Thermosyphons and Heat Pipes: Theory and Applications

Marcia Barbosa Henriques Mantelli

Thermosyphons and Heat Pipes: Theory and Applications

Springer

Marcia Barbosa Henriques Mantelli
Department of Mechanical Engineering
Federal University of Santa Catarina
Florianópolis, Santa Catarina, Brazil

ISBN 978-3-030-62772-0 ISBN 978-3-030-62773-7 (eBook)
https://doi.org/10.1007/978-3-030-62773-7

This Springer imprint is published by the registered company Springer Nature Switzerland AG
The registered company address is: Gewerbestrasse 11, 6330 Cham, Switzerland

I dedicate this book to my most complex and unfinished experiment, from which I learn every day: my family.

To you, Sylvio (my husband), Pedro and Fabio (my sons), Edi (my mother) and especially to Fernando (my father, in memoriam), who, as engineer, inspired me, my brother and sister to follow the intriguing engineering paths.

To God.

Acknowledgments

Representing the Federal University of Santa Catarina, I thank Prof. Sergio Colle, who invited me to the adventure of starting a new laboratory. I am really proud to be a member of the Mechanical Engineering Department of UFSC, Brazil, that hosted my technical visions, allowing me to build a fruitful laboratory.

Thanks to National Institute for Space Research (INPE) and Brazilian Space Agency, which lead my first steps to the heat pipe technology development field.

Special thanks to the Brazilian Industry, especially to Petrobras, Brazilian Oil Company, that trusted our research group, proposing us to be partners of several development projects, especially in the heat pipe industrial application field. These many projects along the years allowed us to get the facilities and knowledge we need.

My special thanks to you, my students, who, along the years, helped in the construction of the knowledge presented in this book. Each one of you is the co-author of a piece of this work. Special thanks to Luis R. Cisterna and João Victor C. Batista, who helped in formatting models and building graphics for this book, being happily available any time of the day, including sleep hours and weekends. My very special thanks also goes to my student Gabriel S. C. Vieira and Larissa Krambeck, for proposing and solving some of the Chap. 6 examples. Thanks to Prof. Fernando H. Milanese and Priscila Gonçalves, who reviewed parts of this book. Also thanks to the Heat Pipe Laboratory (LABTUCAL) staff, here represented by Dra. Kenia W. Milanez (manager) and Dr. Juan P. F. Mera (researcher), Leandro da Silva and Charles Nuernberg (technicians) and Andreza Souza Silva (secretary), who kept the laboratory running smoothly, so that I could dedicate hours to this book.

To all students that worked in the laboratory along these many years, too many to mention, who had very important contributions, my deep thanks.

Above all, I thank God for the family, health, intelligence and strength that I am blessed with.

Contents

Nomenclature

Roman Letter Symbols

a, b, c, d	Position on the liquid column, plate dimension [m]
A	Area, disjoining pressure model constant $[m^2]$
b	Width [m]
B	Disjoining pressure model constant
Bi_C	Modified Biot number
Bo	Bond number
c	Sound velocity $[m/s^2]$
c_p	Specific heat $[J/kg \cdot K]$
C	Dimensionless friction, constant, conductance parameter, correction factor
C_a	Adjustment parameter for vapor velocity profile
C_f	Adjustment parameter for flooding limit
C_0	Vapor distribution coefficient for cylindrical tubes
Co	Dimensionless confinement number
D	Diameter [m]
D+	Non-dimensional diameter
E	Free Helmhotlz energy $[J]$, inernal energy [J]
f	Parameter, Fanning (friction) factor, correction factor
F	Force [N], view factor, filling ratio
g	Gravity$[m/s^2]$
G_n	Geometry non-dimensional parameter
Gr	Grashof number
Gr_c	Corrected Grashof number
h	Convection heat transfer coefficient $[W/m^2K]$, groove height [m]
h_{lv}, h_{vl}	Latent heat, liquid-vapor and vapor-liquid, respectively $[J/kg]$
H	Height $[m]$, hydraulic head [m]

I_0, I_1	Modified Bessel functions of the first kind, order 0 and 1
J	Chilton-Colburn J analogy factor
Ja	Jakob number
J_0, J_1	Bessel functions of the first kind, order 0 and 1
k	Thermal conductivity $[W/m \cdot K]$
k_b	Boltzmann constant [J/K]
K	Permeability $[m^2]$
K_a, K_f	Kinetic energy constant
K_l	Bubble depart parameter
K_{loss}	Localized Pressure drop coefficients
Kn	Knudsen number
Kp	Dimensionless pressure parameter
K_T	Dimensionless boiling parameter
K_0 e K_1	Modified Bessel functions of the second kind, order 0 and 1
L	Turbulence scale [m]
l	Length $[m]$, characteristic length
L_m, M_m, N_m, O_m, P_m Q_m	Superposing solutions for the separation of variables method
m	Mass [kg], adjustable constant, fin convection
$\dot{m}$	Mass flow rate $[kg/s]$
$\dot{m}''$	Mass flux $[kg/m^2 s]$
M	Mach number, molar mass [g/mol], fin parameter
n	Moles number, adjustable constant, number
N	Mesh number $[1/m]$, moles number, number
N_t	Merit number for thermosyphons $[kg/s^{5/2} \cdot K^{3/4}]$
Nu	Nusselt number
p	Pressure $[N/m^2]$
P	Perimeter [m]
p_r	Reduced pressure
Pr	Prandtl number
Pt	Diagonal pitch of tube arrangements [m]
q	Heat rate [W]
q'	Heat rate per unit length [W/m]
q''	Heat flux $[W/m^2]$
$\dot{q}$	Volume generated heat $[W/m^3]$
r	Distance [m], radius [m]
R	Thermal resistance [°C/W], universal constant of the gas $[J/kg \cdot K]$
R'	Thermal resistance per unit length [°C/W·m]
Ra	Rayleigh number
Re	Reynolds number
R_p	Surface roughness [μm]

s	Pitch [m]
S	Crimp factor
t	Time [s], thickness [m]
T	Temperature [K, ℃]
u	Velocity in the axial direction [m/s]
U	Global coefficient of heat transfer [W/m^2·K]
v	Velocity in the radial direction [m/s]
V	Vapor mass flux velocity [m/s], volume [m^3]
$\dot{V}$	Volume rate [m^3/s]
V+	Non-dimensional vapor drift velocity
w	Width [m], eddy velocities [m/s]
We	Weber number
x	Body force term [N]
x,y	Cartesian axis, variable neck radius [m]
X	Vapor quality
Xl, Xt	Longitudinal and transversal pitch for tube arrangement [m]
z	Axial direction, height of the sintering neck [m]

Greek Letter Symbols

α	Void fraction, thermal diffusivity [m^2/s], groove angle
β	Momentum correction factor, coefficient of thermal expansion [1/K], thermal coefficient
β_e, χ_e, ς_e δ_e φ_e ν_e ξ_e	Evaporator geometry dimensions [m]
β_1, β_2	Half-angle of the wire-liquid opening and meniscus curvature
Γ	Fin parameter
$\Gamma_{v,air}$	Mass diffusion coefficient of vapor in air [m^2/s]
γ	Ratio of specific heats, phase change Arckmann factor
δ	Film thickness [m]
ε	Area porosity, emissivity, effectiveness
ε'	Volumetric porosity
ε_r	Surface roughness
ϵ, r_0	Molecule potential force parameters
θ	Inclination angle, solid-liquid contact angle, temperature difference
κ	Compressibility (1/Pa)
λ	Laplace length scale, mean free path
$\lambda_m, \zeta_m, \eta_m, \upsilon_n, \tau_n, \Xi_m$	Eigenvalues
μ	Dynamic viscosity $\left[(\mathrm{N}\cdot\mathrm{s})/\mathrm{m}^2\right]$
η	Efficiency
ν	Cinematic viscosity $\left[\mathrm{m}^2/\mathrm{s}\right]$

ν_e	Evaporator pumping region length [m]
ω	Mass fraction at the liquid-vapor interface, absolute humidity
ξ	Boiling correlation parameter, variable thickness fin parameter
ρ	Density $\left[kg/m^3\right]$
σ	Surface tension [N/m], Stefan-Boltzmann constant
σ_d	Collision distance between molecules $[\dot{A}]$
Σ	Mass diffusion coefficient
τ	Shear stress $\left[N/m^2\right]$
υ	Specific volume $[m^3/kg]$
φ_b	Dimensionless bubble release number
φ_c	Dimensionless fin condenser parameter
φ_e	Evaporator pumping region insulated length [m]
φ_T	Modified dimensionless fin condenser parameter
ϕ	Two-phase multiplier, eigenvalue function
$\Phi(r)$	Molecular potential function
χ	Pool parameter
Ψ	Fluid properties parameter, pressure difference
Ψ_m	Percentage of the evaporator heat used in phase-change
Ω	Martinelli parameter
Ω_m	Mass transfer coeficiente [m/s]

Subscripts and Superscripts

a	Adiabatic section, advancing
act	Active
air	Air
amb	Ambient
atm	Atmosphere
ax	Axial
b	Bubble, boiling, base, bulk
c	Condenser, condensation, casing, cold
C	Characteristic length, characteristic length of condenser
cap	Capillary
cc	Compensation chamber
cd	Condensed
ch	Chamber, channel
col	Column
cool	Cooling
cond	Conduction
conv	Convection
cont	Continuum

cr	Critical
crit	Critical
cs	Cold source
cw	Cooler-wall interface
ct	Cooling tower, continuum
d	Disjoining, dynamic
D	Diameter
db	Dry bulb
e	Evaporator, entrainment
ef	Effective
eq	Equivalent
ev	Evaporated
ex	External
exp	Experimental
f	Flooding, fin, film
g	Gravity, groove
GB	Geyser Boiling
h	Heater, hydraulic, hot
hb	Humid bulb
hem	Hemisphere
hs	Heat sink
i	Interface, ith component, internal, initial
in	In
inf	Inferior
k	Kinetic
l	Liquid
L	Length
lat	Latent
lc	Liquid channel
Le	Evaporator length
ll	Liquid line
lm	Logaritmic mean
loss	Loss
lv	Liquid-vapor
m	Mean
$\dot{m}_L$	Condensate film
max	Maximum
min	Minimum
mix	Mixture
n	Index, nucleation, neck
N	Normal direction, natural convection
NCG	Non condensable gas
Nu	Nusselt
o	Overall
op	Operational

out	Out
p	Pore, pool, pressure, plug, plate
pc	Phase change
pm	Porous media
r	Radial, rescinding
s	Surface, solid, sphere, sonic
sat	Saturation
st	Stagnation point
su	start up
sup	Superior
t	Total, top
tr	Transversal, transition
v	Vapor, viscous
vl	Vapor line
w	Wet, wick, wall, wire, water
wf	Working fluid
x, y, z	Local, direction
X	Body force term
β	Wick plug
δ	Condensate film thickness
σ	Surface tension

Numerical

0	$(x = 0)$, $(t = 0)$
2ϕ	Two-phase flow
∞	Environment, ambient

Chapter 1
Introduction to Thermosyphons and Heat Pipes

Heat pipes and thermosyphons are highly efficient heat transfer devices that use two-phase cycles of fluids as the operating principle. Typical heat pipes and/or thermosyphons consist of an evacuated tube casing, within which a controlled amount of a working fluid is introduced. In heat pipes, capillary forces provided by a wick, are responsible for the movement of the fluid through the device. In thermosyphons, gravity forces do this job. Although heat pipes and thermosyphons may have many different geometries and configurations, they are basically composed of three main regions: evaporator, adiabatic section and condenser. In some applications, the adiabatic section may not be present.

1.1 Working Principles

Figure 1.1 illustrates the physical principles that drive the thermosyphon operation. Heat is delivered to the thermosyphon in the evaporator section, reaching the working fluid, causing liquid–vapor phase change. The generated vapor, due to pressure gradients inside the thermosyphon, crosses the adiabatic region and reaches the condenser, where heat is removed. The vapor contained in the condenser region condenses and the resulting liquid returns to the evaporator by the action of gravity forces. Therefore, the evaporator must be located in a position inferior to the condenser. Due to the surface tension effects between the tube material and the liquid, small rivulets are formed over the tube internal wall.

Heat pipes operates in very similar way, however, the return of the working fluid from the condenser to the evaporator happens due to the capillary forces resulting from the presence of a wick structure located inside the device, as illustrated in Fig. 1.2.

Thermosyphons and heat pipes can also be designed to operate in loops. In this case, internally smooth tubes connect evaporators to condensers, within which only vapor or liquid flows. This arrangement avoids the dragging forces between liquid

M. B. H. Mantelli et al., *Thermosyphons and Heat Pipes: Theory and Applications*,
https://doi.org/10.1007/978-3-030-62773-7_1

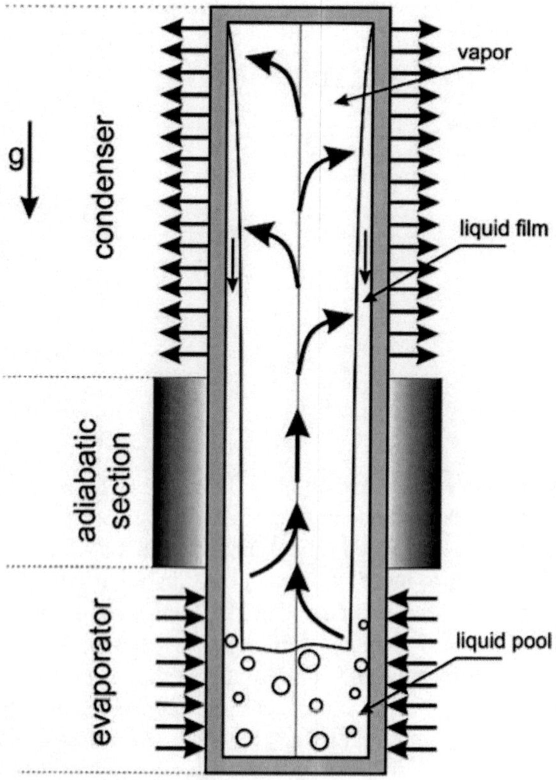

Fig. 1.1 Working physical principles of thermosyphons

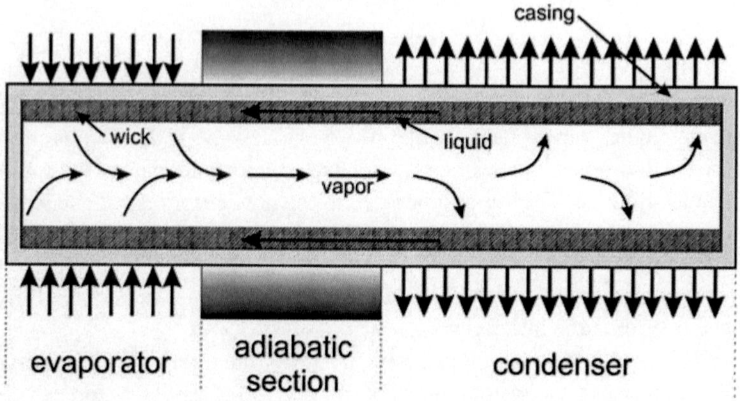

Fig. 1.2 Working physical principles of heat pipes

and vapor due to countercurrent flows. Schematics of loop thermosyphons and loop heat pipes are presented in Fig. 1.3.

In a general sense, the casing and the working fluid are considered the main components of thermosyphons. In heat pipes, besides these two, the wick is another key element. Usually located in the inner tube wall, the wick structure is responsible for the capillary forces that pump the working fluid from the condenser to the evaporator.

Casings can be made of metal, ceramic or other material. Besides, several different liquids can be used as working fluids (liquid nitrogen, water, alcohol, naphthalene, liquid sodium, etc.). Basically, the selection of these materials depends on the device

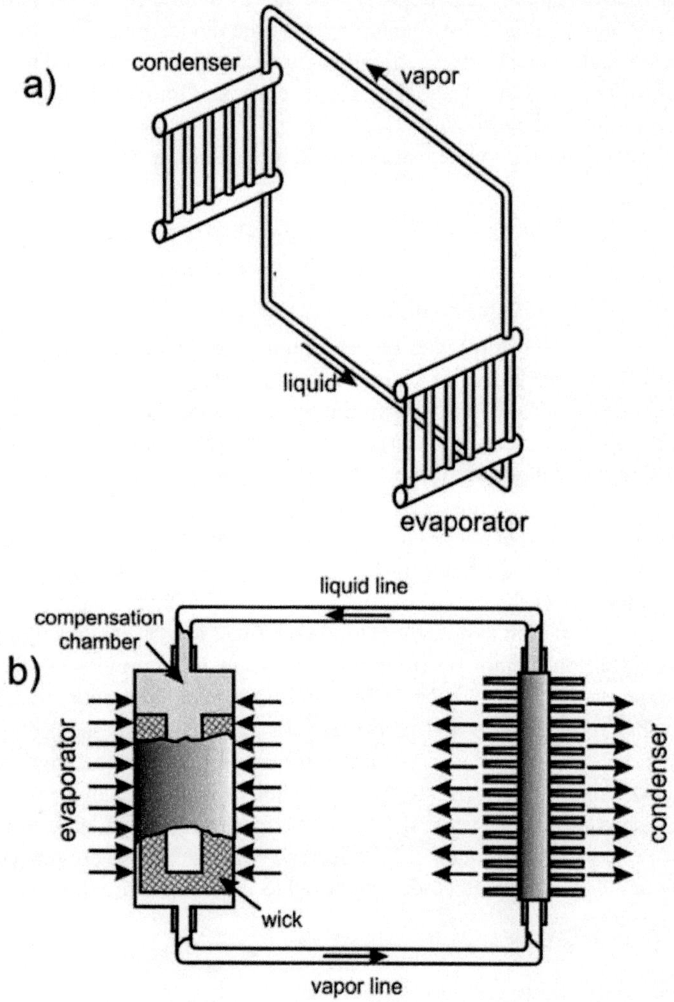

Fig. 1.3 Loop thermosyphons (**a**) and loop heat pipes (**b**)

working temperature and its application. The wick, in heat pipes, can be made of different porous materials such as metal screens, sintered metal powder, longitudinal grooves, corrugated fiber glass, etc.

Usually, the design of thermosyphons or heat pipes starts with the selection of the working fluid, which must operate at the required temperature level. After the working fluid is selected, the casing is chosen. The casing material should be chemically compatible with the working fluid, avoiding the formation of non-condensable gases that would block part of the device, decreasing its performance. For heat pipes, the wick structure must provide the necessary liquid pumping capacity and be compatible with the working fluid and with the casing material.

Among the several configurations of thermosyphons or heat pipes (loop thermosyphon, loop heat pipe, pulsating heat pipe, vapor chamber, etc.) the engineer must select the one more suitable for the application: the device must have the expected thermal performance and be geometrically adaptable. These two-phase technologies are actually versatile, they can operate at basically any temperature level and they can assume multiple shapes.

Design procedures and models are presented in details in this book.

1.2 Applications

Thermosyphon and heat pipes can be used in applications where heat needs to be transferred with high efficiency or where uniform temperatures are required. These devices are able to manage heat in a myriad of equipment, from very small, such as electronic components, up to very large ones, such as oil storage tanks in petroleum refineries. Thermosyphons are more suitable for industrial applications because, as they do not require porous media to operate, they are easy to construct. Therefore, the fabrication costs are much reduced when compared to heat pipes. Besides, they are able to transport at least one order of magnitude more heat than heat pipes. However, as they need the gravity action to operate, the heat sources (evaporators) necessarily are positioned at below positions relative to the heat sinks (condensers). When this condition cannot be fulfilled, such as in microgravity applications or mobile computers, for instance, heat pipes are the correct technology to be applied.

This book is about theories and applications of thermosyphons and heat pipes. The major objective of this book is to organize information so that the engineer is able to select and design the appropriate thermosyphon or heat pipe device for the target application. However, it was not the intention to duplicate available information. Only brief reviews are presented for the material that is extensively reported in the open literature. The interested reader is invited to further explore this material.

1.3 Book Outline

The present book is divided into three major parts.

In **PART I, FUNDAMENTALS AND FABRICATION ASPECTS**, the fundamental operation principles of thermosyphon and heat pipes and the physical phenomena involved are described. Theoretical aspects about condensation in cooled vertical walls, liquid–vapor and vapor–liquid phase change, liquid–solid interface tension, wetting phenomena, disjoining pressures, fluid flow in tubes, boiling, etc., are presented. Thermodynamics aspects are used to construct basic thermal and hydraulic models for thermosyphons and heat pipes. Liquid and vapor pressure drop distribution models are presented. Discussion and modelling of the heat transfer operation limits are shown. Geyser boiling effect, which deeply affects thermosyphon hydraulic and thermal behaviors, are discussed, as well as mechanisms used to avoid it. Basic steady-state one-dimensional models, based on the thermal resistance circuits, are presented and compared with experimental results for both thermosyphons and heat pipes. Correlations for the coefficients of heat transfer, required for the determination of the thermal resistances, are presented. Thermosyphons and heat pipes design methodologies are discussed. Fabrication and experimental test setup design aspects, are also discussed. To illustrate the modelling and design methodology proposed, PART I is concluded with the design and testing of several actual thermosyphons.

In **PART II, SPECIAL THERMOSYPHON AND HEAT PIPE TECHNOLOGIES**, several special configuration of thermosyphons and heat pipes are presented and modelled. First, these devices are classified according to their operational principles. Then, loop thermosyphons, capillary pumped loops (including loop heat pipes), pulsating heat pipes and vapor chambers are discussed in details. Mini thermosyphons and heat pipes technologies and models are also presented, including hybrid heat pipe/thermosyphons, wire-plate heat pipes and hybrid wire-plate/sintered wick heat pipes.

In **PART III, APPLICATIONS**, special attention is given to applications of thermosyphons and heat pipes. The design and construction aspects of thermosyphon heat exchangers, largely employed in the industry, are presented. Several special mini heat pipes and cabinet wall thermosyphon assisted technologies for electronics cooling, including avionics and smart gadget applications, are discussed. The use of thermosyphons for saving water in wet cooling towers, for solar heating of houses, for improving the thermal performance of ovens (vertical and conveyor belt) designed to bake food (bread, biscuits, etc.) or to dry fruits and medicinal herbs, are also shown. Thermosyphon equipment designed for the petroleum industry are also presented, including heaters for oil storage tanks and for natural gas, in gas distribution city gates.

One of the objectives of this book is to collect the necessary information to design new devices or to improve already established industrial process and equipment by the use of thermosyphons and heat pipes. However, the major purpose is to inspire engineers to use this very promising and effective technology to propose new quality and creative thermal solutions for many different industrial problems.

Part I
Fundamentals and Fabrication Aspects

Chapter 2
Physical Principles

The physical principles that drive the operation of thermosyphons and heat pipes are mostly based on the thermal and mechanical behavior of the several interfaces inside these devices, especially the interfaces between the working fluid phases (liquid and vapor) and between the working fluid and the solid surfaces, such as casing, grooves or wicks. Therefore, most of this chapter is devoted to the understanding of these physical principles and to the modeling of heat transfer and forces resulting from these interfaces. In the first part of the chapter, thermosyphon related models are discussed, followed by heat pipe models.

2.1 Nusselt Model for Wall Condensation

Gravity is the major driving force in thermosyphons, responsible for pushing condensate working fluid from the condenser to the evaporator and for dragging the vapor from the evaporator to the condenser, by buoyancy effects.

The heat transfer mechanisms and the thermal behavior of the condenser and evaporator sections of thermosyphons are mostly described based on literature well-known models, such as condensation over cooled walls and pool boiling. In this section, these models are briefly presented, highlighting how they can be adapted to predict the performance of thermosyphons.

Usually, the Nusselt model, briefly treated in this section, is used to predict the condensate formation over internal walls of the condenser sections of thermosyphons.

In general, the vapor condensation over vertical cooled walls, in the presence of saturated vapor, can happen in heterogeneous and homogeneous conditions (Collier and Thome 1994).

Three heterogeneous condensation conditions can be recognized: drop-wise, film-wise and direct contact. Heterogeneous condensation usually happens close to impurities (such as dust), where the saturated vapor suffers a localized pressure drop. For

M. B. H. Mantelli et al., *Thermosyphons and Heat Pipes: Theory and Applications*,
https://doi.org/10.1007/978-3-030-62773-7_2

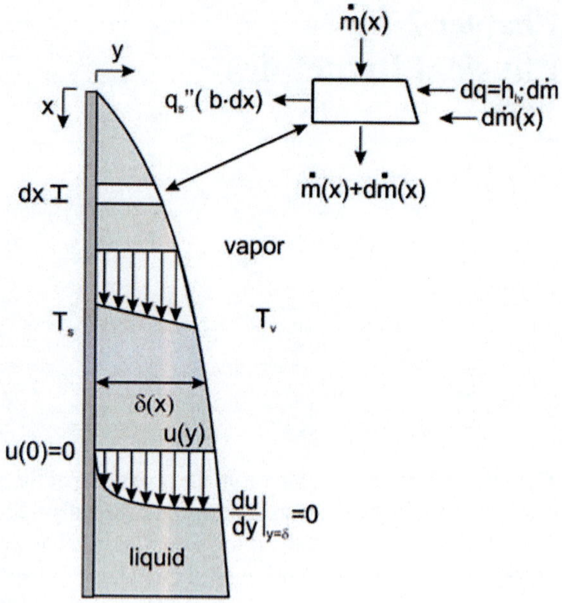

Fig. 2.1 Schematic of condensate film for Nusselt model

thermosyphon equipment design, these models are not usually employed and so they are not described in this text.

A simple model developed by Nusselt (1916) for homogeneous vapor condensation over a vertical cooled wall in contact with a pure single-component saturated vapor, is presented in this Chapter. Stagnant quiescent vapor is considered in contact with a vertical cooled wall, over which the condensate liquid forms a uniform thin liquid film, which flows in downward direction, due to gravity action. Its thickness is zero at the wall top and increases with the descending film. Figure 2.1 shows the schematic of the condensate film over a cooled vertical surface. The following hypothesis are assumed:

- The flow is laminar;
- The vapor is saturated and pure;
- The vapor temperature is uniform;
- The gravity is the only external force acting on the film;
- As the vapor is stagnant, no dragging forces over the film external surface are considered;
- The fluid properties are constant;
- No sensible cooling of the film is considered (negligible when compared to the phase change latent heat);
- The curvature of the liquid–vapor interface is negligible: no tension forces are considered and so the temperature of the interface is determined from the saturation vapor-pressure curve of the fluid.

- Momentum and energy transfer by advection within the condensate film are considered negligible.
- The thermal profile of the liquid film is assumed to be fully developed.

Due to these hypotheses, the temperature profile across the film is linear and the heat transfer is one-dimensional. The heat transfer mechanism is heat conduction through the film, in the direction: vapor–liquid interface to the wall.

The x-momentum conservation equation, for the element depicted in Fig. 2.1, neglecting the inertia (advective) terms, can be written as:

$$\frac{\partial^2 v}{\partial y^2} = \frac{1}{\mu_l}\frac{dp}{dx} - \frac{X}{\mu_l}, \tag{2.1}$$

where v is the velocity in the x direction, $\mu_l$is the dynamic viscosity, X represents the body force term and dp/dx is the free stream pressure gradient in the quiescent vapor region, outside the liquid layer of thickness δ_l, given by $\rho_v g$. Gravity is the body force inside the liquid layer, given by $\rho_l g$. ρ_l and ρ_v are the liquid and vapor densities, respectively and g is the gravity acceleration. Therefore, the last equation becomes:

$$\frac{\partial^2 v}{\partial y^2} = -\frac{g}{\mu_l}(\rho_l - \rho_v). \tag{2.2}$$

which, integrated twice and considering $v(0) = 0$ and $dv/dy|_{y=\delta} = 0$, results in the Nusselt velocity profile:

$$v(y) = \frac{g(\rho_l - \rho_v)\delta_l^2}{\mu_l}\left[\frac{y}{\delta_l} - \frac{1}{2}\left(\frac{y}{\delta_l}\right)^2\right] \tag{2.3}$$

One should note that the velocity given by this expression depends on both coordinates y and x, this last indirectly, as δ_l depends on x.

The heat for the liquid film condensation q is taken from the wall. The energy balance equation, for the elementary region of length dx illustrated at Fig. 2.1, is:

$$dq = h_{lv} d\dot{m} = q_s''\, b\, dx \tag{2.4}$$

where h_{lv} is the latent heat of vaporization, $\dot{m}$ is the mass flux, q_s'' is the heat flux at the surface and b is the wall width. As the temperature profile within the liquid film is considered linear, convection heat transfer coefficient can be expressed as $h = k_l/\delta_l$ and, according to the Fourier law, one has:

$$q_s'' = \frac{k_l(T_{sat} - T_s)}{\delta_l} = \frac{k_l \cdot \Delta T}{\delta_l} \tag{2.5}$$

where k_l is the thermal conductivity of the liquid film, ΔT is the temperature difference between the liquid–vapor saturation temperature (T_{sat}) and the surface temperature T_s.

Therefore, the condensed liquid mass rate, by wall unit width, is given by:

$$\frac{\dot{m}}{b} = \int_0^{\delta_l(x)} \rho_l v(y) dy = \frac{g\rho_l(\rho_l - \rho_v)\delta_l^3}{3\mu_l} \tag{2.6}$$

Substituting Eqs. 2.6 and 2.5 in Eq. 2.4, one gets the following ordinary differential equation:

$$\delta_l^3 d\delta_l = \frac{k_l\mu_l(T_{sat} - T_s)}{g\rho_l(\rho_l - \rho_v)h_{lv}} dx \tag{2.7}$$

which, integrated from $\delta(0) = 0$ to $\delta(x)$, gives:

$$\delta_l(x) = \left[\frac{4k_l\mu_l(T_{sat} - T_s)x}{g\rho_l(\rho_l - \rho_v)h_{lv}}\right]^{1/4} \tag{2.8}$$

Using the same definition of the liquid film convection heat transfer coefficient as before ($h = k_l/\delta_l$), one has:

$$h(x) = \left[\frac{g\rho_l(\rho_l - \rho_v)h_{lv}k_l^3}{4\mu_l(T_{sat} - T_s)x}\right]^{1/4} \tag{2.9}$$

A mean heat transfer coefficient along a characteristic length L along th cooled wall vertical direction can be determined, through the following expression:

$$\overline{h}_L = \frac{1}{L}\int_0^L h\,dx \tag{2.10}$$

Substituting Eq. 2.9 into Eq. 2.10 and performing the integration, the following expression is obtained, for the mean heat transfer coefficient (Kedzierski et al. 2003):

$$\overline{h}_L = 0.943\left[\frac{\rho_l g(\rho_L - \rho_v)h_{lv}k_l^3}{\mu_l(T_{sat} - T_s)L}\right]^{1/4} \tag{2.11}$$

It is interesting to note that $\overline{h}_L = 4/3 \cdot h(L)$. Equation 2.11 is also valid when the heat flux (and not the temperature difference) is regarded as uniform. In this case, q is replaced by the mean heat flux $\overline{q} = k_l\Delta T/L \int_0^L 1/\delta_l(x)dx$ and ΔT is obtained using the uniform wall temperature (Rose 1988).

One should note that Eq. 2.9 presents local heat transfer coefficients. Furthermore, Collier and Thome (1994) suggest that the mean heat transfer coefficient can also be obtained considering a heat balance over the total length of the cooled surface: the total heat transferred along the cooled surface of length L is equal to the energy used for vapor condensation. Therefore, the mean condensation heat transfer coefficient can be given by:

$$\overline{h}_L = \frac{\dot{m}(L)\, h_{lv}}{b\, L\, (T_{sat} - T_s)}. \tag{2.12}$$

Eliminating the temperature difference of this expression by using Eqs. 2.5 and 2.4, one has:

$$\delta_l(x) = \frac{k_l\, \dot{m}(L)\, dx}{\overline{h}_L\, L\, d\dot{m}} \tag{2.13}$$

Inserting this expression in Eq. 2.6 and after some algebraic manipulation, the following differential equation is found:

$$\left(\frac{g\,\rho_l(\rho_l - \rho_v)k_l^3\, b}{3\mu_l\, \dot{m}(L)}\right)^{1/3} \frac{1}{x} dx = \overline{h}_L\, \dot{m}^{1/3}\, d\dot{m} \tag{2.14}$$

which, integrated over the length L, results in the following expression for the mean heat transfer coefficient:

$$\overline{h}_L = 0.925 \left(\frac{g\,\rho_l(\rho_l - \rho_v)k_l^3\, b}{\mu_l\, \dot{m}(L)}\right)^{1/3} \tag{2.15}$$

For thermosyphon design, this expression can be more convenient than Eq. 2.11 , when the wall temperature is not known. Collier and Thome (1994) also highlights that the heat transfer coefficient can be expressed in terms of the Reynolds number of the condensate liquid film, which is defined as:

$$\mathrm{Re}_\delta = \frac{4\dot{m}(L)}{\mu_l\, b}. \tag{2.16}$$

Substituting the Reynolds number expression on Eq. 2.15 , one gets the following equation, based on the mean heat transfer coefficient (Eq. 2.12), valid for $\mathrm{Re}_\delta \leq 30$:

$$\frac{\overline{h}_L}{k_l}\left(\frac{\mu_l^2}{g\,\rho_l(\rho_l - \rho_v)}\right)^{1/3} = 1.47\mathrm{Re}_\delta^{-1/3} \tag{2.17}$$

As $\overline{h}_L = 4/3 \cdot h(L)$, the last expression, in terms of local condensing heat transfer coefficient, can be rewritten as:

$$\frac{h(L)}{k_l}\left(\frac{\mu_l^2}{g\rho_l(\rho_l-\rho_v)}\right)^{1/3} = 1.1\mathrm{Re}_\delta^{-1/3} \tag{2.18}$$

From these last equations, a characteristic length, for the condensation phenomena over a vertical cooled plate, can be defined as:

$$L_C = \frac{\mu_l^2}{g\rho_l(\rho_l-\rho_v)} \tag{2.19}$$

As already mentioned, most of the literature models regarding gravity-driven condensation heat transfer over cooled walls is based on the Nusselt model. However, the actual condensation heat transfer mechanisms can be quite different from the model predictions, due to the influence of several factors that affect the phase change mechanism. Revised Nusselt models are available in the literature and those relevant for thermosyphon applications, are briefly presented next.

Effect of the Condensate Temperature

According to Collier and Thome (1994), Bromley in 1952 and Rohsenow in 1956 extended the Nusselt theory to consider the subcooling and the non-linear distribution of the temperature through the film, showing that the latent heat of vaporization, used in Eq. 2.9 , could be replaced by:

$$h'_{lv} = h_{lv}(1 + 0.68 \cdot Ja_l) \tag{2.20}$$

where Ja_l is the Jakob number, a relative measurement of the degree of subcooling of the liquid film, defined as:

$$Ja_l = \frac{c_{p_l}(T_{sat} - T_s)}{h_{lv}} \tag{2.21}$$

where $c_{pl}$is the specific heat at constant pressure of the condensate liquid film. However, Rose (1988) states that it is not evident that this correction improves significantly the Nusselt model results.

As stated by Collier and Thome (1994), the thermophysical properties of the film should be evaluated for the effective film temperature, using the following expression:

$$\overline{T} = T_s + 0.25(T_{sat} - T_s) \tag{2.22}$$

Minkowycz and Sparrow (1966) suggest the substitution of the number 0.25 to 0.31 in the last equation.

Effect of Vapor Shear Stress

The hypothesis that the interface between laminar falling film is smooth is not always valid. In counter-current annular film flow, ascending vapor causes shear stress over the external surface of the descending film, causing ripples and waves. This is the case of thermosyphons, where the interface shear stress may cause the liquid to absorb some vapor, reducing the velocity and increasing the film thickness. In addition, the convection within the condensation film may affect the heat transfer. Both effects decrease the heat transfer coefficient and were not considered in Nusselt's model.

Many literature papers (Chun and Seban 1971; Fujii and Uehara 1972; Rose 1988) deals with vapor shear stress when vapor and liquid film moves in the same downward directions or in horizontal plates (Koh 1962). As these models cannot be applied directly to a typical thermosyphon as described in Fig. 1.1 these works will not be described here (note that some of these models might be useful for describing the condenser phenomena of some loop thermosyphons).

Koh et al. (1960) studied numerically the two-phase flow problem in laminar condensation over vertical cooled walls, considering that the condensed liquid flow induces motions within the vapor, which, in turn, affects the condensate velocities. In other words, they considered vapor shear forces in their model. They concluded that the effects of the interfacial shear stress can be especially important for the condensation of liquids with very small Pr_l numbers, such as liquid metals.

Chen, in 1961, proposed the following correlation to correct the Nusselt expression to account for the vapor drag effects, which compared with numerical results within 1%:

$$h = h_{Nu}\left[\frac{1 + 0.68\mathrm{Ja_l} + 0.02\frac{\mathrm{Ja_l^2}}{\mathrm{Pr}_l}}{1 + 0.85\frac{\mathrm{Ja_l}}{\mathrm{Pr}_l} - 0.15\frac{\mathrm{Ja_l^2}}{\mathrm{Pr}_l}}\right]^{1/4} \tag{2.23}$$

Pr_l/Ja_l is a dimensionless acceleration effect parameter that takes into account the effect of the convection within the liquid layer, where the Pr_l is the Prandtl number, given by $Pr_l = c_{pl}\mu_l/k_l$. This last equation is valid for $\mathrm{Ja_l} \leq 2$ and for $Ja_l/Pr_l \leq 20$, for liquids with Pr_l larger than one or less than 0.05. Besides, Rose (1998) sates that, in most actual applications, the effect of the vapor shear stress in free convection is very small. Also, if Eq. 2.23 is used, no corrections of the temperature effect in the latent heat of vaporization (use of Eq. 2.20) is necessary.

Effect of Turbulence

As happens to any boundary layer flow, as the Reynolds number increases, a transition from laminar to turbulent flow is expected. In general, a regime of laminar flow is observed over the upper region of the surface, changing to turbulent as the liquid flows to the rear parts of the cooled plate. For the case of a condensate film flowing down the wall, even for low Reynolds numbers, the assumption of viscous flow behavior is questionable, as waves can be easily observed. Experimental data shows that the heat transfer coefficients can be much greater than those predicted by Eq. 2.17. Other literature works such as Kutateladze (1981) used similarity analysis to study the arising waves mechanisms. Carey (1992) suggests, based on the literature, that waves are expected for flows with $\mathrm{Re}_\delta > 33$.

For long vertical surfaces where both regimes can be observed, Collier and Rose (1994) proposed the following correlation obtained from experimental data, valid for $Re_\delta > 2000$:

$$\frac{h_l(x)}{k_l}(L_c)^{1/3} = 0.056 \cdot \mathrm{Re}_\delta^{0.2}\left(\Pr_l\right)^{1/3} \quad (2.24)$$

The idea is to integrate the local heat transfer coefficient using Eq. 2.9 along the plate length up to a Reynolds number Re_f of 2000 and to use Eq. 2.24 for rest of the length. Another expression is available in the literature, valid for $1 \leq \mathrm{Pr_l} \leq 5$ and $Re_\delta > 1400$ (Carey 1992):

$$\overline{\mathrm{Nu_L}} = \frac{\overline{h}}{k_l}\left(\frac{v_l^2}{g}\right)^{1/3} = 0.0131\,\mathrm{Re}_\delta^{1/3} \quad (2.25)$$

where ν_l is the liquid viscosity.

On the other hand, Kutateladze (1963) suggests the use of the following correlation, for the laminar wavy film condensation, for $30 < Re_\delta < 1800$:

$$\overline{\mathrm{Nu_L}} = \frac{\overline{h}}{k_l}\left(\frac{v_l^2}{g}\right)^{1/3} = \frac{\mathrm{Re}_\delta}{1.08\mathrm{Re}_\delta^{1.22} - 5.2} \quad (2.26)$$

Besides, according to Incropera et al. (2014), Labuntsov recommends the use of the following expression when the fluid is in turbulent flow:

$$\overline{\mathrm{Nu_L}} = \frac{\overline{h}}{k_l}\left(\frac{v_l^2}{g}\right)^{1/3} = \frac{\mathrm{Re}_\delta}{8750 + 58\,\mathrm{Pr}_l^{-0.5}\left(\mathrm{Re}_\delta^{0.75} - 253\right)} \quad (2.27)$$

In the present work, special attention is given to counter-current flowing vapor configuration (reflux condensation condition, such as happens in thermosyphon applications), which retards the condensate flow and thickens the condensate layer. Chen et al. (2007) proposed a general expression, applicable to all regimes (laminar-wave or turbulent), based on correlations for quiescent vapor, i.e., gravity dominated film condensation for vertical cooled walls, based on the method of Churchill and Usagi (1974) and on literature correlations, resulting in the following correlation for the local Nusselt number:

$$\mathrm{Nu}_l(x) = \frac{h_l(x)}{\mathrm{k}_l}\left(\frac{v_l^2}{g}\right)^{1/3} = \left[\left(0.31\mathrm{Re_l(x)}^{-1.32} + \frac{\mathrm{Re_l}(x)^{2.4}\mathrm{Pr}_l^{3.9}}{2.37 \times 10^{14}}\right)^{1/3} + \frac{\mathrm{Pr}_l^{1.3}}{771.6}\tau_i^*\right]^{1/2} \quad (2.28)$$

where $Re_l(x)$ is the liquid Reynolds number which varies with the length of the cooled wall and τ_i is the interface shear stress. For annular film condensation inside tubes, the following expression for the vapor phase change dimensionless interfacial shear stress, is proposed:

$$\tau_i^* = \frac{\tau_i}{\rho_l (g \nu_l)^{2/3}} \tag{2.29}$$

The last term of Eq. 2.28 which accounts for the shear stress, may have plus or minus signs, depending on the direction of the vapor motion: plus if in favor, or concurrent, minus if against, or countercurrent to the condensate gravity driven liquid displacement. If the secondary effects are neglected, such as augmentation of the interfacial shear stress due to phase change, the Fanning friction coefficient (Seban and Hodgson 1982) can be used to determine interface shear stress. Chen et al. (2007) proposed the following expression for turbulent flow inside tubes:

$$\tau_i^* = -C\mathrm{Re_l}(x)^{1.8} \tag{2.30}$$

where the dimensionless group C is defined by:

$$C = \frac{0.023}{D^2 g^{2/3}} \frac{\mu_l^{1.133} \mu_v^{0.2}}{\rho_v \rho_l^{0.333}} \tag{2.31}$$

and D is the tube diameter.

Expression 2.28 can be averaged along the length of the tube, as the $Re_l(x)$ varies from zero at the top of the cooled wall (or tube) to some value at position $x = L$, using the expression:

$$\overline{\mathrm{Nu_L}} = \frac{\overline{h}}{k_l}\left(\frac{\nu_l^2}{g}\right)^{1/3} = \mathrm{Re}_l(L)\left(\int\limits_o^{\mathrm{Re}_l(L)} \frac{d\mathrm{Re}_l(x)}{\mathrm{Nu_l(x)}}\right)^{-1} \tag{2.32}$$

The following correlation is, therefore, obtained:

$$\overline{\mathrm{Nu_L}} = \left(\mathrm{Re}_l(L)^{-0.44} + \frac{\mathrm{Re}_l(L)^{0.8}\,\mathrm{Pr}_l^{1.3}}{1.718 \times 10^5} - \frac{C\,\mathrm{Pr}_l^{1.3}\,\mathrm{Re}_l(L)^{1.8}}{7.716 \times 10^4}\right)^{1/2} \tag{2.33}$$

Chen et al. (2007) states that this expression is valid if there is no flooding within the tube, as in this case, no counter-current could be possible. Therefore, the following constraints should be respected:

$$\mathrm{Re_l(x)} \leq D\left(\frac{g}{\nu_l^2}\right)^{1/3}\left(\frac{C_k}{\mathrm{Bo}^{1/4}}\right)^2\left(\frac{\rho_v}{\rho_l}\right)^{1/2} \tag{2.34}$$

where:

$$C_k = \sqrt{3.2}\tanh\left(\frac{\mathrm{Bo}^{1/4}}{2}\right) \tag{2.35}$$

The Bond number is defined by:

$$\mathrm{Bo} = D\left[\frac{(\rho_l - \rho_v)g}{\sigma_l}\right]^{1/2} \tag{2.36}$$

and σ_l is the liquid surface tension. If the vapor density is much smaller than the liquid, the bond number is simplified to $Bo = D(\rho_l g/\sigma_l)^{1/2}$. These authors used literature data and found a good comparison with these correlations.

Other Effects

The presence of non condensable gases in the condensation over vertical cooled walls has been treated in the literature (Collier and Thome 1994). Minkowycz and Sparrow (1966) presented a wide ranged analytical investigation of laminar film condensation, considering the presence of non-condensable gases. Their analytical model included interfacial resistance, superheating, free convection due to temperature and concentration gradients, mass and thermal diffusion and variable properties of the liquid and gas–vapor regions. They demonstrated that the noncondensable gas concentration can affect definitely the heat transfer rate. These models can be actually very useful for applications where non condensable gas is mixtured with working fluid in thermosyphons.

Collier and Thome (1994) also present expressions for the Nusselt vapor condensation over inclined surfaces. In this case, Eqs. 2.9 and 2.11 are slightly modified, to account for the inclination angle θ, resulting in, respectively:

$$h(x) = \left[\frac{g \cdot \sin\theta \cdot \rho_l(\rho_l - \rho_v)h_{lv}k_l^3}{4\mu_l(T_{sat} - T_s)x}\right]^{1/4} \tag{2.37}$$

$$\bar{h}_L = 0.943\left[\frac{\rho_l g \cdot \sin\theta \cdot (\rho_L - \rho_v)h_{lv}k_l^3}{\mu_l(T_{sat} - T_s)L}\right]^{1/4} \tag{2.38}$$

2.2 Liquid–Vapor Phase Change Model

Liquid–Vapor Transitions at Molecular Level.

The interaction between two molecules is usually characterized by a potential function $\Phi(r)$, defined as the energy necessary to bring two infinitely distant molecules to a finite distance r. It depends on the nature of the interaction between molecules and

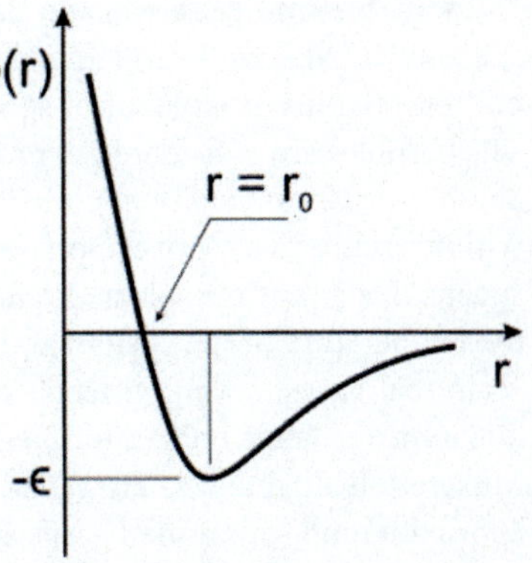

Fig. 2.2 Lennard–Jones potential curve

can be attractive or repulsive. When very close, two molecules usually exert repulsive forces to each other, which increase rapidly as the spacing r decreases. In larger distances, these forces, which are usually characterized as electrostatic, induction and dispersion, became attractive.

According to Carey (1992), Lennard–Jones is one of the most known models for this potential, expressed by:

$$\Phi(r) = 4\varepsilon\left[\left(\frac{r_0}{r}\right)^{12} - \left(\frac{r_0}{r}\right)^{6}\right] \quad (2.39)$$

where the first term at right side represents repulsion and the second the attraction forces. A schematic plot of this expression is shown in Fig. 2.2. The parameter r_0 (distance where the potential is zero) and ε (parameter related to the distance where the potential is minimum: $\varepsilon = -r_{min}$, see Fig. 2.2) vary according to the type of molecules. The variation of the potential implies that energy must be removed to bring close two molecules that are very far apart from each other. However, if two molecules are close enough to be attracted (not so close so that repulsive forces takes place), energy must be supplied to increase the spacing between the molecules. More information about Lennard–Jones potential can be found, among several books in the literature, in Tien and Lienhard (1972).

These energy exchanges are associated with the latent heat of vaporization (where closed molecules are moved apart) or condensation (where distant molecules are moved closer).

The kinetic theory of gases considers that a gas is composed by a very large number of particles moving in random directions, which can collide elastically with one another. Between collisions, it is assumed that the particles move in straight lines. The pressure can be understood as resulting from the force derived from the rate of change of momentum of colliding particles with a wall of a recipient where the gas is confined. The temperature of a gas is directly associated with the random motion of the particles, so that, at absolute zero temperature, the molecules motion ceases completely.

Individual molecules in the gas may travel in different directions, at different speeds. The Maxwell–Boltzmann probability distribution is used to describe gaseous particles (atoms or molecules) speeds of an ideal gas in thermodynamic equilibrium. The particles are considered to move freely inside a stationary container. Their interaction consists of very brief collisions in which they exchange energy and momentum within each other. A particle speed probability distribution indicates the range of speeds that a particle selected randomly is likely to be. The distribution depends on the temperature of the system and on the mass of the particle.

In real gases, several effects, such as van der Waals interactions, vortical flows, quantum exchange interactions, etc., can make their speed distribution different from Maxwell–Boltzmann's. However, the Maxwell speed distribution is an excellent approximation for rarefied gases at ordinary temperatures, which behavior is similar to an ideal gas. More information about kinetic gas theory can be found in the literature (Hirschfelder et al. 1954).

In ideal cases, the Maxwell–Boltzmann speed distribution can be used in the prediction of the kinetic energy distribution of the molecules, resulting in a distribution curve as illustrated in Fig. 2.3. In this figure, the hachured area represents the fraction of molecules that have kinetic energy above the threshold value, represented by the vertical line. The fact that the energy distribution among molecules of liquid and vapor are similar to Maxwell–Boltzmann distribution can be used to explain the relation between the equilibrium pressure and temperature of saturated vapor and liquid. Due to the kinetic energy distribution, even at low temperatures, a fraction of the molecules of the liquid has sufficient energy to escape from the cohesive forces of other liquid molecules. This fraction increases rapidly with temperature. Therefore, it is expected that a liquid with small cohesive energy have a higher vapor pressure

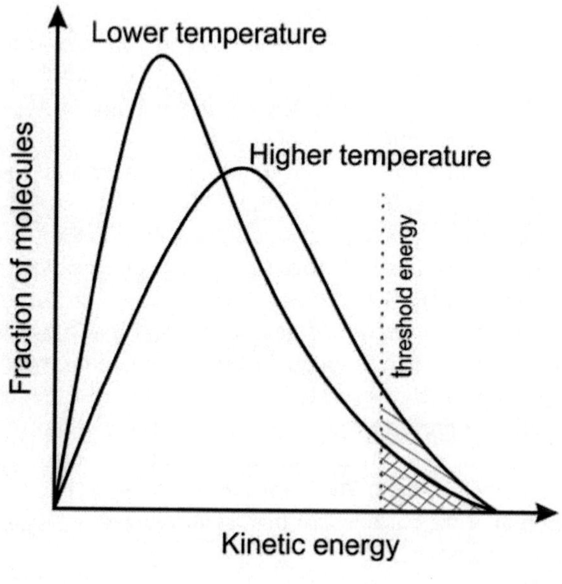

Fig. 2.3 Molecule energy distribution

than other with large cohesive energy. Actually, the latent heat of vaporization is a macroscopic indicator of the cohesive energy of the liquid, so that a liquid with high latent heat of vaporization should have a lower vapor pressure and vice-versa (Carey 1992).

Metastable States

The classical thermodynamics treats the phase change as quasi steady state equilibrium conditions in the saturation conditions. However, phase change happens under non-equilibrium conditions (Carey 1992). Actual processes of vaporization require superheating temperatures, i.e., temperatures above the saturation temperature levels. The same observation can be made about condensation, which can be achieved only after at least part of the vapor is sub-cooled, below the saturation temperature. Therefore, in a saturation curve plot of a pure substance, there are regions where the liquid is superheated or vapor is subcooled. These states are known as metastable.

Figure 2.4 shows the *P*–*v* Clapeyron diagram for a pure substance. The metastable regions are highlighted by the gray areas. An isotherm curve, obtained from the Van der Waals model for real gases, is also plotted. This curve is split in several sections, representing different thermodynamics conditions: section A-B stands for stable sub-cooled liquid, section F-G for stable superheated vapor, B-F (horizontal line) for liquid–vapor equilibrium state, B-C for metastable superheated liquid, E–F for metastable subcooled vapor and CDE represents a region with no phase stability. The metastable regions is located between the saturation curve and the spinodal curve. Phase separation happens for infinitesimal fluctuations in composition and density of substances within the spinodal curve.

Fig. 2.4 Pure substance P–V diagram and Van de Waals curve

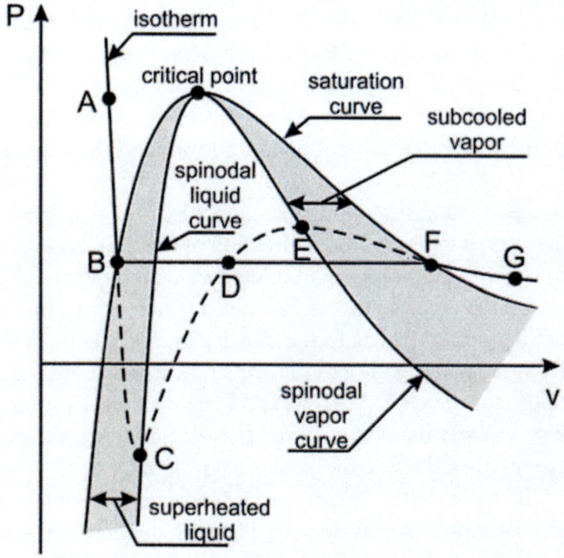

A state is thermodynamic stable when $(\partial P/\partial v)_T < 0$. This means that, keeping the same temperature level, an increase in the pressure necessarily implies in a decrease of volume (or vice-versa). Observing Fig. 2.4, the CDE line does not fulfill the stability criteria, as the pressure increases with the volume and therefore no stability is expected in the region between the saturation and the spinodal curves. However, curves BC and EF follow the phase stability criteria, but are in a state of non-thermodynamic equilibrium, which means that, for the same pressure, there is another state of lower chemical potential. In the metastable regions, any considerable change in the thermodynamic conditions lead to the phase change, as it represents lower chemical potential.

Boiling can be considered a process where a discontinuity of the properties of the substance is observed. When the liquid reaches the saturation condition and there is an infinitesimal change in the environment, which causes a discontinuity of the properties of the substance, a "jump" between the liquid metastable to the saturated vapor states is observed or between the vapor metastable to saturated liquid.

Interfacial Tension

Although, in most two phase (liquid-vapor) interface models, a sharp discontinuity in the fluid composition and properties (such as density) is assumed, actually there is a transition that happens within the thin region centered at this interface. The bulk liquid molecules are closer to each other when compared to those in the liquid–vapor interface region. As the distance between molecules is larger in the interfacial region and considering that energy must be supplied to move molecules apart, it is logical to suppose that the energy per molecule is greater in the interface than in the bulk region. In other words, there is an additional free energy per unit area in the interface (Girifalco and Good 1957).

In the bulk liquid region, the repulsion forces act in each molecule in all directions, so that the resultant force is zero. However, in the regions where the liquid is near to other medium (vapor, liquid or solid), the mean spacing between molecules is larger and the repulsive forces are reduced considerably, so that the forces acting in the molecules are not in equilibrium and the resultant force draws the molecules toward the liquid bulk region. In the interface, the resultant of the radial forces due to the interactions among the molecules is null parallel to the interface (Carey 1992). The resultant force in the direction of the bulk liquid over the liquid–vapor interface makes the drop of a liquid in contact with its vapor to get the spherical shape.

On the other hand, the interface tension can be interpreted as a force per unit area acting perpendicular to the density gradient and parallel to the interface, resulting from the combination of the attractive and repulsive interactions between molecules of the substance. Due to this spontaneous trend to contract, the liquid surface behaves like a rubber membrane under tension and the interface tension can be interpreted as energy stored in the molecules near the interface.

To increase the liquid–vapor interface area, work must be exerted. The surface free energy E associated with this work divided by the surface area A, is denominated as surface tension σ, or (Adamson and Gast 1997):

$$\sigma = \left(\frac{\partial E}{\partial A}\right)_{T,p,n_i} \tag{2.40}$$

which dimension is energy by area or force by unity length. The surface tension depends on the temperature T, pressure p and the number of moles n_i of the gas i component. This equation is valid for interfaces between liquid–gas, liquid–solid and liquid–liquid, this last one for the interaction between two immiscible liquids. The surface tension is a fundamental property that characterizes the surface properties of a fluid. The interface between liquid and gas is a very thin, tridimensional surface and its properties are different from those of the phases that this surface separates.

Wetting Phenomena–Contact Angle

Usually, in heat transfer equipment that makes use of liquid–vapor phase change mechanisms, energy is transferred to the liquid by a wall of a container. Hence, the knowledge of the ability of a liquid to wet a solid surface is of major importance for their design.

The interaction of liquid and solid surfaces depends mainly on the properties and characteristics of the solid wall material (including its surface finishing) and of the liquid. The capacity of the liquid to "wet" the solid surface results from the "affinity" between liquids and solids, a consequence of the attractive forces among the liquid and solid molecules. However, the forces between the liquid and the solid molecules can be repulsive. In this last case, the attraction forces among the liquid molecules make the liquid drop to assume a spherical shape and, in the limit, the liquid–solid contacting area reduces to a single point.

For a drop resting in a solid surface, the angle formed by the liquid–vapor interface and the solid is denominated as "contact angle". The value of this angle is actually a quantification of the liquid–solid wettability. The contact angle is usually weakly dependent of the container shape and of gravity forces. Figure 2.5 illustrates the geometry of a liquid drop over a solid surface in a vapor environment. The three phase (solid, liquid and vapor) line is circular and is denominated as the "contact line". Forces resulting from the surface tensions: σ_{lv} (liquid–vapor), σ_{sl} (solid–liquid) and σ_{sv} (solid–vapor) act over the contact line. The following expression results from horizontal component forces, in equilibrium state:

$$\sigma_{sv} = \sigma_{sl} + \sigma_{lv}\cos\theta \tag{2.41}$$

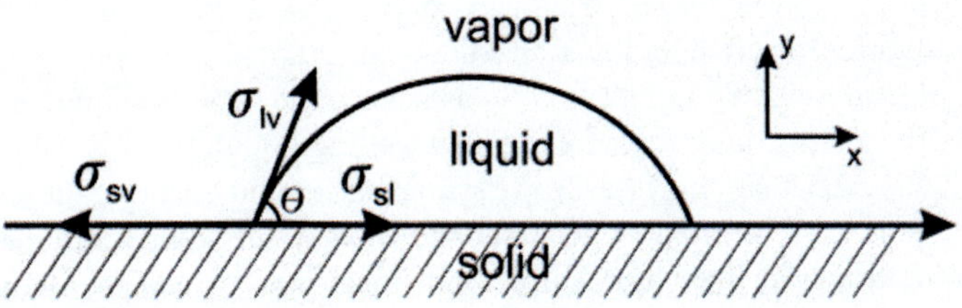

Fig. 2.5 Drop resting on a solid surface

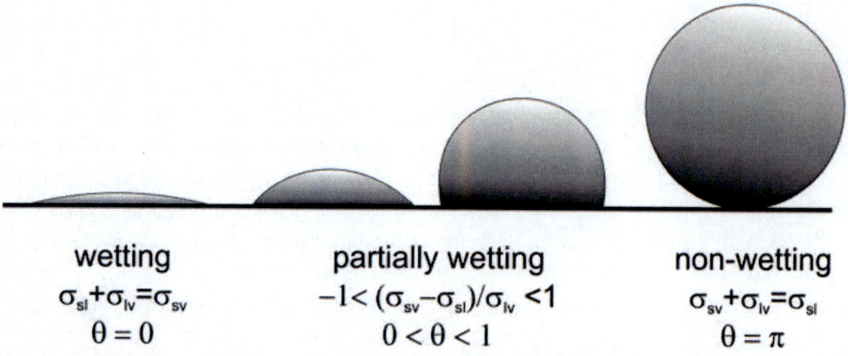

Fig. 2.6 Wetting characteristics of liquid drop over a solid surface

This equation is known as the Neumann Formula or Young Equation (Carey 1992). Isolating the contact angle θ, one gets:

$$\theta = \cos^{-1}\left(\frac{\sigma_{sv} - \sigma_{sl}}{\sigma_{lv}}\right) \tag{2.42}$$

If $\sigma_{sv} > \sigma_{sl}$, the contact angle θ is acute and a liquid wetting condition is perceived. However, if $\sigma_{sv} < \sigma_{sl}$, the contact angle is obtuse and a non-wetting condition is observed. Actually, for a fixed volume liquid drop (considering that no condensation or vaporization is taking place) the increase of the contact angle θ represents a decrease of the solid–liquid interface area. On the other side, the decrease of angle θ results in the liquid spreading, and, as a result, the increasing of the solid–liquid interface area. In the limit, as $\theta \to 0$, the liquid is considered as "full wetting" and spreads all over the solid surface, forming a thin film. For $\theta \to \pi$ the liquid is "non-wetting" and the liquid-surface contact reduces to a single point. Figure 2.6 shows a liquid drop in wetting, non-wetting and partial wetting conditions ($\theta < \theta < \pi$). In wetting liquids, the adhesive forces between liquid and solid are predominant to the cohesive forces among the liquid molecules themselves. In non-wetting fluids, the opposite happens and, if gravity or other field forces are neglected, the liquid drop shape tends to a sphere.

Due to the surface tension effects, the upper surface of a liquid within a container is curved, especially close to the wall regions, forming the so-called meniscus. This curvature can be either concave or convex, depending on the liquid wetting characteristics. Figure 2.7 left side, illustrates the contact angle of a wetting liquid and, right side, of a non-wetting liquid. In this case, the angle between liquid and surface is θ < π for the wetting and θ > π for the non-wetting liquid.

Back to Fig. 2.5 that shows a drop resting on a solid surface, the energy dE (Helmholtz surface free energy) necessary to modify the interface areas, can be determined considering the three surface tensions: σ_{lv}, σ_{sl} and σ_{sv} acting in the contact line, by the expression:

$$dE = \sigma_{lv} dA_{lv} + \sigma_{sl} dA_{sl} + \sigma_{sv} dA_{sv} \tag{2.43}$$

If the liquid–solid area increases by dA_{lv}, the vapor–solid area dA_{sv} decreases by the same amount. Besides, considering that the drop has a truncated hemispherical shape, $dA_{lv} = dA_{sl} \cos\theta$. So, the last expression can be given as:

$$\frac{\partial E}{\partial A_{sl}} = \sigma_{lv} \cos\theta + \sigma_{sl} - \sigma_{sv} \tag{2.44}$$

The meniscus stable form is obtained when the liquid–vapor surface presents the least free energy, or:

$$\frac{\partial E}{\partial A_{sl}} = 0 \tag{2.45}$$

In the case of a liquid drop, considering no other forces, this shape is a sphere (as already noted).

Usually, due to the small order of magnitude and the difficulty of measuring the solid–liquid and liquid–vapor surface tensions, these terms are neglected in the design of thermosyphons and heat pipes and only the liquid–vapor tension is considered.

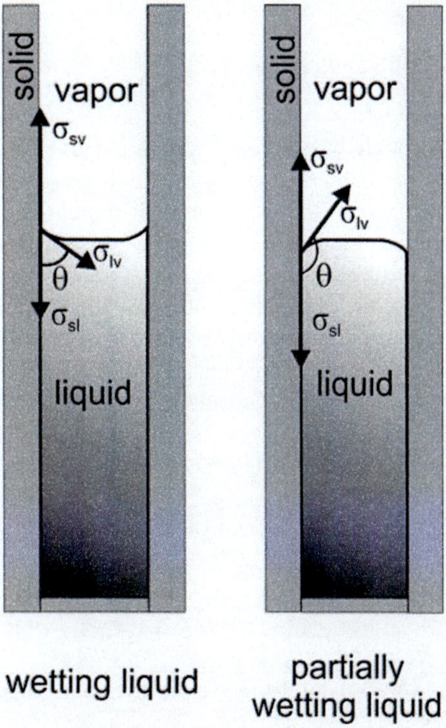

Fig. 2.7 Wetting (left) and partially wetting (right) liquid in tube

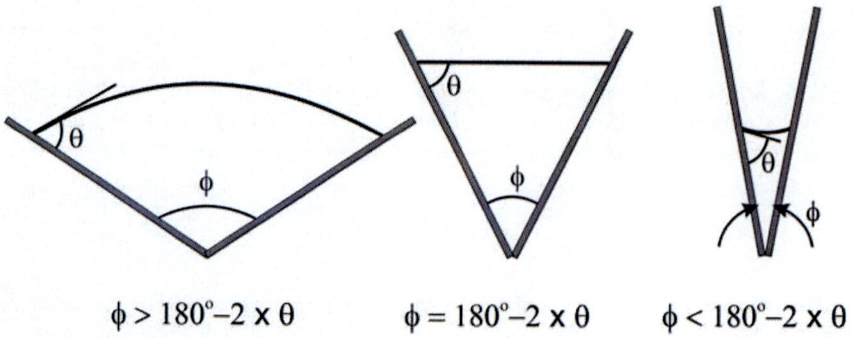

Fig. 2.8 Menisci geometry in a two plate channel with angle φ

This surface tension property is available for several liquids in many books from the literature (Faghri 2016; Peterson 1994; Reay and Kew 2006 among others).

In some thermosyphons and heat pipes, the liquid can be within a V shaped channel, with φ angle (see Fig. 2.8). For a fixed volume of liquid, the shape of the meniscus and of the wall-liquid contact area can be controlled by the φ angle. If the liquid–solid contact angle is θ, the liquid meniscus is concave if $\phi < 180 - 2 \times \theta$ and convex otherwise. Figure 2.8 illustrates these menisci. Usually, especially for heat pipe evaporators, large contact areas between the heated walls and the liquid are desirable.

Capillary Pressure

As described in Chap. 1, the operation principle of thermosyphons and heat pipes is mainly based on an evaporation and condensation cycles, that allows the transfer of very large amounts of heat. To guarantee this cycle, liquid must be continuously supplied to the evaporator. In the case of thermosyphons, the gravity is usually responsible for this task. However, for heat pipes, a porous wick, usually located inside the device, do this job.

For the case of heat pipes, the liquid–vapor phase change takes place in the wick. The evaporation process leads to an increase in the concavity of the meniscus, which, in turn, leads to an unbalance of the surface tension forces between evaporator and condenser menisci, resulting in a capillary pressure. Therefore, the capillary pressure phenomenon is directly related to the meniscus radius of curvature.

Figure 2.9 shows a schematic of a meniscus with curvature radius r_x and r_y in the x and y directions, respectively. A small displacement dz in the z direction causes a deformation of the meniscus area, here represented by a stretch of dx and dy, so that the original area $A = x \cdot y$, changes to:

$$\Delta A = (x + dx)(y + dy) - xy \tag{2.46}$$

Developing the above equation and considering that $dx \cdot dy \approx 0$, then:

$$\Delta A = y \cdot dx + x \cdot dy \tag{2.47}$$

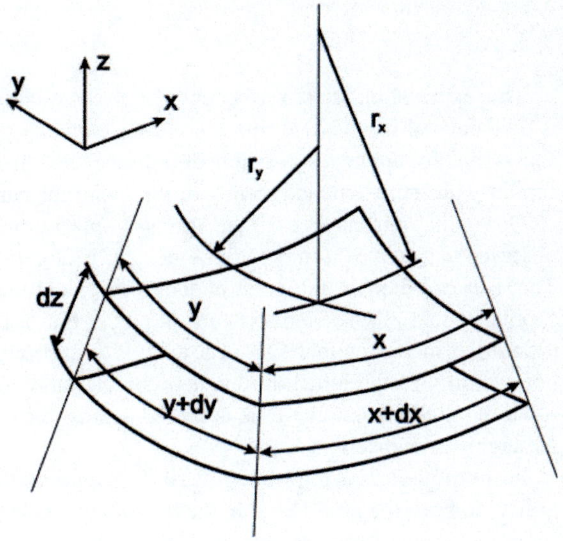

Fig. 2.9 Stretched meniscus geometry with curvature radius r_x and r_y

According to Eq. 2.40, the necessary energy to displace the meniscus surface by *dz* is equal to the work exerted by the resulting normal force over the meniscus area $(\Delta p \cdot xy)$:

$$dE = \sigma(x \cdot dy + y \cdot dx) = \Delta p \cdot xy \cdot dz \tag{2.48}$$

where Δp is the pressure difference between the liquid and vapor regions, separated by the meniscus. Considering similarity of triangles, one gets:

$$\frac{x + dx}{r_x + dz} = \frac{x}{r_x} \text{ and } \frac{y + dy}{r_y + dz} = \frac{y}{r_y} \tag{2.49}$$

Therefore, *dx* and *dy* can be expressed as, respectively:

$$dx = \frac{x \cdot dz}{r_x} \text{ and } dy = \frac{y \cdot dz}{r_y} \tag{2.50}$$

Substituting in Eq. 2.48, one gets:

$$\Delta p \cdot xy \cdot dz = \sigma\left(\frac{xy \cdot dz}{r_x} + \frac{xy \cdot dz}{r_y}\right) \tag{2.51}$$

Δp corresponds to the capillary pressure p_{cap}, due to pressure differences between the liquid and vapor, in both sides of the meniscus and is given by:

$$p_{cap} = \Delta p = \sigma \left(\frac{1}{r_x} + \frac{1}{r_y} \right) \tag{2.52}$$

This expression, known as Young–Laplace equation (Carey 1992; Faghri 2016) is fundamental for the determination of the capillary pressure of a wick structure. It shows that, for applications where the surface tension does not vary significantly, the capillary pressure depends mainly of the meniscus curvature.

One of the consequences of the Young–Laplace equation is that the surface tension observed in a convex liquid–vapor meniscus is larger than that of a concave surface. This fact can explain the phenomenon associated with the penetration of fluid in capillary tubes. Two schematics are shown in Fig. 2.10: in the left, a capillary tube is inserted in a container full of a wetting liquid, while in the right, the liquid is non-wetting. In the left, liquid penetrates the tube up to a height H, while, in the right, the liquid inside the tube is forced against the container fluid and a recession of height -H is observed.

From the left drawing of Fig. 2.10, it is evident that another force acts against gravity to keep the liquid plug in the high level inside the tube of radius r. The same force, but in the opposite direction also acts in the right drawing. Considering a quiescent interface and only the liquid–vapor surface tension, the capillary pressure over the meniscus area (approximated to πr^2) results in a force equal to the sum of

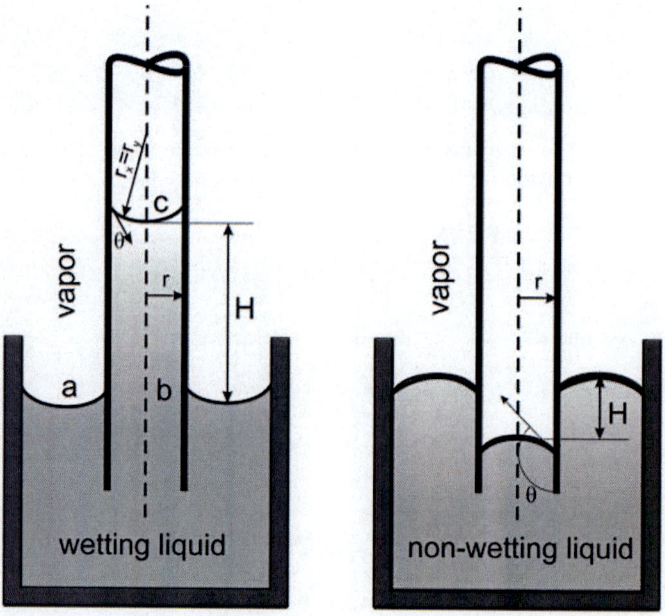

Fig. 2.10 Capillary elevation of liquid in a tube

the forces acting over the solid–liquid contact line, or:

$$p_{cap} \cdot \pi r^2 = 2\pi r \cdot \sigma \cos\theta \tag{2.53}$$

Isolating p_{cap} from the last equation one gets:

$$p_{cap} = \frac{2\sigma}{r} \cos\theta \tag{2.54}$$

and considering that, for a circular tube, the curvature radius r_x and r_y are the same, the radius of the tube and the of meniscus curvature are related by:

$$r_x = r_y = \frac{r}{\cos\theta} \tag{2.55}$$

Point *b* of Fig. 2.10 left side, is located in the basis of the liquid column and is under the action of the hydraulic pressure of the liquid column of height H, of the capillary pressure and of the vapor pressure p_v (which acts at the meniscus at point *c*), i.e.:

$$p_b = \rho_l g H + p_v - p_{cap} \tag{2.56}$$

However, in point *a* of this same sketch, the pressure exerted by vapor is composed of two parcels: column of height *H* and the vapor pressure at point *c*, resulting in:

$$p_a = \rho_v g H + p_v \tag{2.57}$$

As the system is in equilibrium, equating Eqs. 2.56 and 2.57, one has:

$$p_{cap} = (\rho_l - \rho_v) g H = \frac{2\sigma}{r} \cos\theta \tag{2.58}$$

The maximum capillary pressure happens when $\cos\theta = 1$, or $\theta = 0$. In this case, the liquid wets the tube completely. For the non-wetting liquid, $\theta > \pi/2$ and the capillary forces pushes the liquid in the container direction, so that a liquid depression is observed in the tube, as shown in Fig. 2.10 right side. Therefore, for heat pipe applications, working fluids with high wettability are desirable, aiming the effectiveness of the capillary action of the wick.

In heat pipes, due to the evaporation process, the working fluid tend to recede into the wick in the evaporator section and the meniscus radius tends to decrease. The opposite happens on the condenser region, where the excess of liquid makes the radius of the meniscus to increase, tending, in the limit, to infinite. Therefore, in the evaporator, the meniscus pull the liquid from the condenser region though the wick, while, in the condenser the meniscus pushes the liquid to the evaporator region. The combination of both effects results in the capillary pumping of the liquid. Figure 2.11 shows a schematic of the meniscus geometry along a metal screen wicked heat pipe, highlighting the curvature radius of the meniscus in the condenser r_c and evaporator r_e. For the limit case where, in the condenser, the meniscus radius is infinite, only the

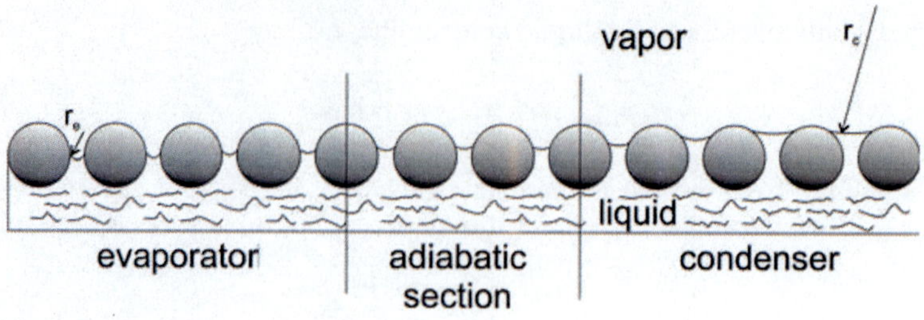

Fig. 2.11 Meniscus radius in a screen wicked heat pipe

evaporator wick is responsible for the capillary pumping, as the capillary pressure provided by the evaporator region is zero, as stated by Eq. 2.52. This is also the same pumping condition for the loop heat pipes, where the wicks are mainly concentrated at the evaporator.

Many different structures can be used as wicks for heat pipes. Schematics of common wicks are shown in Fig. 2.12 (Reay and Kew 2006). The most simple one comprises of a capillary tube and is mostly used to transport condensate from the condenser to the evaporator with low pressure drops. The most usual wicks in electronic cooling applications are homogeneous, such as superposed metal wire screen layers and sintered metal powders, in which the pores are interconnected. Due to their

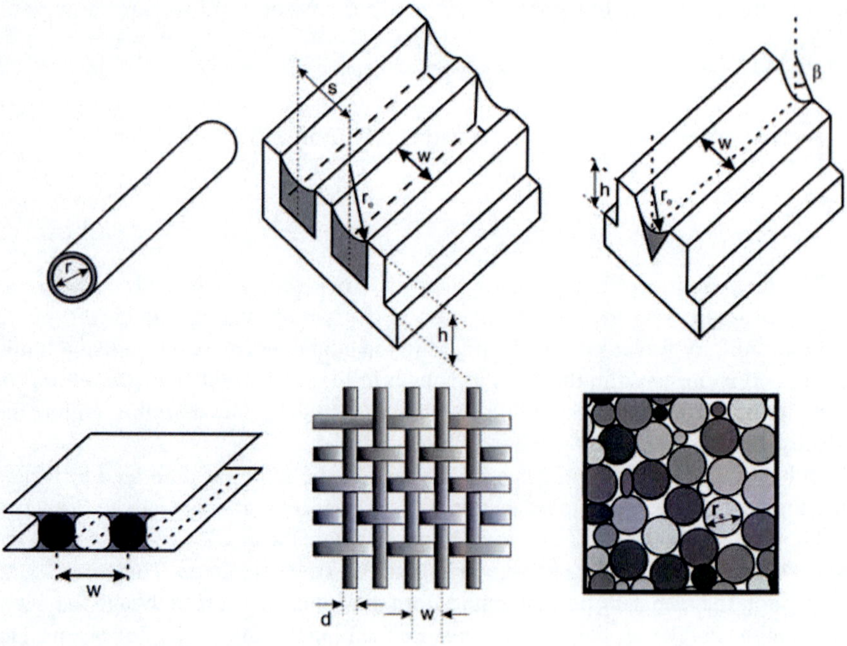

Fig. 2.12 Schematic of usual wicks

small pores, these wicks present high capillary pumping capacity. On the other hand, internal longitudinal grooves, of different cross section geometries, such as rectangular and triangular, are not interconnected. They provide small pressure drops but also small capillary pumping effects. More complex and efficient porous media designs are also proposed, such as those composed by the combination of grooved channels and wire screen wicks, for instance.

For modeling purposes, plain smooth meniscus surfaces are usually considered, as for the contact angle model (Eq. 2.42). Usually, the capillarity of a wick structure is determined experimentally. The concept of maximum capillary pressure for a small diameter (capillary) tube is used to describe the pumping capacity of a wick, by means of an effective meniscus radius (r_{ef}) that characterizes the capillary porous media. The maximum capillary pressure for circular cross section tube is obtained when it is filled with perfect wetting liquid, being given by:

$$p_{cap,max} = \frac{2\sigma}{r_{ef}} \tag{2.59}$$

According to Florez et al. (2017), for a capillary structure made of sintered metal powder, the effective porous radius can be calculated using the concept of hydraulic diameter of channels formed among packed spheres of uniform radius. Considering the powder as composed by spheres of same diameter r_p, these authors showed that, among several packing configurations, the cubic is the one that represents better the actual particle disposal in sintered wicks. Figure 2.13 shows schematics of this packing configuration and of the cross section of the liquid path through the sintered wick. To determine the effective porous radius, the concept of hydraulic diameter is used, where the transversal irregular cross section geometry (see Fig. 2.13 center drawing), is considered as a circular duct of variable cross section radius. The use of the hydraulic diameter concept based on the square root of the area ($D_H = \sqrt{A}$) and of the radius of the liquid flow cross section area proposed by Bahrami et al. (2006), resulted in the duct shown in the right side of Fig. 2.13. Based on this geometry, the following expression for the radius of the liquid channel,r_{lc}, is proposed (Florez 2017):

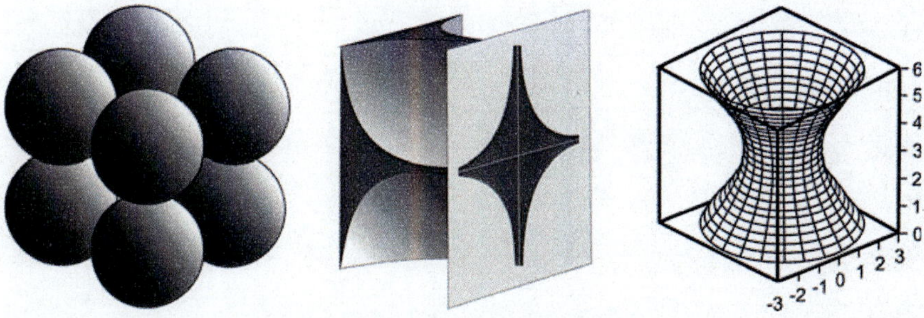

Fig. 2.13 Schematic of the cubic metal sphere packing and liquid channel

$$r_{lc} = \frac{\sqrt{4r_s^2 - \pi r_s^2\left(1 - \frac{(z-r_s)^2}{r_s^2}\right)}}{2} \tag{2.60}$$

where z is the length of the channel, varying from zero to r_s. These authors considered that the largest cross section area is observed at the center of the packing $z = 0$, which radius is actually the same of the metal sphere that composes the porous medium, here represented by the effective radius:

$$r_{ef} = r_s \tag{2.61}$$

These authors also stated that this simple model compared with their data within 25%.

For a grooved capillary structure which cross section channel is a rectangle of width w, the effective meniscus radius can be considered as w (Chi 1976), and the maximum capillary pressure is $P_{max,cap} = 2\sigma/w$. Other expressions for capillary radius of the wick structures shown in Fig. 2.12 are given in Table 2.1.

Many methods are available for measurement of the surface tension (Reay and Kew 2006). For heat pipe applications, special attention is given to the determination of the parameter cos θ, which actually determines the capillary force. Data for surface tension are available in many specialized books. Recently, Faghri (2016) published a very complete compilation of heat pipes and thermosyphon working fluid data that includes surface tension parameters.

Disjoining Pressures

The disjoining pressure, usually a phenomena observed in ultra thin films, can be important in mini or micro heat pipes and thermosyphons or in other devices where liquid is in intimate contact with sharped grooved metal surfaces, such as wire-plate mini heat pipes, pulsating heat pipes, miniature rotating heat pipes, among others.

Table 2.1 Effective capillary radius expressions for several wick strutures

Wick structures	r_{ef}	
Circular cylinder*	r	r: radius
Rectangular groove*	w	w: groove width
Triangular groove*	$w/\cos\beta$	w: groove width β: half angle
Parallel wires*	w	w: wire spacing
Wire screens*	$(w + D_w)/2$	w: wire spacing D_w: wire diameter
Sintered metal powder**	$r_s r_s$	r_s: sphere radius
Packed spheres	$0.41 \cdot r_s$	r_s: sphere radius

Source *Chi, 1976, **Florez et al. (2017)

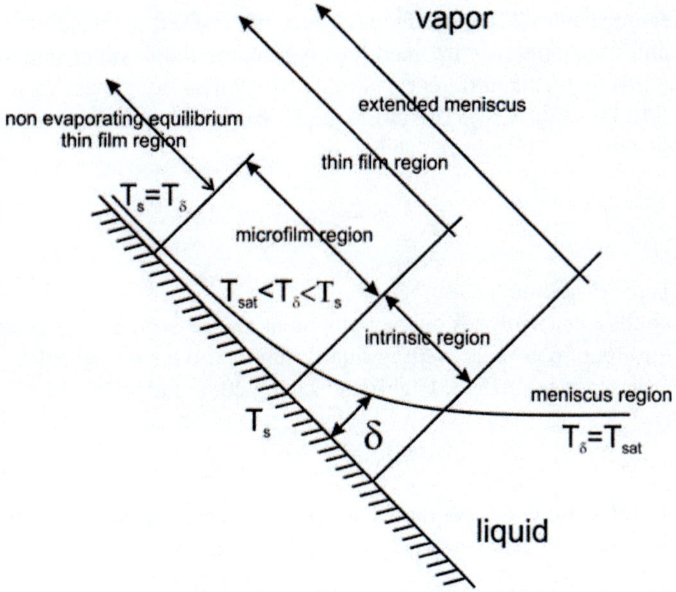

Fig. 2.14 Liquid film over an inclined solid surface

Faghri and Zhang (2006) state that, due to disjoining pressure effect, a liquid with high wetting capability characteristics (polar or non-polar) in contact with a planar solid wall, forms an extended meniscus, which can be divided into three regions: equilibrium thin film, micro film and intrinsic region, as shown in Fig. 2.14. The equilibrium thin film and the microfilm regions actually compose the thin film region.

As stated by Wayner et al. (1975), if the film is sufficiently thin, the van de Waals forces between the solid and the liquid are the predominant ones. Consequently, the intermolecular attractive forces among liquid–liquid and liquid–solid molecules tend to push back the liquid to the liquid film direction, making the liquid tightly adhered to the wall and, consequently, no evaporation is observed. In this case, the liquid vapor interface temperature T_δ is the same of the wall surface, T_s. This adhesion to the wall forces result in a pressure denominated "disjoining pressure".

Virtually, if the surface temperature is larger than the saturation temperature, all the evaporation happens at the microfilm region and the liquid-vapor interface temperature lies between the vapor (saturation) and the wall temperatures: $T_{sat} \leq T_\delta \leq T_s$. The disjoining and capillary pressures significantly affect the shape of the meniscus in this region (Faghri 2016).

Moreover, the surface tension is dominant in the intrinsic region, where the effect of the disjoining pressure can be neglected. Actually, the liquid flow that feeds the evaporating microfilm is pumped by a meniscus pressure gradient, caused by changes of the disjoining pressures as the thickness of the liquid layer increases. Therefore, the larger the meniscus curvature, the larger the meniscus pumping capacity of the intrinsic region (Faghri 2016). As the film becomes thicker, the disjoining pressure effect becomes negligible. In the meniscus region, the liquid vapor interface tends to be planar. Actually, the different meniscus curvature among these regions drives the liquid flow (see Fig. 2.14). Therefore, the pressure difference between vapor and liquid at a vapor–liquid interface can be given by:

$$p_v - p_l = p_{cap} + p_d \tag{2.62}$$

where P_d is the disjoining pressure.

Based on the Lennard Jones interaction potential (see Sect. 2.2.1), the disjoining pressure, for a pure non-polar wetting liquid film over a horizontal solid substrate, is given by (Wayner et al. 1975; Faghri and Zhang 2006; Faghri 2016):

$$p_d(\delta) = -A\delta^{-B} \tag{2.63}$$

where A and B are constants that characterize interactions at the molecular and electrostatic levels (for more information see Potash and Wayner 1972).

Vaporization and Condensation at Liquid–vapor Interfaces

As already observed, the non-equilibrium conditions resulting from the different curvature radius of the liquid–vapor meniscus, in the condenser and in the evaporator regions, are the major driving mechanism for the capillary pressure development within the porous media, which is responsible for the working fluid transportation in heat pipes.

If the meniscus shape in a wick is concave (as happens for most of the heat pipe applications), the attraction between molecules is superior to that of a planar meniscus. This means that the energy necessary to a molecule to leave a concave surface is larger than that of a planar interface. As a result, the density and the vapor pressure over a concave surface are greater. The inverse is observed for convex meniscus.

From a simple hydrostatic pressure balance and based on Eq. 2.58, the saturation pressure difference between a planar and can be given as:

$$\Delta p_{cap} = \frac{2\sigma}{r}\cos\theta\left(\frac{\rho_{sat}}{\rho_l}\right) \tag{2.64}$$

This expression shows that, for very small equivalent porous medium radius (r_{ef}), of about $1\mu m$ or less, this pressure difference can be quite important (around 20 to 30%) of the vapor pressure. However, for large porous, this difference can be negligible (Peterson 1994).

Several other effects can influence the liquid–vapor interface such as the temperature. Temperature rises are typically associated with surface tension decrease. In the limit, when the density of the saturated liquid and vapor are the same, the surface tension tends to zero. The following expression, to correct the temperature influence on the surface tension (in Newtons per meter) for pure water in contact with its vapor, is proposed by the International Association for the Properties of Water and Steam (1994) (see Ghiaasiaan 2008):

$$\sigma = 0.238\left(1 - \frac{T}{T_{cr}}\right)^{1.25}\left[1 - 0.639\left(1 - \frac{T}{T_{cr}}\right)\right] \tag{2.65}$$

with T in Kelvin and where $T_{cr} = 647.15\ K$ is the critical temperature of the water. This equation shows an almost linear relation of the surface tension with temperature, as observed for most liquids. Jaspar (1972) proposed linear curve fittings for many substances, among them several working fluids of heat pipes and thermosyphons.

As stated by Reay and Kew (2006), the following expression, proposed by Fink and Leibowitz, is valid for liquid metals:

$$\sigma = \sigma_0\left(1 - \frac{T}{T_{cr}}\right)^{n} \tag{2.66}$$

where, for sodium for instance,: $\sigma_0 = 0.241\ N/m$, $n = 1.126$ and $T_{cr} = 2503.7\,K$.

The presence of one or more other substances (solute) in the working fluid (solvent) can also have significant influence on the surface tension. Typically, the solute are surface active and decreases the surface tension, as they concentrate at the liquid–vapor interface (Carey 1992). Solutes that highly enrich the liquid–vapor interfaces are called surfactants. A typical case is soap in water.

Another important factor to be considered is the Marangoni effect (Carey 1992), which is observed mainly in zero or close to zero gravity conditions. As the surface tension is a function of both temperature and species concentrations, variations of any of these factors over the liquid–vapor interface can cause motion of the liquid close to the regions where the surface tension is larger. To understand this effect, one can consider, for instance, a vapor bubble in a pool, in microgravity conditions (no natural convection), where one of the pool walls are heated and another one cooled. A close examination shows that the surface tension over the bubble surface portion that is in contact with the colder liquid is larger than that in contact with the hotter liquid. Therefore, the liquid–vapor pressure difference through the coldest interface area is larger than that of the hottest area. As a result, the bubble moves towards the hottest wall direction. This effect provokes a liquid force against the density gradient. In the case of heat pipes operating at zero gravity, a fine vapor layer over a heated wall may be formed due to this effect, decreasing sensibly the heat transfer rate from the solid surface.

Hysteresis Effect

The contact angle hysteresis is an important parameter that should be considered in the design of heat pipes and thermosyphons. This hysteresis is directly related to the variation of the liquid–solid contact angle, as a result of the direction in which forces act over the liquid. A typical example is a drop attached to a vertical wall

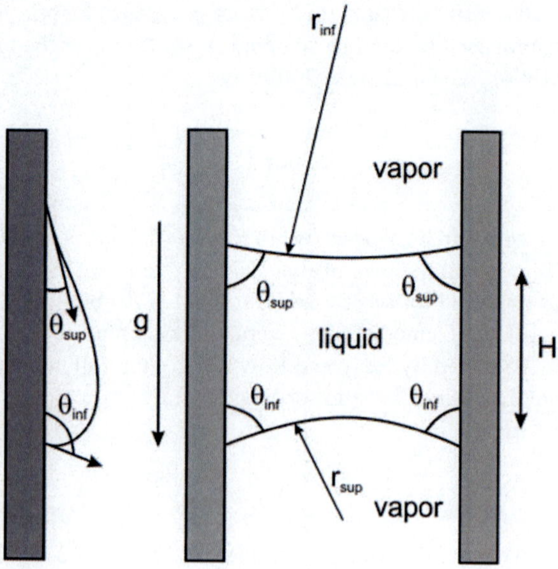

Fig. 2.15 Liquid plug sustained by capillary forces

under the influence of gravity, as illustrated in Fig. 2.15. The affinity between liquid and solid (tension at the solid-liquid interface) makes the liquid drop to be attached to the vertical wall. The action of the gravity force pushes the drop down, decreasing and increasing the superior and inferior contact angles, respectively.

Taking another example, some liquid drops can stand within straws, even when subjected to gravity forces, due to the hysteresis effect. If the liquid is able to wet the straw wall, the hysteresis effect of liquid–solid contact angle due to gravity action causes the contact angle between the liquid drop and the straw to be different in the upper (θ_{sup}), and lower (θ_{inf}) liquid–solid interfaces. Consequently, the superior and inferior liquid–vapor meniscus have different curvature radius, $r_{sup} > r_{inf}$, resulting in an upward surface tension force that acts against downward gravity. The meniscus geometry of a drop within a straw is also shown in Fig. 2.15. Using Eq. 2.59, the superior meniscus capillary pressure, considering that the liquid wets perfectly the solid material ($\cos\theta = 0$) is $2\sigma/r_{sup}$. For the lower meniscus, the capillary pressure is $2\sigma/r_{inf}$. The liquid force, resulting from this capillary pressure unbalance, counterbalances the hydrostatic pressure of the liquid plug of height H, resulting in the following expression, where the vapor density, which is much smaller in comparison with the liquid, is neglected:

$$2\sigma\left(\frac{1}{r_{\text{sup}}} - \frac{1}{r_{\text{inf}}}\right) = \rho_l g H \tag{2.67}$$

In other words, this expression shows that the difference in the radius of curvature is responsible for the force able to support the weight of the liquid plug.

2.3 Fluid Flow

Hagen-Poiseuille Flow

Many models concerning the flow behavior of working fluid in thermosyphons and heat pipes are based on the Hagen-Poiseuille flow model, briefly described in this section.

Figure 2.16 shows a schematic of a steady state laminar flow of an uncompressible fluid in a circular cross section tube with axial velocity u (variable in the r direction), and a differential element where the forces are sketched, where τ_r is adjacent layer shear stress (variable in the r direction). The Hagen-Poiseuille flow hypothesis must be observed: the radial velocity component $v(r)$ and the gradient of the tube axial velocity in the axial direction $\partial u/\partial x$ are both zero. Considering the flux momentum null and that only shear and pressure forces act over the differential element, the following partial differential equation results from a force balance:

$$\frac{d}{dr}(r\tau_r) = r\frac{dp}{dx} \tag{2.68}$$

Applying the Newton law of viscosity $\tau_r = -\mu du/dr$ the last equations turns to:

$$\frac{\mu}{r}\frac{d}{dr}\left(r\frac{du}{dr}\right) = \frac{dp}{dx} \tag{2.69}$$

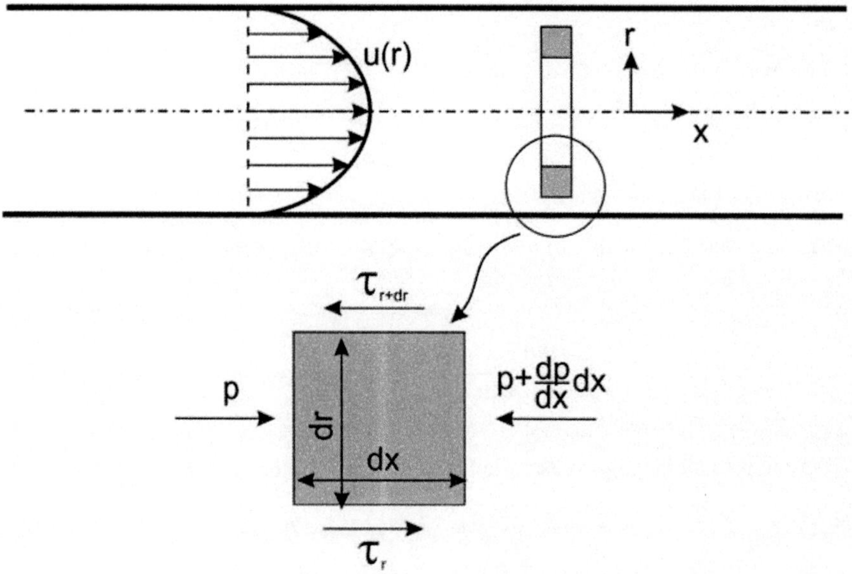

Fig. 2.16 Flow velocity profile in a Hagen-Poiuseuille flow and force balance in a differential fluid element

As the pressure in the axial (x) direction does not depends on r, the last expression can be integrated twice on r to get:

$$u(r) = \frac{1}{\mu}\left(\frac{dp}{dx}\right)\frac{r^2}{4} + C_1 \ln r + C_2 \tag{2.70}$$

Applying the non-slip ($u(r_i) = 0$, where r_i is the internal radius) and symmetry $\left(\partial u/\partial r|_{r=0} = 0\right)$ boundary conditions, one gets:

$$u(r) = -\frac{r_i^2}{4\mu}\left[1 - \left(\frac{r}{r_i}\right)^2\right]\frac{dp}{dx} \tag{2.71}$$

Applying the first order approximation to the pressure derivative $dp/dx \approx (p_2 - p_1)/L$ where L is any length, the axial velocity can be given by:

$$u(r) = \frac{r_i^2}{4\mu}\left[1 - \left(\frac{r}{r_i}\right)^2\right]\frac{p_2 - p_1}{L} \tag{2.72}$$

Therefore, the velocity curve has a parabolic shape, with its maximum value at the center of the tube:$u_{max}(0) = r_i^2/4\mu \cdot (p_2 - p_1)/L$ and so the velocity profile can be written as:

$$u(r) = u_{\max}\left[1 - \left(\frac{r}{r_i}\right)^2\right] \tag{2.73}$$

The mass flow rate can be determined by the expression:

$$\dot{m} = \rho \bar{u} A_{cs} \tag{2.74}$$

where A_{cs} is the tube cross section area and $\bar{u}$ is the mean velocity, which, for an uncompressible fluid with variable velocity, can be determined based on the mass flow rate by the following expression:

$$\bar{u} = \frac{\int_{A_{cs}} \rho u(r,x) dA_{cs}}{\rho A_{cs}} = \frac{2}{r_i^2}\int_0^{r_i} u(r,x) r\, dr \tag{2.75}$$

For the present Hagen-Poiuseuille flow, this mean velocity is given by:

$$\bar{u} = -\frac{r_i^2}{8\mu}\left(\frac{dp}{dx}\right) \cong \frac{r_i^2}{8\mu}\frac{p_2 - p_1}{L} \tag{2.76}$$

So, the pressure gradient can be expressed as:

$$\frac{dp}{dx} = -\frac{8\mu\,\overline{u}}{r_i^2} = \frac{p_2 - p_1}{L} \tag{2.77}$$

Therefore, the mass flow rate (see Eq. 2.74) can be determined as:

$$\dot{m} = \rho\pi r_i^2 \overline{u} = \rho\frac{\pi r_i^4}{8\mu}\left(\frac{dp}{dx}\right) = \rho\frac{\pi r_i^4}{8\mu}\frac{p_2 - p_1}{L}. \tag{2.78}$$

Recalling that the Reynolds number express the relation between inertial and viscous forces, in the present case, one has:

$$\mathrm{Re} = \frac{\rho\overline{u}D_i}{\mu} = \frac{\rho\overline{u}^2}{\mu\frac{\overline{u}}{2r_i}} \tag{2.79}$$

where $D_i = 2r_i$ is the internal diameter of the tube. For Re < 2100, the flow is laminar, for 2100 < Re < 4000, the flow is transition and if Re > 4000, the flow is turbulent. As Re increases, the velocity distribution changes to a flattened pattern, as illustrated by Fig. 2.17.

As discussed by Reay and Kew (2006), the equivalent pressure drop resulting from the kinetic energy of the fluid with velocity $\overline{u}$ is $\rho\overline{u}^2/2$, while the equivalent pressure drop due to the viscous forces is $8\mu\overline{u}L/r_i^2$ (see Eq. 2.77). The ratio between these pressure drops is $\left(r_i^2\rho\overline{u}^2/16\mu\overline{u}L\right) = (Re \cdot r_i/32L)$. This means that, in laminar flows, the equivalent kinetic pressure drop must be smaller than the viscous one, when $(Re \cdot r_i/32L) = 1$, which means that kinetic is equivalent to the viscous energy for $L = Re \cdot r_i/32$.

Turbulent flow–Fanning Equation

The pressure drop due to a turbulent flow is related to the mean velocity by means of the Fanning equation:

$$\frac{p_2 - p_1}{L} = \frac{4}{d}f\frac{1}{2}\rho\overline{u}^2 \tag{2.80}$$

where f is the Fanning factor, related to the Reynolds number given by Blausius Equation (Reay and Kew 2006):

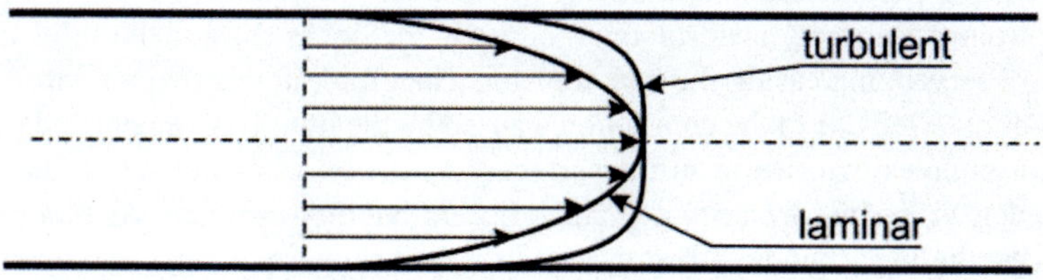

Fig. 2.17 Turbulent and laminar flow

$$f = \frac{0.0791}{\mathrm{Re}^{1/4}} \qquad for \quad 2100 < \mathrm{Re} < 10^5$$
$$f = \frac{16}{\mathrm{Re}} \qquad for \quad \mathrm{Re} < 2100 \tag{2.81}$$

2.4 Boiling

Boiling is the evaporation phenomena that happens in a solid–liquid interface. When the surface temperature is larger than the saturation temperature, in a difference denominated temperature excess (also known as liquid superheat, $\Delta T_e = T_s - T_{sat}$), the formation of bubbles is observed. The bubble vapor growing and detaching dynamics depends on several factors, such as: solid surface finishing, liquid and solid temperatures (levels and excesses) and fluid properties (surface tension, among them). The bubble formation dynamics affects the fluid movement close to the solid surface and so the heat transfer. Essentially, the boiling may happen in pool boiling or forced convection modes. Pool boiling happens when the liquid is quiescent, moved only due to natural convection. In forced convection, the fluid movement, and so the bubble dislocation, are induced by external means, such as pumps. Boiling can occur in subcooled and saturated liquids. If subcooled, the major part of the liquid is in a temperature lower than the saturation and the vapor bubbles, after detaching from the solid surface, condensate within the liquid. In saturated liquid boiling, the liquid is in a temperature level a little higher than the saturation´s and the formed bubbles are pulled out of the solid surface by buoyancy forces in the liquid–vapor interface direction, where vapor is released.

In this section, theories concerning pool and forced convection boiling in saturated liquids are briefly presented, as these phenomena can be observed within thermosyphons and heat pipes.

Pool Boiling

Saturated pool boiling may happen in different regimes, according to the temperature excess: natural convection, nucleate boiling, transition regime and vapor film. The nucleate boiling is the regime where heat is more effectively transferred. Besides, nucleate boiling can be sub-divided into two regimes: isolate bubbles, where bubbles are formed in the nucleate sites and not isolated bubbles (Zuber, 1978). As bubbles depart from the surface, a substantial mixture in the liquid close to the solid surface can be observed, increasing the heat transfer. Therefore, in this regime, most of the heat is transferred due to the convection, caused by the liquid movement and not due to vapor bubbles that rise from the surfaces. As the excess temperature increases, more nucleation sites are activated and adjacent bubbles coalesce. As this process enhances, the vapor tends to rise in jets, increasing the liquid agitation and so the heat transfer, until a maximum heat transfer value is reached (critical flux). However, as more bubbles coalesce, vapor layers tend to be formed over the solid surface, decreasing the heat transfer.

As already observed, the nucleate boiling is the most effective heat transfer boiling regime, reaching very high levels, such as $10^4 W/\left(m^2 \cdot K\right)$ for water (see Incropera et al. 2014), and so it is usually desirable that any heat transfer equipment, where liquid–vapor phase change of a fluid is the major heat transfer mechanism, work in this condition.

Therefore, the boiling phenomenon can be understood as a similar to forced convection over a surface, driven by the ascending bubbles movement and, as for convection, the boiling correlations can be obtained in expressions of the type (Rohsenow 1951):

$$Nu = C_1 \cdot \mathrm{Re}^m \overset{n}{\mathrm{Pr}} \tag{2.82}$$

where *Nu, Re* and *Pr* are the Nusselt, Reynolds and Prandtl numbers, respectively and C_1, *m* and *n* are adjustable constants. A suitable length scale to be used in these non-dimensional parameters can be related to the size of the bubbles that depart from the heated surface, obtained from equating buoyancy and surface tension. Therefore, the bubble diameter is related to the liquid Laplace length scale (Rohsenow and Hartnett 1973, Zuber 1962, and Kutateladze 1961):

$$D_b \propto \lambda_l = \sqrt{\frac{\sigma}{g(\rho_l - \rho_v)}} \tag{2.83}$$

The characteristic velocity of the fluid, in boiling intense movement, can be obtained as the ratio between the distance that the fluid has to move to fill the space left by a departing bubble and the time t_b between consecutive bubble departments, calculated as the time required to the bubble of dimension D_b to be formed by vaporization. Therefore (Incropera et al. 2014):

$$u_b \propto \frac{D_b}{t_b} \propto \frac{D_b}{\left(\frac{\rho_l h_{lv} D_b^3}{q_s'' D_b^2}\right)} \propto \frac{q_s''}{\rho_l h_{lv}} \tag{2.84}$$

Substituting Eqs. 2.83 and 2.84 in Eq. 2.82, the following widely used expression for the surface heat flux, developed by Rohsenow in 1952, is obtained, where the constant *m* was experimentally adjusted to be 2/3 and *C* and *n* depends on the combinations of the liquid and solid (Ghiaasiaan 2008; Collier and Thome 1994, Incropera et al. 2014):

$$q_s'' = \mu_l h_{lv} \left[\frac{g(\rho_l - \rho_v)}{\sigma}\right]^{1/2} \left(\frac{c_{p_l} \Delta T_e}{C_b h_{lv} \,\mathrm{Pr}_l^n}\right)^3. \tag{2.85}$$

For the combination of water and mechanically polished stainless steel, for instance, $C_b = 0.0132$ and $n = 1$. Other liquid-surface combination parameters can be found in the literature (Ghiaasiaan 2008).

Many other researches proposed different correlations for boiling heat transfer (Ghiaasiaan 2008), among them, one can find the correlations of: Foster and Zuber (proposed in 1954), Stephan and Abdelsalam, (proposed in 1980), Copper (proposed in 1984) and Gorenflo (proposed in 1993).

Pool Boiling Critical Heat Flux

The critical heat flux (CHF) represents the peak heat flux in boiling process. It represents the maximum heat flux that a heated surface can bear before the liquid loses the physical contact with the heated surface. Many coolers are designed to operate close to this upper limit, which, especially for safety purposes, must be known, to avoid surface burn-out that may happen if this value is exceed.

Zuber et al. (1963) postulated that critical heat flux is a phenomenon controlled by hydrodynamic capacity of a surface to prevent the development of large dry patches over the heated surface during the transferring of vapor to the surrounding. These researchers proposed the following expression for this limit:

$$q''_{\max} = C h_{lv} \rho_v \left[\frac{\sigma g (\rho_l - \rho_v)}{\rho_v^2} \right]^{1/4} \tag{2.86}$$

where $C = \pi/4 \approx 0.131$ for large surfaces (such as large horizontal cylinders and spheres) and $C = 0.149$ for large horizontal flat plates. The CHF for these and other geometries can be found in Ghiaasiaan (2008).

Forced Convection Boiling.

Depending on the heat input over the walls, the liquid flowing inside heated tubes may reach the boiling regime, with bubble formation over the inner surface of the tube wall. In this case, the growing and detaching of the bubbles from the solid walls are strongly influenced by the liquid flow velocity and differ a lot from the pool boiling hydrodynamics.

In a convection boiling process, as vapor is formed in a heated tube with liquid flow in bottom to top direction, several two-phase patterns may be observed, as the vapor quality increases along the fluid flow (Carey 1992; Collier and Thome 1994; Ghiaasiaan 2008).

Figure 2.18 shows a schematic of these two-phase flow patterns, where the wall is uniformly heated and the incoming liquid is at subcooled state in the bottom entrance. In the rear region, a monophasic forced convection region is observed. Delivering heat over the external tube wall, the wall and liquid temperature increase, reaching the liquid saturation temperature. At this point, liquid starts to vaporize at the wall, in a subcooled flow boiling condition, with high radial temperature gradients. Therefore, bubbles form at the wall, but subcooled liquid still flows in the tube center area and the vapor within the formed bubbles may condense. As the fluid flows, the onset

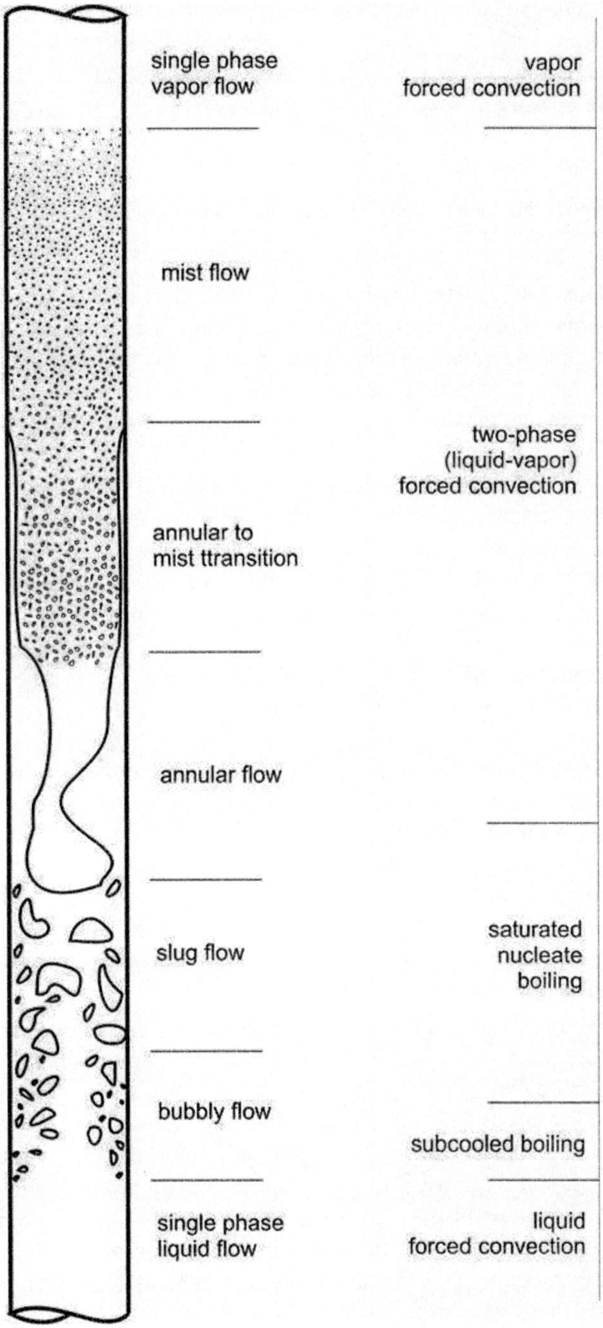

Fig. 2.18 Two-phase flow and boiling regimes in a heated vertical tube

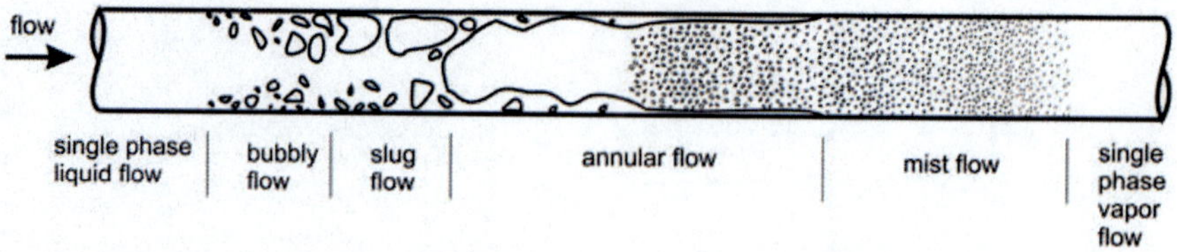

Fig. 2.19 Two-phase flow and boiling regimes in a heated horizontal tube

of more bubbles forms vapor layers over the wall. The temperature of the liquid flowing in the center of the tube increases, to the point in which all the liquid reaches the vaporization temperature. At this stage, bubbles can be observed in any radial position of the tube, determining the onset of the saturated flow boiling region. At this stage, the average fluid velocity increases significantly, due to the density difference between liquid and vapor.

The saturated boiling region can be divided into several sub-regions. The first one is the bubbly flow regime, where loose bubbles can be observed. As the vapor quality increases, individual bubbles coalesce forming vapor slugs, in the slug flow regime. The coalesced bubbles tend to move with higher velocity through the tube center, while a slower liquid layer moves over the walls, reaching the so called annular flow regime. In the sequence, the liquid layer evaporates and a transition region with liquid droplets is observed, followed by a mist regime. Finally, the tube is completely filled with superheated vapor and a second monophasic forced convection region is observed.

The heat transfer coefficient varies significantly (up to one order of magnitude), as the vapor quality increases along the tube. Typically, the lower convection coefficients are located at the forced convection vapor region and the boiling saturated regions present the highest heat transfer coefficient.

Figure 2.19 shows the influence of gravity in the phenomena sketched in Fig. 2.18, observed for tubes operating in horizontal positions. The fluid two-phase flow phenomena for inclined thermosyphons lie between those observed for vertical and horizontal positions.

Several heat transfer correlations can be found in the literature for boiling forced convection (Kandlikar 1990; Kandlikar et al. 1999; Collier and Thome 1994; Carey 1992, Ghiaasiaan 2008).

The reason for discussing some boiling forced convection in this section is that these phenomena are also observed in evaporators of thermosyphons. It worth noting that, in conventional thermosyphons and due to gravity, a liquid film of condensate runs over the internal tube wall from the condenser to the evaporator, causing vapor and liquid counter-current flows in part of the thermosyphon evaporator. In this case, the evaporator phenomena can be somewhat different from the discussions presented here. However, in loop thermosyphons (especially those where the evaporator is composed by heated tubes), the evaporator boiling heat transfer phenomena can be described as presented in this section. Models for the heat transfer at the evaporator region in thermosyphons are discussed later in this book.

2.5 Closure

In this chapter, some multidisciplinary fundamental phenomena, useful for the understanding of thermosyphon and heat pipe operation mechanisms, are described and selected literature models are discussed. The idea is to construct theoretical grounds needed to model the phenomena observed in thermosyphons and heat pipes. In fact, the construction of good tools for designing thermsosyphon and heat pipes demands the understanding of the physical principles of these devices.

References

Adamson, A.W. and Gast, A.P.: Physical Chemistry of Surfaces, 6th edn, John Wiley and Sons (1997)

Bahrami, M., Yovanovich, M.M., Culham, J.R.: Effective Thermal Conductivity of Rough Spherical Packed Beds. Int. J. Heat Mass Transf. **49**, 3691–3701 (2006)

Carey, V.P.: Liquid-vapor Phase-change Phenomena: an Introduction to the Thermophysics of Vaporization and Condensation Process in Heat Transfer Equipment. Series in Chemical and Mechanical Engineering, Taylor and Francis, Hebron (1992)

Chen, S.L., Gerner, F.M., Tien, C.L.: General Film Condensation Correlations. Experimental Heat Transfer: A Journal of Thermal Energy Generation, Transport, Storage, and Conversion **1**, 2 (2007)

Chen, Y, Chien K., Wang, C., Hung, T., Ferng, Y. and Pei, B., Investigations of the Thermal Spreading Effects of Rectangular Conduction Plates and Vapor Chamber, Journal of Electronic Packaging, **129**, 3, 348–356 (2007)

Chi, S.W.: Heat Pipe Theory and Practice A Sourcebook, MacGraw-Hill (1976)

Chun, K.R., Seban, R.A.: Heat Transfer to Evaporating Liquid Films. J. Heat Transfer **93C**, 391–396 (1971)

Churchill, S. W. and Usagi, R., A Standardized Procedure for the Production of Correlations in the Form of a Common Empirical Equation, Ind. Eng. Chem., Fundam., **13** (1), 39–44 (1974)

Collier, J.G., Thome, J.R.: Convective Boiling and Condensation, 3rd edn. Science Publications, Oxford, Oxford (1994)

Faghri, A.: Heat Pipe Science and Technology, 2nd edn, Global Digital Press (2016)

Faghri, A. and Zhang, Y.: Transport Phenomena in Multiphase Systems, Elsevier (2006)

Florez, J.P.M., Chiamulera, M., Mantelli, M.B.H.: Permeability Model of Sintered Porous Media: Analysis and Experiments. Heat Mass Transf. **49**, 1–9 (2017)

Florez, J.P.M., Heat and Mass Transfer Analysis of a Cooper Loop Heat Pipe, Doctoral Thesis, POSMEC, Federal University of Santa Catarina, Brazil (2017)

Fujii, T., Uehara, H.: Laminar Filmwise Condensation on a Vertical Surface. Int. Journal Heat Mass Transfer **15**, 217–233 (1972)

Ghiaasiaan, S.M.: Two-Phase Flow, Boiling, and Condensation, Cambridge University Press (2008)

Girifalco, L. A. and Good, R. J., A Theory for the Estimation of Surface and Interfacial Energies. I. Derivation and Application to Interfacial Tension, J. Phys. Chem., 1957, **61** (7), 904–909 (1957)

Kandlikar, S.G.: A general Correlation for Saturated Two-Phase Flow Boiling Heat Transfer Inside Horizontal and Vertical Tubes. J. Heat Transfer **112**, 219–228 (1990)

Kandlikar, S. G.: In Kandlikar, S. G., Nariai. H.: (eds) Handbook of Phase Change, Boling and Condensation, pp. 331–364, Taylor and Francis, New York (1999)

Kedzierski, M. A., Chato, J. C. and Rabas, T. J., Condensation. In: Bejan, A. and Kraus, A.D. (eds.), Heat Transfer Handbook, pp 719–796. John Wiley & Sons, New Jersey (2003)

Koh, J.C.Y.: Film Condensation in a Forced-Convection Boundary-Layer Flow. Int. Journal Heat Mass Transfer **5**, 941–954 (1962)
Kutateladze, S.S.: Fundamentas of Heat Transfer. Academic Press, New York (1963)
Minkowycz, W.J., Sparrow, E.M.: Condensation Heat Transfer in the Presence of Noncondensables, Interfacial Resistance, Superheating, Variable Properties and Diffusion, Int. J. Heat Mass Transfer **9**, 1125–1144 (1966)
Nusselt, W.: Die Oberllachenkondensation Des Wasserdampfes. Z. VDI **60**, 541–569 (1916)
Peterson, G.P.: Heat Pipes Modeling Testing and Applications, John Wiley and Sons (1994)
Potash, M., Jr., Wayner, P.C., Jr.: Evaporation from a Two-Dimensional Extended Meniscus. J. Heat Mass Transfer **15**, 1851–1863 (1972)
Reay, D.A. and Kew, P.A.: Heat Pipes Theory, Design and Applications, 5th edn, Butterworth-Heinemann (2006)
Rohsenow, W.M.: A Method for Correlating Heat Transfer Data for Surface Boiling of Liquids , Technical Report No. 5, Massachusetts Institute of Technology, (1951)
Rose, J.W.: Fundamentals of Condensation Heat Transfer: Laminar Film Condensation. JSME International Journal **31**(3), 357–375 (1988)
Seban, R.A., Hodgson, J.A.: Laminar Film Condensation in a Tube with Upward Vapor Flow. Int. J. Heat Mass Transfer **25**(9), 1991–1300 (1982)
Stephan, K., Abdelsalam, M.: Heat Transfer Correlations for Natural Convection Boiling. Int. J. Of Heat and Mass Transfer **23**(1), 73–87 (1980)

Chapter 3
Thermal and Hydraulic Models

In this chapter, the thermodynamic behavior of thermosyphon and heat pipes is discussed. Following, classical literature thermosyphon and heat pipe thermal and hydraulic models, used in the design of these devices, are presented.

3.1 Thermodynamic Aspects

The selection of the working fluid for a thermosyphon or heat pipe is one of the most important steps in the design of these devices, as the fluid must perform the expected thermodynamic cycle, according to the operation temperature. Basically, these devices can operate in any temperature and, obviously, the selected working fluid must operate within their melting and boiling points. In cryogenic levels, (5–100 K) working fluids such as helium and hydrogen are usually employed. For low to intermediate temperatures (200–400 K), ammonia, pentane, acetone, alcohols, are some of the used working fluids. For larger temperatures (280–550 K) water is the best working fluid. At little higher temperature levels (400–620 K) naphthalene is the most applied working fluid. Finally, for upper temperature levels (770–1450 K), liquid phase metal working fluids, such as potassium and sodium, are among those commonly used. Figure 3.1 shows a plot of these most used working fluids, as a function of the operation temperature levels, highlighting their melting and boiling points at one atmosphere. One should note that the selection of the working fluid must also take into consideration thermophysical properties such as wettability and surface tension.

The operation mechanisms of thermosyphon and heat pipes are based on the vaporization and condensation of a fluid (working fluid), which process follows a constant volume thermodynamic cycle, as described in the diagrams shown on Fig. 3.2.

After charged and sealed, the heat pipe or thermosyphon working fluid remains in a constant specific volume $\vartheta\left[\mathrm{m^3/kg}\right]$ during operation. If the device working

M. B. H. Mantelli et al., *Thermosyphons and Heat Pipes: Theory and Applications*,
https://doi.org/10.1007/978-3-030-62773-7_3

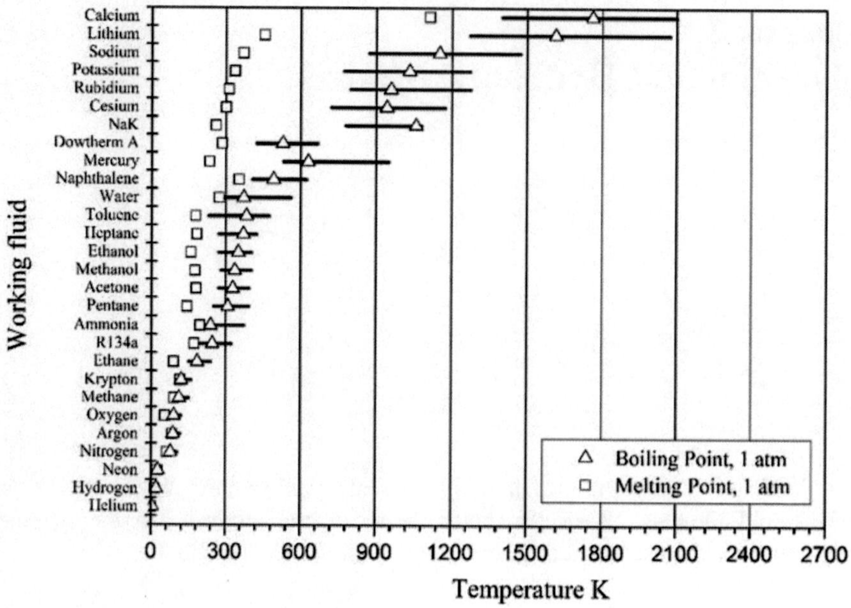

Fig. 3.1 Temperature ranges for working fluids

fluid starts up from liquid and/or vapor states, the total amount of liquid plus vapor masses does not change in any operation conditions. However, as for high temperature (liquid metal working fluids), usually the device starts up from solid state, in vacuum conditions. As the working fluid melts and vaporizes, the sum of liquid and vapor masses changes.

Starting from the condition A (see the pressure-specific volume diagram, Fig. 3.2), the working fluid is in vapor and liquid saturation conditions, however its initial temperature T_i is below the operation temperature level. As heat is introduced, the working fluid temperature increases, until the evaporator and condenser reach the operating temperatures of T_e and T_c, respectively. The complete startup process is represented by A-A' line, in Fig. 3.2. The phase changes in the evaporation and condensation processes happen along the B_l-B_v and C_l-C_v lines, shown in both diagrams of Fig. 3.2. In heat pipes, liquid is retained within the porous medium, where larger pressures are imposed by the capillary structures and, in this position, the liquid is in a sub-cooled liquid state. In thermosyphons, this pressure difference (and therefore the sub-cooling) is attributed to the liquid pool presence, at the bottom of the evaporator: the larger the pool height, the larger the sub-cooling state. The liquid under this thermodynamic state, represented by point D in Fig. 3.2 diagrams, is characterized by the highest pressure and lowest temperature (T_{lc}) level that the working fluid is subjected under steady state operation conditions, and is considered located close to the tube casing wall, in the heat pipe evaporator, or in the bottom

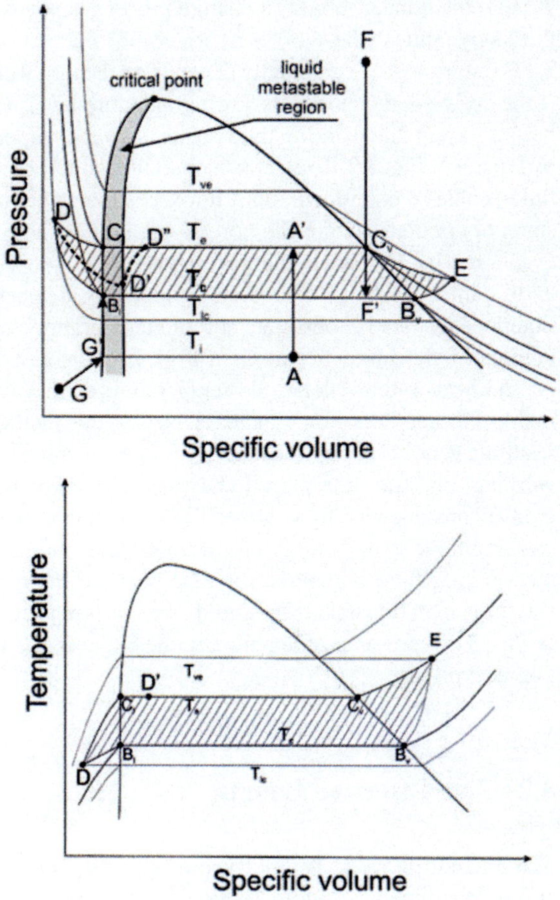

Fig. 3.2 Pressure and temperature as a function of volume for the working fluid

of the pool region, in thermosyphons. However, this hypothesis leads to a thermodynamic inconsistency: as the heat is delivered to the wall and, before it reaches the liquid–vapor interface, crosses the soaked wick structure (for heat pipes) or the pool (for thermosyphons), the liquid necessarily needs to be at higher temperature level than that of the liquid–vapor interface. Therefore, this thermodynamic condition is actually better represented by point D', located in the metastable region of the saturation curve see "Metastable states" section, in Chap. 2).

The liquid warms up (D'-D" process) as it is pumped by the wick (in heat pipes) or it changes of phase (in thermosyphons), forming bubbles in the pool. When the evaporator temperature T_e is reached, phase change process happens along the C_l-C_v line. In the evaporator region, the vapor undergoes a slightly superheating process (C-E). In comparison to the saturation line, point E is characterized by a lower pressure and a higher temperature. However, point E pressure is still above the condenser pressure and this excess causes the displacement of vapor into the condenser direction. In its way, the vapor cools and loses pressure, down to the condensation level,

T_c, and the condenser phase change process happens in the B_v-B_l line, closing the thermodynamic cycle.

The startup process actually depends on the initial conditions of the working fluid. If the device operates at very low temperature levels (such as in cryogenic devices), before startup, the working fluid presents a vapor superheated condition, represented by F point in Fig. 3.2. To establish operation conditions, heat must be removed, so the fluid is able to cool down, until it reaches the saturation temperature and the liquid starts to circulate within the porous medium (in heat pipes) or to be pushed to the evaporator by gravity (in thermosyphons).

In contrast, in high temperature operating devices, the working fluids, usually liquid metals, are in solid state at room temperature and occupies part of the internal volume of the tubes (point G in Fig. 3.2).The rest of the volume is in vacuum. As the heat is introduced, the temperature of the working fluid increases until it melts. Further heat input increase causes the melted solid to evaporate. So, the resulting vapor occupies the empty volume, meaning that the volume occupied by the working fluid (considering all phases) increases and the process cannot be considered at constant specific volume. The G-G' line represents the process in which the working fluid starts from the solid state. After the saturation line is reached and all the working fluid is melted, the heat pipes or thermosyphons operate in the same way of conventional start up liquid state working fluid devices. The hachured areas in Fig. 3.2 represents schematically the operation working cycle of heat pipes and thermosyphons.

3.2 The Effective Length

The amount of working fluid vapor and the liquid masses vary along heat pipes and thermosyphons, due to its evaporation and condensation processes. If heat is uniformly delivered and removed in the evaporator and condenser sections, the rate of vapor production and absorption can be considered uniform. Therefore, the vapor mass flow rate increases linearly within the evaporator region and decreases linearly within the condenser. In the adiabatic region, the vapor mass flow is constant.

At this point, the effective length concept must be presented. Due to the linear natures of the vapor mass increase in the evaporator and decrease in the condenser, the length that the working fluid (in vapor or liquid phase) moves along the heat pipe or thermosyphon can be considered equivalent to the distance between half length of the evaporator and half length of the condenser. Figure 3.3 shows the concept of the effective length of heat pipes and thermosyphons, which is given by the expression:

$$l_{ef} = \frac{l_e}{2} + l_a + \frac{l_c}{2} \tag{3.1}$$

where l_e is the evaporator length, l_c is the condenser length and l_a is the adiabatic length.

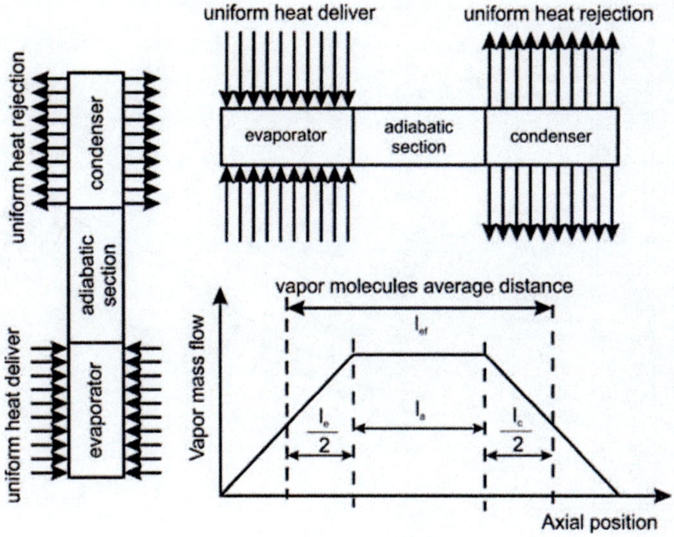

Fig. 3.3 Effective length of thermosyphons and heat pipes

3.3 Liquid Phase Pressure Distribution

The pressure drop within the working fluid liquid phase is a very important parameter to be considered in the design of heat pipe capillary structures that guarantee the appropriate functioning of the device.

However, in thermosyphons, the gravity is responsible for the liquid displacement from the evaporator to the condenser. Actually, gravity is very efficient in transporting the liquid in most thermosyphon geometries and configurations and, therefore, the designer usually is not concerned about this parameter. One exception is for long, very small diameter thermosyphons, where the working fluid vapor bubbles may reach the same dimension of the tube radius, so that liquid is trapped between bubbles, as observed in pulsating heat pipes. Other exception are the loop thermosyphons, where the liquid has to flow within long liquid tube lines: in this case, the simple Haggen Poiseiulle and Fanning models can be used in the device design (see Sect. 2.3). Some discussion of the gravity effect in the heat transfer operation limits (entrainment limits) are presented later in this chapter.

Thermosyphons

Figure 3.4 shows schematically the physical behavior of the working fluid subjected to liquid–vapor phase change within thermosyphons, in vertical and inclined positions. The following regions can be observed: pool boiling evaporation in the evaporator bottom, film evaporation in the evaporator upper region, film with no phase change in the adiabatic region, film with condensation in the condenser region and, finally, only vapor in the top of the device. This last region is especially observed for

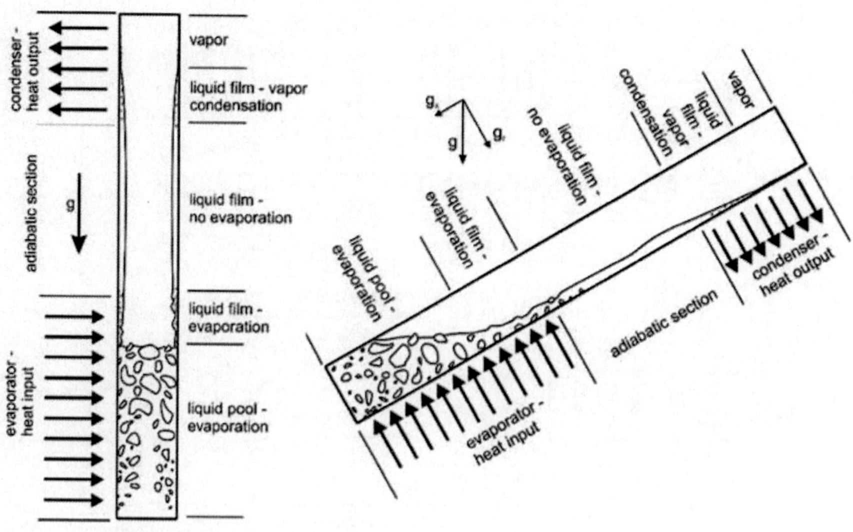

Fig. 3.4 Schematic of liquid film and pool in thermosyphons

long thermosyphons. As the figure shows, in vertical position, the gravity acts only in the thermosyphon longitudinal direction, while, in inclined devices, the gravity actuates in both longitudinal and radial directions and the condensate film also accumulates in the bottom radial region of the tube.

As in most thermosyphon applications, where the gravity is a very efficient liquid displacement mechanism, the designer usually does not need to be concerned about its effect and no liquid pressure distribution models are necessary for designing. However, in some inclined thermosyphon applications, according to the working fluid used (liquid metal for instance) the gravity might play an important role. The pressure distribution can be predicted by determining the gravity force action over the condensate film, using the Nusselt-based models described in Chap. 2. Usually, the molecular interface phase-change mechnisms involving evaporation and condensation are neglected, but they might have a local influence on the liquid pressure drop.

Heat Pipes, Loop Heat Pipes and Capillary Driven Devices.

As already observed in Chap. 2, in a heat pipe, the capillary forces, induced in a liquid by a porous medium, are responsible for the liquid displacement from the condenser to the evaporator.

Wick Structures for Heat Pipes

The fundamental task of a wick structure in heat pipes is to promote the displacement of the liquid from the condenser to the evaporator. In other technologies such as vapor chambers, the wick must assure the spreading of the working fluid along the device, avoiding dry regions. The displacement of the liquid is resulting from the pumping action of the capillary pressure from wicks.

As discussed in Chap. 2, the smaller the wick porous, the higher the capillary pumping capacity of the porous media. However, small wick porous result in media with small permeability (ability of the liquid to flow through a material), resulting in high liquid pressure drops along the heat pipe. Due to this contrast, the literature presents several different wick designs, which try to combine high capillary pumping with low pressure drops.

The most basic wick structures are composed by wrapped screens, grooves and sintered metal powder (more details in the next section), as illustrated in Fig. 3.5. The composed wick structures, also shown in this figure, are designed for special applications such as the thermal control of electronics in satellites. They are more complex and, consequently, more expensive to produce. These special structure designs aim to combine high capillary pumping capacity of the porous media, obtained by the small pore radius, with low liquid pressure drops, by means of free passages for the liquid. Also, these technologies try to avoid the liquid and vapor counter-current flows that cause shear forces. Discussions concerning special heat pipe technologies and wick structures, developed for some special applications, are provided in Part II of this book. The Young–Laplace expression, given by Eq. 2.52, describes the capillary pressure difference between liquid and vapor in a meniscus. As already discussed, the meniscus curvature radius in the evaporator region of an operating heat pipe is smaller than that of the condenser (see Fig. 2.11). This unbalance in the curvature radius causes the capillary pressure that impels the fluid to flow through the porous medium, along the heat pipe. For a well designed heat pipe, the minimum curvature radius location, usually denominated as "dry point", must be at the beginning of

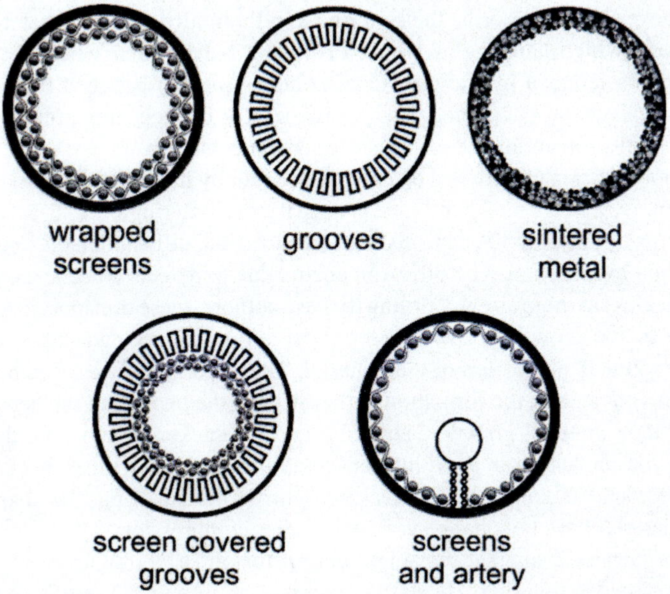

Fig. 3.5 Common wick structures

the evaporator. Likewise, the point where the meniscus radius tends to infinity and therefore the liquid and vapor pressures are nearly the same, denominated as "wet point", is located at the condenser tip.

Using the concept of effective capillary radius and assuming a fluid that wets perfectly a porous material ($\cos\theta = 0$, see Sect. 2.2), the maximum capillary pressure, equated to the difference between the evaporator and condenser capillary pressures ("dry" minus "wet" point), are given by:

$$\Delta p_{cap,max} = \frac{2\sigma}{r_{ef,e}} - \frac{2\sigma}{r_{ef,c}} \tag{3.2}$$

In steady sate conditions, in most heat pipes, the meniscus surface is almost planar, i.e., the curvature radius in the condenser is large (considered as infinite). Therefore, the maximum capillary pressure drop can be expressed as a function of the effective radius of the wick, only at the evaporator, being expressed as:

$$\Delta p_{cap,max} = \frac{2\sigma}{r_{ef,e}} \tag{3.3}$$

Table 2.1 shows capillary radius for several capillary structures. A more complete table can be found in Faghri 2016.

Metal Powder Sintering Process

According to Florez et al. (2014), sintering is a technique used to process particulate materials, aiming among other purposes, the fabrication of porous media. Thümmler and Oberacker (1993) describe the sintering as a thermally activated material transport process, which happens in a bulk of compacted powder, where the particle surface area is reduced by the growth of contacts (necks) between particles. The neck size depends on parameters such as the rate of heating, time, temperature of sintering, particle material and of mass transport mechanisms.

Actually, the grain growth and densification occur by means of the mass transport mechanism. Swinkels and Ashby (1981) classify the sintering process according to the following 5 transport mechanisms: surface diffusion, diffusion in the boundary of the grain and evaporation. The diffusion occurs due to defects (vacancies, gaps and inconsistencies) in the grain. According to these authors, these transport mechanisms contribute to the growth of the neck but only the boundary and lattice diffusion contribute to the densification of the material. The sintering process can be divided in the following stages: the formation and growth of the necks and the densification and particulate material growth. Thümmler and Oberacker (1993) state that micro welding in the contacting areas within particulates form micro bridges (necks), which size depends on the compression pressure. During the sintering, the shape of the sintered material does not change very much after compression.

Ashby (1981) affirms that the superficial diffusion is the dominant mechanism for copper powder sintering. He stated that growing rate of the radius neck, given by $\dot{r}_n$ depends on the temperature and on the physical properties: surface energy, surface diffusion coefficient, energy of activation of the surface diffusion, atomic volume, among others (see Florez et al. 2014). Ashby (1981) affirms that the geometry non-dimensional parameter:

$$G_n = \left[\left(\frac{1}{r_m} - \frac{1}{r_n}\right) + \frac{2}{r_s}\right]\left(1 - \frac{r_n}{r_s}\right) \tag{3.4}$$

is considered as the "driving force" of the sintering process, where r_s is the spherical particle radius, r_n is the neck radius and r_m is the neck meniscus radius (see Fig. 3.6). According to Ashby (1981), the meniscus radius is related to the neck radius through the expression:

$$r_m = \frac{1}{2}\frac{r_n^2}{1 - r_n^2} \tag{3.5}$$

Liquid Pressure Drops

The pressure distribution of the working fluid in its liquid phase in heat pipes is resulting from the combination of viscous and inertial forces. This distribution is not linear due to the effect of the liquid mass injection in the condenser (condensation of vapor) and liquid mass removal in the evaporator (evaporation of liquid). Therefore, the liquid mass flux within the wick increases in the axial direction, from the evaporator to the condenser.

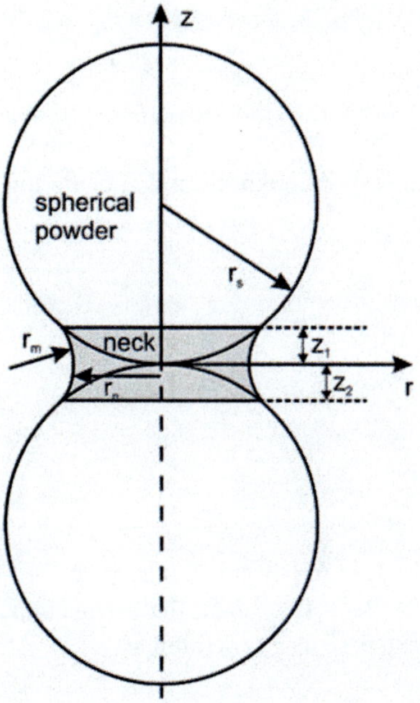

Fig. 3.6 Schematic of the neck formed between spheres in the sintering process of spherical metal powder

While the determination of viscous forces of liquids flowing in arteries and channels (see Chap. 2 for these geometries) is simple and resembles that of a liquid flowing through a tube, the determination of the viscous pressure drops in porous media such as screens and sintered metal, can be quite complex. Neglecting the influence of the dynamic pressures, the liquid pressure gradient, resulting from frictional drag forces and gravity action, is given by Chi (1976):

$$\frac{dp_l}{dx} = -\frac{2\tau_l}{r_h} \pm \rho_l g \sin\theta \tag{3.6}$$

where τ_l is the shear stress in the solid–liquid interface,r_h is the hydraulic radius ($r_h = 2A_l/P_w$) and θ is the heat pipe inclination angle.

The Reynolds number for the working fluid in its liquid phase is defined as:

$$\mathrm{Re}_l = \frac{2r_h\rho_l u_l}{\mu_l} \tag{3.7}$$

The drag coefficient is:

$$f_l = \frac{2\tau_l}{\rho_l u_l^2} \tag{3.8}$$

The liquid velocity depends on the amount of working fluid condensed and so is related to the local heat flux, by the expression:

$$u_l = \frac{q}{\varepsilon A_w \rho_l h_{lv}} \tag{3.9}$$

where A_w is the wick cross section area and ε is the wick porosity, defined as the ratio of working fluid flow area and the wick cross section total area. Combining these last four equations and considering that the heat pipe works at horizontal position ($\theta = 0$), one gets:

$$\frac{dp_l}{dx} = \left(\frac{(f_l \mathrm{Re}_l)\mu_l}{2\varepsilon A_w r_h^2 h_{lv}\rho_l}\right) q \tag{3.10}$$

The parameter permeability of a porous medium (K) is defined as (Chi 1976):

$$K = \frac{2\,\varepsilon\, r_h^2}{f_l \mathrm{Re}_l} \tag{3.11}$$

It is important to note that, for laminar flows, where the laminar friction factor ($f_l Re_l$) is constant, the permeability behaves as a geometric parameter. Therefore, Eq. 3.10 can be written as:

$$\frac{dp_l}{dx} = \left(\frac{\mu_l}{K A_w h_{lv} \rho_l}\right) q \tag{3.12}$$

For circular passages (arteries or tunnel shaped porous media), the hydraulic radius is equal to the radius of the circular cross section area ($r_H = r$), the porosity is one ($\varepsilon = 1$) and, as established by Hagen Poiseuille flow (see Sect. 2.3), $f_l Re_l = 16$. Substituting these values in Eq. 3.11, the permeability is:

$$K = \frac{1}{8} r^2. \tag{3.13}$$

Other values for friction factors can be found in the literature for other passages, such as annular and rectangular ones (Kays 1966), so that Eq. 3.11 can be used to determine the permeability of a specific structure, once the porosity ε and the hydraulic radius r_h are known.

The literature also reports more specific pressure drop models for several heat pipe wicks. Reay and Kew (2006) show that, for a circular tube internally coated by a uniform layer of a homogeneous interconnected porous medium, such as superposed screens layers or sintered metal powder (see Fig. 3.7), the liquid flow cross section area is given by

$$A_w = \pi \left(r_w^2 - r_v^2\right) \varepsilon \tag{3.14}$$

where r_v is the vapor flow region and r_w is the wick external radius (see the external and internal diameters in Fig. 3.7). Considering a Hagen-Poiseuille laminar flow inside the porous medium, where the internal radius r_i is substituted by the effective capillary radius (see Table 2.1) and considering Δp as the pressure drop along the effective length l_{ef}, the liquid mass flow rate ($\dot{m} = \rho \bar{u} A_W$) is determined as:

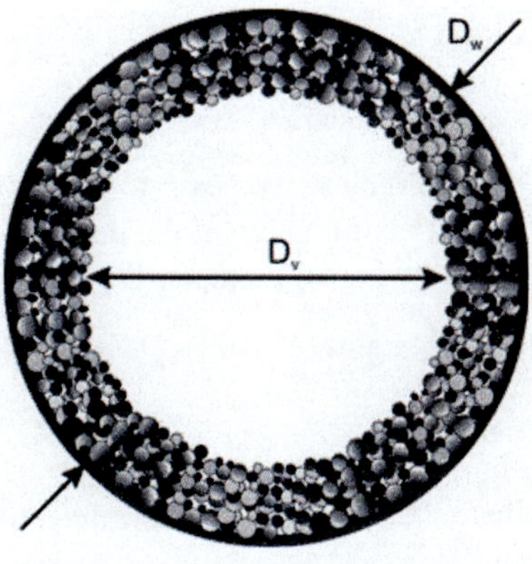

Fig. 3.7 Homogeneous wick recovering the internal surface of a heat pipe

$$\dot{m} = \frac{\pi\left(r_w^2 - r_v^2\right)r_{ef}^2 \varepsilon \rho_l}{8\mu_l}\frac{\Delta p_l}{l_{ef}} \tag{3.15}$$

and $\overline{u}$ is given by Eq. 2.76. Based on the heat used for the working fluid phase change $\left(q = \dot{m}h_{lv}\right)$, the pressure drop of the liquid within homogeneous wicks can be expressed as:

$$\Delta p_l = \frac{8\mu_l\, q\, l_{ef}}{\pi\left(r_w^2 - r_v^2\right)\varepsilon\, r_{ef}^2 \rho_l\, h_{lv}} \tag{3.16}$$

Actually, the liquid phase flow regime is mostly laminar within wicks. However, the fluid does not follow a straight trajectory and so the flow through a circular cross section tube may not be a precise model. Therefore, the Hagen-Poiseuille model need to be corrected. The literature reports (Reay and Kew 2006) that the number 8 can be substituted by coefficients to correct for the actual tortuosity (parameter that characterizes the convoluted pathways of fluid flow through a porous medium). However, these models are of little practical use, as they require the following parameters, which can be hard to get: tortuosity, porosity ε and effective radius r_{ef}.

The Blake-Kozeny equation for the pressure drop in laminar flow of a fluid of viscosity μ through a packed bed of spheres can also be used to predict the pressure drop along a homogeneous porous wick. This expression is:

$$\Delta p_l = \frac{150\mu_l\left(1 - \varepsilon'\right)^2 l_{ef}\, \overline{u}}{D_s^2 \varepsilon'^3} \tag{3.17}$$

where D_s is the diameter of the sphere and ε′ is the volumetric porosity (ratio between the volumes of pores and of the porous medium). This expression can be applied for laminar flow, i.e., when:

$$Re = \frac{\rho \bar{u} d_s}{\mu(1 - \varepsilon')} < 10. \tag{3.18}$$

According to Peterson (1994), Marcus (1972) suggests the following equation for the determination of the permeability of a porous medium composed by screens made of wires of diameter D:

$$K = \frac{D^2 \varepsilon^3}{122(1 - \varepsilon)^2} \tag{3.19}$$

where the porosity ε is determined by the expression:

$$\varepsilon = 1 - \frac{1}{4}\pi S N d \tag{3.20}$$

where N é the mesh number and S is the crimping factor, that takes into account the mesh superposition between layers, being able to vary between 0.95 to 1.05.

Florez et al. (2017) proposed a permeability model based on the liquid flow among the packed spheres of a sintered porous media, using the hydraulic resistance concept. Differently from the others, their model depends only of the average radius of the spherical metal powder. They compared an hydraulic resistance expression, obtained from the application of the Darcy's law, with another one, proposed by Akbariet al. (2010) for converging–diverging laminar fully developed flows within a duct. Equating these equations and using Eq. 2.60 for the radius of the cylindrical duct (equivalent to the irregular circular duct among sintered spheres), the following simple expression, which compared within less than 0.5% with data, was obtained:

$$K = \frac{\pi}{800} r_s^2 \tag{3.21}$$

It is important to note that the porosity is not needed to predict permeability using Florez et al. (2017) model, as it is true for the other models discussed in this section. However the porosity is still an important parameter for other wick design purposes.

Florez et al. (2014), proposed a model to predict the porosity of sintered wick, where the particle average diameter and sintering temperature are the input parameters. As for the effective porous radius and the permeability models developed by this research group, the powder particles are modeled as spheres of uniform radius r_s in cubic arrangement. In this case, the porosity is defined in terms of volume, as the ratio between the empty volume (voids) and the total volume of a tridimensional unit cell in the shape of a cubic box, which edge length is D_s (see dashed lines in Fig. 2.13 that represent the line connecting the center of two adjacent spheres). The void volume is calculated by subtracting, from the total volume of one cell (D_s^3), the solid volume occupied by 8 parts of 1/8 of the sphere volume plus half the volume of the six necks (one neck for each sphere contacts).

According to Florez et al. (2014), the volume of one neck, V_n, is determined as:

$$V_n = \pi \int_{-z_1}^{z_2} r(z)^2 dz - \frac{2}{3}\pi h^2 (3r_s - h) \tag{3.22}$$

where the first term represents the volume of a revolving solid of variable radius r(z) along its longitudinal axis, over the interval $[z_1, z_2]$ (see Fig. 3.6) and the second term represents the two spherical cap that belongs to the spheres of radius r_s, cut by by a plane in $-z_1$, for the lower sphere, and in $+z_2$, for the upper sphere. The parameter h represents the average of these cap heights:

$$h = \frac{z_2 - z_1}{2} \tag{3.23}$$

Through algebraic manipulation, the neck volume is:

$$V_n = 2\pi h r_s\left[\frac{r_s}{2}+\frac{r_s}{2}\left(\frac{r_n+r_m}{r_s+r_m}\right)\right]^2 - \frac{\pi}{3}\left[r_s\left(1-\frac{r_s}{r_s+r_m}\right)\right]^2\left[3r_s - r_s\left(1-\frac{r_s}{r_s+r_m}\right)\right] \quad (3.24)$$

Using this expression and the expressions for the determination of the sphere and cube volumes, one can easily determine the porosity using the Equation:

$$\varepsilon = \frac{D_s^3 - \frac{4}{3}\pi\left(\frac{D_s}{2}\right)^3 - 3V_n}{D_s^3} \quad (3.25)$$

Chi (1994), on the other hand, presented the following simplified expression for permeability of porous media made from sintered metal powders, in which the porosity is an input parameter and the powder is considered made of small spheres of radius r_s:

$$K = \frac{r_s^2\varepsilon^3}{37.5(1-\varepsilon)^2} \quad (3.26)$$

The permeability of a wick of porosity ε formed by closed rectangular grooves can be determined using Eq. 3.11, considering $f_l Re_l = 16$ and $r_h = wh/(w+h)$. The friction factor coefficient $f_l Re_l$ may change from 24, for low values of w/h ratio up to 14, for high ratio value. More information about the variation of the friction factor as a function of w/h can be found in Chi 1976. For open rectangular wicks, the hydraulic radius is $r_h = 2wh/(w+2h)$. The total pressure drop for a tube with n internal longitudinal grooves can be determined adapting Eq. 3.16, resulting in:

$$\Delta p_l = \frac{8\mu_l\, q\, l_{ef}}{\pi\, r_h^4 n \rho_l\, h_{lv}} \quad (3.27)$$

It is important to note that, due to the high velocity of vapor, the shear forces, to some extent, will retain the liquid within the grooves. A way to mitigate this problem is the application of a homogeneous layer over the groove tips (thin screens, for instance). This kind of wick designs are known as composed wicks.

However, the major idea behind the composed wick structures is to ally the capillary pumping capacity of the homogeneous structures (sintered or screens, for instance) with the low pressure drop of channels or tunnels wicks). The homogeneous wick layer must be thin to avoid high pressure drops (predictable using Eq. 3.16). Figure 3.8 shows a schematic of a tunnel-artery composed wick, especially adequate for liquid metal heat pipes, where the two layers of the homogeneous wicks are responsible to transfer condensate along the heat pipe. Reay and Kew (2006) propose the following expression for the pressure drop, when the homogeneous wick thickness w is much smaller than the vapor radius r_v:

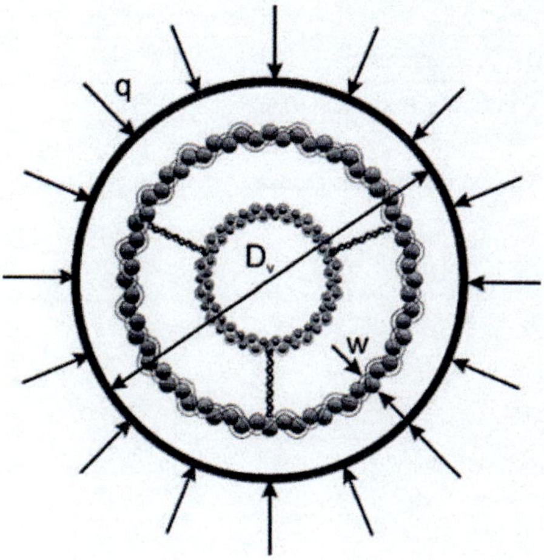

Fig. 3.8 Schematic of a composed wick

$$\Delta p_l = \frac{6\mu_l\, q\, l_{ef}}{\pi\, r_v w^3 \rho_l\, h_{lv}} \tag{3.28}$$

A compilation of permeability of several wicks is presented in Table 3.1.

Another alternative is the use of the Darcy´s model to calculate the pressure drop. Considering the heat flux and all properties as constants, Darcy´s equation can be stated as:

$$\Delta p_l = \frac{\mu_l\, l_{ef} \dot{m}_l}{\rho_l\, K A_w} \tag{3.29}$$

3.4 Vapor-Phase Pressure Distribution

Different from the liquids, gases have the ability to occupy easily available volumes. Therefore, vapor pressure distribution is almost uniform along most thermosyphons and heat pipes and, in many practical applications, vapor pressure drops can be simply neglected. However, the understanding and modeling of the vapor pressure distribution elucidates the hydraulic behavior of thermosyphons and heat pipes, and so, some models are discussed in this section.

The pressure distribution of vapor in thermosyphons and heat pipes is deeply affected by the evaporation and condensation of the working fluid. Actually, most of the models available were developed for heat pipes. As these phenomena are similar for heat pipes and thermosyphons, usually the same models can be applied for both cases.

For devices operating in steady state conditions and subjected to uniform heat supply in the evaporator, the vapor mass flow rate can be considered as increasing

Table 3.1 Permeability of several wick structures

Wick structures	K	
Circular cylinder (arteries or tunnels)	$r^2/8$	r = radius of the liquid passage
Closed rectangular groove	$2\varepsilon r_h^2/fRe$	ε = porosity = w/s w = groove width s = pitch $r_h = wh/(w+h)$
Open rectangular groove	$2\varepsilon r_h^2/fRe$	ε = porosity = w/s w = groove width s = pitch $r_h = 2wh/(w+2h)$
Superposed wire screens	$D^2\varepsilon^3/122(1-\varepsilon)^2$	D = wire diameter $\varepsilon = 1-\pi SNd/4$ S = crimping factor N = mesh number
Packed spheres	$r_s^2\varepsilon^2/37.5(1-\varepsilon)^2$	r_s: sphere radius ε = porosity (depends on the sphere packing)
Sintered metal powder	$\frac{\pi}{800}r_s^2$	r_s: sphere radius
Circular grooves [Brennan and Kroliczek (1979) *	$0.0221\frac{D_g^{2.2}(2\pi-\alpha+sen(\alpha))^{2.1}}{w^{0.2}(2\pi-\alpha)^2}$	D_g: groove diameter α: groove angle w: groove width

Obs.: the porosity of the circular grooved surface is $\frac{D_g[\alpha/2-1/2sin(\alpha)]}{2s[1-cos(\alpha/2)]}$ (Brennan and Krolicek 1979).

linearly. If the heat removal is also uniform, the vapor mass flow rate, in the condenser, can be considered as decreasing uniformly, while it remains constant in the adiabatic section, as shown in Fig. 3.3. Therefore, due to the conservation of mass principle and for usual devices with constant vapor cross section area, as vapor is injected linearly in the evaporator, its axial velocity also increases linearly. On the other hand, for the same reason, the vapor velocity decreases linearly in the condenser.

Neglecting any viscous forces and the gravity effects, Bernoulli´s equation for vapor flow, can be written as:

$$\frac{u^2}{2}+\frac{p}{\rho}=\text{constant} \tag{3.30}$$

Therefore, the increase in the velocity (and so increase of the kinetic energy) results in a decrease of the pressure and vice versa. If the velocity increases linearly with the distance, the pressure drop decreases under a square function of space.

In ideal devices, the pressure drop resulting from the variation of the kinetic energy in the evaporator, can be recovered in the condenser. In Fig. 3.9, the full line illustrates this vapor pressure behavior. There are experimental evidences of this kinetic pressure drop recovery, especially for high density working fluids, as

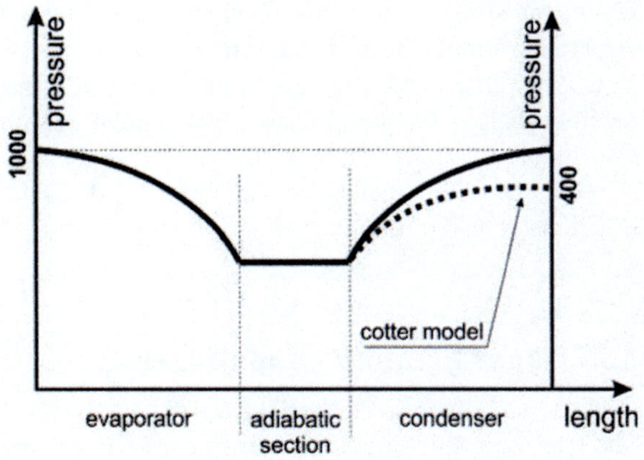

Fig. 3.9 Vapor pressure distribution in thermosyphons and heat pipes

liquid metals. However, more realistic models consider the partial recuperation of the pressure, as shown in Fig. 3.9, dashed line.

However, vapor pressure drops due to the viscous forces are always observed in thermosyphons and heat pipes. The viscous forces act against the vapor flow direction and tend to reduce the vapor velocity. Therefore, due to this effect, the pressure drops decreases monotonically from the evaporator to the condenser. The rate of vapor pressure variation with distance depends on the vapor velocity: the larger the velocities, the larger the shear stresses. Hence, as the vapor velocity changes along the device, the pressure variation also changes. Figure 3.10 illustrates the pressure distributions due to the sum of the kinetic and shear pressure drops.

In thermosyphons, the condensed liquid film, moving in downward direction due to gravity effects, suffers, at the liquid–vapor interface, the shear stress forces of the vapor in upward direction. In the limit, vapor can drag some liquid to the condenser

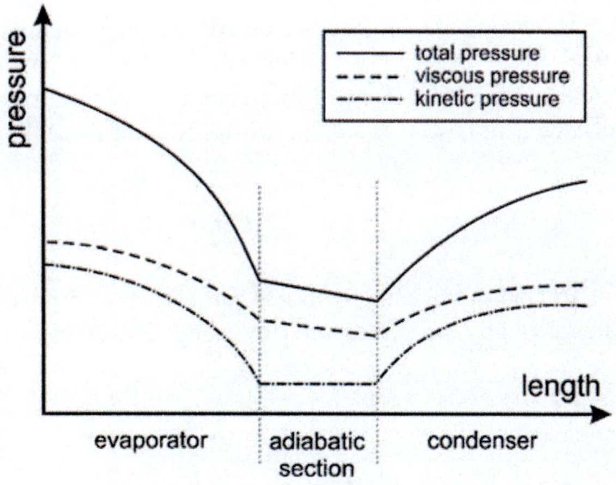

Fig. 3.10 Parcels for the vapor pressure distributions

region, hindering the thermosyphon operation. These phenomena is described in the "Operation Limits" section, later in this book.

On the other hand, in conventional heat pipes, the solid material of the porous medium shelters the liquid phase and the meniscus, so that the vapor shear stresses over the liquid phase can be negligible. Nevertheless, the wick shear stress effects over the vapor flows can be important for the design of some devices, such as long, liquid metal heat pipes.

3.5 Vapor Pressure Drop Models

The total vapor phase pressure drops can be considered as the sum of two parcels: inertial (kinetic, $\Delta p'$) and viscous ($\Delta p''$). In the evaporator, the kinetic parcel must be able to change the radial to axial velocity direction of the working fluid just evaporated and to accelerate the vapor flow to reach the end of the condenser region. The inertia term has a positive sign in the evaporator and negative in the condenser and, in the limit, they can cancel each other. The viscous parcel must overcome the vapor flow viscous forces. Besides, the total pressure drop can be taken as the sum of three terms: evaporator (Δp_{ve}), adiabatic section (Δp_{va}) and condenser (Δp_{vc}).

A good hypothesis is to consider the mass flux in the adiabatic section as unidimensional in the axial direction, while, in the evaporator and condenser, the fluxes are axial and radial, due to the evaporation and condensation effects.

At this point, it is interesting to define the following radial Reynolds number, which takes into consideration the radial vapor velocity v_v in the interface with the wick (Reay and Kew 2006):

$$\mathrm{Re}_r = \frac{\rho_v v_v r_v}{\mu_v} \tag{3.31}$$

The Reynolds number is positive (center direction of the circular cross section area) in the evaporator and negative in the condenser. In most applications, this number varies from 0.1 to 100. Related to the injection and removal of vapor per unit of tube length ($d\dot{m}_v/dx$), the Reynolds number can be expressed as:

$$\mathrm{Re}_r = \frac{1}{2\pi\mu_v}\frac{d\dot{m}_v}{dx} \tag{3.32}$$

The radial and axial Reynolds number, for uniform evaporation and condensation, are related by the expression (Reay and Kew 2006):

$$\mathrm{Re}_r = \frac{\mathrm{Re}}{4}\frac{r_v}{x} \tag{3.33}$$

where x is the distance from the beginning of evaporator or condenser. In Chap. 2, it was established, for the Hagen-Poiseuille laminar flow, that the pressure drop, due to the viscous forces, is comparable to the kinetic energy pressure drop, for the equivalent length $L = Re \cdot r_v/32$. If x is substituted by L, Eq. 3.33 results in:

$$\mathrm{Re}_r = 8 \tag{3.34}$$

Therefore, if $Re_r > 8$, the radial direction of the vapor flow must be considered in modelling. Otherwise, unidimensional models can be applied.

Simplified one-dimensional model.

Vapor pressure distribution models can be complex to be formulated due to the compressibility of the gas and to the addition or removal of vapor in the evaporator or condenser sections. However, in this section a simplified model, where the flow is considered unidimensional, is proposed. In steady state conditions, the vapor and liquid mass flow crossing the adiabatic region is the same. However, due to the large density difference between these phases, the vapor velocity is much larger than the liquid. Therefore, besides the pressure drop due to the shear forces in vapor, dynamic pressure variations due to velocity changes must be considered, even for the unidimensional simple model. While the liquid flow is usually considered laminar, a good hypothesis for heat pipes where the liquid flows within wicks and for some thermosyphon operation conditions, the vapor flow can be either laminar or turbulent.

In the simplified model, the vapor flow is treated as an uncompressible fluid. This means that the velocity of flux u is small when compared to the sound velocity c, with the Mach number lower than 0.3 ($M < 0.3$). In this case, the pressure drop in vapor (Δp_v) is small when compared to the average vapor pressure (p_v). This hypothesis is not always observed during the startup process or for liquid metal working fluid devices.

According to Reay and Kew (2006), if the mass flow per unit cross section area is $\rho_v u$, the corresponding momentum flux per unit length is $\rho_v u \cdot u$. This energy is provided by the kinetic pressure gradient, in the evaporator section:

$$\Delta p'_{ve} = \rho_v u^2 \tag{3.35}$$

For laminar flows, the pressure drop due to viscous forces can be estimated based on the Hagen-Poiseuille model. For the evaporator, considering the rate of vapor mass entering the evaporator per unit length ($d\dot{m}_v/dx$) constant, which means that the injection (or removal) of vapor is linear, Eq. 2.78 can be integrated along the evaporator length, resulting in:

$$\Delta p''_{ve} = \frac{8\mu_v \dot{m}_v}{\rho_v \pi r_v^4} \frac{l_e}{2} \tag{3.36}$$

where $\dot{m}_v$ is the vapor mass flow that leaves the evaporator section (at $x = l_e$). Therefore, the evaporator total pressure drop is:

$$\Delta p_{ve} = \rho_v u^2 + \frac{8\mu_v \dot{m}_v}{\rho_v \pi r_v^4} \frac{l_e}{2} \tag{3.37}$$

The condenser must be treated in the same way, but in this case, the axial moment must be dissipated as the vapor velocity decreases and the kinetic term is negative (there is a recovery of the inertial energy) and the total pressure drop is due only to viscous forces. However, in actual applications the kinetic pressure cannot be fully recovered.

In the adiabatic region, as there is no evaporation and condensation, only viscous terms influences the pressure drop, which, for laminar flows, can be determined using the Hagen-Poiseuille model and, for turbulent flows, using the Fanning equation, through the following expressions:

$$\Delta p_{va} = \frac{8\mu_v \dot{m}_v}{\rho_v \pi r_v^4} l_a, \text{ for } Re < 2100 \tag{3.38}$$

$$\Delta p_{va} = \frac{2}{r_v} f \frac{1}{2} \rho_v u^2 l_a, \text{ for } Re > 2100 \tag{3.39}$$

where, according to Eq. 2.81:

$$f = \frac{0.0791}{\text{Re}^{1/4}} \tag{3.40}$$

Adding the three terms, one has, for laminar flow and no recuperation of kinetic pressure:

$$\Delta p_v = \Delta p_{ve} + \Delta p_{ve} + \Delta p_{vc} = \rho_v u^2 + \frac{8\mu_v \dot{m}_v}{\rho_v \pi r_v^4} \left(\frac{l_e + l_c}{2} + l_a \right) \tag{3.41}$$

For total kinetic pressure recuperation:

$$\Delta p_v = \Delta p_{ve} + \Delta p_{ve} + \Delta p_{vc} = \frac{8\mu_v \dot{m}_v}{\rho_v \pi r_v^4} \left(\frac{l_e + l_c}{2} + l_a \right) \tag{3.42}$$

Chisholm e Chi unidimensional model Chisholm and Chi (1976) developed a model based on the momentum conservation equation applied to a vapor control volume in the evaporator region, subjected to viscous and pressure forces, as shown in Fig. 3.11, which is expressed as:

$$\sum F = \frac{d}{dt}(m u_v) \tag{3.43}$$

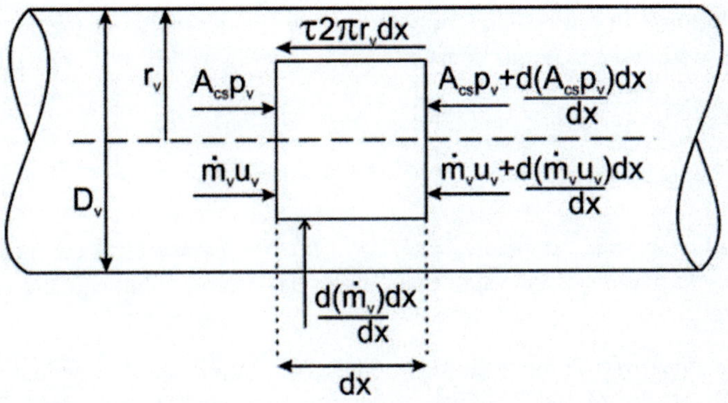

Fig. 3.11 Control volume for one-dimensional model

where m is mass (kg). Using the Reynolds mass transport theorem and neglecting the gravity effects, one has, for steady state conditions:

$$\frac{d}{dt}(mu) = \frac{d}{dx}\int\limits_{A_{cs}} \rho_v u_v^2(r)dA_{cs}dx \tag{3.44}$$

The forces acting over the control volume are in equilibrium, according to the force balance:

$$\sum F = -\frac{d(A_{cs}p_v)}{dx}dx - \tau_v 2\pi r_v dx \tag{3.45}$$

Equating Eqs. 3.44 and 3.45, one gets:

$$-\frac{d(A_{cs}p_v)}{dx}dx - \tau_v 2\pi r_v dx = \frac{d}{dx}\int\limits_{A_{cs}} \rho_v u_v^2(r)dA_{cs}dx \tag{3.46}$$

Developing the derivative term and isolating the derivative of pressure in respect to x, one gets:

$$\frac{dp_v}{dx} = -\frac{\tau_v 2\pi r_v}{A_{cs}} - \frac{1}{A_{cs}}\frac{d}{dx}\int\limits_{A_{cs}} \rho_v u_v^2(r)dA_{cs} \tag{3.47}$$

The second term of the right hand side represents the momentum rate that crosses the cross section area A_{cs}, which is not the same as the momentum calculated using the average values of $\dot{m}_v$ and $\overline{u}_v$. The following factor to correct the momentum is introduced: $\beta = \int_{A_{cs}} \rho_v u_v^2(r)dA_{cs}/\rho_v\overline{u}_v^2(r)A_{cs}$.

Substituting β in the last equation and taking the derivative of the second right term, considering ρ and A_{cs} as constants and assuming dβ/dx ≈ 0, one has:

$$\frac{dp_v}{dx} = -\frac{(f_v\mathrm{Re}_v)\mu_v\dot{m}_v}{2A_{cs}r_v^2\rho_v} - \beta\frac{2\dot{m}_v}{\rho_v A_{cs}^2}\frac{d\dot{m}_v}{dx} \tag{3.48}$$

where: $Re_v = 2r_v\rho_v\overline{u}_v/\mu_v$ and $\tau_v = 1/2 f_v\rho_v\overline{u}_v^2$. The parameter β compensates for the velocity variations of the vapor that crosses the control volume. If the velocity is constant, $\beta = 1$.

Cotter two-dimensional model In the model given by Eq. 3.42, a velocity profile, completely developed and full kinetic pressure recovery are considered. However, some authors propose more precise models, adopting less restrictive hypothesis. According to Reay and Kew (2006), Cotter in 1965, considering that the velocity profiles change along the axial direction in heat pipes and thermosyphons, recommended the following expression for the vapor pressure gradient, based on the Yuan e Finkelstein model for a cylindrical porous wall duct, where vapor is uniformly injected:

$$\frac{dp_v}{dx} = \frac{8\mu_v\dot{m}_v}{\rho_v\pi r_v^4}\left(1 + \frac{3}{4}\mathrm{Re}_r - \frac{11}{270}\mathrm{Re}_r^2\right) \tag{3.49}$$

One should note that the radial mass injection (bi-dimensional effect) is taken into consideration in the last expression by the use of the radial Reynolds number (Eq. 3.31). For $Re_r \ll 1$ and considering uniform mas injection rate along the length, the last equation can be integrated along the evaporator length, resulting in the expression for the evaporator pressure drop, which is similar to Eq. 3.36:

$$\Delta p_{ve} = \frac{4\mu_v l_e q}{\pi\rho_v r_v^4 h_{lv}} \tag{3.50}$$

For $Re_r \gg 1$, Cotter used the pressure gradient proposed by Yuan e Finkelstein for a flow between parallel plane walls with vapor injection, resulting in the following expression:

$$\Delta p_{ve} = \frac{\dot{m}_v^2}{8\rho_v r_v^4} = \frac{\left(\rho_v\pi r_v^2 u\right)^2}{8\rho_v r_v^4} = \frac{\pi^2}{8}\rho_v u^2 \approx 1.23\cdot\rho_v u^2 \tag{3.51}$$

similar to Eq. 3.35 prediction, showing that inertia effects take the major role in this flow condition.

However, for the condenser region, Cotter proposed a different velocity profile, so that the vapor pressure recovery can be estimated by:

$$\Delta p_{vc} = -\frac{4}{\pi^2}\frac{\dot{m}_v^2}{8\rho_v r_v^4} \tag{3.52}$$

which is also equivalent to Eq. 3.35, modified by the constant $-4/\pi^2 0.4$, meaning that, according to Cotter, around 40% of the pressure drop due to kinetic losses is recovered.

In the adiabatic section, fully developed laminar flow is considered, so that Eq. 3.38 can be used. Therefore, the total pressure drops expression proposed by Cotter for the whole tube is:

$$\Delta p_v = \left(1 - \frac{4}{\pi^2}\right)\frac{\dot{m}_v^2}{8\rho_v r_v^4} + \frac{8\mu_v \dot{m}_v}{\rho_v \pi r_v^4} l_a \tag{3.53}$$

3.6 Gravity Pressures

In addition to the pressure drops due to kinetic, viscous and capillary forces (this last only for heat pipes), the gravity action over the liquid and vapor phases of the working fluid, also results in pressure. The gravity effect is the driving force for conventional thermosyphons, however, it can be negligible in heat pipes, especially those operating in horizontal positions.

Usually, for the devices operating in leaning positons (with the evaporator located below the condenser), considering the bottom to top direction, an decrease in the liquid phase pressure and an increase in the vapor phase pressure, due to gravity, is observed. In this case, the gravity pressure has two components: axial (in the direction of the tube main axis) and normal. The normal component is always observed along the inclined length, being larger for tubes in horizontal position; however, it is not found in vertical devices.

Usually, the pressure drop due to gravity in the vapor phase of the working fluid is neglected, both for thermosyphons or heat pipes. It can be of some importance for high density working fluids.

In thermosyphons, gravity is responsible for driving the condensate formed in the condenser to the evaporator direction and so a vertical component of gravity force must always be present. In actual applications, for straight thermosyphons, an inclination of at least a 7° with the horizontal position must be provided, to guarantee its operation. However, only the pressure due to the liquid pool height at the bottom of the device, can be an important pressure component, especially for thermosyphons with large filling ratios (ratio of the working fluid to the evaporator volumes). The gravity pressure effects can be negligible over the liquid film region, although gravity is responsible for the liquid film displacement towards the evaporator direction (see Nusselt model discussed in Chap. 2). One exception might be the use of high density working fluids, such as liquid metals.

In heat pipes, the gravity force is substituted by the capillary action and the pressure drops due to gravity, although always present (at least in the normal direction) usually can be neglected.

Figure 3.12 illustrates a leaning heat pipe or thermosyphon. If the inclination angle relative to the horizontal position is ψ, the pressure drop due to the liquid column, in the axial direction along the tube is:

$$\Delta p_{gL} = \rho_l g l \sin\psi \tag{3.54}$$

while the gravity force along the normal direction is:

$$\Delta p_{gN} = \rho_l g D \cos\psi \tag{3.55}$$

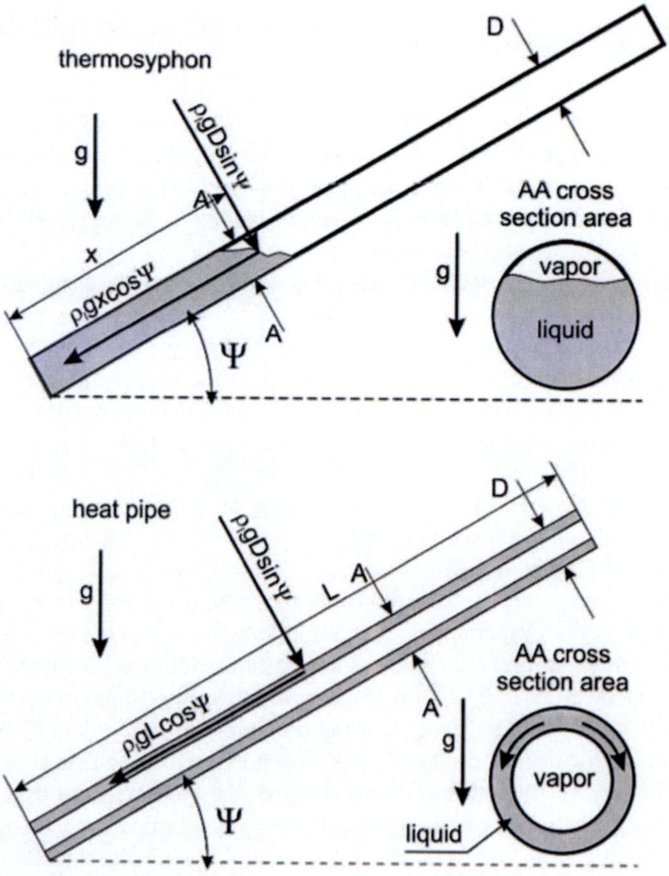

Fig. 3.12 Gravity pressure drops

3.7 Liquid and Vapor Pressure Distribution Curves

The schematic plot of the total pressure distribution curves for liquid and vapor phases in a typical heat pipe in horizontal position, highlighting the contribution of each pressure drop component, can be seen in Fig. 3.13. The effect of gravity in the vapor is neglected in this plot. The pressure drops due to the transition of phases at the interfaces (liquid to vapor and vapor to liquid) were also neglected. Besides, the effect of the inertia terms in the liquid phase is not considered, as the viscous forces, in most heat pipes, play the major role. As already observed, the capillary forces, which are the result of the deformation of the working fluid meniscus in the liquid vapor interface, being variable along the length, is responsible for the liquid movement through the device. In this plot it is highlighted the order of magnitude of the capillary forces necessary to operate the heat pipe, which, in fact, must overcome the sum of all pressure drop components. If the wick is not able to provide this capillary pressure, the device does not work.

When the inertia is the major component of the total pressure drop along the heat pipe, the shape of the pressure distribution curves tends to that of a parabola. On the other hand, when the viscous forces plays the major role, the pressure distribution curves tends to a linear distribution. Besides, a well designed heat pipe should require the smallest capillary pressure possible to operate. This condition is achieved when the largest liquid pressure in the condenser is located at the end of the condenser and has the same pressure magnitude of the vapor at this same point. This point is identified as the "wet point" (already mentioned). Figure 3.14 shows two vapor curves representing the vapor pressure distribution for a well designed heat pipe: one for vapor flow mainly subjected to inertia forces and the other for vapor flow affected primary by viscous forces. The liquid curve is the same for both vapor pressure distributions as the liquid is usually subjected only to viscous forces.

Figure 3.15 shows a schematic of the pressure distribution of vapor and liquid flows in thermosyphons. The vapor pressure distribution is similar to that of heat pipes. The kinetic effects are expected to be more pronounced, as the viscous forces

Fig. 3.13 Heat pipe pressure curves

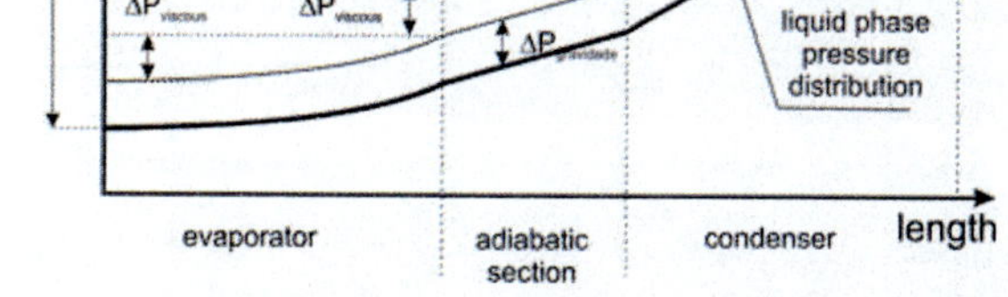

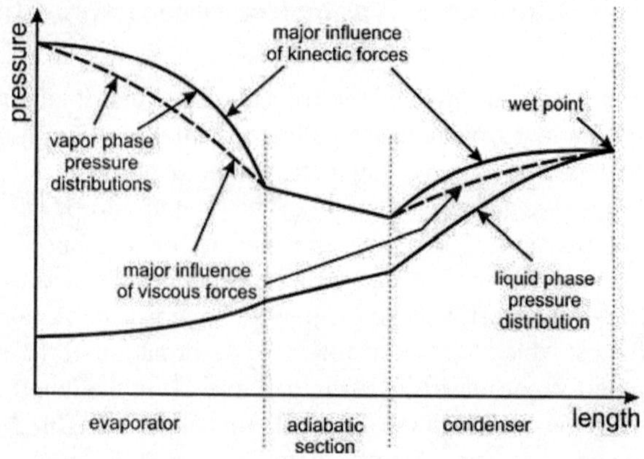

Fig. 3.14 Heat pipe pressure distribution curves, for vapor subjected mainly to inertia and to viscous forces effects

in thermosyphons, resulting from the interaction of the liquid flowing down over the internal wall and the vapor going up, are usually weak. However the pressure distribution in the liquid is different, as the pressure drops are primarily due to the action of gravity and the viscous forces can be considered negligible. In the condenser region, the liquid pressure distribution curves is very close to that of the vapor. As the condensate film thickness (and its mass) increases in its path to the evaporator, the pressure resulting from gravity force increases as well. In the pool region, the mass of liquid increases sharply with the concentration of liquid mass in the thermosyphons basis.

As already observed, in thermosyphons, instead of using capillary effects, the liquid is pumped back from the condenser to the evaporator by gravity action and the vapor is conducted to the condenser by buoyancy effects. In general, the gravity is much more efficient as driving force than the capillary forces and so thermosyphons usually present lower thermal resistances than heat pipes.

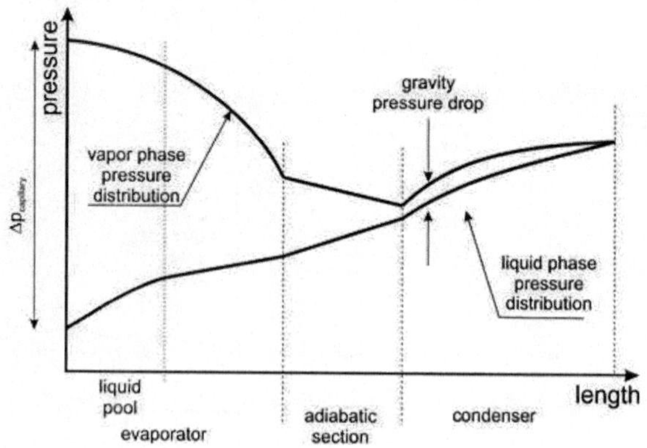

Fig. 3.15 Thermosyphon pressure curves

It should be noted that Figs. 3.13, 3.14 and 3.15 show schematic plots. For conventional application, for both thermosyphons and heat pipes, the liquid pressure drops are much larger than those in vapor, so that the vapor pressure drops can be neglected.

3.8 Operational Limits

Thermosyphons and heat pipes may not operate in some circumstances, known as operational limits. These limitations may indicate both that the device cannot transport heat at all or that, in some cases, it is not able to transfer heat efficiently. These operational limits must be known a priori, so that the engineer can avoid designing equipment working in these conditions.

One should note that, in conventional thermosyphons, the working fluid liquid phase flows over the internal surfaces of the tubes, in counter-current flow with the vapor phase. On the other hand, in heat pipes, the liquid flows within wick structures. Besides, gravity is the driving force for thermosyphons while capillary pumping is responsible for liquid displacements within heat pipes. Therefore, as the devices operate in different conditions, their operational limitations are attributed to different causes. However, some of the operational limits are analogous and are predicted by similar models, as discussed in this section.

Thermosyphons

One of the most usual symptoms that indicates that a thermosyphon heat transport limit is achieved, is a sudden increase of the evaporator temperature (or of one region of the evaporator), resulting in an increase of the overall thermal resistance of the device. This temperature increase is usually denominated as “dry out”.

Usually, two mechanisms may lead to the dryout phenomenon in thermosyphons. First, if the working fluid volume is small, i.e., there may not be enough fluid to guarantee the continuous circulation of the vapor and liquid phases within the device, even for small radial heat fluxes. In this case, the condensed liquid evaporates in its way down from the condenser to the evaporator and liquid phase is not able to reach the evaporator base, resulting in the dry out of inferior regions of the evaporator.

The second mechanism happens for high heat fluxes, when the entrainment limit is achieved, i.e., in the special conditions when the vapor dragging forces over the liquid film is so strong that condensate cannot reach the evaporator bottom or the liquid pool. In this case, an initial dryout region may grow to a larger area, denominated as “dry patch”. Usually, in vertical thermosyphons, the dryout conditions are reached before the formation of dry patches, while the opposite happens for inclined tubes, as gravity pushes liquid from the upper to the bottom radial positions. However, both dryout and dry patch regions increase the thermal resistance of thermosyphons. Besides, the onset of dryout can be intermittent, depending on the operation conditions, as thermosyphons subjected to high heat fluxes can present oscillatory behaviors, such as the one outlined in Fig. 3.16.

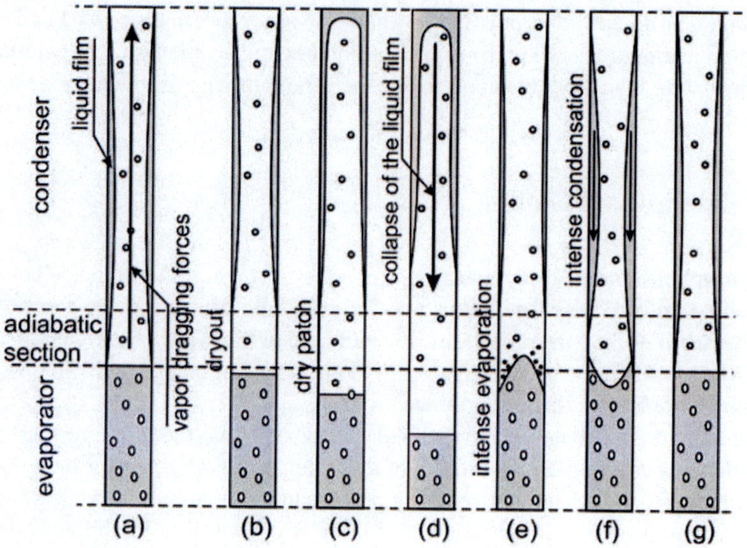

Fig. 3.16 Oscillation effect in thermosyphons with high filling ratios and high power inputs

The phenomenon described in Fig. 3.16 is specially observed for vertical thermosyphons, with medium to high filling ratios, large aspect ratios (ratio of the length to the diameter) and subjected to high heat fluxes. Figure 3.16a shows the thermosyphon operating in its normal conditions. If the power input is high, the liquid in the thinner inferior regions of the condensate layer may evaporate, causing the onset of a dryout region (Fig. 3.16b). Depending on the heat power, the dryout area may increase, forming a dry patch (Fig. 3.16c) as, in addition to the evaporation of the film, the dragging forces caused by the high volume of the vapor formed may force the liquid film back to the condenser region. This liquid accumulates in the condenser until the gravity acting over the volume is able to push the liquid film back to the evaporator area, collapsing the liquid film (Fig. 3.16d). The collapsed film finds a high temperature surface in the superior region of the evaporator and a violent evaporation takes place (Fig. 3.16d). The formed vapor finds a free of liquid condenser area and efficient vapor condensation takes place. Gravity returns the liquid to the evaporator (Fig. 3.16f) and the device returns to the initial state, represented by Fig. 3.16a, closing the cycle. The oscillation usually happens during the transition of the heat transfer operation modes of the thermsoyphons, such as in startups. However, persistent oscillations may happen, for special operation conditions.

One should note that the phenomena described in Fig. 3.16 is similar but not the same of the geyser boiling effect, commonly observed in thermosyphons, which deserves a special subsection, to be presented later in this chapter.

For devices operating in very low temperature levels, the vapor pressure difference between evaporator and condenser can be extremely small. In some cases, the viscous forces over the vapor can be larger than the pressure gradient caused by the temperature fields. When this happens, the vapor pressure gradients may not be enough to overcome the drag forces and vapor stops. This state of insufficient vapor in the condenser or of no vapor flow at all is known as "viscous limit". This very low vapor pressure level is frequently observed in cryogenic thermosyphons or heat pipes, in devices with very long condensers or in devices in which working fluid starts up from the solid state.

The heat transfer limit phenomena are usually modelled using conservation of energy equations and appropriate state equations, where the vapor is considered a perfect gas. In viscous regime, the inertia terms are considered negligible and the viscous heat transfer limit is obtained considering that pressure at the end of condenser reaches the zero level. In the inertia regime, the viscous forces are small compared to the inertia forces, and the heat transfer is limited by the chock wave formation (sonic limit).

Viscous Limit

Viscous limit expression can be obtained using the following hypothesis: vapor is an isotherm perfect gas, flowing in laminar regime within a long cylindrical tube. Heat is inserted and removed perpendicularly to the tube axis. The equation of state, for an isothermal perfect gas at a constant temperature T_o is:

$$\frac{p}{\rho} = \frac{p_0}{\rho_0} = \frac{RT_0}{M} \tag{3.56}$$

where R is the universal gas constant, M is the average molar mass and the subscript 0 refers to the evaporator initial position. It is also considered that the energy necessary to increase the vapor temperature is small when compared to the energy necessary to change the phase of the working fluid (evaporation). Besides, considering that the latent heat of vaporization does not depend on the temperature and that the device is perfectly insulated, the axial heat flux, in steady sate conditions, can be determined from the expression:

$$q'' = \rho_v \overline{u}_v h_{lv} \tag{3.57}$$

As discussed in Chap. 2, the fluid velocity profile in viscous flow regime in a circular cross section tube, neglecting the influence of the inertia forces, is parabolic (see Eq. 2.72), and the pressure gradient is given by Eq. 2.77. However, in the inertia regime, experimental observations show that the velocity assumes a profile proportional to a cosine function ($cos\left(2\pi r_i^2/D_i^2\right)$), getting flatter as it moves into the evaporator exit direction.

The maximum heat that can be transferred in the axial direction due to the viscous limits are related to the minimum pressure drop necessary to make the vapor to move along the device (Faghri 2016). To obtain an expression to determine this limit, Eqs. 2.77, 3.56 and 3.57 can be combined to obtain:

$$p\frac{dp}{dx} = -\frac{32\mu_v}{D_v^2}\frac{p_0 q''}{h_{lv}\rho_0} \tag{3.58}$$

In this case, the subscript "0" means the beginning of the evaporator ($x = 0$). Integrating along the device length, one has:

$$p_0^2 - p_L^2 = \frac{64\mu_v}{D_v^2}\frac{p_0}{h_{lv}\rho_0}\int_0^l q''(x)dx \tag{3.59}$$

where p_L is the pressure at the end of the device ($x = L$). The effective length for variable heat flux can be determined as:

$$l_{ef} = \frac{1}{q''_{\max,v}}\int_0^l q''(x)dx \tag{3.60}$$

where q''_{max} is the maximum axial heat flux that the device is able to transfer due to viscous forces, when the pressure at $x = l$ is p_L. The combinations of Eqs. 3.60 and 3.59 results in:

$$q''_{\max,v} = \frac{D_v^2 h_{lv}\rho_0 p_0}{64\mu_l l_{ef}}\left(1 - \frac{p_L^2}{p_o^2}\right) \tag{3.61}$$

The viscous limit is obtained by setting p_L to zero, or:

$$q''_{\max,v} = \frac{D_v^2 h_{lv}\rho_0 p_0}{64\mu_l l_{ef}} \tag{3.62}$$

Predictions using this expression have been compared favorably with experimental data, showing that, actually, the vapor flow can be limited by the difference between the vapor pressures at the evaporator exit and its minimum at the condenser.

A practical criterion between the total vapor pressure drop and the absolute pressure is proposed to avoid the viscous limit:

$$\frac{\Delta p_v}{p_v} < 0.1 \tag{3.63}$$

Sonic Limit

Concerning to the sonic limit, the behavior of heat pipes and thermosyphons are similar. Differently from the other heat transfer limits, the sonic limit is actually an upper heat transfer boundary and does not result in the dry-out or in the total failure of the device. If this limit is exceeded, an increase of the temperature gradient along the tube is observed and the isothermal characteristic of the device is no longer perceived.

Kemme (1969) showed in his experiment for liquid metal heat pipes that the vapor could accelerate in the evaporator section up to the sonic velocity by addition of heat (thus, of vapor mass) in the evaporator. After the vapor reached the sonic velocity at the end of the evaporator, the vapor velocity could not increase any further with the decrease of the condenser temperature and so the evaporator temperature remains the same. Actually, the sonic limit establishes that the maximum heat transfer capacity of the device happens when the vapor reaches the sonic velocity.

An expression can be obtained for the sonic limit, adopting the hypothesis of compressible mass flux in a constant cross section duct with the addition and removal of vapor, using convergent-divergent nozzle theory. Figure 3.3 illustrates the vapor mass flux distribution, considering uniform vapor generation and condensation along the evaporator and condenser sections, respectively. Similar vapor mass flux distributions can be found in Fig. 3.17 (Busse 1973), for a device with no adiabatic section, considering no viscous forces and constant temperatures at the beginning of the evaporator and at the end of condenser. The numbers 1 to 5 represent the vapor mass flux curves resulting from increasing levels of transported heat power. Up to curve number 2, one observes a decrease of the absolute pressure along the evaporator and a subsequent total recuperation in the condenser. However, if more heat

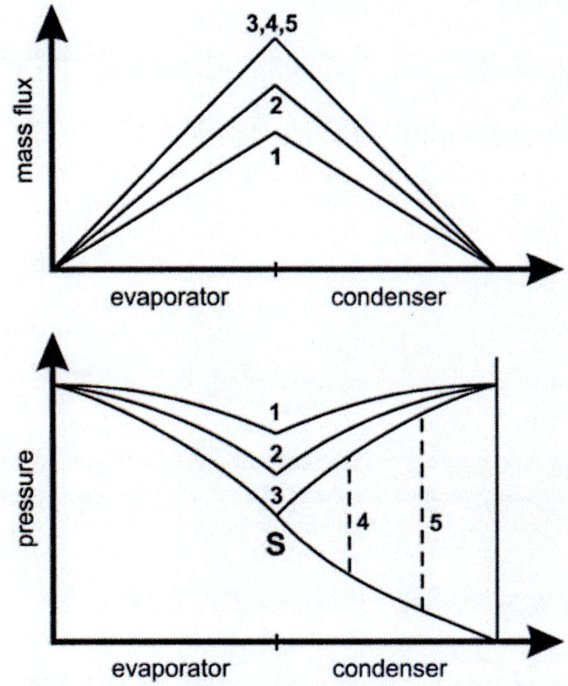

Fig. 3.17 Mass flux and pressure distributions along a thermosyphon or heat pipe

is delivered, the vapor velocity increases, reaching the sonic level (curve 3), represented by letter S in the plot. Kemme (1969) shows that, if more heat is removed from the condenser, the condenser temperature and its absolute pressure decreases, not affecting, however, the pressure level at the evaporator, as shown in this figure. Actually, at the low pressure region in the condenser, the vapor reaches supersonic velocities. Due to the viscous forces, the pressure distribution curve is continuous, however, considering an ideal model with no dissipation forces, the pressure recover along the condenser takes the form of a shock wave front, represented by the vertical dashed lines in Fig. 3.17 (Busse 1973).

One expression for the sonic limit can be obtained for a one-dimensional vapor flow, considering that the vapor properties follows the perfect gas law and that the inertia terms dominate, i.e., the viscous forces are ignored. These limits are reasonable for devices operating in low densities and high velocities. Therefore, by the ideal gas law, one has:

$$\frac{p_{st}}{\rho_{st} T_{st}} = \frac{p_{le}}{\rho_{le} T_{le}} \tag{3.64}$$

where the subscripts "*st*" refer to the stagnation point, position where the vapor velocity is zero and the subscript "*le*" to the vapor at the end of evaporator (onset of the adiabatic or condenser section). The energy conservation equation for the vapor, for adiabatic steady state regime, with no gravitational forces, is (see Bird 1960):

$$h_{st} = h_{Le} + \frac{u_{le}^2}{2} = c_p T_{Le} + \frac{u_{le}^2}{2} = c_p T_{st} \tag{3.65}$$

Equation 3.65 can be rewritten as:

$$\frac{T_{st}}{T_{le}} = 1 + \frac{u_{le}^2}{2c_p T_{le}} \tag{3.66}$$

Besides, the sonic velocity is expressed as;

$$u_s^2 = \gamma_v R T_v = \frac{c_p}{c_v} R T_v \tag{3.67}$$

where $\gamma_v = c_p/c_v$ is the ratio of pressure to volume specific heats of the vapor and R is the gas constant. As $R = c_p - c_v$, from Eq. 3.67, one has $C_p T_v = u_s^2/(\gamma - 1)$. Therefore, Eq. 3.65 can be rewritten as:

$$\frac{u_s^2}{\gamma - 1} + \frac{u_v^2}{2} = \frac{u_{s,st}^2}{\gamma - 1} \tag{3.68}$$

From this equation, applying the definition of Mach number (ratio between the vapor and sonic velocities, $Ma = u_v/u_s$) and Eqs. 3.66 and 3.67, the ratio between the sonic velocities at the stagnation and local temperatures is:

$$\frac{u_{s,st}^2}{u_s^2} = \frac{T_{st}}{T_v} = 1 + \frac{\gamma - 1}{2} Ma^2 \tag{3.69}$$

The pressure difference along the evaporator and the stagnation point is given by:

$$p_{le} - p_{st} = \rho_v u_v^2 \tag{3.70}$$

Dividing the last expression by p_{Le}, using the equation of state for low pressure gas ($p_v = \rho_v RT$) and substituting the Mach number definition, one has:

$$\frac{p_{st}}{p_{le}} = 1 + \gamma Ma^2 \tag{3.71}$$

Combining the above equations, the density ratio (see Eq. 3.64) can be expressed as:

$$\frac{\rho_{st}}{\rho_{le}} = \frac{1 + \gamma Ma^2}{1 + \frac{\gamma - 1}{2} Ma^2} \tag{3.72}$$

Besides, the vapor axial mass flux depends on the heat transferred (see Eq. 2.78) by:

$$\dot{m}_v'' = \rho_v u_v = \rho_v\, Ma\, \sqrt{\gamma R T_v} = \frac{q}{A_{cs} h_{lv}} \tag{3.73}$$

Substituting the expressions for T_v and ρ_v in the last equation, one has:

$$q = \frac{A_{cs} h_{lv} \rho_{st} \sqrt{\gamma R T_{st}} Ma \sqrt{1 + \frac{\gamma - 1}{2} Ma^2}}{1 + \gamma Ma^2} \tag{3.74}$$

The sonic limit happens when the vapor velocity at the end of evaporator reaches the sonic velocity, i.e., when $Ma = 1$. Therefore, the maximum heat that the device can transport due to the sonic limit is (Peterson 1994) is:

$$q_{\max,s} = A_{cs} h_{lv} \rho_{st} \sqrt{\frac{\gamma R T_{st}}{2(\gamma + 1)}} \tag{3.75}$$

The ratio between the actual and the max heat transported is:

$$\frac{q}{q_{\max,s}} = \frac{Ma \sqrt{2(\gamma + 1)\left(1 + \frac{\gamma - 1}{2} Ma^2\right)}}{1 + \gamma Ma^2} \tag{3.76}$$

The local temperature at the liquid–vapor interface can be determined using Eq. 3.69, with the Mach number given by Eq. 3.73. In most applications, the wall temperature distribution can be considered equal to the liquid–vapor interface temperatures.

These equations were developed for the evaporator (where heat is inserted) but can also be used for the condenser (where heat is removed). However, while the Mach number is necessarily lower or equal to one in the evaporator, it can be larger than one in the condenser. For condenser vapor fluxes characterized by Mach numbers much greater than one, the actual and predicted (using these models) temperature distributions may be very different. In these conditions, one must add the temperature gradient due to radial conductions in the wall and in the porous media to the liquid–vapor interface temperature. In the condenser, the radial conduction temperature difference must be subtracted from the temperature predicted by this model.

An alternative procedure to determine the sonic limit was proposed by Busse (1973). For laminar unidimensional vapor flux, the Navier–Stokes equation takes the form:

$$\frac{d\overline{p_v}}{dx} = \frac{d}{dx}\overline{\rho_v u_v^2} \tag{3.77}$$

where $\overline{p_v}$, $\overline{\rho_v}$ and $\overline{u_v}$ are the average vapor pressure, vapor density and vapor velocity, over the cross section area. After integrating this equation and using the non-slip condition at the contour, one gets:

$$\overline{p}_0 - \overline{p}_v = \overline{\rho_v u_v^2} = C_a \cdot \rho_v u_v^2 \tag{3.78}$$

where $\overline{p}$ is the average pressure at the evaporator cross section and C_a is the coefficient that adjusts the real velocity profile from the parabolic (Haggen-Poiuseuille) profile. The combination of this last expression with the continuity equation and ideal gas relations, results in the following expression for the maximum heat that vapor can transfer for a specific pressure condition:

$$q_{\max,s} = h_{lv}\left(\frac{\rho_0 p_0}{A_{cs}}\right)^{1/2}\left[\frac{p}{p_0}\left(1 - \frac{p}{p_0}\right)\right]^{1/2} \tag{3.79}$$

The sonic limit is achieved when pressure drops do not cause any change in the average heat transferred, i.e., when $dq_{max}/dp = 0$. By using Eq. 3.79 to obtain this derivative and using appropriate C_a values, Busse (1973) proposed the following expression for the sonic limit:

$$\overline{q}_{\max,s} = 0.474 \cdot h_{lv}(\rho_0 p_0)^{1/2} \tag{3.80}$$

Transition Between Viscous and Sonic Limits

Especially during the startup, the best way to avoid the viscous limit is to increase the heat flux applied to the evaporator. This procedure causes the increase of the vapor temperature and so of the vapor pressure, up to the point where the pressure drop along the vapor flow exceeds the viscous force pressures. The major concern about this procedure is that, at low absolute pressure levels, temperature increases may lead quickly to sudden increases in the vapor velocities, so that the vapor can easily reach the sonic level.

The temperature levels, for which the viscous pressure differences are around the same magnitude of the pressure drops due to inertia forces, can be obtained theoretically. Equation 3.62 shows that the heat transfer limits due to viscous forces depend on the product: $\rho_0 p_0$ while Eq. 3.80 shows that the sonic limit depends on the parameter $\sqrt{\rho_0 p_0}$.When $\rho_0 p_0$ is small, the viscous forces dominate while, when this term is large, the inertia terms dominate. As the parameters ρ_0 and p_0 defines the temperature at the beginning of the evaporator, T_0, the boundary between the viscous and sonic limit can be determined equating Eqs. 3.62 and 3.80 and finding which temperature corresponds to that specific physical condition. The definition of this temperature may be of major importance for high temperature thermosyphons operating with liquid metal as working fluid.

Entrainment/Flooding Limits

The interaction between the vapor and liquid, in countercurrent flows, may induce a viscous force in the liquid–vapor interface that can prevent the liquid return from the condenser to the evaporator. When this happens, the entrainment limit was achieved.

Actually, the increase of the heat flux results in the increase in the vapor velocity, which, in turn, may cause the instability of the liquid flow, causing the formation of wavelets over the tube internal surface. If the interfacial shear forces due to the vapor interaction with the liquid is larger than the forces resulting from the liquid surface tension, liquid droplets may be dragged by the vapor flow and pushed to the condenser region, which than accumulates working fluid. The droplet dragging and the consequent excess of liquid (flooding) in the condenser results in several drawbacks. First, it decreases the condenser heat transfer coefficient, as the heat has to cross a thicker liquid layer before reaching the condenser vapor–liquid interface. Second, it decreases the amount of liquid in the evaporator and so less working fluid is available to evaporate. Third, it promotes the circulation of fluid in liquid state that do not change of phase within the device, decreasing the thermosyphon performance. The entrainment/flooding limit is especially important for thermosyphons with small aspect ratios (ratio between diameter of the tube and length).

Several correlations are available for the prediction of this limit, some of them are presented here. In 1973, Sakhuja proposed the following semi-empirical correlation for the prediction of the entrainment/flooding limit, based on the hydrostatic and inertia forces (Nguyen-Chi and Groll 1981):

$$q_{\max,f} = A\frac{C_f^2 h_{lv}\sqrt{gD_i(\rho_l - \rho_v)\rho_v}}{\left[1 + (\rho_v/\rho_l)^{1/4}\right]^2} \tag{3.81}$$

Where A is the cross section area and C_f is an empirical non-dimensional constant obtained as a function of the thermophysical properties of the fluid and is, for the major part of applications, between 0.7 and 1 (Faghri 2016). This correlation is applied for one-dimensional, uncompressible fluxes of one component, two-phase fluids, considering that the thickness of the liquid film is constant and in steady sate conditions.

It is expected that fluids with high surface tension parameters require more energy to have droplets removed off the liquid film. Tien and Chung (1979) proposed the following expression for the flooding limit, similar to Eq. 3.81, which included the surface tension parameter:

$$q_{\max,f} = A\frac{C_f^2 h_{lv}[\sigma g(\rho_l - \rho_v)]^{\frac{1}{4}}}{\left(\rho_l^{-1/4} + \rho_v^{-1/4}\right)^2} \tag{3.82}$$

where C_f, for water as working fluid, is:

$$C_f = (3.2)^{1/2}\tanh\left(0.5 \cdot Bo^{1/4}\right) \tag{3.83}$$

and, for other working fluids (Faghri 2016), given by:

$$C_f = \left(\frac{p_l}{p_v}\right)^{0.14}\tanh^2\left(Bo^{1/4}\right) \tag{3.84}$$

with the Bond number defined as in Eq. 2.36.

Nguyen-Chi and Groll (1981) proposed the following correction factor $f(\theta)$ to account for the influence of the inclination angle:

$$f(\theta) = \left[\frac{\theta}{180^o} + \sqrt{\sin 2\theta}\right]^{0.65} \tag{3.85}$$

Imura et al. (1983) suggested another correlation for the maximum heat flux that compares with most of literature data within 30% of differrence:

$$q''_{\max,f} = 0.64\left(\frac{\rho_l}{\rho_v}\right)^{0.13}\left(\frac{D_i}{4l_e}\right)h_{lv}\left[\sigma g\rho_v^2(\rho_l - \rho_v)\right]^{\frac{1}{4}} \tag{3.86}$$

Boiling Limit

Boiling limits are usually observed in thermosyphons charged with high volumes of working fluid, when subjected to high heat fluxes in the evaporator section. It happens during the transition of nucleate to film boiling, i.e., when de device is subjected to the evaporation critical heat flux, where the vapor bubbles generated over the heated surface in the liquid pool coalesce, forming a vapor film. Due to the low thermal conductivity of the vapor, the vapor film isolates thermally the tube wall. In this case, the wall temperature increases and may reach the melting point of the tube material, damaging the thermosyphon.

The boiling limit is also known as "burn-out limit" and, according to Groll and Rosler (1992), can be determined from the following equation developed for a liquid pool and therefore adequate for application to the lower region of the liquid film evaporator:

$$q_{\max,b,e} = 0.12 \cdot h_{lv} \rho_v^{\frac{1}{2}} [\sigma g (\rho_l - \rho_v)]^{\frac{1}{4}} \tag{3.87}$$

The above equation therefore is more appropriate to the rear regions of the evaporator. One way to avoid the dryout due to the boiling effect is to increase the external evaporator area, by increasing the length of the evaporator region and/or the diameter of the thermosyphon tube, so that the heat flux decreases. Usually, long tubes with long evaporators are not in risk of reaching the boiling limit, however, the entrainment limit can be easily reached.

In 1976 Gorbis and Saychenkov (Faghri 2016) also proposed another correlation for the maximum heat flux due to the boiling limit. This correlation is valid for inclination angles between vertical position (0°) and 86° (relative to the vertical position), filling ratios (working fluid volume to the evaporator volume) between 0.029 and 0.6, and non-condensable gas ratio (non-condensable gas volume to the condenser volume) between 0.006 and 1:

$$q_{\max,b,e} = q_{crit} C^2 \left[0.4 + 0.012 r_i \sqrt{\frac{g(\rho_l - \rho_v)}{\sigma}} \right]^2 \tag{3.88}$$

where q_{crit} is the pool boiling critical heat flux, given by:

$$q_{crit} = 0.142 \sqrt{\rho_v} [g \sigma (\rho_l - \rho_v)]^{\frac{1}{4}} \tag{3.89}$$

and the constant C is determined by the equation:

$$C = \xi \left(\frac{D_i}{l_c} \right)^{-0.44} \left(\frac{D_i}{l_e} \right)^{0.55} \left(\frac{V_l}{V_e} \right)^{n'} \tag{3.90}$$

where $\xi = 0.538$ and $n' = 0.13$, for $V_L/V_e \leq 0.35$ and $\xi = 3.54$ and $n' = -0.37$, for $V_L/V_e \geq 0.35$.

Pressure Drop Limit for Loop Thermosyphons

As for other devices treated in this book, the flow of working fluid inside a loop thermosyphon is associated with pressure drops. The larger the heat transfer rates, the larger are the working fluid mass flow rates, the fluid (vapor and liquid) velocities and, consequently, the pressure drops. For passive gravity assisted systems, a hydraulic head H, i.e., the positive difference between the level of the condensate in the liquid return tube and in the evaporator, must compensate the pressure differences due to the working fluid flow pressure drops. Figure 3.18 presents a schematic of the working fluid flow inside a loop thermosyphon, highlighting the overhead level. The maximum overhead H_{max} is actually determined by the loop thermosyphon geometry,

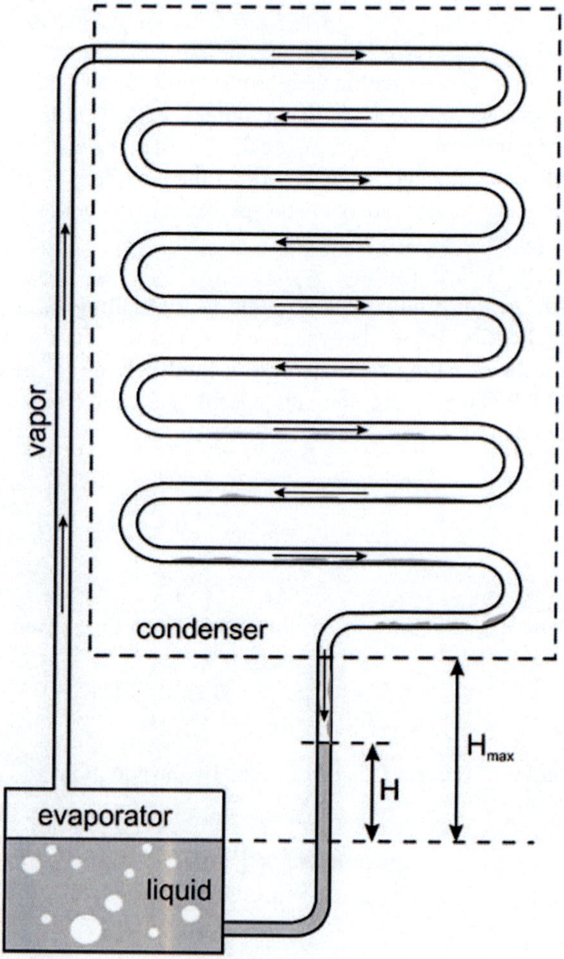

Fig. 3.18 Hydraulic head H in a loop thermosyphon

as also seen in Fig. 3.18. If the evaporator power input is too high, the vapor pressure over the evaporator upper level pushes the liquid to the condenser region, blocking partially the condenser, even leading to the failure of the system. In this condition, the condensation capacity of the device is decreased, and so, the overall thermal resistance increases, as the portion of the condenser that is filled with condensate no longer operates in two-phase heat transfer mode.

The pressure drop limit for thermosyphons is defined as the maximum heat that the device is able to transport, before liquid enters the condenser region. This limit is especially important for when the condenser is a long tube, such as in steam coil shape condensers, as shown in Fig. 3.18. Milanez and Mantelli, in 2010, developed a model for determining this heat transfer limit, based on the literature correlations for viscous fluid flow pressure drop, which is briefly presented.

The working fluid flow total pressure drop (Δp_t) is related to the hydraulic head H though the expression:

$$\Delta p_t = (\rho_l - \rho_v) g H \tag{3.91}$$

The pressure drops due to momentum variations at the liquid vapor interfaces are neglected. The total pressure drop Δp_t results from the summation of the pressure drops: evaporator Δp_e, vapor line Δp_{vl}, condenser Δp_c and liquid line Δp_{ll}:

$$\Delta p_t = \Delta p_e + \Delta p_{vl} + \Delta p_c + \Delta p_{ll} \tag{3.92}$$

It is considered that only vapor phase is present in the vapor line (good thermal insulation) and that the condenser is operating in the limit ($H = H_{max}$), i.e., the liquid line is completely filled with liquid (single phase). Therefore, the vapor and liquid line pressure drops can be determined using the Fanning equation (Eq. 2.80) here reproduced:

$$\Delta p = f \frac{l_{eq}}{D_i} \frac{\rho u^2}{2} \tag{3.93}$$

where l_{eq} is the summation of the tube length with the equivalent length of the bends and other accessories of the circuit. For smooth pipes and turbulent flows, the friction factor f for single flows can be computed as (Fox et al. 1999):

$$f = \frac{0.316}{\mathrm{Re}^{\frac{1}{4}}} \tag{3.94}$$

The other two pressure drop parcels (Eq. 3.92) are more complex to determine due to the two-phase nature of the flow. The literature (Collier and Thome 1994; Wallis 1969; Carey 1992) presents two classical models for the pressure drop in two-phase flows: homogeneous and separated. In the homogeneous model, the two phases flow at the same velocity, while, in the separated model, they flow at different velocities, but with a non-zero relative velocity between phases. The literature shows that the separated model presents more precise results. Therefore, the separated model is described.

According to Collier and Thome (1994), an expression for the local pressure gradient for the liquid phase is:

$$-\frac{dp}{dx}=\frac{1}{\kappa}\left\{\begin{array}{l}\varphi_l^2\left[\frac{2f_l\,\dot{m}''^2(1-X)^2}{\rho_l D_i}\right]+[(1-\alpha)\rho_l+\alpha\rho_v]g\sin\theta+\\ \dot{m}''^2\frac{dX}{dx}\left\{\left[\frac{2X}{\rho_v\alpha}-\frac{2(1-X)}{\rho_v(1-\alpha)}\right]+\frac{d\alpha}{dX}\left[\frac{(1-X)^2}{\rho_v(1-\alpha)^2}-\frac{X^2}{\rho_v\alpha^2}\right]\right\}\end{array}\right\} \quad (3.95)$$

where κ is the vapor compressibility, which, for water, is approximately 1 (Carey 1992) and f_l is the liquid friction coefficient, calculated as:

$$f_l=0.079\left[\frac{\dot{m}''(1-X)D_i}{\mu_l}\right]^{-\frac{1}{4}} \quad (3.96)$$

The Eq. 3.95 can also be used to determine the pressure drop for the vapor phase, however, the vapor friction coefficient must be determined from:

$$f_v=0.079\left[\frac{\dot{m}''D_i}{\mu_l}\right]^{-\frac{1}{4}} \quad (3.97)$$

In Eq. 3.95, φ_l is the two phase multiplier that can be determined by several models available in the literature. Milanez and Mantelli (2010) showed that the following model, proposed by Wallis (1969), presents the best comparison with data:

$$\varphi_l=\left[1+\left(\frac{1}{\Omega}\right)^{\frac{16}{19}}\right]^{\frac{19}{16}} \quad (3.98)$$

where Ω, the Martinelli parameter, is the ratio between: the pressure drop of the liquid, considering only the presence of the liquid phase, and the pressure drop of the vapor, considering only the presence of the vapor phase, i.e., (Carey 1992):

$$\Omega^2=\frac{\left(\frac{dp}{dz}\right)_l}{\left(\frac{dp}{dz}\right)_v} \quad (3.99)$$

Heat Pipes

As for thermosyphons, the heat transfer limit can be considered reached, observing temperature increases, especially in the evaporator region. From all the limits, the most strict one is the capillary.

Capillary Limit

The capillary limit is achieved when the porous media, in a capillary driven heat transfer device, is not able to provide enough liquid pumping capacity and the liquid cannot move from the condenser to the evaporator. The physical phenomena that drive the capillary action were described in Chap. 2. Figure 2.11 illustrate the liquid–vapor meniscus along a heat pipe, highlighting that meniscus radii in the liquid–vapor interface, in the evaporator, is much smaller than those in the condenser. This means that the major pumping capacity of the porous media is concentrated in the evaporator. Therefore, many devices, such as loop heat pipes, are designed considering that only the evaporator porous media contributes for the liquid transportation pumping head.

The "pumping active region" of a media is between the points were the meniscus (liquid–vapor interface) presents the smaller and the larger radius. For a well designed capillary media in a conventional heat pipe, this points should be located as far as possible to each other within the device, preferably at the edges of the evaporator and condenser. Therefore, the heat pipe works properly if the capillary pressure drop between the larger and smaller meniscus radii in a porous medium is able to overpass all pressure drops along the liquid path in the device, i.e.:

$$\Delta p_{cap} \geq \int_{L_{ef}} \frac{\partial p_v}{\partial x} dx + \int_{L_{ef}} \frac{\partial p_l}{\partial x} dx + \Delta p_{h,N} + \Delta p_{h,ax} \tag{3.100}$$

where $\partial p_v/\partial x$ and $\partial p_l/\partial x$ represent the summation of the inertia and viscous pressure drops of the vapor and liquid phases and $\Delta p_{h,N}$ and $\Delta p_{h,ax}$ determines the hydraulic normal and axial pressure drops, respectively. The liquid–vapor phase-change pressure drops are neglected in this equation. All the models and expressions necessary to determine the capillary pressure limit are presented in Chap. 2.

Viscous Limit

The same conditions discussed for thermosyphons are necessary for heat pipes to reach the viscous limits. Usually they happen at low temperature levels, when, according to the working fluid used, the vapor pressure differences can be extremely low, such as in heat pipes operating at cryogenic temperature. Also, they can be observed for very long condensers or for devices that start up from solid state conditions (liquid metal heat pipes). Equation 3.62 can be applied for the determination of the viscous limits in heat pipes.

Sonic Limit

As for thermosyphons, the behavior of a compressible mass flow in a constant cross section area of a heat pipe, with addition and subtraction of vapor, is analogous to a compressible convergent-divergent nozzle. Equation 3.80 can be used for the prediction of the sonic limit. Also, the same observations for the transition between sonic and viscous limits are valid for heat pipes.

Entrainment Limit

In heat pipes, the working fluid in its liquid phase flows within porous media, which prevents the vapor shear forces from removing droplets and dragging them to the condenser. However, in some special conditions, the pull-off of droplets may happen. In the limit, too much liquid can be pushed to the condenser direction, preventing the evaporator to generate the necessary volume of vapor for the proper functioning of the device. When this happens, the entrainment limit (also known as flooding limit) is reached. A necessary condition is that the wick has to be saturated with working fluid.

As the velocity of vapor is of several orders of magnitude larger than that of the liquid, only the vapor shear force over the liquid is considered in the modelling of the entrainment limit. Therefore, the shear force, proportional to the dynamic pressure (constant C_1), is given by (Faghri 2016):

$$F_{lv} = C_1 \frac{\rho_v \overline{u}_v^2 A_{lv,p}}{2} \tag{3.101}$$

where $A_{lv,p}$ is the liquid vapor interface area of an individual wick pore and $\overline{u}_v$ is the mean axial vapor velocity. On the other hand, the surface tension force, which maintains the liquid in the wick, can be determined through the expression:

$$F_\sigma = C_2 \sigma P \tag{3.102}$$

where P is the wetted perimeter of the individual pore and C_2 is a constant.

Cotter, in 1997 (Peterson 1994), proposed the use of the Weber number (We) to determine the onset of the liquid dragging phenomenon in heat pipes. This parameter is defined as the ratio between the liquid–vapor viscous (Eq. 3.102) and interface dragging forces (Eq. 3.101). The heat pipe entrainment limit is reached when *We* reaches the unitary value, i.e.:

$$We = \frac{C_1 \rho_v \overline{u}_v^2 A_{lv,p}}{2 C_2 \sigma P} = 1 \tag{3.103}$$

Defining a hydraulic radius of the wick pore as: $r_{H,w} = 2A_{lv,p}/P$ (Busse and Kemme 1980), and using the experimental value of $C_1/C_2 \approx 8$, obtained from the literature (Faghri 2016), Eq. 3.103 can rewritten as:

$$We = \frac{2 \rho_v \overline{u}_v^2 r_{H,w}}{\sigma} = 1 \tag{3.104}$$

In this condition, the maximum axial heat power input due to the entrainment limit can be determined as (Wright 1970):

$$q_{\max,e} = A_v h_{lv} \left(\frac{\sigma \rho_v}{2 r_{H,w}} \right)^{\frac{1}{2}} \tag{3.105}$$

where, for screen wicks, the hydraulic dimeter is $r_{H,w} = 0.5w$ (w is the wire spacing), for axial grooves, $r_{H,w} = w$ (w is the groove width) and, for packed spheres, $r_{H,w} = 0.250D$ (D is the diameter of the spheres).

It is important to note that, besides some attempts (Kim and Peterson, 1994), the literature does not report any conclusive experimental and theoretical study of the entrainment limits for heat pipes. Some researchers even do not think that this limit is relevant for the heat pipe performance, as the porous media protects the working fluid in its path along the device, as reports Faghri (2016).

Boiling Limit

The boiling limit is directly related to the capacity of the bubble formation in the working fluid. Depending on the heat flux delivered to the heat pipe in the evaporator section, bubbles can be formed and remain trapped within the wick, blocking the liquid return from the condenser to the evaporator, resulting in early dryout.

Different from others, this limit is associated to the radial application of heat and not to the axial heat transportation. As for thermosyphons, the determination of the heat transfer limit is based on the nucleate boiling theory, which englobes two different physical phenomena: formation of bubbles and their growth and collapse.

The bubble formation depends on the number and size of the nucleation sites in the solid surface. For the onset of bubbles from a heated surface, a temperature difference between the heated wall and the working fluid, also known as liquid superheat (see Sect. 2.4), must be provided. This temperature difference is usually given in terms of the heat flux by the equation:

$$q_{\max,b} = \frac{k_{ef}}{T_s} \Delta T_{cr} \tag{3.106}$$

where ΔT_{cr} is the liquid superheat temperature difference $(T_s - T_{sat})$, given by (Marcus 1972):

$$\Delta T_{cr} = \left(\frac{T_{sat}}{h_{lv} \rho_v} \right) \left(\frac{2\sigma}{r_n} - \Delta p_{\max,i} \right) \tag{3.107}$$

where r_n is the radius of the nucleation site, which values, according to Peterson (1994) lies in the range $2.54 \cdot 10^{-5}$ to $2.54 \cdot 10^{-7}$ m, for heat pipe conventional metallic case materials. However, these expressions under-predict the superheat temperature, probably because the heat pipe fabrication cleaning process may prevent the presence of adsorbed gases in the nucleation sites over the boiling surfaces, increasing the superheat temperature. In Eq. 3.106, k_{ef} is the wick structure effective thermal conductivity, which includes the liquid and porous media, as discussed in the next section.

The growth and collapse of a bubble over a flat surface depends on the liquid differences of temperature and pressure between the wall and the liquid–vapor interface. These parameters depend on the vapor pressure and on the liquid surface tension.

As proposed by Chi (1976), two major forces acts over a bubble: the pressure and the surface tension. Therefore, the force balance in the bubble-surface interface, considering a spherical bubble, can be given by the expression:

$$\pi r_b^2 \left(p_{s,w} - p_{lv} + p_{cap}\right) = 2\pi r_b \sigma_l \tag{3.108}$$

where $p_{s,w}$ is the liquid–vapor saturation pressure of the working fluid in the tube-wick structure interface and p_{cap} the capillary pressure in this same position.

Clausius–Clapeyron equation (perfect gas) can be applied to relate temperature and density along the saturation (liquid–vapor) line as:

$$\frac{dp}{dT} = \frac{h_{lv}}{(\upsilon_v - \upsilon_l)T_{lv}} \tag{3.109}$$

As the specific volume of gas is much larger than that of the liquid, it can be removed from this equation. Discretizing the pressure–temperature gradient, it is possible to considerer that:

$$p_{s,w} - p_{lv} \approx \left(T_{s,w} - T_{w,v}\right)\frac{dP}{dT} \tag{3.110}$$

where $T_{s,w}$ is the is the liquid–vapor saturation temperature of the working fluid in the tube-wick structure interface and $T_{w,v}$ is the liquid temperature at the wick-vapor interface. Therefore $\left(T_{s,w} - T_{w,v}\right)$ is the temperature drop across the wick structure at the evaporator section which, combining the last three equations, can be given as:

$$\left(T_{s,w} - T_{w,v}\right) = \frac{T_{lv}}{\rho_v h_{lv}}\left(\frac{2\sigma_l}{r_b} - p_{cap}\right) \tag{3.111}$$

This temperature difference in the flooded wick structure can also be determined using Fourier equation:

$$\left(T_{s,w} - T_{w,v}\right) = \frac{q \ln(r_i/r_v)}{2\pi l_e k_{ef}} \tag{3.112}$$

where q represents the necessary energy to keep vapor bubbles with radius r_b stable. However, in many application, the bubble radius can the considered as equal to the nucleation site radius $r_b \approx r_n$. Equating the last two equations and solving for q, the following expression comes out:

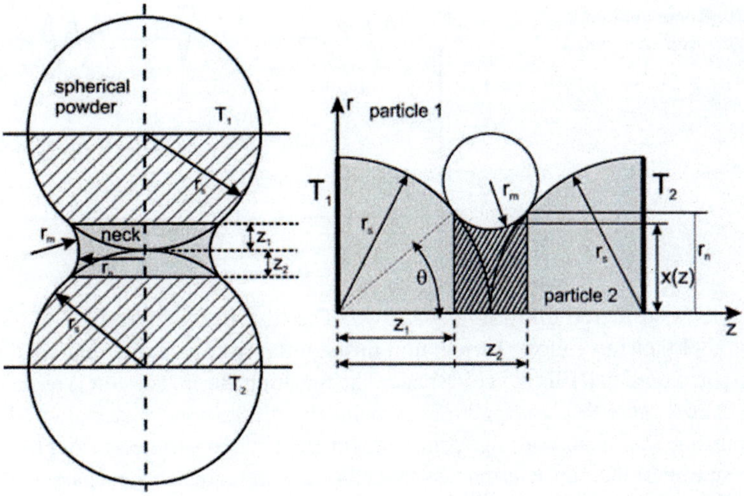

Fig. 3.19 Solid material in a unit cell formed by two metal spheres in contact

$$q_{\max,b} = \frac{2\pi l_e k_{ef} T_{lv}}{\rho_v h_{lv} \ln(r_i/r_v)} \left(\frac{2\sigma_l}{r_n} - p_{cap} \right) \tag{3.113}$$

The analysis of the last equation shows that, if the heat input is smaller than stablished by the last equation, the bubbles collapse, but if the heat input is larger, bubbles are formed and grow within the wick structures, blocking the liquid displacement along the porous media. Therefore, Eq. 3.113 can be used to determine the boiling limit for devices with wick capillary structures.

Effective Thermal Conductivity

Florez et al. (2013) developed an effective thermal conductivity model for a sintered wick, based on two contacting metal powder spheres, as shown in the schematics displayed in Fig. 3.19. A unit cell, formed by two metal hemispheres of radius r_s, (average metal particle radius) represents the whole porous media. The actual porous media geometry can be reproduced by stacking several of these cells in a vertical arrangement and replicating this pile in a horizontal array, forming a three-dimensional structure. The cubic arrangement is assumed for the sintered wick models: effective radius, permeability and porosity.

The hemispheres are joined by means of a neck, formed during the sintering process (see Fig. 3.6). The voids among adjacent spheres are considered filled with stagnant liquid. No heat exchange between the solids and the liquid and no convection within the liquid layer are considered, as the space occupied by the liquid is very small.

The analogy between thermal and electrical circuits is used to model the effective thermal conductivity of this elementary cell, resulting in the thermal circuit shown in Fig. 3.20. The temperatures T_1 and T_2 of the circular central area of the contacting

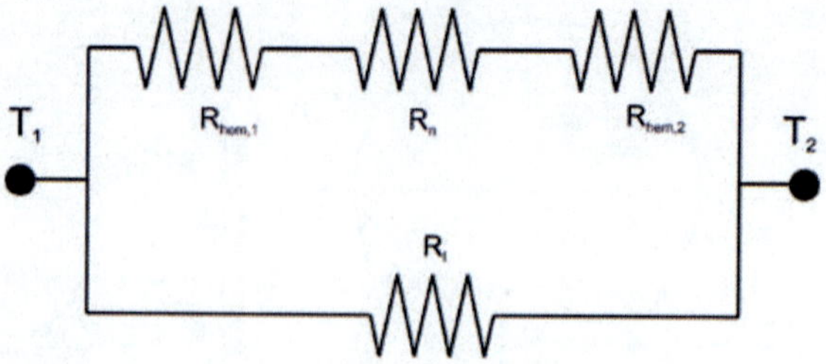

Fig. 3.20 Thermal circuit for two hemispheres joined by neck

spheres are considered uniform and known. The heat flux between T_1 and T_2 has two parallel paths to follow: conduction through the solid material and conduction through the liquid that fills the void space. The resulting thermal circuit is represented in Fig. 3.20, where the upper path is associated with the solid material conduction heat transfer. The heat, coming from hemisphere 1, is constricted through to neck area, resulting in the constriction resistance, $R_{hem,1}$. In sequence, heat passes through the neck, by the associated conduction resistance R_n. Finally, heat is spread in hemisphere 2, through the thermal resistance $R_{hem,2}$. The other parallel path is associated to the heat flowing through the static fluid that surrounds the spheres, represented by a unique resistance, R_l. Therefore, the overall thermal resistance is:

$$R_t = \frac{1}{\frac{1}{R_{hem,1}+R_n+R_{hem,2}} + \frac{1}{R_l}} \tag{3.114}$$

Experimental data showed that the liquid resistance is too high when compared to the solid material resistances, being removed from the circuit, resulting in:

$$R_t = R_{hem,1} + R_n + R_{hem,2} = R_n + R_s \tag{3.115}$$

where R_s, the thermal conduction resistance of a sphere, includes both hemispheres.

According to Florez et al. (2014), the effective thermal conductivity of the unit cell can be estimated equating the conduction heat transferred using the linear approximation of the Fourier law and the thermal resistance definition, i.e.:

$$k_{ef} A \frac{(T_2 - T_1)}{L} = \frac{(T_2 - T_1)}{R_t} \tag{3.116}$$

which gives:

$$k_{ef} = \frac{l}{AR_t} \tag{3.117}$$

The neck and sphere thermal resistances need to be determined for the estimation of R_t and so, of the effective thermal conductivity.

The one-dimensional Fourier Equation can also be employed for the calculation of the thermal resistances of the neck region, assuming that que heat transferred in the z direction, q_z, is constant:

$$q_z = -kA(z)\frac{dT}{dz} \tag{3.118}$$

This expression can be integrated, resulting in (see Fig. 3.19).

$$q_z \int_{z_1}^{z_2} \frac{dz}{A(z)} = -k \int_{T(z_1)}^{T(z_2)} dT \tag{3.119}$$

By rearranging this expression, it results in:

$$\frac{q_z}{T(z_2) - T(z_1)} = \frac{-k}{\int_{z_1}^{z_2} \frac{dz}{A(z)}} \tag{3.120}$$

The left hand side term of the previous equation represents the inverse of neck thermal resistance, therefore:

$$R_n = \frac{\int_{z_1}^{z_2} \frac{dz}{A(z)}}{-k} \tag{3.121}$$

As shown in Fig. 3.19, both spheres have the same radius. Nominating θ as the angle between the center of the sphere and the neck height and considering the neck symmetrical to its center, the following expression is obtained for the neck radius (Florez et al. 2014):

$$x(z) = \sqrt{r_m^2 - \{z - [(r + r_m)\cos\theta]\}^2} + r_m + r_n \tag{3.122}$$

where x is the variable for the neck radius (see Fig. 3.19). Therefore, the thermal resistance of the neck solid phase becomes:

$$R_p = \frac{1}{\pi k_s} \int_{z_1}^{z_2} \frac{dz}{x(z)^2} \tag{3.123}$$

Substituting $x(z)$ in the last equation and performing the integration, the following expression is obtained:

$$R_p = \frac{8r_s\left(1 - \frac{r_s}{r_s + r_m}\right)}{\pi k_s\left[r_n + r_s\left(\frac{r_n + r_m}{r_s + r_m}\right)\right]^2} \tag{3.124}$$

As mentioned, heat is considered to be uniformly inputted to the cell by the left flat circular area of radius r_s maintained at temperature T_1 and removed through the right circular area maintained at temperature T_2 (see Fig. 3.19, right side). After reaching the hot hemisphere, the heat is constricted to the neck region, cross the neck and reaches the cold hemisphere. Within the cold hemisphere, the heat spreads, reaching uniformly its central area. The lateral walls of the elementary cell are considered insulated. The solid sphere constriction thermal resistances are obtained from the temperature distribution solution, given in terms of a series, of the heat equation in spherical coordinates (function of directions r and φ), proposed by Yovanovich et al. (1978). These authors studied the constriction and spreading of the heat transferred along a sphere subjected to concentrated heat input and output, over opposite areas. In the present model, the expressions proposed for the constriction and spreading thermal resistances are the same, so that two hemispheres sum up to a sphere. Bahrami et al. (2006) presented a simplified expression for Yovanovich et al. (1978) solution, valid for a specific range of contact angles and Florez (2017) improved this expression, for a wider range of the θ contact angle, resulting in the following equation for the constriction resistance:

$$R_{hem,1} = R_{hem,2} = \frac{0.56}{k_s r_n} \tag{3.125}$$

The substitution of Eqs. 3.124 and 3.125 in the overall thermal resistance, Eq. 3.115, results in:

$$R_t = \frac{0.56}{k_s r_n} + \frac{8r_s\left(1 - \frac{r_s}{r_s + r_m}\right)}{\pi k_s\left[r_n + r_s\left(\frac{r_n + r_m}{r_s + r_m}\right)\right]^2} \tag{3.126}$$

Form this expression it is possible to obtain the effective thermal conductivity.

Alternatively, another expression for the total thermal resistance, based on the porosity and defined in terms of areas, is:

$$R_t = \frac{2r_s(1 - \varepsilon)}{k_{ef} A_s} \tag{3.127}$$

where A_s is the total solid area in the unit cell. Matching Eqs. 3.126 and 3.127, a non-dimensional effective thermal conductivity of the sintered porous can be determined:

Table 3.2 Effective thermal conductivity of several wick structures (Chi 1976)

Wick structures	k_{ef}
Porous media and liquid in series	$\frac{k_l k_w}{\varepsilon k_w + k_l(1-\varepsilon)}$
Porous media and liquid in parallel	$\varepsilon k_l + k_w(1-\varepsilon)$
Superposed wire screens	$\frac{k_l[(k_l+k_w)-(1-\varepsilon)(k_l-k_w)]}{(k_l+k_w)+(1-\varepsilon)(k_l-k_w)}$
Packed spheres	$\frac{k_l[(2k_l+k_w)-2(1-\varepsilon)(k_l-k_w)]}{(2k_l+k_w)+(1-\varepsilon)(k_l-k_w)}$
Rectangular grooves	$\frac{w_f k_l k_w h + w k_l(0.185 w_f k_w + h k_l)}{(w+w_f)(0.185 w_f k_l + h k_l)}$
Circular grooves	$kef = \varepsilon kl + (1-\varepsilon)\cdot\left(\frac{k_l k_w D_g}{0.185(k_w)(s-w)+k_l D_g}\right)$

Where: k_l is the liquid thermal conductivity, k_w is the wick material thermal conductivity, w is the groove width (see Fig. 2.11), w_f is the groove fin thickness (thickness of the wall between grooves), h is the groove depth (see Fig. 2.11), α is the opening angle [rad], s is the pitch between grooves, D_g is the groove diameter

$$k_{ef} = \frac{2(1-\varepsilon)}{\pi r_s} k_s \left(\frac{0.56}{r_n} + \frac{8 r_s \left(1 - \frac{r_s}{r_s + r_m}\right)}{\pi \left[r_n + r_s \left(\frac{r_n + r_m}{r_s + r_m} \right) \right]^2} \right)^{-1} \tag{3.128}$$

The porosity can be calculated from the radius of the metallic powder from which the wick is fabricated, as discussed in the Liquid Pressure Drops section.

Table 3.2, adapted from Chi (1976), presents several other expressions for the determination of the effective thermal conductivity for several different wick structures.

3.9 Geyser Boiling Phenomena in Thermosyphons

When designing an equipment assisted by thermosyphons, the designer must consider possible periodic temperature oscillations, which happens under some especial operation conditions, usually during the start-up period, especially for thermosyphons with large aspect ratios (length over diameter). These temperature oscillations are actually resulting from a phenomenon known as Geyser Boiling, caused by the formation and burst of large bubbles in the evaporator.

Literature experimental work (Pabón et al. 2019) shows that, usually, the Geyser Boling effect results from the very fast growth of a single vapor bubble, formed in a nucleation site in the evaporator internal wall, such as a small cavity or wall imperfection, as shown in the schematic of Fig. 3.21a. The bubble grows very fast, reaching the liquid–vapor interface (Figs. 3.21b–d). According to Pabón et al. (2019), the elapsed time for vapor bubble growth in Geyser Boiling conditions is very short, typically less than 50 ms (miliseconds), depending on the thermosyphon design parameters.

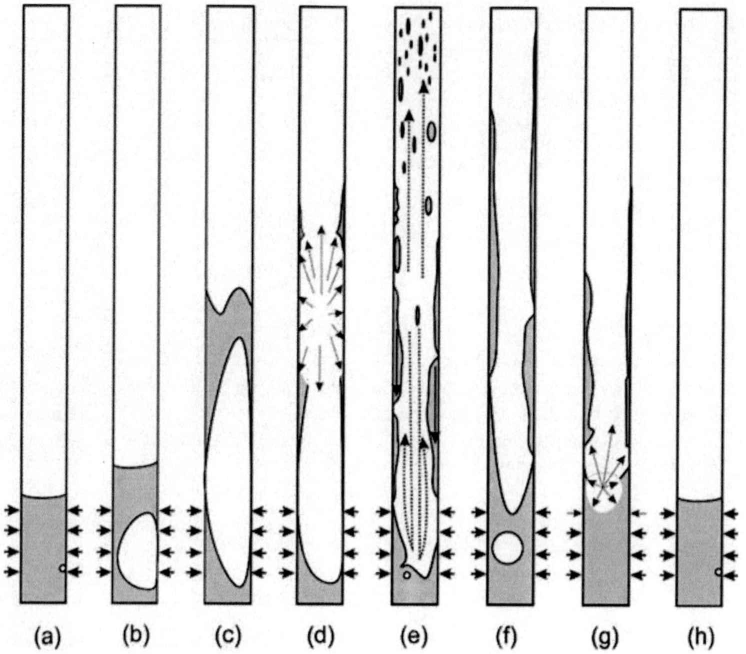

Fig. 3.21 Geyser Boiling phenomenon in thermosyphons

The high amount of heat used for the liquid–vapor phase change that provides the vapor to the bubble, during its growing process, is absorbed from the liquid pool and from the evaporator wall, resulting in a very fast evaporator temperature drop.

If the tube has a small diameter, the bubble radius can reach quickly the tube radius and the boiling process can be considered confined. Actually, the pressures inside the bubble in vertical confinement may be larger than those observed in pools heated by a horizontal surface. During the bubble grow process, some liquid is pushed upwards, causing an increase in the liquid pool height. As it reaches the liquid–vapor interface, the bubble bursts, spreading a hot, high velocity mist, as the vapor contained in the bubble core, due to the sudden pressure decrease, immediately converts to very small liquid droplets, which reach the condenser wall in a spray heating condition. It is well recognized that hot spray is a very effective heating mode, and so the condenser wall temperature increase almost instantaneously. Therefore, during the vapor bubble burst, the condenser temperature rises at almost the same instant and rate that the evaporator temperature drops. Figure 3.22 illustrates the evaporator and condenser temperature behaviors for a thermosyphon operating under Geyser Boiling conditions.

Besides, the high vapor pressure release due to the bubble burst causes an intense movement of the working fluid within the thermosyphon, as illustrated in Fig. 3.21e. During the liquid re-accommodation period due to gravity action, small bubbles can be trapped within the pool (see Fig. 3.21f), usually close to the liquid–vapor

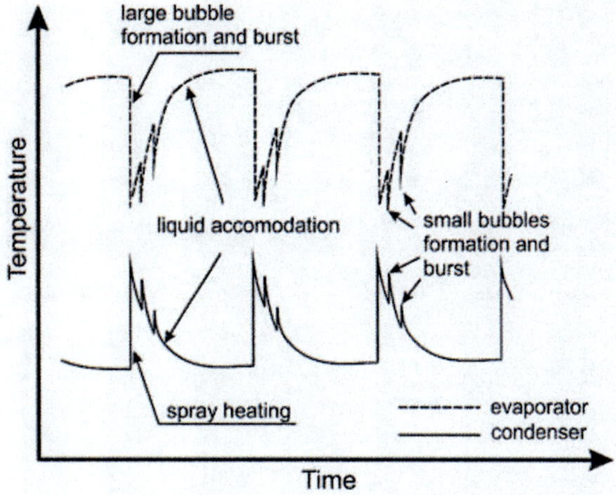

Fig. 3.22 Evaporator and condenser temperature as a function of time during the formation and burst of bubbles in Geyser Boiling

interface, playing the role of new vapor nucleation points, which grow quickly to bubbles. These bubbles rapidly reach the liquid–vapor interface and burst, however with smaller energy. These smaller bubbles causes smaller temperature variations on the evaporator and condenser. After all the entrapped bubbles are burst, the liquid "slowly" accommodates again and the cycle restarts (see Fig. 3.21h).

Geyser Boiling is more likely to happen during start-up processes. However, this phenomenon can be avoided by increasing the evaporator heat input, which makes the boiling more intense, increasing the frequency of the bubble formation. In this condition, the bubbles have energy to detach from the walls before they grow large, not having enough time to accumulate energy. The free bubbles move to the vapor–liquid interface, where they burst gently. In the limit, the fully developed boiling regime is reached. Another way to avoid the Geyser Boiling is by the appropriate choice of the thermosyphon geometry, by selecting, for instance, larger diameter tubes.

Working fluid behavior under Geyser boiling in a thermosyphon was observed by Pabon et al. (2019) in a glass-water thermosyphon. The glass tube had 20 mm of internal diameter and 400 mm of total length. The thermosyphon adiabatic section temperature was around 43 °C, resulting in the vapor pressure of about 103 Pa. Heat was externally delivered to the thermosyphon by an electrical cartridge, through a concentrated small lateral external area, with the objective of a unique nucleation site. The idea of this test configuration was to guarantee the formation of a single bubble, which can be easily visualized. Figures 3.23 and 3.24 were taken using a high velocity camera with 4000 frames per second. In Fig. 3.23 the liquid level, for the thermosyphon in rest conditions, is only 40 mm higher than the heat source, while, in Fig. 3.24 the liquid is 83 mm above the heater position. Bellow each picture, the time in which the picture was taken is shown. The arrows indicate the points where bubbles start: the onset of the larger bubbles is shown in the first frame of each figure, while the entrapped bubbles are shown in the other frames. From these photographs, it is possible to visualize the several steps of the working fluid behavior

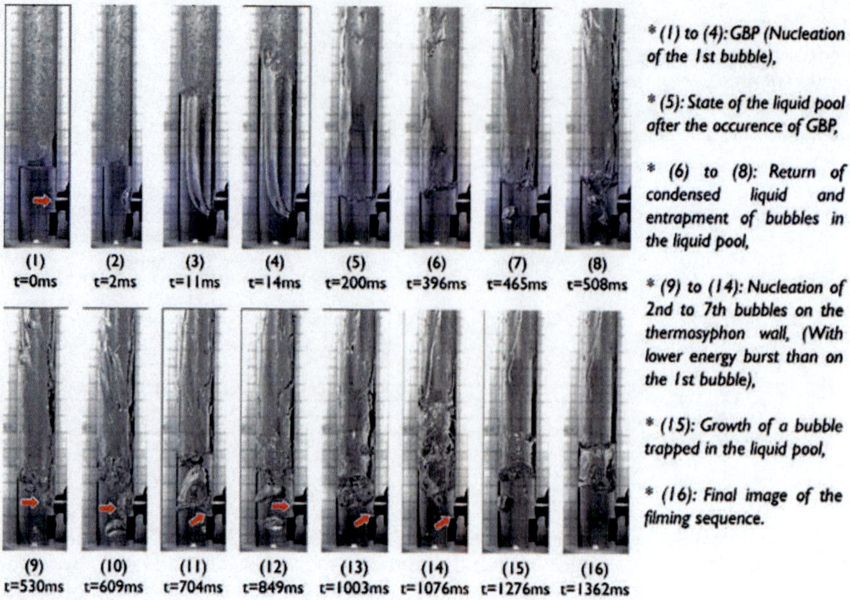

Fig. 3.23 Photographs of a glass-water thermosyphon where the volume of liquid height at rest is little higher height of the heat source

as schematized in Fig. 3.21. These images show clearly the bubble growth and rise in the liquid pool, starting from a single nucleation site, until its burst. Swirling in the liquid film in its return from the condenser is observed, a consequence of the concentrated lateral heating mode. Although the images shown in this book are from the same set of images presented in Pabon et al. (2019), Figs. 3.23 and 3.24 show different images from those published.

Therefore, the interested reader in visualization of this phenomenon, is advised to see Pabon et al. (2019).

Comparing Figs. 3.23 and 3.24 it is possible to note that higher filling ratios result in taller bubbles that demand larger time to burst. Higher vapor pressures are developed within larger bubbles. When the bubbles burst, larger temperature drops and rises, in the evaporator and condenser, are expected.

Pabon et al. (2019) also measured the vapor temperature and pressure in the condenser region. Based on their data, plots against time are shown in Fig. 3.25, for the thermosyphon operating as in Fig. 3.23 conditions. Readings from two thermocouples installed in the wall are shown, one closer to the evaporator and the other in the middle of the condenser. As expected, the temperature close to the evaporator is higher. However, it is clearly observed the almost perfect correlation between the condenser pressure and temperatures, for the thermosyphon operating under Geyser Boling conditions.

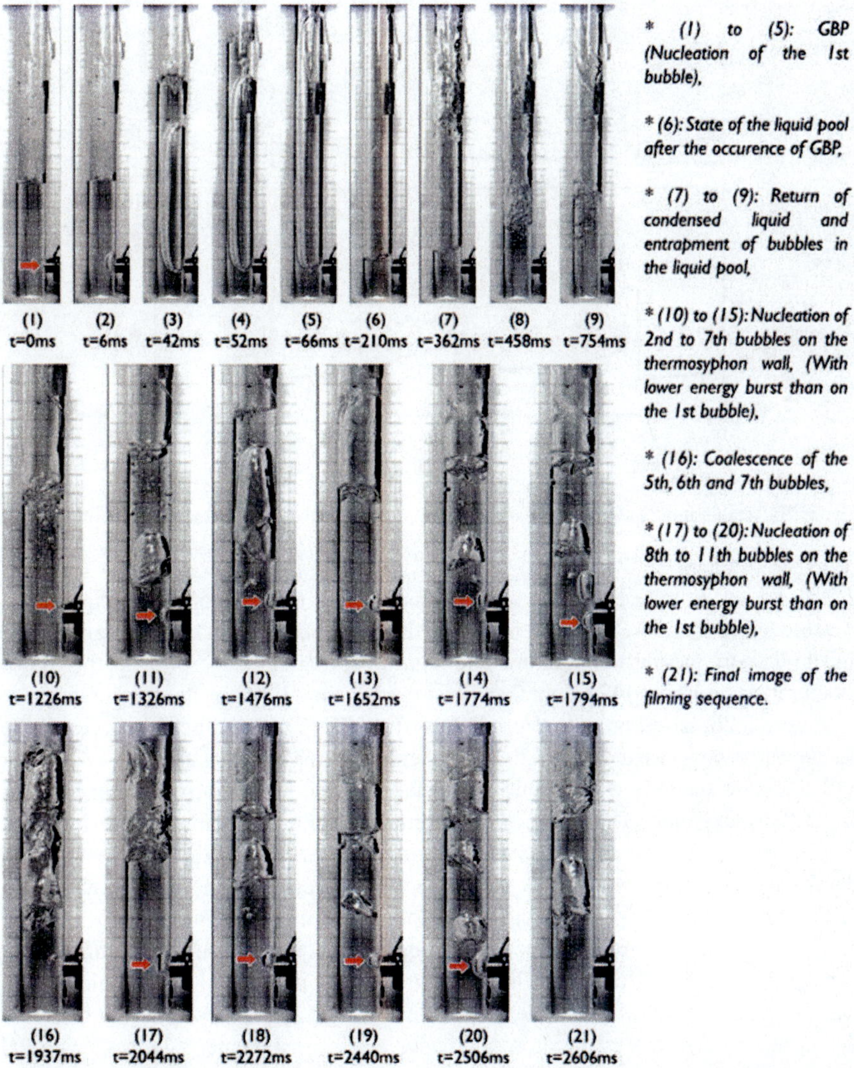

Fig. 3.24 Photographs of a glass-water thermosyphon in Geyser Boiling where the liquid height at rest is 83 mm above the heat source

In practical applications, the Geyser Boiling phenomenon can be recognized by the observation of the typical temperature–time behavior of the evaporator and condenser regions, suchas illustrated in Fig. 3.22.

The bubble burst also causes audible and recognizable sounds accompanied by vibrations, as treated by Tecchio et al. (2017), in their work. When Geyser Boiling is observed during the start-up period, the first bubble bursts are followed by loud sounds and vibrations. As the liquid pool heats up, the bubble burst frequency increases, while

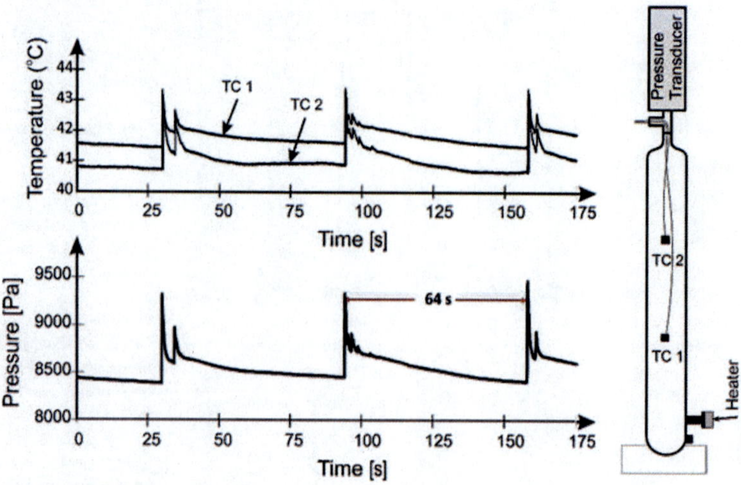

Fig. 3.25 Temperature and pressure as a function of time for the glass water thermosyphon

the sound and vibration continuously decrease, until the developed boiling regime is achieved and almost no noise and vibrations are observed. The lower amplitude and higher frequency temperature and sound oscillations superpose to the major temperature and sound behaviors, as illustrated in Fig. 3.22, for the temperature.

Figure 3.26, based on data from Pabon et al. (2019), shows actual curves of the condenser vapor pressure and of the mean temperatures of the evaporator, condenser and adiabatic sections, as a function of time, obtained for a relatively long copper–water thermosyphon. The copper tube external and internal diameters were 22 and 20 mm and the total length was of 920 mm. The evaporator, adiabatic section and condenser lengths were 200, 90 and 630 mm, respectively. The filling ratio (ratio of the volumes of working fluid to the evaporator) was 66%. The power input was 150 W. The temperature curve behaviors suggest that the device is operating under Geyser Boiling conditions (see Fig. 3.22). As already mentioned, the bubble growth and burst is a very fast phenomenon, presented by a close to vertical temperature decrease curve in the evaporator, followed, almost instantaneously, by a vertical temperature increase in the condenser. After the bubble burst, all temperature curves steadily tend to constant values, with some perturbations due to the burst of smaller bubbles.

In the present case, due to the long narrow geometry of the thermosyphon and because the condenser is longer than the evaporator, the adiabatic section temperature follows closely the condenser temperature behavior. The amplitude of the temperature variation of the evaporator is around 8 °C, while it is less than 2 °C in the condenser. This difference is due to the friction that the wall imposes over the vapor flow through the lengthy condenser, damping temperature variation effects. The pressure curve, also shown in Fig. 3.26, follows very closely the condenser

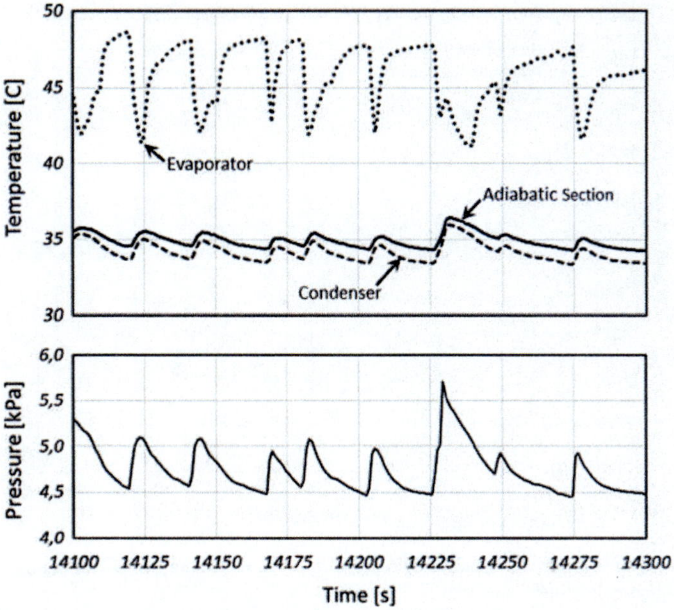

Fig. 3.26 Temperature and condenser pressure as a function of time for an actual copper–water thermosyphon working in Geyser Boiling conditions

temperature, as the pressure sensor is located at the end of condenser. The Geyser Boiling cycle frequency, for both temperature and pressure, is around 0.04 s^1.

Figure 3.27 shows similar temperature plots for a high temperature thermosyphon, made of stainless steel tube and liquid metal sodium as the working fluid, as presented by Rodrigues (2019). Evaporator, adiabatic section and condenser curves are presented, for a thermosyphon with 350 mm of total length and 25.4 mm and 21.5 of external and internal diameters, with 100, 50 and 200 mm of evaporator, adiabatic section and condenser lengths, respectively. The thermosyphon is filled at ambient temperature with 36.25 g of 99.8% pure sodium, resulting in a filling ratio of 127%. As seen from this plot, the evaporator and condenser temperature variation amplitudes are of around 250 °C and 300 °C, much higher than those observed in Fig. 3.26, for the copper–water thermosyphon. However, the Geyser Boiling frequency, of $2{\cdot}10^{-3}$ s^{-1} in this case, is much smaller than that observed for water. In fact, the thermal behavior of the liquid metal working fluid explains this difference. The surface tension and the molecular weight of liquid sodium is much larger than those of water. Therefore, the hydraulic and tension forces acting over the bubble liquid–vapor interface are able to hold much higher vapor pressures, so that, when the bubble disrupts, the sodium vapor pressure drop is much larger and so are the outburst effects. This vapor eruption also results in high vibration and noise. For this reason, Geyser Boiling must be avoided during actual applications of liquid metal thermosyphons.

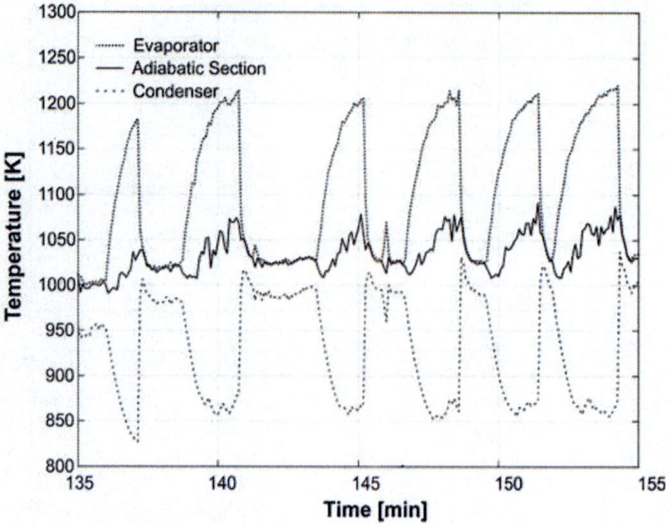

Fig. 3.27 Temperatures and pressure as a function of time for an actual copper–water thermosyphon working in Geyser Boiling conditions

Loop Thermosyphons

Although this phenomenon is more likely to be observed in thermosyphons, loop thermosyphons may also experience Geyser Boiling. Actually, the loop thermosyphon geometry organizes better the vapor and liquid flows within the device, preventing the counter-current fluxes, which, in many configurations, is enough to avoid the Geyser Boiling.

Figure 3.28 illustrates the thermal behavior of a loop thermosyphon under Geyser Boiling conditions. According to the geometry and the heat input level, after the liquid reaches superheated conditions, large bubbles almost instantaneously forms and burst over some nucleation sites located at the wall. The pressure variations due to the bubble ruptures "pump" vapor through the vapor line up to the condenser, which warms up quickly. Pressure drops are expected in the vapor line, which tend to attenuate the temperature variations of the condenser section. This means that the Geyser Boling effect over the condenser can be reduced by the appropriate design of the loop thermosyphon and/or by controlling the power input level.

Simas (2017) studied the thermal behavior of a copper loop thermosyphon, designed for cooling electro/electronic devices, which geometry is presented schematically in Fig. 3.29. It basically consists of a flat evaporator of $190 \times 190 \times 10\ mm^3$, with square, 3 mm thick, front and back walls. So, the evaporator internal volume is around 650 ml. The evaporator has internal fins to promote the appropriate mechanical strength to afford the internal vapor pressure during operation. The condenser consists of a heat exchanger with a 10 turn spiraled cooper pipe of 12.7 mm of external diameter and 0.88 mm of wall thickness (where working fluid vapor flows), which is located within a cylindrical copper case of 100 mm of diameter

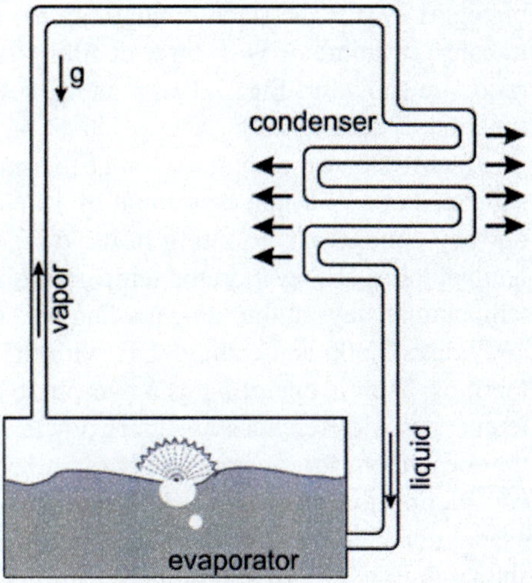

Fig. 3.28 Schematic of Geyser Boiling for loop thermosyphons

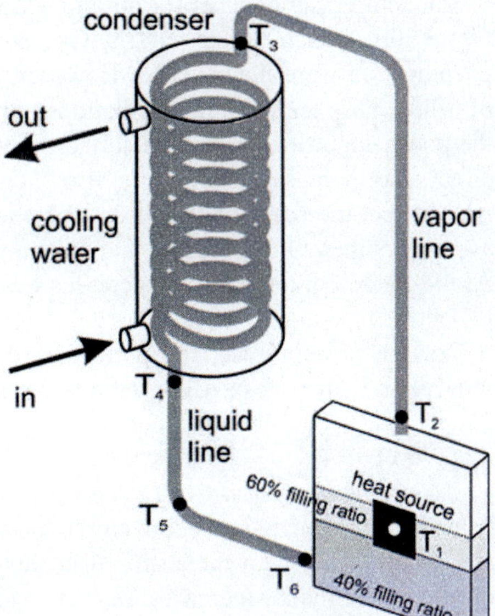

Fig. 3.29 Schematic of the loop thermosyphon studied by Simas (2017)

and 335 mm of length. An internal copper helix that follows the spirals improves the heat exchange within the device. The liquid and vapor lines, of approximately 320 and 680 mm of length, respectively, are made of the same tube as the evaporator.

Several tests were conducted, as depicted in Simas (2017). Among them, the results obtained with the device in the vertical position with a heat source located in the center of the evaporator (see Fig. 3.29) are discussed. Plots of temperatures against time, for the thermocouples T_1 to T_6, which position is depicted in the schematic of Fig. 3.29, for the three configurations: no working fluid (pure conduction heat transfer), evaporator with 60% of filling ratio and evaporator with 40% of filling ratio, are shown in Fig. 3.30a, b and c, respectively, in curves of temperature as a function of time.

Figure 3.30a were obtained for the thermosyphon operating with no working fluid, subjected to a constant heat input of 150 W. In this case, the heat was exchanged only by conduction, resulting in the dissociation of the evaporator and condenser temperatures: the evaporator temperature increases steadily, while the condenser temperature stays at the same cooling water level of the condenser heat exchanger.

Figures 3.30b and c show that, with the presence of the working fluid, the loop thermosyphon is operating as a two-phase device, as the evaporator and condenser temperatures differences are much lower than those observed for the empty loop thermosyphon, for a same power input level. No dry out was noticed as expected, considering that, for both filling ratios (60 and 40%), the liquid level within the evaporator reaches at least the tail position of the electrical resistance block, as shown in Fig. 3.29 (it should be remembered that the liquid level in the evaporator raises with the presence of bubbles).

The temperature curves for 60% of filling ratio show no major oscillations, (see Fig. 3.30b), which means that the Geyser Boiling conditions for the onset of the phenomenon were not reached. However, in Fig. 3.30c, for the device with 40% of filling ratio, temperature oscillations typical of Geyser Boiling are observed. As the power input increases (and so the evaporator temperature), the Geyser Boiling effect also increases. This figure also shows that the loop thermosyphon presents two distinct thermal performances. For power inputs above 350 W, the evaporator average temperature decreases and the temperature oscillation frequency increases. As the condenser temperature increases steadily, the overall thermal resistance tends to decrease.

For the present case, the performance (thermal resistance) of the loop thermosyphon with 40% of filling ratio is better than that of 60%, which shows larger temperature differences between the evaporator and condenser, for the same power input level.

According to Simas (2017), loop thermosyphons may present a very unstable behaviors. In general sense, the amplitude of the Geyser Boiling temperature oscillations decreases with increasing heat flux at the evaporator surface, whereas the oscillation frequency increases. Decreasing the filling ratio, increases the probability of Geyser Boiling occurrence. Actually, the gravity action over the high volume of liquid above the bubble nucleation site in the evaporator, results in high confinement conditions, in which the liquid may not reach the necessary overheating for the onset of boiling. In this case, boiling conditions may be reached by increasing the power inputs. Besides, small bubbles eventually formed may collapse within the liquid, disappearing, as vapor turns to liquid before reaching the liquid–vapor interface. When boiling is not present, the evaporation happens only in the liquid–vapor interface, however with much larger thermal resistance. For this reason, boiling regime is always desirable for a better thermal performance of the device.

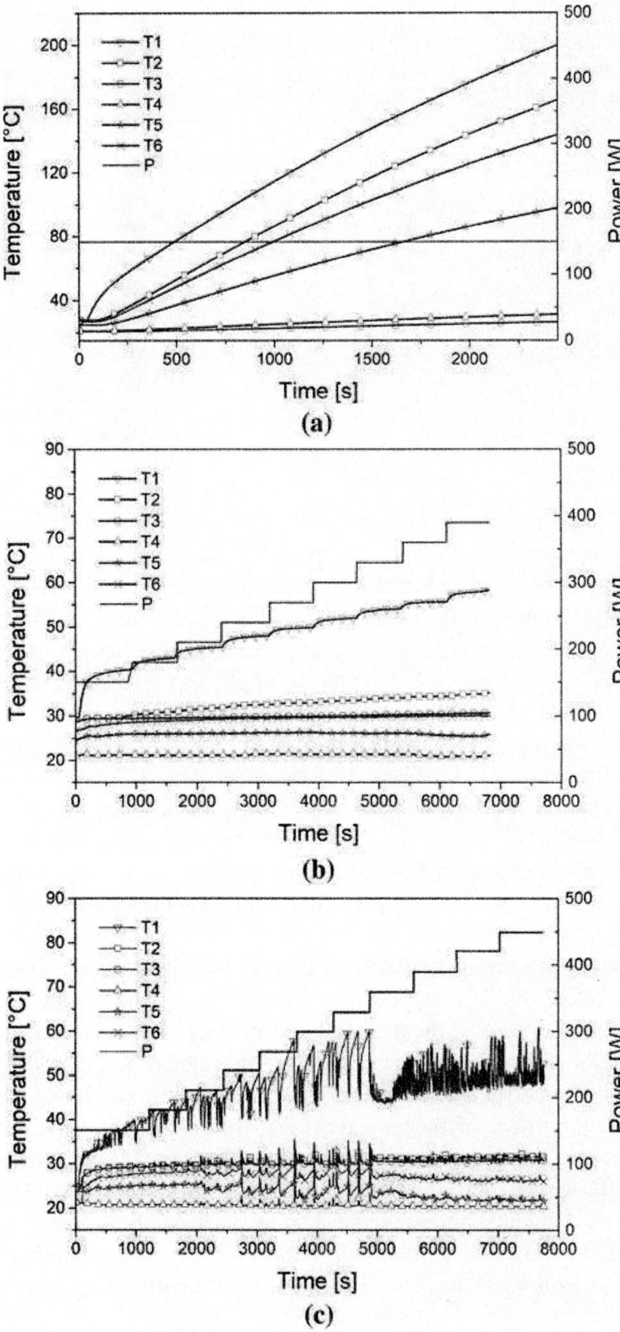

Fig. 3.30 Temperature against time for loop thermosyphon (a) no working fluid, (b) 60% and (c) 40% of working fluid filling ratio

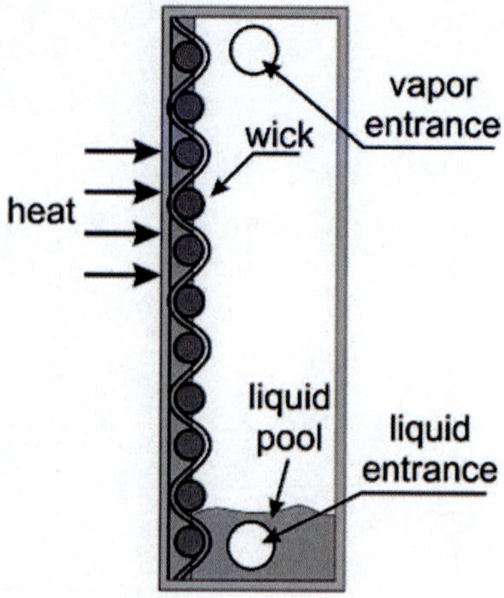

Fig. 3.31 Schematic of the of cross section of Fig. 3.29 cross section with internal wall of the heated surface heated by a wick

As already discussed, in most applications, Geyser Boiling effect is not desirable due to vibration and noise. One way to mitigate this effect is to design thicker evaporators, which may not be convenient, for instance, in the cooling of electro/electronic equipment applications. Simas (2017) propose a solution to eliminate the Geyser Boiling, and, additionally, reduce the overall thermal resistance of the device. A screen wick (or any porous media) should be applied in the heated internal face of the evaporator. Due to the capillary effect of the porous media, liquid is spread over the active surface of the evaporator, avoiding the formation of pools. The liquid–vapor phase change, in this case, happens in the meniscus of the wick and no large bubbles are formed. Figure 3.31 show a cross section view of an evaporator, in which the heated wall is internally covered by a wick. The temperature as function of time for the loop thermosyphon with 40% of filling ratio without screen and with screen are shown in Figs. 3.32a (the same of Fig. 3.30c) and Fig. 3.32b, respectively. Comparing both plots it is clearly noted that the Geyser Boiling was completely eliminated. The overall thermal resistance reduction was also quite evident, as the evaporator and condenser temperature difference decreased for the same power input: for instance for the power input of 450 W, the temperature difference dropped from around 30 to about 10 °C and the thermal resistance reached the value of 0.055 °C/W.

The wick structure over the internal surface of heated wall of a loop thermosyphon evaporator has the effect of pushing the liquid (even against gravity), spreading it over the surface, so that the heat source can be located in any position of the evaporator wall, without the risk of dry-out. Simas (2017), installed the same heat source (same area and heat power) in the right up position of the evaporator external wall. Figure 3.33 show the plots of the temperatures for the evaporator without (Fig. 3.33a) and with wick structure (Fig. 3.33b), both with filling ratios of 40%.

Figure 3.33a, relative to the non-wicked evaporator, shows that the device is not working (dry-out) as the evaporator temperature keeps rising with the power input increase, while the condenser temperature is detached from the evaporator, controlled by the cooling water. However, the device works very well with the presence of the

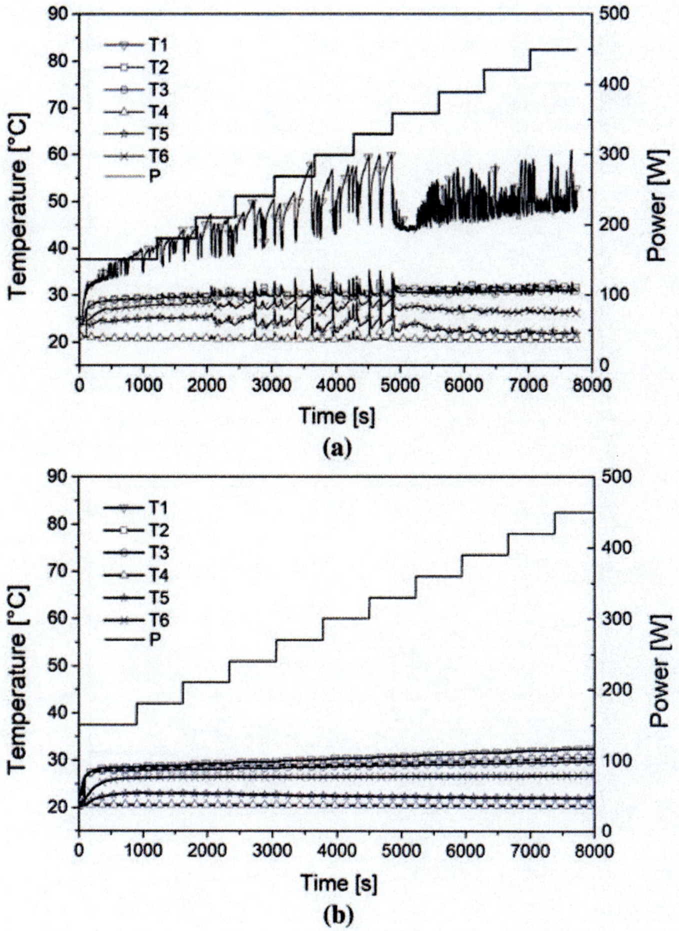

Fig. 3.32 Temperature against time for the Fig. 3.29 loop thermosyphon with 40% of filling ratio: **a** with no wick structure **b** with wick structure installed over the heated internal wall of evaporator

wick, with the overall thermal resistance almost as low as that observed for the central heat source (see Fig. 3.32b). As for the Fig. 3.32b, Fig. 3.33b do not show any noticeable temperature oscillations due to Geyser Boiling, which is completely eliminated in both cases.

3.10 Closure

In this chapter, the physical models regarding the several phenomena that drive the behaviors of thermosyphons and heat pipes are discussed. The major available

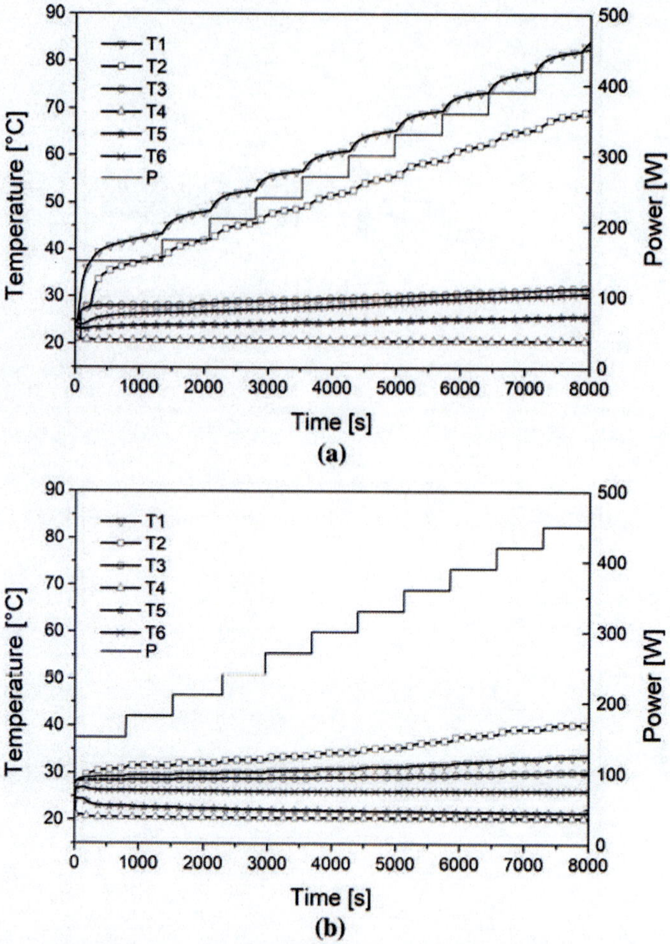

Fig. 3.33 Temperature against time for Fig. 3.29 loop thermosyphon with 40% of filling ratio, with the heat source located in the up-right position: (a) without wick structure and (b) with wick structure

thermal and hydraulic theoretical models in the literature are presented and effective properties are defined, so that the literature state of the art is established. A deep discussion about the Geyser Boiling phenomena observed in the different thermosyphon configurations is provided. Therefore, the main purpose of this chapter, which is to make available analytical tools for the design of such devices, is fulfilled.

References

Akbari, M., Siston, D., Bahrami, M.: Laminar Fully Developed Flow in Periodically Converging-Diverging Microtubes. Heat Transfer Eng. **31**(8), 628–634 (2010)

Bahrami, M., Yovanovich, M.M., Culham, J.R.: Effective Thermal Conductivity of Rough Spherical Packed Beds. Int. J. Heat Mass Transf. **49**, 3691–3701 (2006)

Brennan, P. J., and. Kroliczek, J., Heat Pipe Design Handbook. Towson, NASA Goddard Space Flight Center (1979)

Busse, C.A.: Theory of the Ultimate Heat Transfer Limit of Cylindrical Heat Pipes. Int. Journal Heat Mass Transfer **16**, 169–186 (1973)

Busse, C.A., Kemme, J.E.: Dry-out Phenomena in Gravity-Assist heat Pipes with Capillary Flow. Int. Journal Heat Mass Transfer **23**, 643–654 (1980)

Carey, V.P.: Liquid-vapor Phase-change Phenomena: an Introduction to the Thermophysics of Vaporization and Condensation Process in Heat Transfer Equipment. Series in Chemical and Mechanical Engineering, Taylor and Francis, Hebron (1992)

Chi, S.W.: Heat Pipe Theory and Practice A Sourcebook, MacGraw-Hill (1976)

Collier, J.G., Thome, J.R.: Convective Boiling and Condensation, 3rd edn. Science Publications, Oxford, Oxford (1994)

Faghri, A.: Heat Pipe Science and Technology, 2nd edn, Global Digital Press (2016)

Florez, J.P.M., Chiamulera, M., Mantelli, M.B.H.: Permeability Model of Sintered Porous Media: Analysis and Experiments. Heat Mass Transf. **49**, 1–9 (2017)

Florez, J.P.M., Mantelli, M.B.H., Nuernberg, G.V.: Effective thermal conductivity of sintered porous media: Model and experimental validation. Int. Journal Heat Mass Transfer **66**, 868–878 (2013)

Florez, J.P.M., Mantelli, M. B. H. ; Nuernberg, G. V. and Milanez, F. H., Powder Geometry Based Models for Sintered Media Porosity and Effective Thermal Conductivity, Journal of Thermophysics and Heat Transfer, **28** (3), 507–517 (2014) Thümmler, F. and Oberacker, R.: An Introduction to Powder Metallurgy, The Institute of Materials, London (1993)

Florez, J.P.M., Heat and Mass Transfer Analysis of a Cooper Loop Heat Pipe, Doctoral Thesis, POSMEC, Federal University of Santa Catarina, Brazil (2017)

Fox, R, W., Pritchard, P. J. and McDonald, A. T., Introduction to Fluid Mechanics, 4th Ed., McGraw-Hill (1999)

Groll, M., Rosler, S.: Operation Principles and Performance of Heat Pipes and Closed Two-Phase Thermosyphons. Journal of Non-Equilibrium Thermodynamic **17**, 91–151 (1992)

Imura, H., Sasaguchi, K., Kozai, H., Numata, S.: Critical Heat Flux in a Closed Two-Phase Thermosyphon. Int. J. Heat Mass Transfer **26**(8), 1181–1188 (1983)

Kays, W.M.: Convective Heat and Mass Transfer. McGraw-Hill Book Company, New York (1966)

Kemme, J.E.: Ultimate Heat-Pipe Performance. IEEE Transactions of Electron Devices **8**, 717–723 (1969)

Marcus, B. D.: Theory and Design of Variable Conductance Heat Pipes, Report No. NASA CR-2018, National Aeronautics and Space Administration, Washington, D.C. (1972)

Milanez, F.H., Mantelli, M.B.H.: Heat Transfer Limit Due to Pressure Drop of a Two-Phase Loop Thermosyphon. Heat Pipe Science and Technology, an International Journal **3**, 237–250 (2010)

Nguyen-Chi, H., Groll, M.: Entrainment or Flooding Limit in a Closed Two-Phase Thermosyphon. Heat Recovery Systems **1**, 275–286 (1981)

Pabón, N.Y.L., Florez, J.P.M., Vieira, G.S.C., Mantelli, M.B.H.: Visualization and Experimental Analysis of Geyser Boiling Phenomena in Two-Phase Thermosyphons. Int. Journal Heat Mass Transfer **141**, 876–890 (2019)

Peterson, G.P.: Heat Pipes Modeling Testing and Applications, John Wiley and Sons (1994)

Reay, D.A. and Kew, P.A.: Heat Pipes Theory, Design and Applications, 5th edn, Butterworth-Heinemann (2006)

Swinkels, F.B., Ashby, M.F.: A Second Report on Sintering Diagrams. Acta Metall. **29**, 259–281 (1981)

Tecchio, C., Oliveira, J. L. G, Paiva, K. V., Mantelli, M. B. H., Gandolfi, R. and Ribeiro, L. G. S., Geyser Boiling Phenomenon in a Two-Phase Closed Loop-Thermosyphons, Int. J. Heat Mass Transfer, **111**, 29–40 (2017)

Tien, C.L. and Lienhardt, J.H.: Statistical Thermodynamics, Hemisphere Publishing Corporation, (1979)

Wallis, G.B.: One-Dimensional Two-Phase Flow, McGraw-Hill (1969)

Wright, P.E., ICICLE Feasibility Study, Final Report NASA Contract No. NAS5–21039, Camden (1970)

Yovanovich, M. M. ; Schneider, G. E. and Tien, C. H., Thermal Resistance of Hollow Spheres Subjected to Arbitrary Flux Over Their Poles, 2nd AIAA/ASME Thermophysics and Heat Transfer Conference, Palo Alto, California, 120–134 (1978)

Chapter 4
Design of Thermosyphons and Heat Pipes

Design methodologies for thermosyphons and heat pipes are presented in this chapter. Steady state operation condition hypothesis and the analogy between electrical and thermal circuits are employed in the present models.

The flow chart presented in Fig. 4.1 shows the main steps necessary for the design of thermosyphons and heat pipes, which are briefly described next.

The two major parameters to be known a priori for the design of thermosyphons or heat pipes are the physical dimensions (geometry) available for the installation of the thermosyphons and the major thermal requirements, consisting of expected heat transport capacity and/or the operation temperature levels. Based on this information, the "first guess" of the technology/geometry to be used (straight tube, loop, tree configuration, etc.) is proposed, along with the most basic design parameters: working fluid and case material.

For usual thermosyphon and heat pipes design procedures, the most important parameter needed for the working fluid selection is the operation temperature level. In the sequence, the material of the casing and of the wick (for thermosyphons) is chosen: they must be compatible with the working fluid, to avoid their chemical reaction that could generate the non-condensable gases that accumulate in the condenser highest regions, blocking part of the device and decreasing its thermal performance.

The temperature operation level is, actually, a consequence of the heat transport capacity and vice-versa. This compromise between temperature (difference) and the heat power transported can be represented by the thermal resistance concept, defined as:

$$R = \frac{\Delta T}{q} \tag{4.1}$$

The overall thermal resistance of a thermosyphon or heat pipe is a function of the device geometry and can be obtained through analytical models. For steady-state conditions, resistance networks are usually employed. In parallel, the relevant

M. B. H. Mantelli et al., *Thermosyphons and Heat Pipes: Theory and Applications*,
https://doi.org/10.1007/978-3-030-62773-7_4

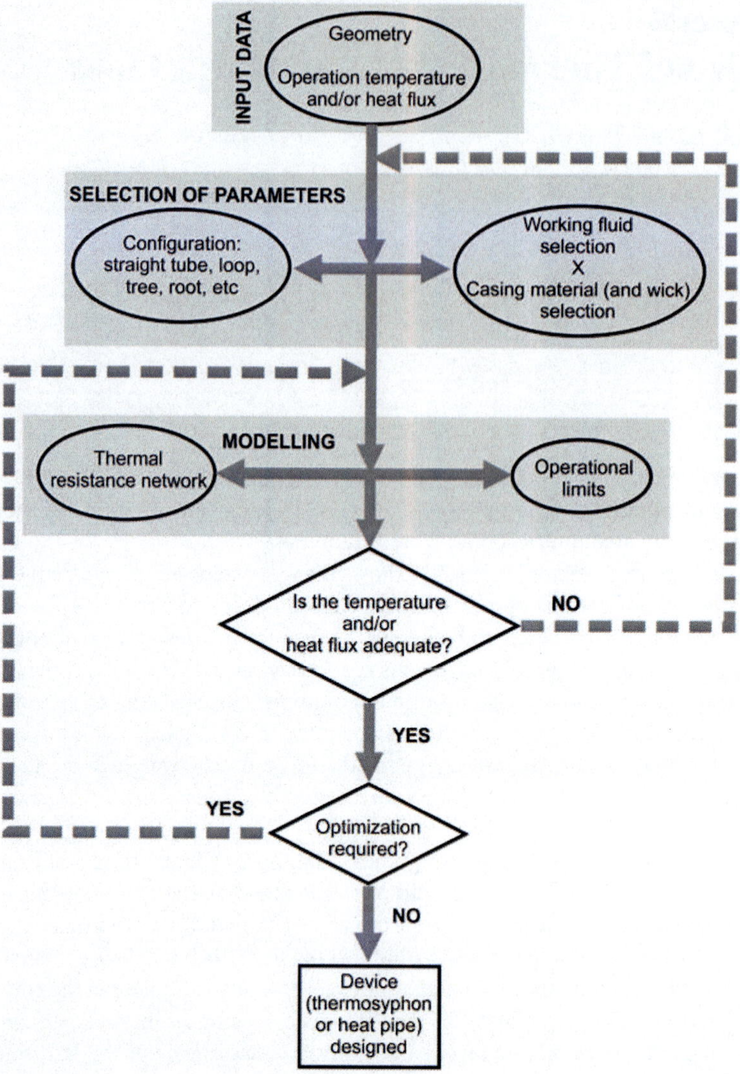

Fig. 4.1 Flowchart for thermosyphon design methodology

operation heat transfer limits (see Chap. 3) must be checked for the range of operation conditions.

Obviously, if the device is not able to transfer the amount of heat and/or to keep the temperature of the device under controlled levels, a new configuration must be proposed and the overall thermal resistance as well as the heat transfer capacity of the new device must be determined again. Optimization procedures for the selection of more suitable devices (lower costs, lower volumes, etc.) can be applied before the thermosyphon or heat pipe are considered finally designed.

4.1 Thermosyphons

Resistance Network

As for other thermal applications, the analogy between electrical and thermal circuits can be used as a modelling tool for thermosyphons. This technique is very simple and powerful for devices that present close to one-dimensional behaviors and operate in steady-state conditions, such as thermosyphons.

The overall thermal resistance can be interpreted as the "difficulty of the thermosyphon to transport heat". The larger the thermal resistance, the larger is the temperature difference (thermal "potential") required to transfer a certain amount of heat. Considering the thermosyphon operating in steady state conditions, an equivalent thermal circuit (analogous to an electric circuit), such as the one presented in Fig. 4.2, must be constructed.

External Thermal Resistances

In Fig. 4.2, heat is delivered to the thermosyphon in evaporator region and removed by the condenser, through their external surface walls. Depending on the heat source or sink, the external heat transfer can happen by several mechanisms: convection,

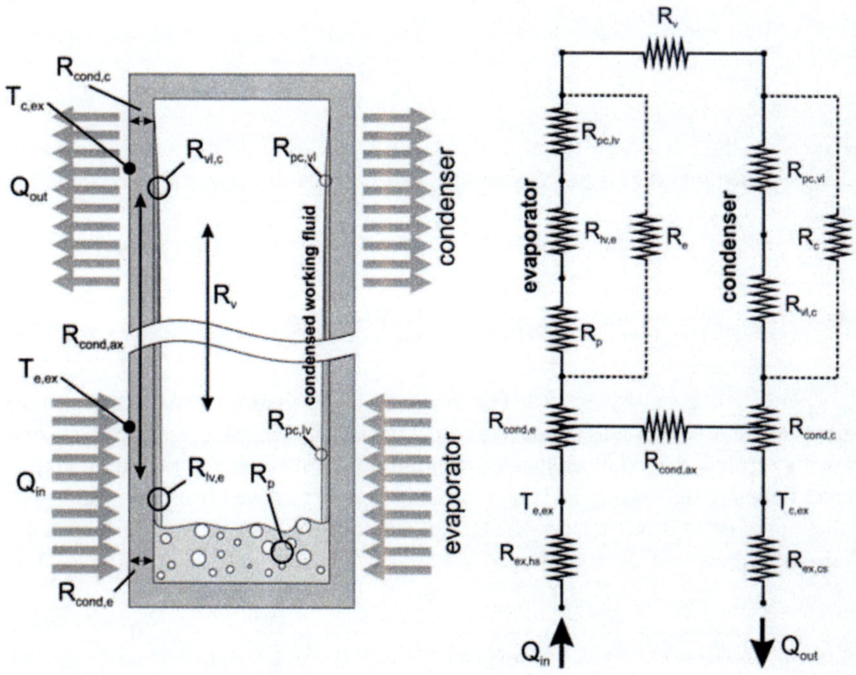

Fig. 4.2 Thermosyphon thermal circuit

conduction (by contact between surfaces), radiation, phase change, etc., or a combination of them. Several models, available in the literature, can be used to predict this heat exchange mechanism, which can be associated to a heat transfer coefficient: $h_{ex,e}$ for the heat source (evaporator) and $h_{ex,c}$ for the heat sink (condenser). The following equations can be used to estimate the external evaporator and condenser thermal resistances, respectively:

$$R_{ex,hs} = \frac{1}{h_{ex,e}A_e} \tag{4.2}$$

$$R_{ex,cs} = \frac{1}{h_{ex,c}A_c} \tag{4.3}$$

where A_e and A_c are the evaporator and condenser external areas.

Conductive Thermal Resistances

Once it reached the evaporator, the heat has two conduction parallel paths through the tube case material: axial, along the tube length, and radial, in the tube center direction. For a cylindrical tube, the radial conduction thermal resistances can be given by:

$$R_{cond,e} = \frac{\ln\left(r_{ex}/r_i\right)}{2\pi k l_e} \tag{4.4}$$

where l_e is the evaporator length and r_{ex} and r_i are the external and internal tube radius, respectively. This expression can also be used to determine the condenser radial resistance, $R_{cond,c}$ substituting l_e by l_c, if the condenser geometry is the same of the evaporator. The axial resistance can be obtained by the expression:

$$R_{cond,ax} = \frac{l_{ef}}{k\pi\left(r_{ex}^2 - r_i^2\right)} \tag{4.5}$$

where l_{ef} is given by Eq. 3.1. For tubes, the axial thermal resistance is usually around three orders of magnitude larger that the radial resistance; therefore, the heat flowing through the axial resistance is negligible when compared to that flowing in the radial direction and so the axial resistance can be removed from the circuit. The following criteria is proposed by ESDU (1983) for the elimination of this resistance from the circuit:

$$\frac{R_{cond,ax}}{R_{cond,e} + R_e + R_{pc,lv} + R_v + R_{pc,vl} + R_c + R_{cond,c}} > 20 \tag{4.6}$$

Thermal Resistance of the Evaporator Pool

In the next step, the heat reaches the bottom internal volume of the evaporator, where the working fluid in liquid state forms a small pool. To predict the thermal resistance of the pool, R_p, equations similar to Eqs. 4.2 and 4.3 must be applied, once the coefficient of heat transfer is known. Literature correlations can be used to predict the these coefficients, depending on the pool heat transfer conditions.

Basically, the liquid in the pool may present three regimes: natural convection, nucleate boiling or a combination of both.

Several correlations can be used to predict the coefficient of heat transfer as discussed by Mantelli (1999), who compared 14 of those correlations, developed for pools in general or specific for thermosyphons, with thermosyphon experimental data, concluding that El-Genk and Saber (1998a, b) correlation and Imura (1979) presented the best comparison.

El Genk and Saber (1998a) proposes the following parameter to determine the pool regime:

$$\chi = \Psi \, Ra^{0.35} Pr_l^{0.35} Kp^{0.7} \mathrm{Re}_v^{0.7} \tag{4.7}$$

where Pr_l is the Prandtl number ($Pr_l = \mu c_p/k$), Ψ is the working fluid property parameter given by:

$$\Psi = \left(\frac{\rho_v}{\rho_l}\right)^{0.4} \left[\left((p\nu_l/\sigma)\left(\rho_l^2/(\sigma g(\rho_l - \rho_v))\right) \right)^{1/4} \right]^{1/4} \tag{4.8}$$

Ra is the Rayleigh number defined as:

$$Ra = \left(\beta g D_i^4 q_e''\right) / (k_l \alpha_l \nu_l) \tag{4.9}$$

where β is the thermal expansion coefficient and $\alpha_l$the liquid thermal diffusivity. q_e'' is the heat flux applied over the evaporator wall. The non dimensional pressure parameter *Kp,* which involves the Laplace length scale (λ, see Eq. 2.83), and the Reynolds number are given by:

$$Kp = \frac{p\lambda}{\sigma} = \frac{p\sqrt{\sigma/g(\rho_l - \rho_v)}}{\sigma} \quad \text{and} \quad \mathrm{Re}_v = \frac{q_e''\lambda}{\rho_l h_{lv} \nu_l} \tag{4.10}$$

According to El Genk and Saber (1998b), natural convection is observed when $X < 10^6$, for which the following expression for Nu_p for the pool $\left(Nu_p = h_p D_i / k_l\right)$ is valid:

$$Nu_{p,N} = 0.475\, Ra^{0.35}\left(\frac{\lambda}{D_i}\right)^{0.58} \quad for \quad \chi < 10^6 \tag{4.11}$$

However, if X > 2.1 · 10^7, nucleate pool boiling is observed. El-Genk and Saber (1998b) proposed a correction factor for the classical boiling correlation of Kutateladze of 1952, i.e.:

$$Nu_{p,b} = (1 + 4.95\Psi)Nu_{ku} \quad for \quad \chi > 2.1 \cdot 10^7 \tag{4.12}$$

where the Nusselt number correlation of Kutateladze, defined as $Nu_{ku} = h_{ku} \cdot \lambda / k_l$, is given by Kaminaga et al. 1992 and El-Genk and Saber 1998a:

$$Nu_{ku} = 6.95x10^{-4}\, \mathrm{Pr}^{0.35}\left(\frac{q_e''\lambda}{\rho_v h_{lv} \nu_l}\right)^{0.7}\left(\frac{p\lambda}{\sigma}\right)^{0.7} \tag{4.13}$$

Combined convection is observed for $10^6 < X < 2.1 \cdot 10^7$; for this range, El-Genk and Saber 1998a propose the use of the expression:

$$Nu_p = \left(Nu_{p,N}^4 + Nu_{p,b}^4\right)^{1/4} \quad for \quad 10^{-6} < \chi < 2.1 \cdot 10^7 \tag{4.14}$$

The correlation of Imura (1979) is one of the most used (Groll and Rösler 1992) and Faghri (2016), being given by:

$$Nu_{p,b} = 0.32\left(\frac{k_l}{D_i}\right)\left(\frac{\rho_l^{0.65} k_l^{0.3} c_{p,l}^{0.7} g^{0.2}}{\rho_v^{0.25} h_{lv}^{0.4} \mu_l^{0.1}}\right)\left(\frac{p}{p_{atm}}\right)^{0.23} q_e''^{0.4} \tag{4.15}$$

A recent study on boiling heat transfer was developed by Kiyomura et al. (2017), where the following correlation was proposed, based on an extended database of experimental results, for unconfined conditions:

$$Nu_{p,b} = 154\left[\left(\frac{c_{p,l} T_{sat}}{h_{lv}}\right)^{1.72}\left(\frac{c_{p,l} \cdot \mu_l}{k_l}\right)^{-0.34}\left(\frac{D_{b,dt} \cdot q''}{\mu_l h_{lv}}\right)^{0.62}\left(\frac{13}{\sigma}\right)^{-0.05}\right] \tag{4.16}$$

where $D_{b,dt}$ is the detachment diameter of vapor bubbles. This correlation was developed based on the Buckingham π theorem with the most relevant parameters for the pool boiling, using a regression analysis of the experimental database. The authors affirm that this correlation can predict 91.8% of experimental data from the literature, with an error of ±30%.

Thermal Resistance of the Evaporator Film

The thermal resistance of the film in the evaporator section, i.e. $R_{lv,e}$ (see Fig. 4.2), can be determined by expressions similar to Eqs. 4.2 and 4.3, once the evaporation coefficient of heat transfer of the liquid film over the evaporator internal surface is known. The prediction of such coefficient can be done by literature correlations that are based on the classical Nusselt model, for condensation over vertical walls (see Sect. 2.1), however with the heat flux in the inverse direction (heat is supplied to the wall). Mantelli (1999) amd Faghri (2016) compared several literature correlations with data from thermosyphons. A simple correlation for the Nusselt number in this region can be obtained from Eq. 2.15, in the form (Gross1992):

$$\text{Nu}_{lv,e} = 0.925 \cdot \text{Re}_{lv,e}^{-1/3} \tag{4.17}$$

where the Nusselt $Nu_{lv,e}$ and Reynolds $Re_{lv,e}$ numbers area defined as, respectively:

$$\text{Nu}_{lv,e} = \frac{h_e}{k_l}\left[\frac{g}{\nu_l^2}\frac{(\rho_l - \rho_v)}{\rho_l}\right]^{-1/3} \tag{4.18}$$

$$\text{Re}_{lv,e} = 4\frac{q_e}{\pi \cdot D_i \cdot \mu_l \cdot h_{lv}} \tag{4.19}$$

However, the heat transfer coefficients calculated by these equations presents higher values than those observed experimentally, as in actual thermosyphons, where the liquid film is broken and transformed into small rivulets, leaving the rest of the tube dry. Gross (1992) suggests the following correction factors for Eq. 4.17, which takes into consideration the diameter of the tube:

$$f_D = 1 - 0.67\frac{D_i}{20}, \quad for \quad 6 < D_i < 20\,\text{mm} \tag{4.20}$$

$$f_D = 0.33, \quad for \quad D_i > 20\,\text{mm} \tag{4.21}$$

This correction factor is not applicable if the thermosyphon tube has small diameter, or operate in inclined position or is subjected to small heat power input ($q_e'' < 200\,\text{W/m}^2$).

Besides, for some boiling conditions, the liquid splash effect, due to the vapor bubbles that burst abruptly on liquid–vapor interface in the evaporator, may increase the heat transfer inside the evaporator region. El-Genk and Saber (1988b) considered boiling in the descendent liquid film in the evaporator region, proposing several correlations that depends on the following non-dimensional parameter:

$$\eta = \frac{\text{Re}_v^2 Kp^2 \text{Re}_{lv,e,x}}{\text{Pr}_l} \tag{4.22}$$

where Re_v is determined according to Eq. 4.11 and the evaporator local film Reynolds number is defined as:

$$\mathrm{Re}_{lv,e,x} = 4\frac{q''_{\hat{e}}(l_e - x)}{\mu_l \cdot h_{lv}} \tag{4.23}$$

where x = 0 at the interface between evaporator and adiabatic section.

For laminar convection, $\eta < 10^9$, the following correlation can be used to predict the evaporator film Nusselt number (that varies with x):

$$\mathrm{Nu}_{lv,e,x} = \left(\frac{4}{3}\right)^{1/3} \cdot \left(\mathrm{Re}_{lv,e,x}\right)^{-1/3}, \quad \eta < 10^9 \tag{4.24}$$

For film nucleate boiling, $\eta > 2.7 \cdot 10^{10}$, El-Genk and Saber (1988b) present the following correlation:

$$\mathrm{Nu}_{lv,e,N} = 1.155 \cdot 10^{-3} \cdot \mathrm{Nu}_f^{0.3} Pr_l^{0.35} Kp^{0.7} \mathrm{Re}_v^{0.7}, \quad \eta > 2.7 \cdot 10^{10} \tag{4.25}$$

where:

$$\mathrm{Nu}_f = \frac{h_{lv,e} \cdot L_f}{k_l} \tag{4.26}$$

and where the liquid film thickness scale L_f is given by:

$$L_f = \left(\frac{\mu_l^2}{g\rho_l(\rho_l - \rho_v)}\right)^{1/3} \tag{4.27}$$

For the combined convection over the liquid film, $10^9 < \eta < 2.7 \cdot 10^{10}$, the following expression is suggested:

$$\mathrm{Nu}_{lv,e} = \left[\mathrm{Nu}_{lv,e,x}^3 + \mathrm{Nu}_{lv,e,N}^3\right]^{1/3}, \quad 10^9 < \eta < 2.7 \cdot 10^{10} \tag{4.28}$$

Knowing the Nusselt number, the heat transfer coefficients can be obtained and so the $R_{lv,e}$ thermal resistance. ESDU (1981) also used a more simple expression of the determination of this resistance, given by:

$$R_{lv,e} = \frac{0.235\, q_e^{1/3}}{D_i^{4/3} g^{1/3} l_e \left(\frac{h_{lv} k_l^3 \rho_l^2}{\mu_l}\right)^{1/3}} \tag{4.29}$$

Combined Evaporator Thermal Resistance

The evaporator overall thermal resistance, R_e, is actually a combination of the pool and the falling film resistances, which depends of the height of the liquid pool, l_p, through the expression:

$$R_e = l_p R_p + \left(1 - l_p\right) R_{lv,e} \tag{4.30}$$

In actual operation conditions, the liquid pool height is, at one hand, increased due to the presence of bubbles within the pool (especially if the evaporator experiences pool boiling), and, on the other hand, decreased due to the accumulation of liquid in the films that form along the internal thermosyphon walls. El-Genk and Saber (1997 and 1998b) developed a model for the non-dimensional height parameter ($H_p = l_p / l_e$), where l_p is the actual length of the pool, given by:

$$H_p = \frac{4m_{wf}}{\pi \cdot l_e \cdot D_i^2 \cdot \rho_{m,p}} \tag{4.31}$$

where the mean pool density is:

$$\rho_{m,p} = \alpha_m \rho_v + (1 - \alpha_m)\rho_l \tag{4.32}$$

and the mean void fraction can be calculated by:

$$\alpha_m = \frac{1}{l_e - l_f} \int_{l_f}^{l_e} \alpha(x) dx \tag{4.33}$$

where l_f is the length of the film in the evaporator section. The void fraction $\alpha(x)$ varies along the evaporator length. In Eq. 4.33, x = *0* is located at the interface between evaporator and adiabatic section and $x = l_e$ in the bottom of the thermosyphon, so that this integration is over the expanded poll height of the evaporator. El-Genk and Saber (1997 and 1998b) propose the following expression to estimate $\alpha(x)$:

$$\alpha(x) = \frac{V_v(x)}{C_0 \cdot V_v(x) + V^{+}\left(\sigma g(\rho_l - \rho_v) / \rho_l^2\right)^{1/4}} \tag{4.34}$$

where the vapor distribution coefficient, for cylindrical tubes, is expressed as:

$$C_0 = 1.2 - 0.2\sqrt{\rho_v / \rho_l} \tag{4.35}$$

and the velocity of the vapor mass flux is determined from an energy balance on the pool, resulting in the expression:

$$V_v(x) = \frac{4q_e''}{\rho_v \cdot D_i \cdot h_{lv}}(l_e - x) \tag{4.36}$$

The non-dimensional vapor drift velocity V^+ is determined based on the film Nusselt number (Nu_f, see Eqs. 4.25 and 4.26) and on the non-dimensional hydraulic diameter:

$$D_H^+ = \frac{D_i}{\sqrt{\sigma/(g(\rho_l - \rho_v))}}, \tag{4.37}$$

using the Kataoka and Ishii (1987) expression, valid for low viscosity fluids, $Nu_f < 2.5 \cdot 10^{-3}$:

$$V^+ = 0.0019 D_H^{0.809}\left(\frac{\rho_l}{\rho_v}\right)^{0.157} Nu_{f,p}^{-0.562}, \quad D_H < 30 \tag{4.38}$$

$$V^+ = 0.03\left(\frac{\rho_l}{\rho_v}\right)^{0.157} Nu_f^{-0.562}, \quad D_H > 30 \tag{4.39}$$

and for $Nu_f > 2.5 \cdot 10^{-3}$:

$$V^+ = 0.92\left(\frac{\rho_l}{\rho_v}\right)^{0.157}, \quad D_H > 30 \tag{4.40}$$

One should note that three hypothesis can be made for the height of the liquid pool. First, the pool is considered undisturbed and its height is obtained from the filling ratio knowledge. Second, the liquid pool height is equivalent to the evaporator length $(H_p = 1)$. Third, film is considered and the height of the pool is determined through iteration process using Eq. 30 to 40, considering: $l_f = l_e - l_p$. The physical properties of the working fluid are calculated at the average pool temperature.

For some applications, such as thermosyphons in which the working fluid is a liquid metal or those operating at low temperature levels, the hydraulic pressure in the pool may influence considerably the saturation temperature in the evaporator and must be taken into account. In this case, the following expression is proposed for the pressure prediction:

$$p_p = p_v + \rho_l\, g\, F\, l_e \sin\theta \tag{4.41}$$

where F is the filling ratio, i.e., the ratio of the working fluid volume to the evaporator volume ($F = V_{wf}/V_e$) and θ is the operation inclination angle. In this case, a temperature gradient can be observed within the evaporator. The temperature in the pool base Tp is considered as the saturation temperature at p_p. Assuming linear temperature increase with the liquid depth, the average temperature in the evaporator is:

$$\overline{T} = T_v(1 - F) + \frac{T_v + T_p}{2} F \quad (4.42)$$

Therefore, the average temperature difference due to the hydrostatic pressure is:

$$\Delta T_H = \overline{T} - T_v = \frac{T_p - T_v}{2} F \quad (4.43)$$

For most of the operation conditions, ΔT_H is very small and neglected. Besides, for liquid metal thermosyphons, this difference maybe be important for low temperature applications (low vapor pressures).

Liquid–vapor Phase Change and Vapor Thermal Resistances

The thermal resistances associated to the evaporation and condensation in the liquid–vapor interface in thermosyphons, $R_{pc,lv}$ and $R_{pc,vl}$, are usually neglected (the order of magnitude is around 10^{-5} °C/W) and is not considered in this text. However, the physical principles behind these phenomena are briefly discussed in Sect. 2.2.

The thermal resistance associated with the vapor displacement along the thermosyphon, R_v, is also very small (order of magnitude 10^{-8} °C/W), and is neglected. The vapor pressure distribution model developed in Sect. 3.4 can be used for the thermal resistance estimative, if necessary.

As these resistances are associated in series in the thermal circuit of Fig. 4.2, they can be removed from the network, causing a very small error in the overall thermal resistance prediction.

Thermal Resistance of the Condenser Film

As for the external thermal resistances (Eqs. 4.2 and 4.3), the condenser coefficients of heat transfer must be known for the determination of the $R_{vl,c}$ thermal resistance. For the majority of the thermosyphon applications, the descending liquid film flow is in turbulent or in the laminar to turbulent transition regime, where small rivulets are observed over the inner condenser surface. In these regimes, the heat transfer are larger than the laminar ones. However, Nusselt based models are used to predict the condensation process.

Mantelli (1999) and Faghri (2016) compared several literature correlations for the condensation coefficient of heat transfer in thermosyphons. Differences of up to two orders of magnitude in the predictions are observed.

Gross (1992a) presented an extensive work about condensation in internal wall of thermosyphons. For the thermosyphon operating in vertical position, the flow is considered one-dimensional and the Reynolds number of the film is defined as:

$$\mathrm{Re}_{f,c} = \frac{u_f \delta_f}{\nu_l} = \frac{q_c}{\pi D_i \mu_l h_{lv}} \tag{4.44}$$

where u_f and δ_f are, respectively, the velocity and the thickness of the condensed liquid in the downward direction. If $Re_{f,e} < 5$, the film flow is considered laminar with a smooth vapor-side interface. In this case, heat is transferred by pure conduction across the film. If $Re_{f,e} \approx 5$, rivulets are formed as the dumping effects of the viscosity and surface tension are not able to suppress wavy structures. The heat transfer across the film rate is increased due to the reduction of the effective thickness of the laminar film. An even higher heat transfer rate can be obtained for dropwise condensation, which happens when there is a strong cohesion between the working fluid and the tube wall material.

The film Reynolds number and the condenser modified Nusselt number, given by:

$$\mathrm{Nu}_c^* = \frac{h_c \delta_v}{k_l} = \frac{h_e \delta_l}{k_l} \cdot \left(\frac{g \delta_f^3}{\nu_l^2} \cdot \frac{\rho_l - \rho_v}{\rho_l} \right)^{-1/3} \tag{4.45}$$

can be, for laminar flows, correlated using a variation of the Nusselt model, for a constant temperature difference between the saturation and the wall temperatures and for quiescent vapor. This expression is:

$$\mathrm{Nu}_c^* = 0.925 \mathrm{Re}_{f,e}^{-1/3} \tag{4.46}$$

For thermosyphon working in inclined position and laminar flow, Gross (1992b) suggests the use of Eq. 4.46, with the Nusselt number, corrected as:

$$\mathrm{Re}_{f,e,\theta} = f_\theta \cdot \mathrm{Re}_{f,e} \tag{4.47}$$

where $f_\theta = 1$, for vertical tubes and $f_\theta = 2.87(D_i/l \sin\theta)$ for $\theta > 10°$. For $0 \le \theta \le 75°$, Gross (1992a), based on Semena e Kiselev work (1978), proposes the expression:

$$\frac{\mathrm{Nu}_{c,\theta}^*}{\mathrm{Nu}_c^*} = 1 + 0.0074 \left(\mathrm{sen}\theta \cdot {}^{l}/_{D_i} \right)^{0.7} \tag{4.48}$$

Gross (1992a) proposes the use of Uehara et al. correlation for laminar-wavy flow, i.e., for $2 < Re_{f,e,\theta} < 1333 \cdot Pr_l^{-0.96}$ (where $Pr_l = \mu \cdot c_p / k_l$):

$$\mathrm{Nu}_c^* = 0.884 \mathrm{Re}_{f,e,\theta}^{-1/4} \tag{4.49}$$

For the turbulent range, $Re_{f,e,\theta} > 1333 \cdot Pr_l^{-0.96}$, the following correlation is proposed:

$$\mathrm{Nu}^* = 0.044 Pr_l^{2/5} \mathrm{Re}_{f,e,\theta}^{1/6} \tag{4.50}$$

Gross (1992a) compared, according to the film flow conditions, the last three correlations (Eqs. 4.47, 4.48 and 4.49) with a large number of data obtained from the literature, proposing the following expression that compares with about 50% of the literature data within ±10%:

$$\mathrm{Nu}_c^* = \left[\left(f_p \cdot \mathrm{Nu}_{c,Eq.4.\ 43}^*\right)^2 + \left(Nu_{c,Eq.4.47}^*\right)^2\right]^{\frac{1}{2}} \tag{4.51}$$

where $f_p = 1/\left(1 - 0.63(p/p_{cr})^{3.3}\right)$ is a pressure correction factor, which approximates to unity when $p/p_{cr} < 0.3$, where p_{cr} is the critical pressure.

Mantelli et al. (1999) compared several literature correlations with data from a water-steel thermosyphon and concluded that the Kaminaga (1992) correlation presented the best comparison. It is given by:

$$\mathrm{Nu}_c \equiv \frac{h_c \cdot D_i}{k_l} = 25 \mathrm{Re}_c^{0.25} Pr_l^{0.4} \tag{4.52}$$

with the Reynolds defined as in Eq. 4.19. These authors recommend the use of this correlation when it is not possible to recognize the liquid film flow regime within the thermosyphon.

ESDU (1981) proposes the use of the following expression, similar to Eq. 4.29, for the condenser thermal resistance $R_{vl,c}$, also recommended by Groll and Rosler (1992), regardless of the film flow regime:

$$\mathrm{R}_c = 0.235 \frac{q_c^{1/3}}{D_i^{4/3} g^{1/3} l_c \left(h_{lv} k_l^3 \rho_l^2 / \mu_l\right)^{1/3}} \tag{4.53}$$

Overall Thermal Resistance

The overall thermal resistance is obtained from the combination of the individual thermal resistances. Removing the phase change, vapor flow, axial wall conduction resistances from the thermal circuit, considering that the radial conduction resistances of evaporator and condenser are the same and including external thermal resistances, the overall resistance is:

$$R_o = R_{ex,hs} + 2 \cdot R_{cond} + R_e + R_c + R_{ex,cs} \tag{4.54}$$

4.2 Heat Pipes

As for thermosyphons, the capacity of transferring heat of heat pipes can be modelled by means of thermal resistance circuits. As the physical phenomena that drive heat pipes are not the same as thermosyphons, some of the thermal resistances are predicted using different expressions. Figure 4.3 shows the thermal resistance network for steady sate conditions, valid for the heat transfer in straight heat pipes, where the internal wall is covered by a wick structure (porous media).

Heat is inserted and removed in the heat pipe by some kind of external heat transfer mechanisms. Actually, the external thermal resistances are very important, frequently determining the temperature levels of the device. The external thermal resistances depends on how the heat pipes are installed and may be represented by convection, radiation or thermal contact resistances. Often, these resistances are larger than the overall heat pipe thermal resistance.

In this text, the external heat transfer mechanisms are modelled by convection thermal resistances. The heat source (evaporator) and cold sink (condenser) resistances $R_{ex.hs}$ and $R_{ex,cs}$ can be calculated using Eqs. 4.2 and 4.3, where $h_{ex,e}$ and $h_{ex,c}$ are predicted using some of the many available correlations of the literature.

After reaching the evaporator surface, which is at temperature $T_{e,ex}$, the heat has two conduction paths: across the radial direction of the tube, $R_{cond,e}$, or along the tube axial direction, $R_{cond,ax}$. Similarly, heat reaches the condenser through two conduction routs: radially, crossing the $R_{cond,c}$ and axially through $R_{cond,ax}$. The axial conduction thermal resistances are usually orders of magnitude larger than the radial ones. As both are associated in parallel, in most models, the axial thermal resistances can be removed from the circuit. These conduction resistances are determined from Eqs. 4.4 and 4.5.

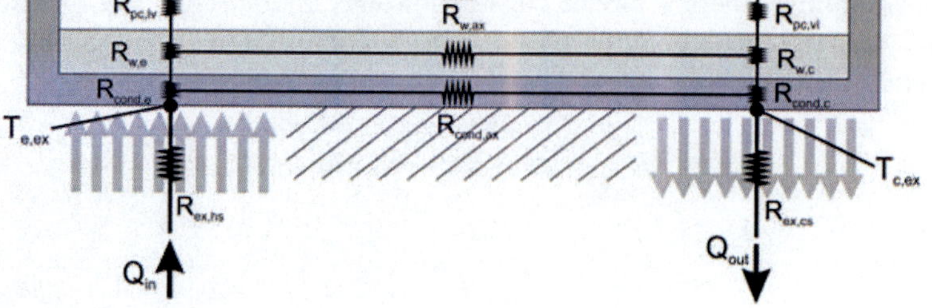

Fig. 4.3 Heat pipe thermal circuit

As for thermosyphons, usually the thermal resistances associated to the evaporation and condensation in the liquid–vapor interface, $R_{pc,lv}$ and $R_{pc,vl}$, are neglected. The thermal resistance associated with the vapor displacement along the thermosyphon, R_v, is also very small and usually negligible. Vapor pressure distribution models can be used for the thermal resistance estimative, if needed. As these three resistances are associated in series, they can be removed from the network, with small error on the overall thermal resistance.

Wick Structure Thermal Resistances

To obtain the overall resistance, thermal resistances concerning the wick structure of the heat pipe need to be determined. Usually, the contact resistance between the wick structure and the tube wall is neglected. Referring to Fig. 4.3, as for the tube wall conduction resistances, once the heat reaches the wick structure, it follows two conduction paths: radial ($R_{w,e}$ and $R_{w,c}$) and axial ($R_{w,ax}$). Expressions similar to Eqs. 4.4 and 4.5 can be used in the determination of the radial and axial conduction thermal resistances, substituting the thermal conductivity by an effective thermal conductivity that takes into consideration the combination of the wick material and the working fluid, which can be determined using Table 3.2 expressions.

4.3 Heat Pipe Selection of Design Parameters

The selection of the working fluid and of the casing material is one of the key thermosyphon and heat pipes design parameters. Basically, the working fluid is selected based on the operation temperature, while the casing (tubes) must be selected considering the chemical compatibility with the working fluid and with the environment where the device is to be installed. Aspects such as geometry, reliability and safety must also be taken into consideration. Practical working fluid and casing material selection criteria are discussed in the sequence.

Working Fluid

The working fluid must be able to fulfill the thermodynamics aspects discussed in Sect. 3.1. Therefore, the selection of the working fluid must be based mainly on the operation temperature. Several candidates are listed in Fig. 3.1.

As already observed, the working fluid can only operate between the critical and the triple point. Above the critical temperature, the vapor cannot be liquefied, no matter how much pressure is applied. Beyond the triple point, the fluid is in equilibrium between solid and vapor. Actually, the fluid operates in ranges much narrower than these.

The liquid mass is also an important parameter, as it determines the specific volume of the working fluid, which is turn, establishes the highest operation temperature. Figure 4.4 shows the schematic of a pressure against specific volume (p x v) diagram of a pure substance, highlighting the location of the critical point and of the triple line. This figure also shows a constant volume heating process in a thermosyphon or

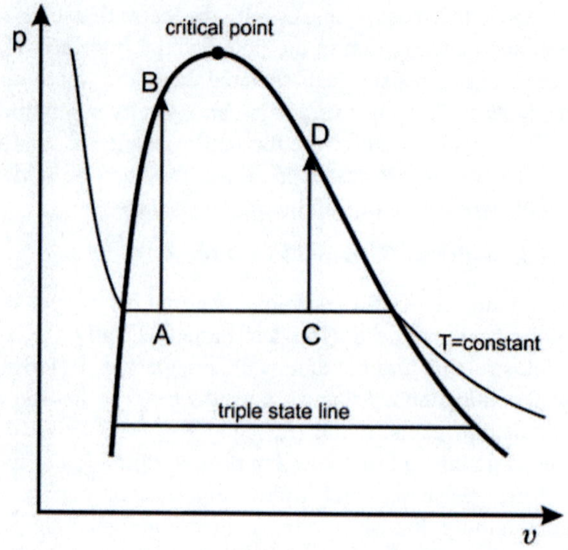

Fig. 4.4 Typical $p \times v$ diagram of a pure substance

heat pipe. If the working fluid specific volume is lower than that of the critical point (point A), adding heat increases the saturation temperature and pressure (represented by line AB) but decreases the vapor quality (ratio of the mass of vapor to the total mass). On the other hand, if the specific volume is larger than that of the critical point, (point C), adding heat increases the vapor quality (line CD). Therefore, ultimately, the working fluid mass determines the device maximum operation temperatures (points B or D) and the amount of vapor as well.

Besides, the following aspects are also important:

- Chemical compatibility between the working fluid and the case material (including porous media, if heat pipes are being designed),
- Chemical stability
- Toxicity
- Stability
- Thermal conductivities
- Latent heat
- Wettability between working fluid and casing
- Surface tension
- Viscosity, etc.

According to Mantelli (2013), for industrial applications in which the operating temperature range is 300 to 600 K, including: heat exchangers, furnaces, electronic cooling devices, etc., water is usually a very good working fluid, since it is stable, non-toxic and present high latent heat. Due to this last characteristic, only small volumes of water is enough to produce large vapor volume and transfer large amount of heat. The main concern about water is the vapor pressure, which can augment expressively with the temperature increase. In these cases, the casing material and geometry must be designed to stand the high vapor pressures. Ammonia, although considered a very interesting fluid for low temperature ranges, has its application limited because it is considered toxic. Besides, Freon and other similar refrigeration fluids are considered harmful to the environment and in many applications must also be avoided by law.

Mantelli (2013) also states that, based on its thermal properties, mercury is one of the best working fluids, for medium temperature levels. However, due to its toxicity, it is condemned for most applications. As discussed by Mantelli et al. (2010), many heat exchangers operate at the temperature ranges in which mercury would be recommended. To replace mercury, naphthalene is being used as working fluid.

Sodium is an interesting working fluid for temperature levels above 800 K, but it requires care in its manipulation as it is highly reactive with the water present in humid air. Highly inflammable gases (Mantelli et al. 2016) can be released in this reaction, with potential of explosion. Potassium is also reactive with water and may form explosive mixtures with air, at room temperature.

The thermosyphon number of merit (Nt) is a parameter which relates working fluid properties with its maximum heat power capacity and is used to classify working fluids for thermosyphons. It is expressed as (Reay and Key 2006):

$$N_t = \left(\frac{h_{lv} k_l^3 \rho_l^2}{\mu_l}\right)^{\frac{1}{4}} \tag{4.55}$$

The thermophysical properties which composes the merit number are dependent on the temperature level. The higher is N_t, the best is the working fluid for thermosyphons.

A similar merit number for working fluids can be used for heat pipes (N_{hp}). In this case, the surface tension is an important property, which is introduced to the expression, given by:

$$N_{hp} = \frac{\rho_l \sigma_l h_{lv}}{\mu_l} \tag{4.56}$$

Working fluid impurities, such as the presence of non-condensable gases, may decrease the thermal performance of thermosyphons. These gases may be already mixed within the working fluid when the liquid is charged or they can be the result of chemical reactions, after the working fluid and the casing metal are brought to contact during the liquid charging process and/or during thermosyphon operation. Non-condensable gases are pushed to the upper end of the condenser region of the thermosyphon during operation, blocking part of the condenser and reducing the thermal performance of the device. The impurities can also decrease the viscosity of the working fluid.

Casing Material

As observed before, the casing material must be designed to bear the internal vapor pressures. Besides, it must be compatible with ambient where the device is to be applied and with the working fluid. This means that the working fluid and casing material should not react chemically, so to avoid the generation of non-condensable gases, that would accumulate in the condenser upper region. The most obvious symptom of the presence of non-condensable gases is the cold tip of thermosyphons, observed in the rear condenser regions. No working fluid phase change happens in these regions occupied by the non-condensable gases, which can be considered adiabatic as these gases have insulation properties.

Table 4.1, based on informations presented at Reay and Kew (2006) and Faghri (2016), shows a list of common compatible casing materials and working fluids.

Most of the compatibility studies were conducted in the literature to supply data for space applications, where heat pipes are usually applied for the thermal management of electro/electronic equipment and where the requirements of durability and safety are severe. The combination of water and steel (carbon or stainless) for instance, is considered incompatible for space applications but is currently applied in many industrial applications, such as thermosyphon heat exchangers. Water-steel thermosyphons present low cost and excellent thermal performance and, although eventually non-condensable gases are formed, their production rate is small, reducing slowly the thermal performance of the device. In fact, it is quite probable that the equipment undergoes regular maintenance due to other industrial operational conditions, much sooner than the low performance of the thermosyphon is detected.

Mantelli et al. (2010) studied the influence of the presence of non-condensable gases in the performance of a thermosyphon by analyzing the temperature distribution in a naphthalene carbon steel thermosyphon, where a controlled amount of argon (non-condensable gas) was inserted. Figure 4.5 shows the plots of the temperature distribution along the axial length, where it can be observed that the volume occupied by the non-condensable gases increases with the increasing amount of non-condensable gases, however decreases with increasing the thermosyphon temperature. This happens because, at higher temperatures, the vapor pressure increases and the non-condensable gases are squeezed in the condenser upper region, occupying lower volume.

As mentioned before, the thermosyphon casing material must resist the mechanical forces caused by the pressurized vapor, considering the operation temperature. High vapor pressures demands thicker walls. The water vapor pressure for temperatures above 350 °C, for instance, can reach really high pressures, demanding appropriate material and casing geometries (cylindrical tubes are more resistant to higher pressures than parallelepipeds, for instance).

Besides, the tube material must be easily welded to avoid leakages, ensuring the integrity of the system during operational conditions. While stainless steel can be employed for thermosyphons operating at temperatures up to 800 ºC, other carbon steels are recommended for higher temperatures, up to 1000 °C. At temperatures over 850 °C, ceramics can be considered, as this material presents high mechanical

Table 4.1 Working fluid and casing material compatibility list

	Recommended	Not recommended
Ammonia	Aluminum Steel Nickel Stainless steel	Copper
Acetone	Copper Silica Aluminum Stainless steel[1]	
Methanol	Copper Stainless steel Carbon steel Silica	Aluminum
Mercury	Stainless steel	Nickel Inconel Titanium Niobium
Water	Copper Monel Silica[1,2] Nickel[1,2] Stainless steel[1,2] Carbon steel[1,2]	Stainless steel[1,2] Carbon steel[1,2] Aluminum Silica Inconel Nickel
Dowtherm A	Copper Silica Stainless steel[2]	
Naphtalene	Carbon steel Stainless steel	
Potassium	Stainless steel Inconel	Titanium
Sodium	Stainless Steel Inconel	Titanium
Silver	Tungsten Tantalum	Rhenium

[1]Considered compatible for some authors and incompatible for others
[2]Recommended with caution

strengths at high temperatures, high resistance to corrosion and erosion, and have low cost. The main drawback is that ceramics can be fragile and have low thermal conductivities.

In industrial applications, cost is a major concern and so commercial tubes must be preferably employed. As mentioned before, some moderate incompatibility between tube and working fluid can be tolerated, especially when the degradation of the thermosyphons thermal performance is much slower than actual routine maintenance periods.

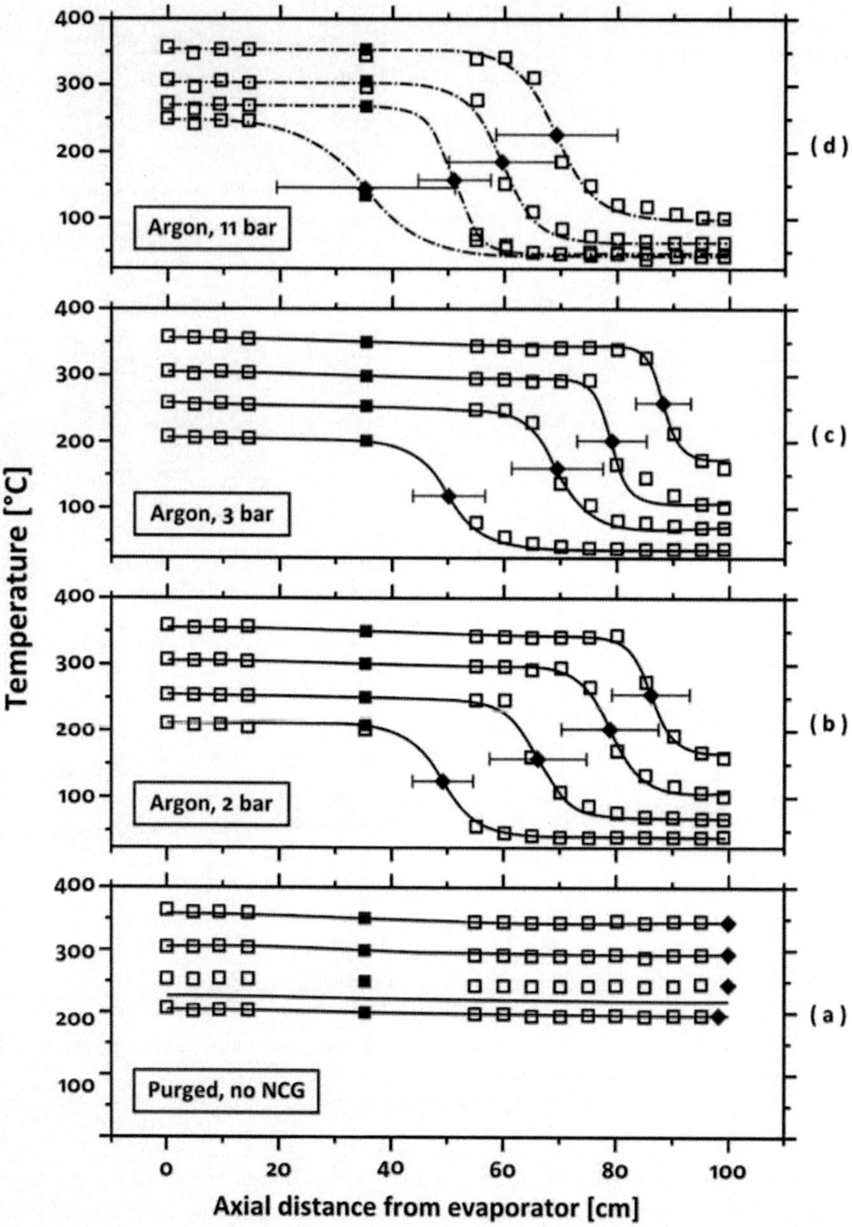

Fig. 4.5 Temperature distribution for naphthalene thermosyphons with the presence of controlled amount of non-condensable gases

Wick Structures for Heat Pipes

Figure 3.5 shows the common wick structure for heat pipes, including wrapped screens, sintered metal powder and groves. Advanced structures are also shown in this figure. As commented, these wick structures have the purpose of guaranteeing the working fluid distribution along the tube and the return of condensed fluid from the condenser to the evaporator. To provide the necessary capillary pumping, the meniscus radius in the liquid–vapor interface must be small, attainable by reducing the wick porous dimensions. However, small porous cause high liquid pressure drops along the tube length. Non-homogeneous wicks are developed in an attempt to fulfill both requirements: larger porous are provided close to the tube wall and smaller porous close to the liquid–vapor interface, to provide the capillary pumping.

Figure 4.6 shows schematically the most common wick structures: wrapped screens, sintered metal powder, sintered metal fibers and groves. More about modeling of the thermal and hydraulic behavior of these wicks can be found in the Chaps. 2 and 3.6.

Wrapped screen wick structures are composed of several layers of screens. The screens can be made of metal, such as copper, stainless steel, bronze, monel, etc. The screens are made of thin wires which are braid in different meshes that can vary from 50 to 400. Sometimes, thicker screens are used to hold fine screens against the tube wall, using their "spring" effect. Spot welding are also used for fixing screens.

Grooved wick structure heat pipes are made by internal extrusion of metal (cooper, aluminum, etc.) tubes. Usually, the grooves are axial, but radial grooves can also be used, to guarantee that the tube is wet radially.

Sintered porous media are obtained by the partial fusion of loose metal powder (cooper or nickel, for instance) that is heated to around 60 to 80% of the metal melting temperature. Ceramic porous media can be applied to heat pipes with very chemical reactive working fluids. One application are the high temperature heat pipes where litium is used as the working fluid. Also chemical protection can be obtained by the application of a layer by vapor deposition of a refractory metal over the tube internal wall. This layer must present chemical compatibility with alkaline metal at high temperature.

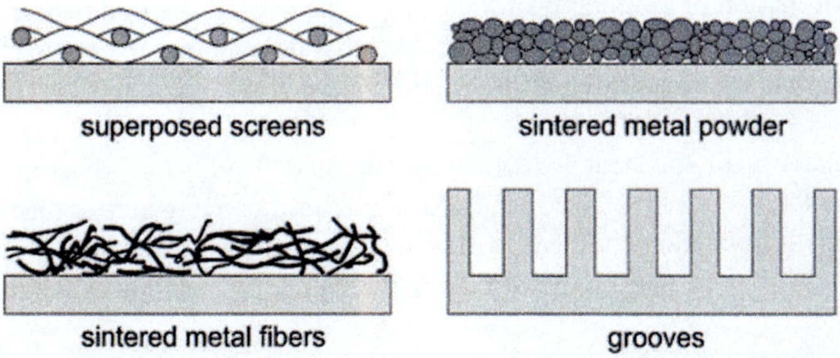

Fig. 4.6 Major wick structures

Metallic foams as well as ceramic or metal fibers are in crescent use, especially for applications where the case is not cylindrical. Foam can be made of nickel, stainless steel and copper, among other materials, while the fibers are available in stainless steel and amorphous silica woven cloth (Refrasil).

4.4 Design Methodology

The following design methodology is proposed for thermosyphons and heat pipes and is based on Mantelli (2013).

In designing thermosyphons or heat pipes, the first major step consists in the determination of all thermal resistances of the thermal circuits presented in Fig. 4.2 and 4.3. After, all the operational limits (according to Chapter 3) must also be determined. Based on this information, the total heat power that the device is able to transport is calculated, which is compared with the operational limits. If the heat to be transported is within the operational limits, optimization procedures, aiming the reduction of cost, improvement of heat transport capacity, of the geometry, etc., can be applied. Testing may be necessary for exquisite thermosyphon designs.

The following steps are proposed to be followed, considering that the thermosyphon or the heat pipe is exchanging heat with a source and sink by convection.

1. Design parameter specification: lengths, tube diameters, inclination angle, evaporator and condenser external areas, external evaporator and condenser convection coefficients (obtained from literature correlations and models), source and sink temperatures, filling ratio and thermal conductivity of the casing metal.
2. Determination of $R_{ext,hs}$, $R_{cond,e}$, $R_{cond,c}$, $R_{ext,cs}$.
3. Using the resistance network, estimative of the vapor temperature.
4. Collection of the following working fluid properties, considering T_v as the saturated temperature: p_v, ρ_l, ρ_v, h_{lv}, μ_l, μ_v, σ, k_l e c_{pl}.
5. Determination of the pressure in the pool basis (Eq. 4.41), to obtain the saturated temperature for p_p. Determination of ΔT_H using Eq. 4.43.
6. Determination of the total temperature difference.
7. Determination of the total heat load (using the overall thermal resistance).
8. Determination of the thermal resistances that depend on the heat load.
9. Determination of the overall thermal resistance and recalculation of the heat load.
10. Comparison of the heat load obtained in step 9 with that of step 7. If the difference is not acceptable, return to step 8 until convergence is reached.
11. Verification whether the heat load is within the operational limits. If not, the thermosyphon must be redesigned and the whole process started again.

4.5 Closure

In this chapter, procedures for designing thermsosyphon and heat pipes are presented. Models for steady-state, one-dimensional heat transfer are discussed. The analogy between electrical and thermal circuits are used for modelling the devices. Literature correlations, appropriate to estimate the coefficients of heat transfer needed for the determination of the thermal resistances of the circuits are presented. Compatibility between casing material and working fluid is discussed. For heat pipes, the compatibility between fluid and wick structure must also be considered. Some of the most common porous media structures are discussed. The methodology proposed in this chapter is applied to design actual thermosyphons and heat pipes in Chapter 6.

References

ESDU, Heat Pipes - Performance of Two-Phase Closed Thermosyphons. Engineering Sciences Data Unit 81038, London (1981)

El-Genk, M.S., Saber, H.H.: Heat Transfer Correlations for Small, Uniformly Heat Liquid Pools. Int. Journal Heat Mass Transfer **41**(2), 261–274 (1998)

El-Genk, M. S. and Saber, H. H., Operation Envelope for Closed, Two-Phase Thermosyphons, 10th International Heat Pipe Conference, Stuttgart, Germany (1997)

El-Genk, M. S. and Saber, H. H., Thermal Conductance of Evaporator Section of Closed Two-Phase Thermosyphons (CTPTs), 7th AIAA/ASME Joint Thermophysics and Heat Transfer Conference, **3**, 99–106, (1998b)

Faghri, A.: Heat Pipe Science and Technology, 2nd edn, Global Digital Press (2016)

Groll, M., Rosler, S.: Operation Principles and Performance of Heat Pipes and Closed Two-Phase Thermosyphons. Journal of Non-Equilibrium Thermodynamic **17**, 91–151 (1992)

Gross, U.: Falling Film Evaporation Inside a Closed Thermosyphon, 8th International Heat Pipe Conference. Beijing, China (1992a)

Gross, U.: Reflux Condensation Heat Transfer Inside a Closed Thermosyphon. Int. J. Heat Mass Transfer **35**(2), 279–294 (1992b)

Imura, H., Kusuda, H., Ogata, J.I., Myiazaki, T., Sakamoto, N.: Heat Transfer in Two-Phase Thermosyphons. Transactions of Japan Society of Mechanical Engineers, Series B **45**(393), 712–722 (1979)

Kaminaga, F., Okamoto, Y., Suzuki, T.: Study of Boiling on Heat Transfer Correlation in a Closed Two-Phase Thermosyphon, 8th International Heat Pipe Conference. Beijing, China (1992)

Kataoka, I., Ishii, M.: Drift Flux Model for Large Diameter Pipe and New Correlation for Pool Void Fraction. Int. J. Heat Mass Transfer **30**(9), 1927–1939 (1987)

Kutateladze, S. S., Heat Transfer During Condensation and Boiling, State Scientific And Technical Publishing House Of Literature On Machinery. 2nd ed., Russia, (1952)

Mantelli, M.B.H., Carvalho, R.D.M., Colle, S., Moraes, D.U.C.: Study of Closed Two-Phase Thermosyphons for Bakery Oven Applications, 33th National Heat Transfer Conference. Albuquerque, New Mexico (1999)

Mantelli, M.B.H., Uhlmann, T.W., Manzoni, M., Marengo, M., Eskilson, P.: Experimental Tests on Sodium Thermosyphons, Joint 18th International Heat Pipe Conference and 12th International Heat Pipe Symposium. Jeju, South Korea (2016)

Mantelli, M.B.H., Ângelo, W.B., Borges, T.: Performance of Naphthalene Thermosyphons with Non-Condensable Gases: Theoretical Study and Comparison with Data. Int. Journal of Heat and Mass Transfer **53**, 3414–3428 (2010)

Mantelli, M. B. H.: Thermosyphon Technology for Industrial Applications. In: Vasiliev, L.L. and Kakaç, S., Heat Pipes and Solid Sorption Transformations - Fundamentals and Practical Application, 411–464, CRC Press, Florida (2013)

Reay, D.A. and Kew, P.A.: Heat Pipes Theory, Design and Applications, 5th edn, Butterworth-Heinemann (2006)

Chapter 5
Fabrication and Testing

In this chapter, fabrication process for thermosyphons and heat pipes are described. Besides, experimental facilities for testing these devices are proposed and discussed.

5.1 Fabrication

Roughly speaking, casings for thermosyphons or heat pipes are basically composed of a metal tube closed in its extremities. When the tube diameter is not too large, the closing can be by smashing the tube. However, most commonly, two closing lids are used. These lids are made of circular flat plates, in the shape of coins, welded in the tube tip borders. In one of the circular lids, a small tube, known as umbilical, used for internal vacuum and charging procedures, is welded. The external diameter of these umbilical tubes is typically ¼ in.

Mantelli (2013) proposes the use of four different types of closing lids, illustrated in Fig. 5.1, designed to be welded to the tube. In Fig. 5.1a, the closing lid has exactly the same diameter of the tube. The welding is done over the external surfaces and a small protuberance results from the process. The lids of Fig. 5.1a and b have mostly the same dimensions, their major difference is in a small groove which is machined in the welding area, to accommodate the material excess. This geometry provides a better external finishing of the device but is more subjected to vacuum leakages. In Fig. 5.1c, the external diameter of the lid is a little smaller than the internal tube diameter, so that the lid fits inside the tube. Figure 5.1d shows a similar to Fig. 5.1c lid geometry, but with a groove, provided to accommodate the welding protuberance. For these two last configurations, the welding is performed inside the tube, along the lid perimeter. This process can be hard to perform, especially for small diameter tubes, as it is difficult to keep the lid in the correct position during the welding, however the results in terms of leakage can be rewarding.

The TIG welding, suitable for small metal pieces such as the thermosyphon components, is recommended. This procedure is relatively clean and provides the

M. B. H. Mantelli et al., *Thermosyphons and Heat Pipes: Theory and Applications*,
https://doi.org/10.1007/978-3-030-62773-7_5

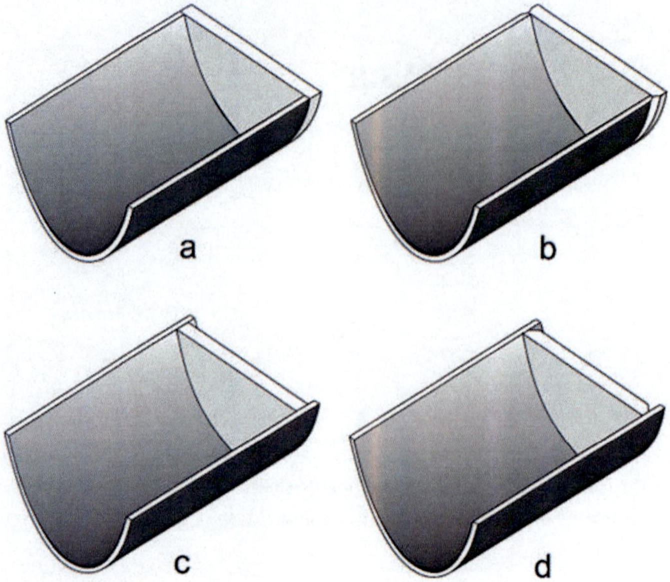

Fig. 5.1 Geometry of the thermosyphon closing lids

necessary leak-tightness and mechanical strength. The other arc-welding processes, which may retain gases in bubbles, are considered dirty, damaging the quality of the vacuum.

Diffusion bonding

Flat plate PHPs (discussed in Chap. 5) are especially difficult to produce. In these devices, the channels, within which the working fluid flows, are internal and usual machining process are impractical. Internal channels in flat plate PHPs can be formed by the union of flat plates, where grooves are machined in, at least, one of the plates. One of the main advantages of this heat pipe configuration is that internal surfaces may be accessed and tailored before being joined. In operation, these devices must be able to stand for possible high working fluid vapor pressures and, according to their application, still be thin. Therefore, the technique selected to fabricate flat plate PHPs must produce devices that fulfill all these requirements.

In the literature, many flat plate PHPs were constructed for visualization of the working fluid behavior and so the closing plate is made of glass (Kim and Kim 2018). In this case, glues can be used for lateral seal of the channels. Brazing of the plates is also a used technique (Ayel et al. 2015), although some brazing adding material can be displaced to the channel, modifying the internal geometry. The use additive manufacture technique, such as laser power bed fusion, has been investigated (Ibrahim et al. 2017). High temperatures involving the fabrication process may modify the crystal structure of metals, affecting the material properties, with possible effects on the channel geometry. These effects are increased when pressure is employed in the fabrication process.

Diffusion bonding is one of the joining techniques that have been lately explored for the fabrication of flat plates PHPs (Jang et al. 2019; Soo et al. 2019; Betancur et al. 2018; Facin et al. 2018). Unlike conventional welding processes, the diffusion bonding is considered a solid-state union process, which, when adequately performed, preserves the base metal microstructure and properties at the joint interface. Also, no addition (filler) materials are necessary. As drawbacks, more preparation (better control of roughness and cleanness) are required to the surfaces to be joined, the thermal cycles are time consuming and the cost of the equipment (furnace) used for the diffusion bonding is quite high (Rusnaldy 2001). Diffusion bonding technique was used for the fabrication of wire-plate heat pipes, as described by Paiva and Mantelli (2015a, b). Hu et al. (2013) also used copper powder sintering and diffusion bonding process to fabricate heat pipes.

The diffusion bonding method developed by Mortean et al. (2016) to manufacture cut-plate compact heat exchangers, can also be employed to fabricate devices with complex inner geometries with relatively ease. Their method consists of cutting flat plates, with a water jet machine, to produce machined plates, which are alternately stacked with non-machined flat plates. This stacking process creates rectangular cross section channels (Moreira Jr. and Mantelli 2019).

The diffusion bonding process allows for the coalescence of surfaces in intimate contact, by the submission to a uniaxial high pressure and high temperature, in a vacuum atmosphere, for a determined period of time. According to Schwartz (1969) and Kazakov (1985), in these conditions, the interdiffusion of atoms across the surfaces happens after the plastic deformation of surface asperities (Lee 2012). Kielhorn et al. (2001) and Lehrheuer (1993) describe the diffusion bonding process as consisting basically of three major stages, illustrated in Fig. 5.2. First, with the application of pressure and heat, the long wavelength undulations of contacting surfaces are leveled by plastic deformation. In the next stage, the joined areas grow quickly by creek, resulting from edge dislocations or thermally-activated vacancy diffusion. In the onset of the third stage, the union line is composed of large isolated vacancies that are closed by similar to sintering process means. This last stage is very affected by the vacancy geometry and involves additional mechanisms of diffusion.

Therefore, the appropriate combination of the parameters: temperature, pressing load and holding time, is quite important for the success of the union and varies according to the materials to be bonded, surface finishing and the geometry of the plates. The number, the thickness of the layers and the initial microstructure of the materials also influence the quality of the bonding interface and are directly associated with possible corrosion or other undesirable effects.

For devices with internal channels, such as in flat PHPs, non-optimized diffusion bonding parameters, i.e. high pressure, high temperature levels or even too long periods of time, may cause excessive and irreversible mechanical deformation of the geometry of the channel cross section. Plastic deformation (when the yield stress limit is overpassed) or creep (cold flow due to diffusion in vacancies) may happen or both. Especially the creep process depends on time. That deformation may decrease

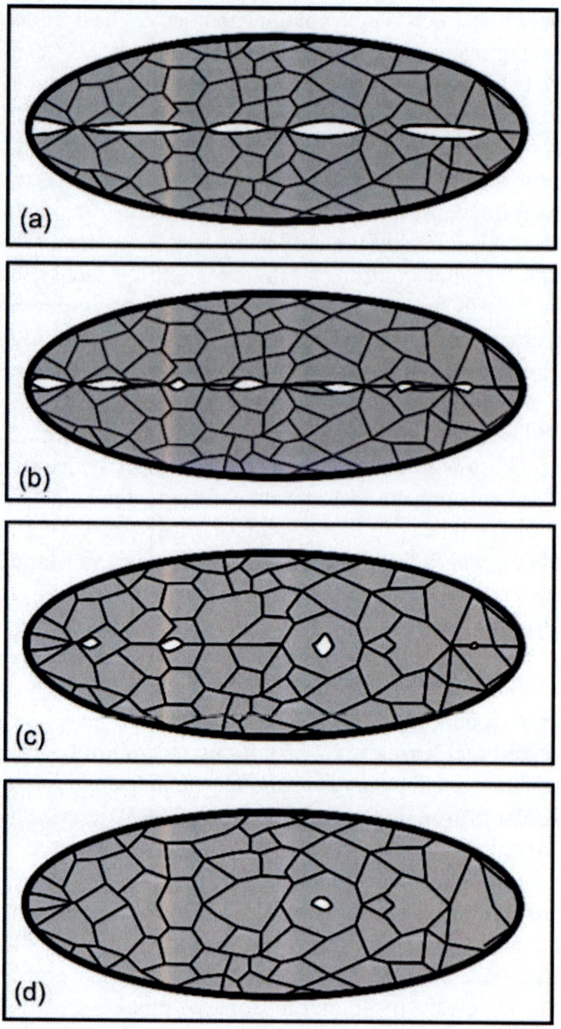

Fig. 5.2 Schematic of the diffusion bonding

the flatness of the external surfaces, which, in the limit, may collapse, blocking some channels. According to Mylavarapu et al. (2012), an appropriate diffusion bonding procedure keeps the channel in acceptable deformation levels. As illustrated in Fig. 5.3 (see Betancur et al. 2020), the circular cross section channels may reach oval shape and the thin walls of the rectangular cross section channels may acquire a "barrel shape" after subjected to high pressures during the fabrication process.

Gietzelt et al. (2016) proposed the following set of parameters for the diffusion bonding of pure oxygen-free copper (OF-Cu): temperature of 850 °C, bearing pressure of 2 MPa and four hours of elapsed time. These authors claim that, especially in thin-walled micro-structures, perfect grain growth across the bonding plate was observed, probably because of the concentration of bearing pressure, while, in

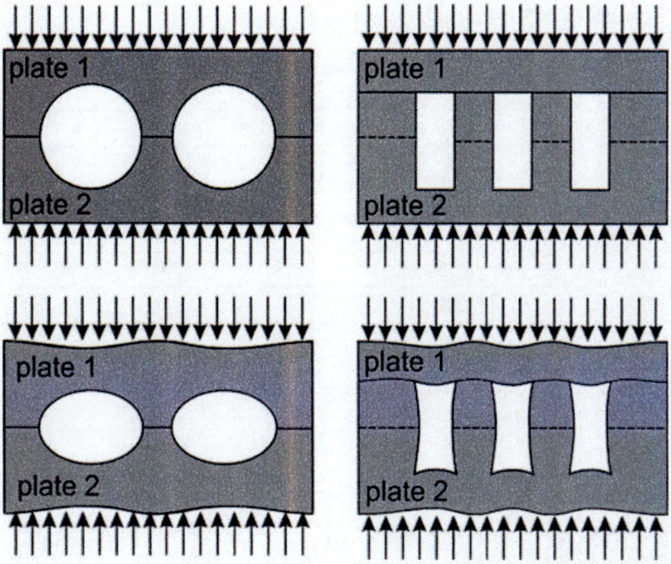

Fig. 5.3 Not deformed (upper schematics) and deformed geometries (lower schematics) due to fabrication process

the larger areas, the grain growth was not so good, as the pressure applied could be insufficient to deform asperities and cause the pores to be filled by the mass diffusion.

Mortean (2014) evaluated experimentally the diffusion bonding process for the fabrication of copper compact heat exchangers, using the following parameters: temperatures of 750, 850 and 950 °C, pressures of 2.5 MPa, 4 MPa and 21 MPa and an ellapsed time of 60 min. This researcher observed that samples of the joined surface interfaces diffusion bonded at a temperature of 750 °C, when subjected to tension tests, ruptured at the interface for pressures ranging from 4 to 21 MPa. However, for temperatures of 850 and 950 °C, these samples ruptured at the base material. Therefore, in terms of mechanical strengths of the bonded surfaces, the temperature level is a much more important parameter than the pressure, as indicated by literature (Kazakov 1985). The plot of Fig. 5.3 shows the thermal diffusion bonding cycles that Betancur et al. (2020) successfully used for constructing copper pulsating heat pipes.

Nuernberg et al. (2017) and Mortean et al. (2019) performed another experimental study of the bonding parameters for stainless steel 316 L, where samples, produced by the combination of parameters shown in Table 5.1, were produced and mechanically tested according to the Standard ASTM E8. Stacked plate samples were diffusion bonded under a high vacuum of at least 5.10^{-5} mbar, initially heated up to 600 °C at a rate of 5 °C/min and kept at this level for 60 min, for temperature homogenization. Then, samples were heated up to the bonding temperature level, as indicated in Table 5.1, at a rate of 2,5 °C/min. The pressure was applied and maintained during the wole bonding time (according to Fig. 5.4).

Table. 5.1 Stainless steel 304 L bonding parameters

Sample	Temperature (°C)	Pressure (MPa)	Time (min)	Mechanical strength (MPa)
S0	*As received*			
S1	945	8.75	105	466
S2	945	7.5	105	521
S3	945	10	105	481
S4	945	8.75	32	402
S5	945	8.75	179	430
S6	1040	9.51	60	565
S7	1040	9.51	150	551
S8	1040	7.98	150	555
S9	1100	8.75	105	524

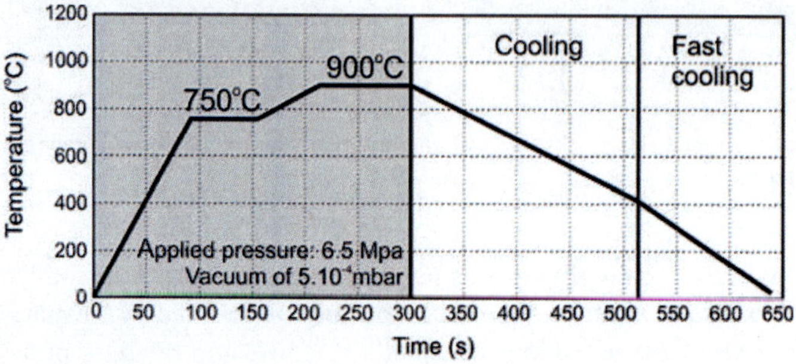

Fig. 5.4 Diffusion bonding cycle

The temperature of 1040 °C, used in S6 (9,51 MPa, 60 min), S7 (9,51 MPa, 150 min) and S8 (7,98 MPa, 150 min) set of parameters, was considered adequate for the diffusion bonding of the austenitic stainless steel 316 L.

5.2 Cleaning Process

The cleaning of the components must be performed before any welding procedure. Two cleaning procedures are usually adopted (Mantelli 2013): primary, performed before the major welding processes; and secondary, done after most the thermosyphon is fabricated and before the working fluid is charged.

The primary cleaning has the objective of removing the protection waxes (used by the tube manufacturer to protect the tubes against rust during the storage and transportation) and other gross impurities found in commercial tubes. Abrasive blasting, using alumina microspheres jets can be employed in this cleaning process.

The secondary cleaning (Mantelli 2013) is performed after the machining and lid welding. In this process, the tube is completely filled with acetone or trichloroethylene, and left in an ultrasonic bath, for about 15 min, to remove the internal tube wall impurities. After this first cleaning, the solvent is removed from the device, which is filled with isopropyl alcohol. The tube is emptied again and left to dry. The drying process can be accelerated by heating the tube by means of a blowtorch.

The process may be simplified for stainless steel tubes, because the commercial tubes are usually cleaner than carbon steels, for instance. In this case, the thermosyphon is first machined and welded before the secondary cleaning process is performed, followed by the vacuum and charging procedures.

Copper tubes are cleaned by immersing the parts into a 10% sulfuric acid. After, the cleaned material is rinsed in water for around 10 min and left to dry. Reay and Kew (2006) propose the cleaning of nickel tubes by immersing them into an acid solution with 25% of nitric acid.

The fabrication of liquid metal thermosyphons and heat pipes involves meticulous cleaning processes, as the presence of small amounts of non-condensable gases can affect very much their thermal behaviors, especially for low temperature levels.

5.3 Out-Gassing of Tube Materials and Working Fluids

Out-gassing of the tube and working fluid is important for some applications, to remove non-condensable gasses absorbed in the metal or in the fluid that could be released during the operation of the thermosyphon or heat pipe. This is especially important for devices operating at high temperatures. Outgassing is performed by subjecting the tube to vacuum, after the cleaning and before the charging processes.

It is desirable to buy the working fluid as clean as possible from the market. If the fluid is not clean enough, it must be purified before being inserted in the thermosyphon or heat pipe. Distillation is a process highly employed for fluids that operate at low temperature levels such as acetone, methanol and ammonia, as the presence of water may cause chemical incompatibilities between the casing metal and the working fluid.

Usually, the out-gassing of fluids working in temperatures up to 200 °C is realized by freezing the fluid, followed by the application of vacuum to remove the not frozen gasses previously absorbed in the liquid. The liquid is than thawed and the process is repeated many times, depending on the working fluid cleaning standards required.

Of course, the more sophisticated is the working fluid cleaning process, more expensive is the device fabrication costs. However, for water-steel thermosyphons for industrial applications such as heat exchangers, the working fluid treatment can be quite simple: first the water is distilled and then out-gassed by means of a vacuum pump.

5.4 Charging Procedures

After welding and cleaning and before charging with the working fluid, the thermosyphon casing must be subjected to vacuum leakage tests. Small leakages can compromise the operational behavior of thermosyphons, as non-condensable gases can intrude the device. The leak tests must be performed by means of very sensible leak detector equipment, such as a helium mass spectrometer. If leakages are detected, repairs must be done or even the casing must be rejected.

The thermosyphon casing must also be tested for the mechanical strength, by filling the tube with pressurized water, at least 150% above the operating pressures. After the pressurized water test, the tube is evacuated and dry. The working fluid is then charged through the umbilical tube.

Usual Charging Processes

Two processes can be applied for charging liquid state working fluids at the room temperature: electrical heating and vacuum pumping, described in the sequence.

Charging by Electrical Heating

The following equipment (Mantelli 2013) are necessary for applying the charging of thermosyphons or heat pipes by electrical heating: electrical power source, electrical resistance, glass beaker, hose, valve and connections. Figure 5.5 presents a schematic of this setup.

After cleaned, the tube is completely filled with the already treated working fluid. A hose connects the tube and glass beaker, which is also filled with a controlled amount of working fluid. A valve connects the umbilical tube to one of the hose ends, with the other extreme introduced inside the beaker. The electrical resistance, installed in the evaporator, heats slowly and continuously the tube, allowing enough time for the small air bubbles trapped in the internal walls to escape to the beaker and then to the environment. The heating process proceeds until working fluid boiling takes place. The formed vapor is expelled to the beaker, which is cold, and the vapor condenses again. When the vapor pressure is not enough to expel the working fluid to the beaker, the valve is closed until pressure grows again. The overpressure must be avoided, to control possible damages of the fragile parts of the filling rig.

After some time, the working fluid is almost completely removed from the tube, when the valve is closed and the electrical resistance is turned off. The tube cools down and the internal pressure decreases. The valve is then opened again, and the working fluid is sucked back to the tube from the beaker. Care must be taken so that air is not introduced in the tube again.

The process of heating up and cooling down is repeated several times (five or six) until air bubbles are not observed during the slow heating process. At this stage, the heater is turned on for the last time to remove the excess of working fluid, so that the appropriate amount is left in the tube. Then, the valve is closed and the hose removed.

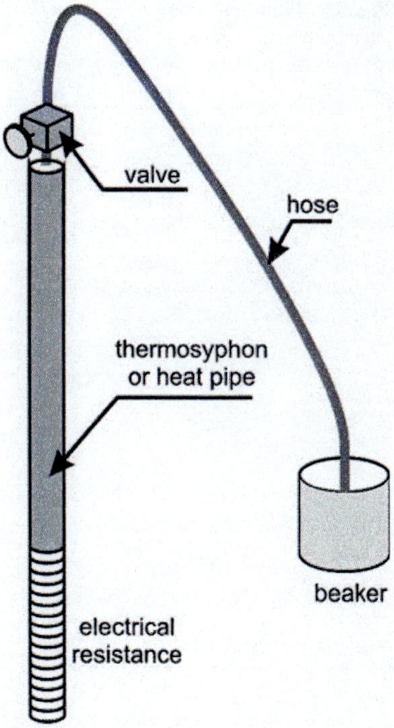

Fig. 5.5 Working fluid filling rig by electrical heating

The amount of working fluid is controlled through the comparison between the total liquid volume (volume of the liquid within the thermosyphon and the beaker in the beginning of the charging process) and the volume left in the beaker after charging.

This process has the advantage of requiring a simple apparatus, with low operational cost. On the other hand, it presents the following disadvantages: time consuming, difficulties to guarantee the working fluid cleanliness and to control the fluid charging volume, as some of the working fluid vapor does not condense in the beaker and is lost to the environment, or is trapped in the hose and valve.

Charging by Vacuum

The following equipment (Mantelli 2013) are necessary for charging thermosyphons or heat pipes by vacuum: vacuum pump, vacuum gauge, beaker, valves, hoses and connections. A schematic of the setup is shown in Fig. 5.6.

After the tube is cleaned, its umbilical is connected to a vacuum pump by opening valves V1 and V3. The vacuum system is turned on until a vacuum of the order of 10^{-3} Torr is reached. Then valve V1 and V3 are closed. Then valve V2, which is connected to the bottom of the beaker, is opened, filling the tubing connecting the beaker and the thermosyphon with the working fluid, by the action of gravity. It is important that no air bubbles are within the connecting tubes and that the beaker is filled with more working fluid than necessary for charging the device. The valve

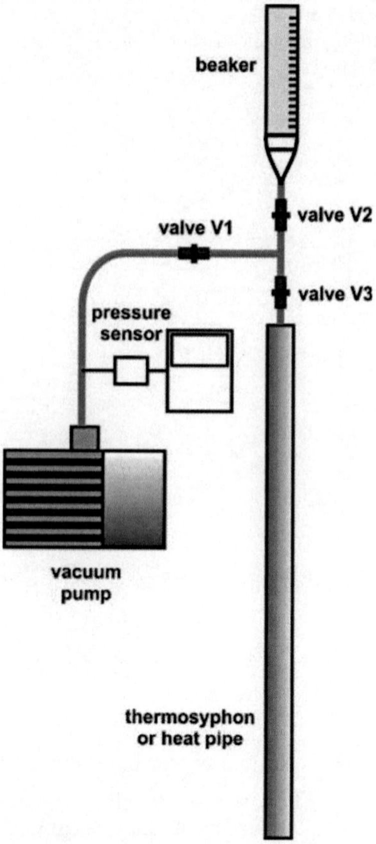

Fig. 5.6 Filling rig using vacuum pump system

V3 is then open and, in the sequence, valve V2 is also opened, but slowly, so that the empty casing tube, in vacuum, sucks the working fluid until its total volume is charged, as measured through the scale of the beaker. The valve V3 is closed and the casing, together with valve V3 is removed from the setup for umbilical sealing.

The advantage of this procedure is its simplicity, quick operation and the precise control of volume of working fluid to be inserted inside the device. The main disadvantage is the need of a complete filling rig, with the associated cost. However, for industrial serial production, the cost of the set-up is diluted along the time.

Between these methods, the vacuum procedure is recommended due to better volume control and fast charging of the working fluid. The heating procedure can be employed when a vacuum system is not available, as happens in actual industrial applications, when the device has to be charged "in situ".

Umbilical Sealing

After the filling processes, the umbilical tube has to be sealed. The sealing is performed in two steps. First, the umbilical is pinched to provide a mechanical

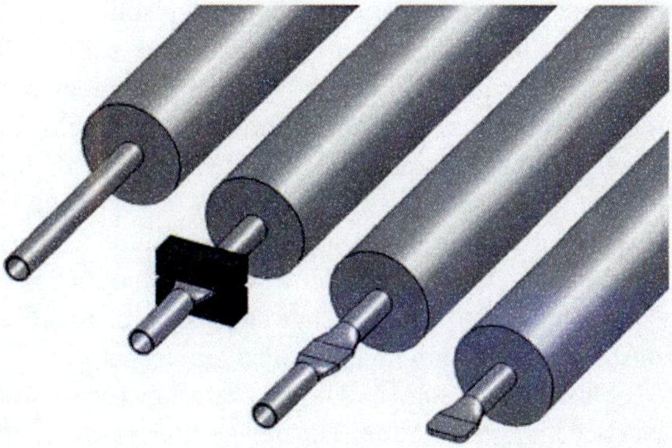

Fig. 5.7 Sealing process including pinching, cutting and welding

sealing, by means of a hydraulic press (Mantelli 2013), which smashes the umbilical between two metal bars. Care must be taken so that the metal bars do not cut the filling tube: for this, the contact area between the pressing bars and tube must be the largest possible. After the pinching, the tube is cut and welded by the usual process (TIG). Figure 5.7 shows the metal bars and the sealing process.

5.5 Sodium Charging—High Temperature Devices

As stated by Cisterna et al. (2020a), extreme care must be taken for the manipulation of high temperature thermosyphons and heat pipes, especially those that operate with sodium as working fluid. Sodium has a large potential for explosion when in contact with the moisture of the atmospheric air.

Sodium hydroxide is formed over the external surfaces of pure sodium as a result of the reaction with the moisture of the air. The clean sodium is silver shiny, while the sodium hydroxide is darker, with an opaque appearance. The impurity layer has a melting and a boiling points around 1.6 and 3.3 times higher than the properties of pure sodium. Therefore, the presence of impurity may require large wall superheat temperatures to initiate the boiling process (start-up). The impurities may generate non-condensable gases (NCG) which accumulates at the end of the condenser, decreasing the useful heat transfer area and so the heat transfer performance of the device.

Sodium hydroxide surface layer may be formed during the charging procedure. According to Reid et al. (2002) and Reay et al. (2016), the best thermosyphon performance is obtained with distilled sodium. However, the charging apparatus is complex and costly, demanding strict security requirements. Distillation consumes significant amount of time, water and energy. Besides, after the charging procedure, the sodium remnants must be carefully removed from the charger apparatus, a quite difficult task (Qu and Duan 2011).

The charging can also be performed in solid state. In this case, a controlled atmosphere chamber is necessary to handle the sodium. This is a low cost procedure, although the sodium cleaning is not very efficient and impurities may be left in the working fluid, resulting in the formation of a cold tip, more evidenced at low heat power inputs (Uhlmann et al. 2016; Mantelli et al. 2017 and Wu et al. 2017).

Finally, the charging procedure may be accomplished with sodium in liquid state. This thermosyphon shows a good performance, as observed by the resulting homogeneous temperature profiles. However, the charging apparatus may be more expensive when compared to the solid state charging process. Besides, the sodium manipulation must be carried out with controlled atmosphere.

Following, the solid and liquid charging processes are described, due to their good cost/benefit good relation. The device casing is fabricated and cleaned, following the same basic fabrication and cleaning process described in this chapter.

Solid State Sodium Charging

The present procedure for solid charging of sodium in thermosyphons or heat pipes is described in details in Uhlmann et al. (2016). This charging process does not require a specific filling rig.

The commercialized sodium is usually stored in flasks filled with kerosene, to avoid its contact with the air. A little larger amount than the volume of sodium to be inserted in the thermosyphon or heat pipe is placed in a Kitasato flask, which is vacuum pumped. This Kitasako flask and the tools necessary for the charging, including: metal tube (usually Inconel), knife, silicon hoses, pressure scissors, scale and a tube diameter adapter, all previously cleaned, where taken into a glove box, which is filled with an inert gas (argon). Inside the glove box, the flask is opened and the external opaque sodium layer was removed with the knife, with the objective of removing the surface impurities.

In the next step, the sodium is cut and small pieces and rolled as a spaghetti. These pieces are weighted and inserted in the metallic tube. The charged tube was then closed with silicon gloves, connected to the tube by a tube adapter, which is smashed by pressure scissor, before being removed from the glove box. Tube heating by electrical resistance blocks, to temperatures up to 98 °C, is necessary during the charging procedure to avoid the sodium to be stacked inside the tube, hindering the sodium loading.

Outside the glove box, the tube is connected to the vacuum pump with care, to avoid air to enter into the system. The tube is vacuumed again, up to 10^{-6} mbar, to remove all gases which could eventually enter in the device. The silicon hose is closed with the pressure scissor and the tube is smashed in a loading press in a two-step process, as described in the last section: first a load of 16 ton is applied with a plane mold and, second, a load for 10 ton is applied with a wedge format mold. In order to close the device and cut remaining parts of the tube, a TIG welding with a current of 100 A is used.

Liquid state sodium charging

Figure 5.8, adapted from Cisterna et al. (2020a, b), shows a schematic of the rig

Fig. 5.8 Schematic of the sodium charging rig

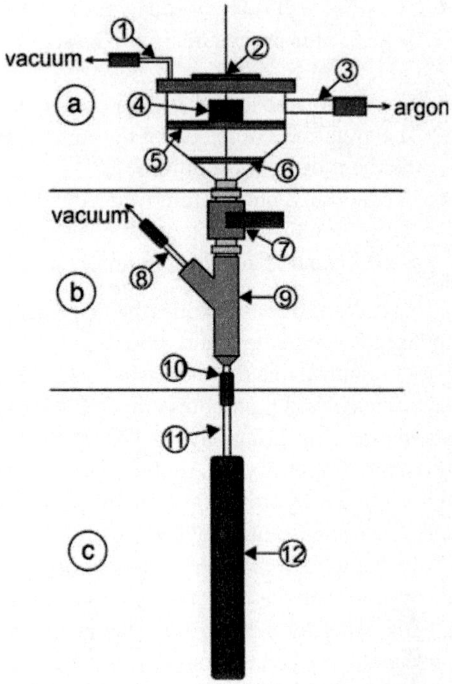

used for the liquid charging of a thermosyphon or heat pipe, which is divided into three main parts: sodium reservoir, Y tube and device (thermosyphon or heat pipe). Initially, the sodium in solid state (4) is placed inside the reservoir over the copper grid (5), by removing the glass visor (2). The glass visor is sealed against leakages, maintaining vacuum during the charging procedure, by means of a flange and o-rings. Next, the system is evacuated to 10^{-1} mbar and then pressurized with argon, to up to 1.1 bar. This procedure is repeated twice in order to guarantee air concentration levels below 10 p.p.m. Then, the following steps must be applied:

- The ball valve (7) and silicone hoses (Kocher clamps) of the umbilical (1), (3) are (8) closed.
- The sodium reservoir is heated with electric heaters, at a temperature at 250 °C, until the sodium melts. The glass visor is used to verify that the sodium is completely molten.
- The Y-tube and the thermosyphon are located inside an oven that is heated to up to 200 °C.
- The reservoir is pressurize with argon, which is inserted through the umbilical, to reach the pressure of up to 1.1 bar (3).
- The Y-tube and the thermosyphon are evacuated to 10^{-1} mbar through the umbilical connection (8).

- The silicone hose is closed by means of a Kocher clamps (8) and the ball valve (7) is open. This procedure drives the sodium from the reservoir to the thermosyphon (12).
- The silicone hose (10) is then closed with Kocher clamps.
- The system is cooled down to room temperature to ensure that the sodium inside the thermosyphon solidifies.
- The thermosyphon is removed from the oven. The silicone hose (10) with Kocher clamps are disconnected and connect to a high vacuum diffuser pump. Vacuum of 10^{-6} mbar is achieved to finally perform the thermosyphon closing process.

Actually, the copper grid is the first barrier that separates liquid sodium and sodium hydroxide, preventing this impurity to enter into the device. While being heated, pure sodium expands, increasing its volume until the sodium hydroxide external layer breaks and liquid flows, by gravity, to the lower regions of the reservoir. Some small pieces of sodium hydroxide are eventually dragged by the liquid sodium to the bottom of the reservoir. Another barrier, formed by a stainless steel net (6) prevents these smaller hydroxide pieces from entering the thermosyphon. Figure 5.9 shows pictures of the sodium hydroxide removed from the working fluid.

To improve even further the quality of the device, a purging should be done after the charging. In this case, two umbilical tubes are required: one umbilical for charging and the other for purging. The charging is performed with a procedure similar to the one just described, except that another step is taken: before removing the device from the charger, the whole system must be pressurized with argon, to up to 1.1 bar. The objective is to ensure that, during the closing process of charging, should a leak occurs, argon would leak out rather than air into the thermosyphon. Umbilical evacuation is done through the silicone hose (8), until the absolute pressure inside the thermosyphon reaches approximately 10^{-6} mbar. Then, the evaporator is heated by an induction furnace, until the temperature at the tip of the condenser reaches 880 °C. That ensures that the sodium vapor pressure is larger than atmospheric (1 bar, sodium saturation pressure of 880 °C). Thus, any remaining NCG inside the thermosyphon is pushed by the sodium vapor to the NCG reservoir (6), which is cooled by convection and radiation to the ambient. At this point, valve (10) is closed and the thermosyphon

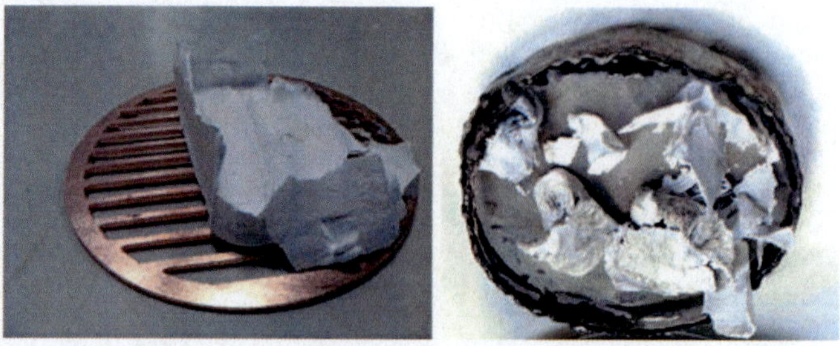

Fig. 5.9 Sodium hydroxide deposited on the copper grid after the charging process (left) and net with smaller impurities (right)

charging procedure is finished. As discussed by Cisterna et al. (2020b), for a uniform heat removal from the condenser, the temperature distribution of the thermosyphon is also uniform, showing that this procedure results in devices with the same quality of those obtained from complex distillation processes.

5.6 Thermal Testing

As thermosyphons and heat pipes are driven by complex physical principles, usually their behavior is difficult to predict analytically. For this reason, in some cases, experimental evaluations may be the only way to assure their viability to a specific application. Usually, experimental tests aim to measure the thermosyphon thermal resistances and/or temperature distributions. Tests are specially recommended when: devices work close to their heat transfer limits, new geometries are proposed, unusual working fluids and/or casing materials are selected or new applications are envisioned. Lastly, but not less important, experimental tests are very useful for validating mathematical models.

A well-succeeded experiment needs careful design. The experimental set up must be able to reproduce operational conditions and so information such as: heat power to be transferred, temperature level, temperature distribution, etc., must be considered for planning the experiment.

The operation conditions of small devices, working in moderate temperature ranges, are not difficult to reproduce in laboratory. On the other hand, testing large equipment, such as heat exchangers, demands the construction of large experimental rigs. These setups may be expensive to construct.

However, it is interesting to note that the external heat source and sink coefficients of heat transfer are usually the dominant resistances of the thermosyphon or heat pipe thermal circuits and, for this reason, their determination must be the crucial parameter to be measured. This means that, not necessarily, the geometry of the device needs to be perfectly reproduced, but rather the operational parameters.

A thermal performance measurement apparatus must include a controlled heat power source, responsible for delivering heat to the evaporator, and a controlled heat sink, to remove heat from the condenser. The power source is usually responsible for the heat input, which is determined from electrical voltage and current readings. On the other hand, the cooler usually consists of a small heat exchanger where cold liquid, controlled by a thermal bath, removes the heat power transferred by the device, while it keeps the condenser temperature at desirable levels. The heat removal is usually measured by the product of the difference between the input and output cooling liquid temperatures and the cooling liquid mass flow rate. In steady state conditions, the heat delivered by the electrical resistance must be equal to the thermal energy removed by the cooling liquid. Heat losses must be carefully computed in the energy balance. When water is the cooling liquid, high mass flow rates must be avoided as, due to the characteristics of the water, the input and output temperature difference could be very small, increasing the measurement uncertainty. The quality of the measurement data would improve considerably with the calibration of the thermocouples used for temperature measurement.

Another important experimental verification is the temperature distribution along the device. Temperature measurements are usually made by means of thermocouples installed over the casing external surface. In some tests, internal temperatures need to be accessed. These internal measurements are difficult to perform as the presence of thermocouples in the liquid and/or vapor flows may change their hydrodynamic behavior affecting the measurements. Besides, some thermosyphon and/or heat pipes operate under high internal pressures and holes in the casing wall, to accommodate the sensors, can cause leakages.

Data acquisition systems are used to read the temperatures, voltages and currents. The data are stored in computers. Several available softwares, to control the experimental parameters such as: voltage and/or electrical current (heat power input), cooling water temperature readings, time, steady state conditions, etc., are available and should be used.

Figure 5.10 presents a schematic of an experimental setup designed for testing the thermal performance of thermosyphons, including all major components. The same system could be employed to heat pipes, if the setup components are disposed in

Fig. 5.10 Schematic of an experimental set up for straight thermosyphon testing

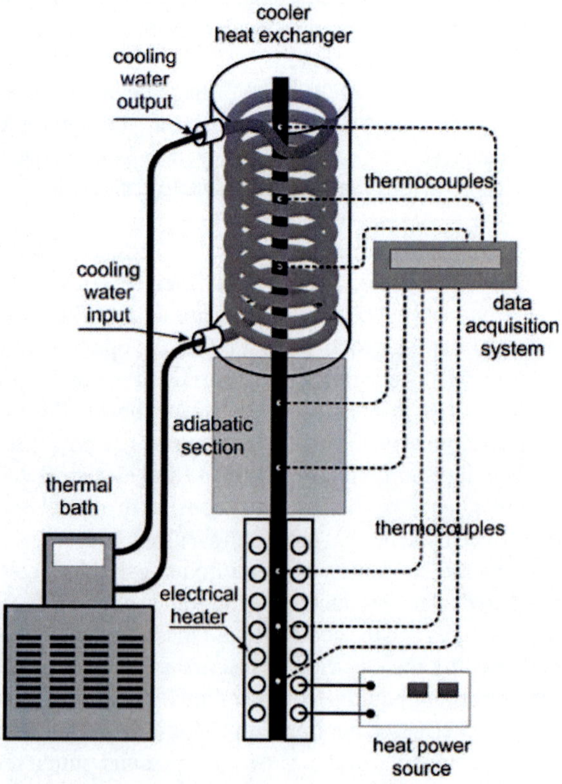

the horizontal position. Actually, this same apparatus could also be used for testing devices operating in any inclination.

Steady state and transient tests can be conducted, depending on the objectives of the experimental work. The steady state criteria (usually an acceptable temperature variation within a reasonable time) must be clearly established, for the testing parameters selected (commonly the power input and thermal bath temperature). The experimental data results can be presented and discussed in several forms, being usual the use of graphics relating the device thermal resistance and/or the temperature with time, length or power input. Transient temperatures may also be important to report, as a function of time.

5.7 Closure

In this chapter, two major subjects were treated: thermosyphon and heat pipe fabrication techniques and testing procedures.

First, some fabrication aspects of thermosyphons and heat pipes were discussed and the major steps to produce high quality, free of non-condensable gases devices are presented. Two charging techniques are described: by electrical resistance heating or by vacuum. Special attention is given for the charging procedures involving liquid metal (sodium, for instance) thermosyphons or heat pipes.

Second, the major aspects concerning the design of testing apparatus, highlighting the most important sections (heater and cooler) are presented and discussed. Furthermore, suggestions of how to present and discuss the obtained data is given.

References

Ayel, V., Areneo, L., Scalambra, A., Mameli, M., Romestant, C., Piteau, A., Marengo, M., Filippeschi, S., Bertin, Y.: Experimental Study of a Closed Loop Flat Plate Pulsating Heat Pipe Under a Varying Gravity Force. Int. J. Therm. Sci. **96**, 23–34 (2015)

Betancur, L, Hulse, P., Melian, I., Mantelli. M.B.H.: Diffusion Bonded Pulsating Heat Pipes: Fabrication Study and New Channel Proposal, Journal of the Brazilian Society of Mechanical Sciences and Engineering, Springer (2020)

Betancur, L.A., Facin, A., Gonçalves, P., Paiva, K., Mantelli, M.B.H., Nuernberg, G.: Study of Flat Plate Closed Loop Pulsating Heat Pipes With Alternating Porous Media, Joint 19th International Heat Pipe Conference and 13th IHPS International Heat Pipe Symposium, Pisa, Italy (2018)

Cisterna, L.H.R., Vitto, G., Cardoso, M.C.K., Fronza, E.L., Mantelli, M.B.H., Milanez, F.H.: Charging Procedures: Effects on High Temperature Sodium Thermosyphon Performance. Journal of the Brazilian Society of Mechanical Sciences and Engineering **42**, 416 (2020a)

Cisterna, L. H. R., Fronza, E. L., Cardoso, M. C. K., Milanese, F. H., Mantelli, M.B.H.: Biot Number for Cold Tip Prediction in Sodium Two-Phase Thermosyphons. Int. J. Heat Mass Transf. **165**, 120699, 1–15 (2020b)

Facin, A., Betancur, L., Mantelli, M., Mera, J.P.F.,Coutinho, B. H.: Influence of Channel Geometry on Diffusion Bonded Flat Plat Pulsating Heat Pipes, 19th International Heat Pipe Conference and 13th International Heat Pipe Symposium, Pisa, Italy (2018)

Gietzelt, T., Toth, V. and Huell, A.: Diffusion Bonding: Influence of Process Parameters and Material Microstructure, IntechOpen (2016)

Hu, R., Guo,T., Zhu, X., Liu, S., Luo, X.: A Small Flat-Plate Vapor Chamber Fabricated by Copper Powder Sintering and Diffusion Bonding for Cooling Electronic Packages, in: Proc. - Electron. Components Technol. Conf. (2013)

Ibrahim, O.T., Monroe, J.G., Thompson, S.M., Shamsaei, N., Bilheux, H., Elwany, A., Bian, L.: An Investigation of a Multi-Layered Oscillating Heat Pipe Additively Manufactured From Ti-6Al-4V Powder. Int. J. Heat Mass Transf. **108**, 1036–1047 (2017)

Jang, D.S., Kim, D., Hong, S.H., Kim, Y.: Comparative Thermal Performance Evaluation Between Ultrathin Flat Plate Pulsating Heat Pipe and Graphite Sheet for Mobile Electronic Devices at Various Operating Conditions. Appl. Therm. Eng. **149**, 1427–1434 (2019)

Kazakov, N.F.: Diffusion Bonding of Materials, Pergamon Press (1985)

Kielhorn, W.H., Adonyi, Y., Holdren, R. L., Horrocks, R.C., Nissley, N. E.: Survey of Joinning, Cutting and Allied Processes. In: Jenney, C. L. and O'Brien, A. (eds) Welding Handbook Volume 1: Welding Science and Technology, 9th Edn, American Welding Society (2001)

Kim, W., Kim, S.J.: Effect of Reentrant Cavities on the Thermal Performance of a Pulsating Heat Pipe. Appl. Therm. Eng. **133**, 61–69 (2018)

Lee, H. S., Diffusion bonding of metal alloys in aerospace and other applications, In: Chaturvedi, M. C. Welding and Joining of Aerospace Materials, Woodhead Publishing (2012)

Lehrheuer, W.: High-temperature Solid-State Welding, In: Olson, D.L., Siewert;, T.A., Liu S., Edwards, S. L. (eds), ASM International, **6** (1993)

Mantelli, M.B.H.: Thermosyphon Technology for Industrial Applications. In: Vasiliev, L.L. and Kakaç, S., (eds.)Heat Pipes and Solid Sorption Transformations—Fundamentals and Practical Application, 411–464, CRC Press, Florida (2013)

Mantelli, M. B. H., Uhlmann, T., Cisterna, L. H. R., Experimental Study of a Sodium Two-Phase Thermosyphon, 9th World Conference on Experimental Heat Transfer, Fluid Mechanics and Thermodynamics, Iguazu Falls, Brazil (2017)

Moreira, A.A., Jr., Mantelli, M.B.H.: Thermal Performance of a Novel Flat Thermosyphon For Avionics Thermal Management. Energy Convers. Manage. **202**, 112219 (2019)

Mortean, M.V.V., Cisterna, L.H.R., Paiva, K.V., Mantelli, M.B.H.: Development of Diffusion Welded Compact Heat Exchanger Technology. Appl. Therm. Eng. **93**, 995–1005 (2016)

Mortean, M. V. V.: Desenvolvimento de tecnologias de recheios para trocadores de calor compactos soldados por difusão, M.Sc. Thesis. Florianópolis: Federal University of Santa Catarina (2014)

Mortean, M. V. V., Mateus, L. B., Rosinski, G., Mantelli, M. B. H.: Estudo dos Parâmetros de União por Difusão do Aço Inoxidável AISI 316L, Revista Soldagem e Inspeção, **24,** (2019)

Mylavarapu, S.K., Sun, X., Christensen, R.N., Unocic, R.R., Glosup, R.E., Patterson, M.W.: Fabrication and Design Aspects of High-Temperature Compact Diffusion Bonded Heat Exchangers. Nucl. Eng. Des. **249**, 49–56 (2012)

Nuernbrg, G., Rosinski, G., Gonçalves, P., Mortean, M., Gonçalves e Silva, R. H., Monteiro A., Mantelli, M.B.H., Mateus, L.: 316L Stainless Steel Diffusion Bonding Optimized Parameters, 24th ABCM International Congress of Mechanical Engineering, Curitiba (2017)

Paiva, K.V., Mantelli, M.B.H.: Theoretical Thermal Study of Wire-Plate Mini Heat Pipes. Int. J. Heat Mass Transf. **83**, 146–163 (2015a)

Paiva, K.V., Mantelli, M.B.H.: Wire-Plate and Sintered Hybrid Heat Pipes: Model and Experiments. Int. J. Therm. Sci. **93**, 36–51 (2015b)

Qu, W., Duan, Y. N.: Fabrication and Performance of High and Super High Temperature Heat Pipes, 10 th International Heat Pipes Symposītum, Taipei, Taiwan (2011)

Reay, D.A., Kew, P.A.: Heat Pipes Theory, Design and Applications, 5th edn, Butterworth-Heinemann (2006)

Reid, R.S., Sena, J.F., Martinez, A.L.: Heat-Pipe Development for Advanced Energy Transport Concepts. Los Alamos National Lab, NM (US) (2002)

Rusnaldy, R.: Diffusion Bonding: An Advanced of Material Process. Rotasi, Indonesia **3**, 23–27 (2001)

Schwartz, M.M.: Modern Metal Joining Techniques. John Wiley & Sons, New York (1969)
Soo, D., Kim, D., Ho, S., Kim, Y.: Comparative Thermal Performance Evaluation Between Ultrathin Flat Plate Pulsating Heat Pipe and Graphite Sheet for Mobile Electronic Devices at Various Operating Conditions. Appl. Therm. Eng. **149**, 1427–1434 (2019)

Chapter 6
Application of Models to Selected Cases

In this chapter, selected design cases of thermosyphons and heat pipes are presented, with the objective of illustrating the use of the models discussed in Chaps. 2–4 in the analysis and design of these devices. Aspects such as the prediction of the working fluid pressure distributions along the devices, the estimation of their heat transfer operational limits and the determination of the overall thermal resistances are discussed. When available, theoretical results are confronted with experimental data. It is important to note that it is not claimed that these selected cases are optimized in any aspect. Actually, the concept of optimization may have different connotations, according to the application of the device.

Modelling

The following hypothesis are considered for this chapter models:

- Steady state conditions.
- The mass flow rate in the condenser and evaporator varies linearly, as the heat is uniformly supplied and rejected in these sections.
- In the adiabatic section, the working fluid mass flow rate is constant.
- The interaction between the liquid and gas phases at the interfaces is neglected.

6.1 Ordinary Geometry Thermosyphon

It is supposed that the proposed thermosyphon, which characteristics are shown in Table 6.1, is designed for an industrial application. Heat is applied to the evaporator and is removed from the condenser by forced convection of cooling water at controlled temperature. For this geometry and temperature, the heat load to be transported and the operational limits must be determined. The vapor pressure distribution is required and so the overall thermal resistance.

M. B. H. Mantelli et al., *Thermosyphons and Heat Pipes: Theory and Applications*,
https://doi.org/10.1007/978-3-030-62773-7_6

Table. 6.1 Thermo-syphon charac-teristics

Thermosyphon parameters	
Inner Diameter [mm]	20.2
Outer Diameter [mm]	22.0
Evaporator Length [mm]	200
Adiabatic Section Length [mm]	90
Condenser Length [mm]	600
Working fluid	Deionized water
Filling ratio [% of evaporator volume]	66
Operation position	Vertical
Wall material	Copper
Wall temperature (adiabatic section)	$T_w = 40\,^{\circ}\mathrm{C}$
Evaporator	Constant heat flux, $T_{w,e} = 50\,^{\circ}\mathrm{C}$
Condenser	Forced convection of water $T_{ex.c} = 20\,^{\circ}\mathrm{C}$, $V_{ex.c} = 0.0132\,\mathrm{m/s}$

The design methodology proposed in Sect. 4.4 is used for the determination of the heat transport capacity of the thermosyphon. The resistance network shown in Fig. 4.2 is used for the present case.

Step 1. The design parameters are specified (see Table 6.1) and the thermophysical properties of the tube material (copper) and working fluid (water) at their operating temperature ($T_o = 70\,^{\circ}\mathrm{C}$) are obtained from literature tables.

Step 2. The external and conduction resistances are determined. Considering that the heat flux at the evaporator external surface is known, the external heat source resistance is considered as zero. The cold source thermal resistance is determined considering forced convection of water in cross current over the external surface of a cylinder, using the well-known expression of Churchill and Bernstein (Incropera and de Witt 2008):

$$\overline{Nu_D} = 0.3 + \frac{0.62 \cdot Re_D^{1/2} \cdot Pr^{1/3}}{\left[1 + (0.4/Pr)^{2/3}\right]^{1/4}} \left[1 + \left(\frac{Re_D}{282000}\right)^{5/8}\right]^{4/5} \tag{6.1}$$

where the Reynolds number is:

$$Re_D = \frac{\rho_{ex,c} \cdot u_{ex,c} \cdot D_{ex}}{\mu_{ex,c}} \tag{6.2}$$

and:

$$\overline{h}_{ex,c} = \frac{\overline{Nu}_D.k_{ex,c}}{D_{ex}} \tag{6.3}$$

For the present case, $\overline{Nu}_D = 20.45$ and $\overline{h}_{ex,c} = 572.72\,\mathrm{W/m^2K}$. Therefore, the external thermal resistance, calculated from Eq. 4.3 is $R_{ex,cs} = 0.0421\,\mathrm{K/W}$.

The radial conduction thermal resistances in evaporator and condenser, according to Eq. 4.4, are $R_{cond,e} = 1.718 \cdot 10^{-4}\,\mathrm{K/W}$ and $R_{cond,c} = 5.726 \cdot 10^{-5}\,\mathrm{K/W}$, respectively. The effective length is determined by Eq. 3.1. The axial thermal resistance is $R_{cond,ax} = l_{ef}/k\pi\left[(r_{ex})^2 - (r_i)^2\right] = 20.77\,\mathrm{K/W}$. The conduction resistances are associated in parallel. As the axial resistance is of several orders of magnitude larger than the radial resistance, the axial resistance can be removed from the circuit with negligible error.

Step 3. The vapor temperature is estimated by the resistance network of Fig. 4.2. Primarily, the heat load is estimated, according to the expression:

$$q = \frac{T_{ex,e} - T_{ex,c}}{R_{ex,hs} + R_{cond,e} + R_{cond,c} + R_{ex,cs}} \tag{6.4}$$

resulting, in the present case, in $\Delta T = 29.34$

The evaporator temperature is, in this first guess, considered the same of the adiabatic section temperature. Therefore, the first approximation of heat load to be transferred is $q = 708.7W$.

The vapor temperature can be determined from the thermal circuit of Fig. 4.2, taking only the resistance path of the condenser branch and the estimated power input. As a first approximation, the thermal resistance of the condenser film can be neglected ($R_{vl,c} = 0$). The following expression can be used:

$$T_v = T_{ex,c} + q(R_{vl,c} + R_{cond,c} + R_{ex,cs}) \tag{6.5}$$

resulting in $T_v = 49.88^oC$.

Step 4. The working fluid thermophysical properties are obtained from literature tables, considering that the vapor is at saturated state, at temperature T_v,

Step 5. Using Eq. 4.41, the pressure in the bottom of the pool and the saturated temperature for this pressure are predicted, resulting, for the present case, in: P_p = 13,557.4 Pa and $T_p = 51.89^oC$. Using Eq. 4.43, the average temperature difference, due to the hydrostatic pressure, can be determined resulting in: $\Delta T_H = 0.66^oC$.

Step 6. The total temperature difference can be determined by the expression:

$$\Delta T = T_{ex,e} - T_{ex,c} - \Delta T_H \tag{6.6}$$

Step 7. The total heat load of the thermosyphon can be estimated again by Eq. 6.4, resulting in $q = 683.0W$.

Step 8. The thermal resistances, which input data is the heat load, are determined in this step. Using the correlation given by Eq. 4.15, the following expression can be used to determine the thermal resistance of the evaporator pool:

$$R_p = \left\{ 0.32 \left[\frac{\rho_l^{0.65} k_l^{0.3} c_{pl}^{0.7} g^{0.2}}{\rho_v^{0.25} h_{lv}^{0.4} \mu_l^{0.1}} \right] (p_v / p_{atm})^{0.3} (\pi D_i l_e)^{0.6} q_e^{\prime\prime 0.4} \right\}^{-1} \tag{6.7}$$

where $q"_e = q/\pi D_i l_e$. Substituting the correct thermophysical parameters, the thermal resistance of the evaporator pool is $R_p = 0.01268\ K/W$. The thermal resistance of the evaporator film is determined in the sequence, using Eq. 4.29, resulting in $R_{lv,e} = 0.03635\ K/W$.

The overall evaporator thermal resistance is composed by the pool and the film resistances, through Eq. 4.30, resulting in $R_e = 0.02073\ K/W$.

At this step, the liquid–vapor phase change and vapor thermal resistances, $R_{pc,lv}$, $R_{pc,vl}$, R_v, are neglected. The condensation thermal resistance is determined using Eq. 4.53, resulting in $R_c = 0.008401\ K/W$.

The total thermal resistance is $R_o = 0.07146\ K/W$, and the heat load is $q = 410.5\ W$

Step 10. In this step, the heat load obtained is Step 9 is compared with that of Step 7. If the difference between them is not acceptable, the heat is calculated again in Step 8.

Step 8 first recalculation:

$$R_p = 0.01563\ K/W$$
$$R_{lv,e} = 0.03053\ K/W$$
$$R_e = 0.0207\ K/W$$
$$R_c = 0.007056\ K/W$$

Step 9 first recalculation:

$$R_o = 0.07009\ K/W$$
$$q = 418.6\ W$$

Step 8 second recalculation:

$$R_p = 0.01551\ K/W$$
$$R_{lv,e} = 0.03073\ K/W$$
$$R_e = 0.02069\ K/W$$
$$R_c = 0.007102\ K/W$$

Step 9 second recalculation:

$$R_o = 0.07012\, K/W$$
$$q = 418.4\, W$$

The comparison of this power with the one obtained in the first recalculation, $q = 418.6\ W$, shows that the convergence can be considered achieved.

Step 11. In this step, the operational limits are calculated and compared with the predicted heat transfer capacity of the device. For that, the thermal proprieties should be determined for an operating temperature of 49.88 °C.

The **viscous limit** is determined by the Eq. 3.62, which, after substituting the appropriated data, results in $q_{max,v} = 950.351 KW$.

The **sonic limit** is calculated by Eq. 3.80, resulting in $q_{max,s} = 11.531 KW$.

The **entrainment/flooding limit** is determined by Eq. 3.86, with the outcome: $q_{max,f} = 2.412 KW$.

The **boiling limit** for a thermosyphon, considering only the rear region of evaporator, can be calculated using Eq. 3.87, resulting in $q_{max,b} = 5.285 KW$.

Therefore, the heat load that this thermosyphon is able to transfer (418.4 W) does not exceed any of the operational limits, showing that this device is appropriate for any application that does not surpass this heat transfer rate. Among the operational limits, the most limiting one is the entrainment, which presents the lowest value.

The heat transfer limits, determined using the above mentioned equations, for the device operating between 20 to 140 °C is presented in Fig. 6.1, as a function of the operating temperature.

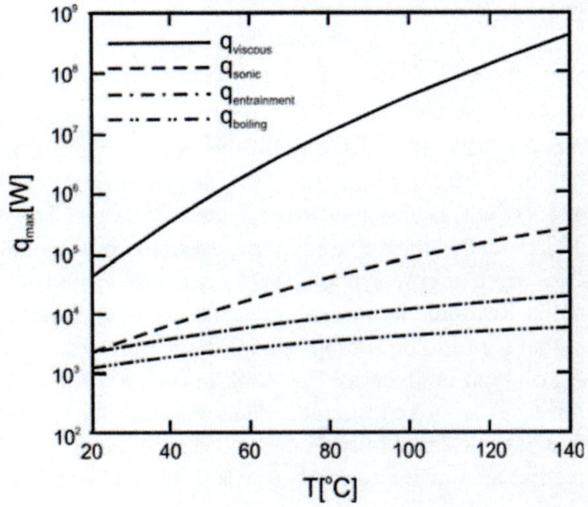

Fig. 6.1 Ordinary geometry thermosyphon operational heat transfer limits

Pressure Distribution

The hypothesis listed in the beginning of this section, where the vapor flow is assumed as linearly variable along each section of the tube, leads to the following expression for the evaporator, adiabatic and condenser sections, respectively:

$$\begin{cases} \frac{d\dot{m}_{v,e}}{dx} = \frac{\dot{m}}{l_e}, & 0 \leq x < l_e \\ \frac{d\dot{m}_v}{dx} = 0, & l_e \leq x < l_e + l_a \\ \frac{d\dot{m}_{v,c}}{dx} = -\frac{\dot{m}}{l_c}, & l_e + l_a \leq x < l_t \end{cases} \tag{6.9}$$

Performing the integration of the last three equations along the evaporator, condenser and adiabatic section lengths, respectively, and expressing the term $\dot{m}$ in terms of the heat input and the liquid–vapor latent heat, the following set of expressions results:

$$\dot{m}_v(x) = \begin{cases} \frac{q}{l_e h_{lv}} x, & 0 \leq x < l_e \\ \frac{q}{h_{lv}}, & l_e \leq x < l_e + l_a \\ \frac{q}{l_c h_{lv}}(l - x), & l_e + l_a \leq x < l_t \end{cases} \tag{6.10}$$

where l_t is $l_e + l_a + l_c$.

To select the appropriated model for predicting the pressure distribution, the radial Reynolds ($Re_{r,e}$, see Eq. 3.31) and Mach ($Ma = v_v/c$) numbers must be determined, resulting in:

$$\mathrm{Re}_{r,e} = \frac{1}{2\pi\mu_v}\frac{d\dot{m}_{v,e}}{dx} = \frac{1}{2\pi\mu_v}\frac{q}{l_e h_{lv}} = 14.26 \tag{6.11}$$

$$Ma_c = \frac{v_{v,e}}{c} = \frac{\mathrm{Re}_{r,e}\mu_v}{\rho_v r_v c} = 1.27 \cdot 10^{-4} \tag{6.12}$$

According to Chap. 3, Chisholm and Chi one-dimensional model can be used if $Re_{r.c} < 8$ and $Ma < 0.3$. While the second condition is by far fulfilled, the Reynolds number is slightly larger than 8. The power input used in this calculation was $q = 418.4W$ but, as presented in the sequence, the tested power input varied between 0 and 3500 W and so the Reynolds number varied from very low to high values. As the major concerns in thermosyphon operation are associated to lower power inputs, where the radial Reynolds is lower than 8, the use of the above mentioned one dimensional model seems to be appropriate for the present problem. The resulting vapor pressure distribution is given by Eq. 3.48, in which $f\,Re = 16$ (see Eq. 2.81, for laminar flows). Also, given the tube slenderness, it is reasonable to assume that the vapor flow is considered fully developed (thus, $\beta = 1$). The term A_{cs} corresponds to cross section area of the tube ($A_{cs} = \pi D_i^2/4$) and r_v to the vapor radius (which can be approximated as $r_v \cong r_i$). The thermal properties are considered constant and only dependent of x according to Eqs. 6.9 and 6.10. The pressure distributions are obtained for each section, as following.

Evaporator section $0 \leq x \leq l_e$

Substituting Eq. 6.10 in Eq. 3.48 and integrating along the evaporator length, one has:

$$\int_0^x \frac{dp_v}{dx}dx = -\int_0^x \left[\frac{(f_v \mathrm{Re}_v)\mu_v \dot{m}_v}{2A_{cs}r_v^2\rho_v} + \beta\frac{2\dot{m}_v}{\rho_v A_{cs}^2}\frac{d\dot{m}_v}{dx}\right]dx \tag{6.13}$$

As the fluid properties are assumed constant over the thermosyphon length, they can be removed from the integral. The term $\dot{m}_v(x)$ is then expressed in function of the x variable. Consequently, this results in:

$$p_{v,e}(x) = p_0 - \left\{\frac{q}{l_e h_{lv}}\frac{x^2}{2}\left[\frac{(f_v \mathrm{Re}_v)\mu_v}{2A_{cs}r_v^2\rho_v} + \beta\frac{2}{\rho_v A_{cs}^2}\frac{q}{l_e h_{lv}}\right]\right\} \tag{6.14}$$

Adiabatic section $l_e. \leq x \leq l_e + l_a$

The vapor flow in the adiabatic section is assumed constant, which means that $d\dot{m}_v/dx = 0$ and the integration of Eq. 3.48 along the adiabatic section length leads to:

$$\int_{l_e}^x \frac{dp_v}{dx}dx = -\int_{l_e}^x \frac{(f_v \mathrm{Re}_v)\mu_v \dot{m}_v}{2A_{cs}r_v^2\rho_v}dx \tag{6.15}$$

Substituting $\dot{m}_v = q/h_{lv}$, the expression for the pressure variation along the adiabatic section is obtained:

$$p_{v,a}(x) = p_v(l_e) - \frac{(f_v \mathrm{Re}_v)\mu_v}{2A_{cs}r_v^2\rho_v}\frac{q}{h_{lv}}(x - l_e) \tag{6.16}$$

Condenser Section $l_e. + l_a \leq x \leq l$

Similarly to the last two thermosyphon regions, the expression for the vapor pressure variation in the condenser is obtained, following the same steps for the evaporator. $\dot{m}_v(x)$ is expressed as a function of x and of the project parameters. So Eq. 3.48 can be written as:

$$\int_{L_e+L_a}^x \frac{dp_v}{dx}dx = -\int_{l_e+l_a}^x \left[\frac{(f_v \mathrm{Re}_v)\mu_v}{2A_{cs}r_v^2\rho_v}\frac{q}{l_c h_{lv}}(l-x) - \beta\frac{2}{\rho_v A_{cs}^2}\left(\frac{q}{L_c h_{lv}}\right)^2(l-x)\right]dx \tag{6.17}$$

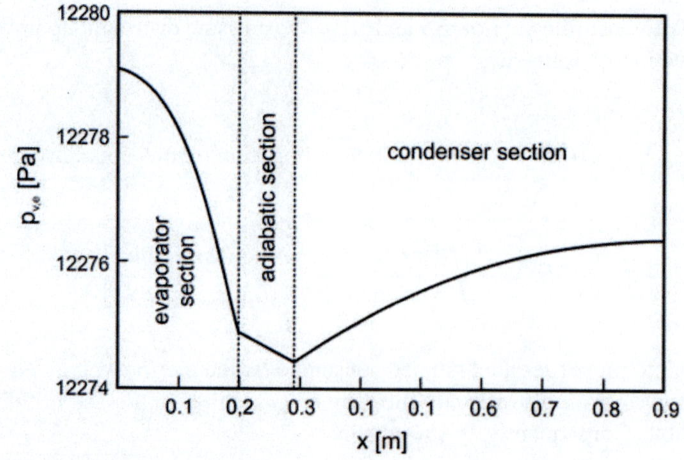

Fig. 6.2 Vapor pressure distribution along the Odinary geometry thermosyphon

Performing the integration and rearranging terms, the following equation is obtained:

$$p_{v,c}(x) = p_{v,a}(l_e + l_a) - \frac{q}{L_c h_{lv}}\left[l_t x - \frac{x^2}{2} - l_t(l_e + l_a) + \frac{(l_e+l_a)^2}{2}\right]\left[\frac{(f_v \mathrm{Re}_v)\mu_v}{2A_{cs} r_v^2 \rho_v} - \beta \frac{2}{\rho_v A_{cs}^2} \frac{q}{l_e h_{lv}}\right] \tag{6.18}$$

These equations are implemented for the data of Table 6.1. The plot of the vapor pressure distribution along the device is shown in Fig. 6.2. The pressure variation along the device is so small that, in practical situations, it can be neglected.

The working fluid liquid phase pressure distribution depends only of the gravity and it is not shown.

Experimental Validation

The thermosyphon described in Table 6.1 was constructed and tested in an apparatus schematically shown in Fig. 6.3, similar to the one depicted in Fig. 5.5. Thermocouples were distributed along the thermosyphon: three in the evaporator, three in the adiabatic section and eight in the condenser. One thermocouple also monitored the cooling water temperature. The heat load varied from 20 to 3500 W, in variable steps. Selected temperature readings represent typical temperature behavior of the evaporator, condenser and adiabatic sections, as shown in Fig. 6.4. The pressure response to the power input is also shown in this graphic.

From Fig. 6.4, for power inputs up to around 150 W, the condenser and the cooling water temperatures are very close to each other. This happens because vapor did not reach the condenser region and so the thermosyphon is not operating properly. The device start to operate at power input levels around 150 W, when the vapor is able to extend along the condenser length, observed by the temperature increase. However, at power input levels of up to 500 W, the device operates under Geyser

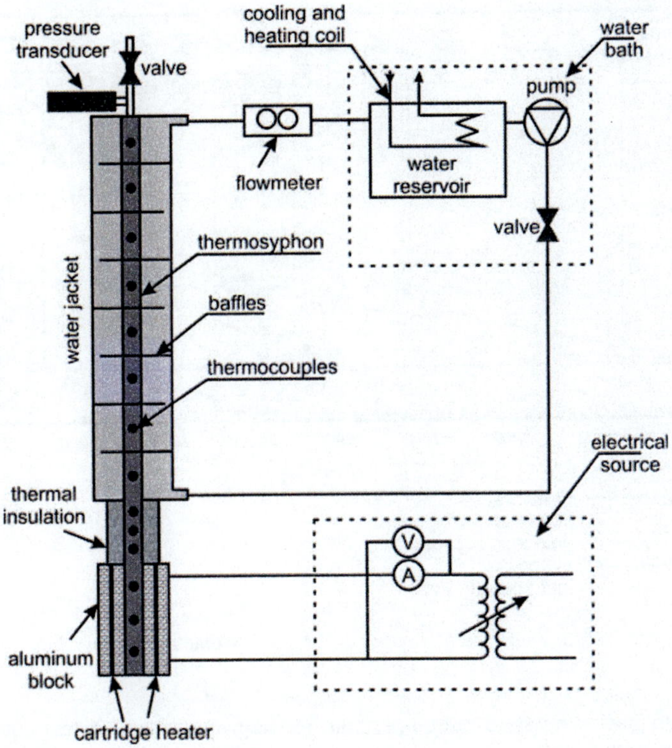

Fig. 6.3 Diagram of the experimental apparatus

boiling regime, observable by strong wall temperature oscillations (see Sect. 3.8 for discussion of Geyser boiling phenomena in thermosyphons). As the power input levels increase further, the temperature oscillations decreases, as well as the difference between the evaporator and condenser temperatures. This means that the device works better at high power inputs, i.e., the thermal resistance decreases, as the power input increases. The pressure curve shows that the pressure increases with the power input, as expected.

Figure 6.5 shows a comparison between the theoretical predictions and the experimental data, as a function of the power input, for the thermosyphon overall thermal resistances. As the wall temperatures are considered known, the external thermal resistances are not accounted for in this plot. The vertical bars represent the experimental uncertainties. The lower plot shows a "zoom" for the power varying from zero to 600 W. In this same power range, the device starts-up and Geyser boiling phenomenon starts and terminates. The comparison between data and model in this range is not good because the models do not take into consideration the Geyser boiling phenomenon.

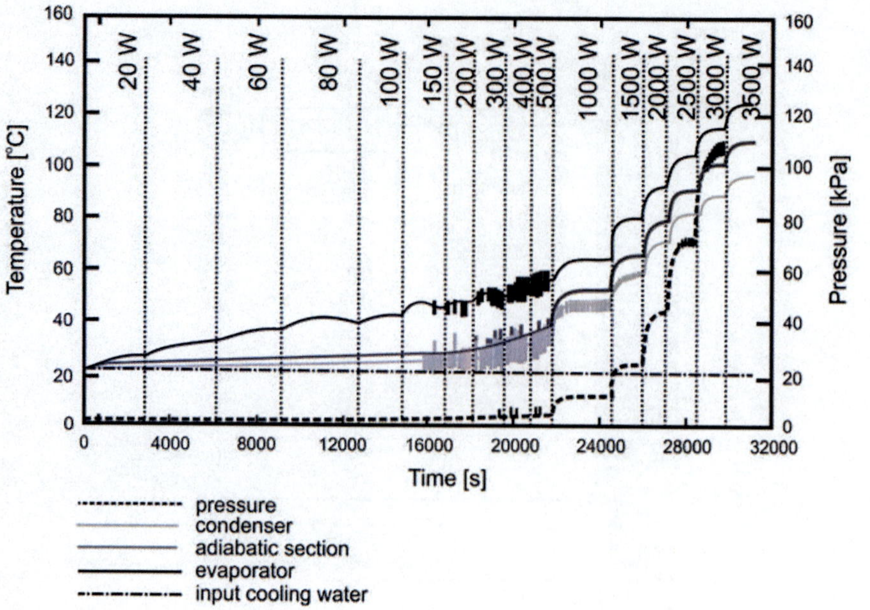

Fig. 6.4 Temperature as a function of time for the ordinary geometry thermosyphon

For high power inputs, the comparison between model and data is quite nice, with the model slightly overpredicting the thermal resistances. This overprediction actually goes in the favor of the safety of the device operation.

6.2 Small Diameter Thermosyphon

The phase change phenomena in small diameter thermosyphons may be quite different from the ordinary geometry devices. For the sake of design example, in this section, a small-diameter copper–water two-phase thermosyphon was experimentally designed, constructed and tested. The geometrical and operation parameters of the thermosyphon are given in Table 6.2. A similar to the Fig. 6.2 apparatus was used in its tests. Temperatures were measured by means of thermocouples attached to the tube surface. A pressure transducer was installed in the topmost portion of the condenser section. Two power input testing conditions are discussed.

Vapor Pressure Distribution

The same procedure applied for the ordinary geometry thermosyphon described in the last section is used here. To verify whether the one-dimensional Chisholm and Chi model can be applied, the Reynolds numbers and Mach numbers must be calculated. Taking the higher power input case (worst case), these calculations results in:

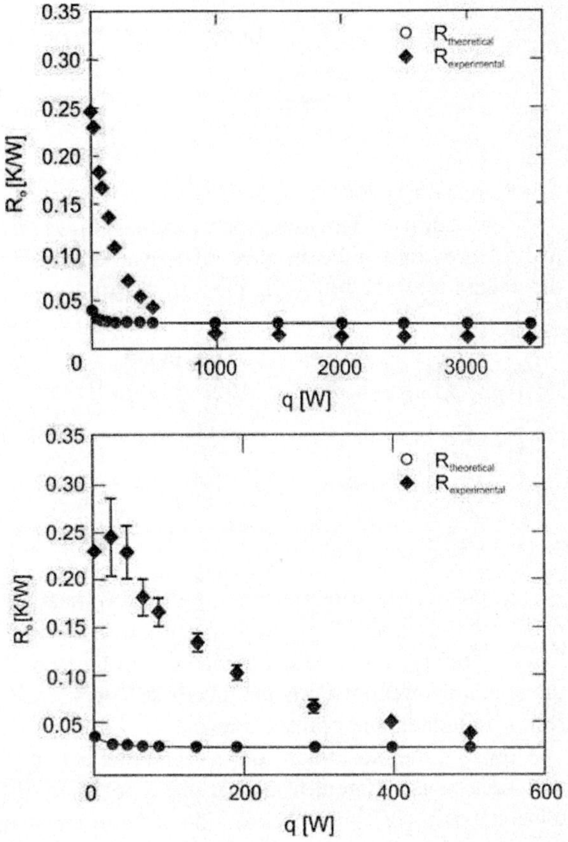

Fig. 6.5 Theoretical and experimental thermal resistances as a function of the power input for the ordinary geometry thermosyphon: second graphic shows a zoom for power input varying from zero to 600 W

Table. 6.2 Geometrical parameters and operation conditions of small dimater thermosyphon

Thermosyphon parameters		
Inner Diameter [mm]	4.77	
Outer Diameter [mm]	6.35	
Evaporator Length [mm]	200	
Adiabatic Section Length [mm]	90	
Condenser Length [mm]	600	
Working fluid	Deionized water	
Filling ratio [% of evaporator volume]	90	
Operation position	Vertical	
Wall material	Copper	
Wall temperature (adiabatic section)	$T_w = 40\,^{\circ}\mathrm{C}$	
Input power [W]	40.7	80.2
Vapor temperature [°C]	32.6	36.4
Pressure [Pa]	4914	6044

$$\mathrm{Re}_{r,c} = \frac{1}{2\pi\mu_v}\frac{d\dot{m}_{v,e}}{dx} = \frac{1}{2\pi\mu_v}\frac{q}{l_e h_{lv}} = 2.59$$

$$Ma_c = \frac{v_{v,e}}{c} = \frac{\mathrm{Re}_{r,e}\mu_v}{\rho_v r_v c} = 6.0 \cdot 10^{-4}$$

which are quite beyond the limits (Re < 8 and Ma < 0.3) showing that one-dimensional models can be used. Obtaining the vapor mass flow rate expressions, substituting them in the expressions of the distance (x) derivative of pressure (Eq. 3.48) and performing the integrations, the following pressure distribution is obtained:

$$p_v(x) = \begin{cases} p_{v,e}(x) = p_0 - \left\{\frac{q}{l_e h_{lv}}\frac{x^2}{2}\left[\frac{(f_v\mathrm{Re}_v)\mu_v}{2A_{cs}r_v^2\rho_v} + \beta\frac{2}{\rho_v A_{cs}^2}\frac{q}{l_e h_{lv}}\right]\right\}, \quad for\ 0 < x < l_e \\ p_{v,a}(x) = p_{v,e}(l_e) - \frac{(f_v\mathrm{Re}_v)\mu_v}{2A_{cs}r_v^2\rho_v}\frac{q}{h_{lv}}(x - l_e), \quad for\ l_e < x < l_e + l_a \\ p_{v,c}(x) = p_{v,a}(l_e + l_a) - \frac{q}{L_c h_{lv}}\left[l_t x - \frac{x^2}{2} - l_t(l_e + l_a) + \frac{(l_e+l_a)^2}{2}\right] \\ \cdot\left[\frac{(f_v\mathrm{Re}_v)\mu_v}{2A_{cs}r_v^2\rho_v} - \beta\frac{2}{\rho_v A_{cs}^2}\frac{q}{l_e h_{lv}}\right], \quad for\ l_e + l_a < x < l_t \end{cases} \tag{6.19}$$

The pressure distribution plots, for the two heat power operation conditions, are presented in Fig. 6.6. Comparing this plot with the one shown in Fig. 6.2, one can see that the pressure level is smaller than that of the wider tube thermosyphon, although the pressure variation along the tube is still very small. Besides, higher input power results in higher vapor pressure levels.

However, the most interesting observation is that the concavity of the condenser region curve is different from that of Fig. 6.2 and of Figs. 3.13, 3.14 and 3.15. To understand this behavior, the complicated pressure distribution curve can be simplified, considering a second degree equation of the form: $p_v(x) = ax^2 + bx + c$, which terms, by comparison with the expression for the condenser (Eq. 6.19), are:

$$a = \frac{q}{2l_c h_{lv}}\left[\frac{(f_v\mathrm{Re}_v)\mu_v}{2A_{cs}r_v^2\rho_v} - \beta\frac{2}{\rho_v A_{cs}^2}\frac{q}{l_e h_{lv}}\right] \tag{6.20}$$

$$b = -l_t\frac{q}{l_c h_{lv}}\left[\frac{(f_v\mathrm{Re}_v)\mu_v}{2A_{cs}r_v^2\rho_v} - \beta\frac{2}{\rho_v A_{cs}^2}\frac{q}{l_e h_{lv}}\right] \tag{6.21}$$

$$c = \frac{q}{l_c h_{lv}}\left[\frac{(f_v\mathrm{Re}_v)\mu_v}{2A_{cs}r_v^2\rho_v} - \beta\frac{2}{\rho_v A_{cs}^2}\frac{q}{l_e h_{lv}}\right]\left[l_t(l_e + l_a) + \frac{(l_e + l_a)^2}{2}\right] + p_{v,a}(l_e + l_a) \tag{6.22}$$

If a is positive, the concavity is up and vice-versa. Also, according to the parameters a, b and c and the position x, the pressure can increase or decrease along the condenser length. It is interesting to note that both parameters a and b are heavily dependent on the viscous forces $(fRe)\mu_v/(2A_{cs}r_v^2\rho_v))$ and on the momentum rate $(2\beta q/(\rho_v A_{cs}^2 l_c h_{lv}))$. Both terms depends on the cross section area, while the momentum rate also depends on the vapor production/consumption rate $(q/h_{lv}l_c)$.

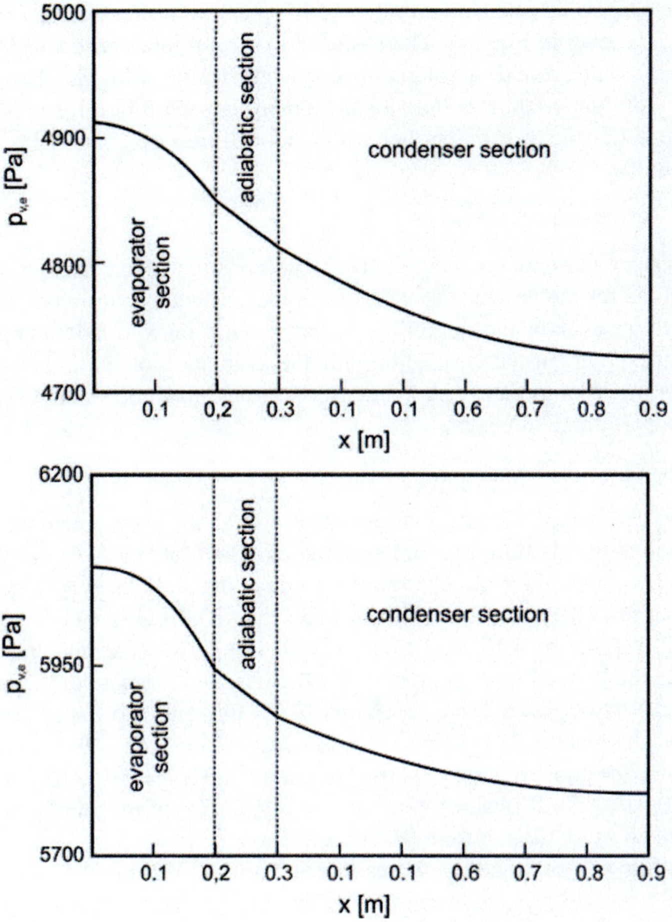

Fig. 6.6 Small diameter thermosyphon vapor pressure distribution for the higher and lower power input

The a and b coefficients have opposite signals, so that the sum $ax^2 + bx + c$ is positive or negative depending on which term (viscous or momentum rate) supplants the other, defining the pressure behavior curve.

The derivative of the $p_v(x)$ with respect to x, in the condenser section is:

$$\frac{dp_{v,c}}{dx} = -\frac{q}{l_c h_{lv}}\left[\frac{(f_v \mathrm{Re}_v)\mu_v}{2A_{cs} r_v^2 \rho_v} - \beta \frac{2}{\rho_v A_{cs}^2} \frac{q}{l_c h_{lv}}\right](l_t - x) \tag{6.23}$$

A positive value of the derivative at the position $x = l_e + l_a$ means that the momentum rate is bigger than the viscous force rate, at the entrance of the condenser region. Therefore, there will be a pressure recovery along the condenser section, as for the Fig. 6.2 pressure curve. However, if the derivative is negative, the viscous forces rate surpass the momentum rate and the pressure decreases even further, as for the plots shown in Fig. 6.6. Therefore, as expected, tubes with small diameter presents higher pressure drops along the condenser length, with the viscous forces playing a more important role than for the larger diameter tubes. It should be kept in mind that, obviously, different behaviors can be observed considering different working fluids.

Thermal Resistance Network

The equivalent thermal resistance network is constructed for the present thermosyphon, as for the previous case. As the vapor pressure drops along this small diameter thermosyphon may not follow the vapor behavior expected for larger diameter thermosyphons, usual correlations for the estimation of the internal thermal resistances in the evaporator and condenser sections may not be the best and so, studies are performed in this regard.

Evaporator Thermal Resistance

As observed in Chap. 4, the liquid–vapor phase change in the evaporator section is a combination of pool boiling and film evaporation. Therefore the evaporator thermal resistance is composed by two resistances: pool boiling and film evaporation.

From the several correlations available (see Chap. 4), El-Genk and Saber (1998a and 1998b, Eq. 4.7 to 4.15) and Imura (1992, Eq. 4.15) were implemented and validated in the light of the experimental results. The pool evaporator thermal resistances, for the two power input cases, using the El-Genk and Saber (1998a and 1998b) correlation are: $R_p = 8.91 \cdot 10^{-2}\ K/W$ and $R_p = 6.63 \cdot 10^{-2}\ K/W$. Using Imura model, the pool evaporator thermal resistance are $R_p = 6.16 \cdot 10^{-6}\ K/W$ and $R_p = 4.50 \cdot 10^{-6}\ K/W$, values which are of three order of magnitude lower than those obtained using El-Genk and Saber correlation.

For the film evaporation thermal resistance, Gross (2002, Eq. 4.17 to 4.19) and Groll and Rösler (1992, Eq. 4.29) correlations were used. The use of Gross correlation resulted in film thermal resistances of $R_{lv,e} = 4.09 \cdot 10^{-2}\ K/W$ and of $R_{lv,e} = 4.97 \cdot 10^{-2}\ K/W$ for both power input levels, while the use of Groll and Rösler correlation results in $R_{lv,e} = 3.79 \cdot 10^{-2}\ K/W$ and $R_{lv,e} = 4.60 \cdot 10^{-2}\ K/W$. These results were combined using Eq. 4.30.

In actual operation conditions, the height of the liquid pool experiences, in one hand an increase, due to the presence of bubbles within the pool (especially if the evaporator experiences pool boiling), and, in the other hand, a decrease, due to the accumulation of liquid in the condensate films along the internal thermosyphon walls. Actually, this height determines the balance between pool and film resistances in the estimation of the evaporator thermal resistance. As mentioned in Chap. 4, El-Genk and Saber (1997 and 1998b) developed a model for the non-dimensional height parameter ($H_p = l_p/l_e$), where l_p is the actual length of the pool, expressed by the set of equations: Eqs. 4.31 to 4.40.

To determine the pool height using these equations, both Nu_f (see Eq. 4.26) and $D_H{}^+$ values are determined, resulting in: $Nu_f = 2.30 \cdot 10^{-7}$ and $D_H^+ = 1.77$ for the lower power input level, and $Nu_f = 1.95 \cdot 10^{-7}$ and $D_H^+ = 1.78$, for the higher power. These parameters are used for the determination of the vapor drift velocity V^+ (Eqs. 4.38 to 4.40). Besides, Eq. 4.34 can be integrated to obtain the void fraction over the length interval $l_f < x < l_e$, where l_f is the liquid film length.

The film height l_f can be estimated based on three possible hypotheses. First, the liquid pool height is assumed undisturbed and the film length is $l = l_e(1 - F))$, where F is the filling ratio. Second, it can be assumed that, due to the presence of bubbles, the liquid pool actually occupies the whole length of the condenser ($H_p = 1$). Third, the liquid pool height is disturbed and the liquid film length $l_f = l_e - l_p$ must be obtained by iterations, using the set of Eqs. 4.30 to 4.40.

In the present case, the boiling correlations, film correlations and pool height hypotheses were combined to generate models that were compared with data. Calculation shown that the liquid pool height might be actually much larger than expected, with the H_p parameter larger than the unity. Therefore, it is assumed that the pool boiling height is the whole evaporator length, with the film resistance discarded. In the case of small diameter thermosyphons, it is not difficult to imagine that bubbles in the evaporator section can grow to occupy the whole diameter, pushing the liquid column upwards, thus increasing the evaporator pool height.

Condenser Film Thermal Resistance

For the calculation of the condenser film thermal resistance, two correlations are used: Kaminaga (1992) given by Eq. 4.52 and ESDU (1981) given by Eq. 4.53. The use of Kaminaga's correlation results in thermal resistances of $R_c = 1.14 \cdot 10^{-2}\ K/W$ and $R_c = 9.70 \cdot 10^{-3}\ K/W$, for both power input levels. On the other hand, using ESDU correlations, one has the thermal resistances of $R_c = 8.60 \cdot 10^{-3}\ K/W$ and of $R_c = 1.04 \cdot 10^{-2}\ K/W$ for the two power input levels.

Overall Thermal Resistance

Finally, the overall thermal resistance can be calculated by the combination of the thermal resistances. Because they are very small, the phase-change, vapor flow and the axial wall conduction resistances were removed from the circuit. If the external wall temperatures in the evaporator and condenser are known, the overall thermal resistance can be determined as the sum of the following resistances (see Eq. 4.54):

$$R_o = 2R_{cond} + R_e + R_c \tag{6.24}$$

Operational Limits

The following thermosyphon operational limits were determined:

- Viscous limit (Eq. 3.62);
- Sonic limit (Eq. 3.80);
- Entrainment/flooding limit (Eq. 3.86);
- Boiling limit (Eq. 3.87).

It should be noted that, in order to calculate the maximum heat input power, the calculated limits have to be multiplied by areas, to which the phenomenon apply.

Table. 6.3 Opera-tional limits for small diameter thermosyphon

Power Input [W]	$q_{max,v}[W]$	$\bar{q}_{max,s}[W]$	$q_{max,f}[W]$	$q_{max,b}[W]$
40.7	519	269	101	835
80.2	798	330	108	918

Cross section areas A_{cs} apply for the viscous and sonic operation limit equations while the evaporator inner surface area ($A_e = \pi D_i l_e$) apply for the boiling and entrainment limits. Table 6.3 shows these limits. Among all the theoretical limits, the entrainment/flooding is the lower one, as expected, since the direct contact of the countercurrent flows of liquid and vapor is increased in small diameter thermosyphons. All the other limits are distant from the entrainment limit by a wide margin.

Comparison Between Model and Data

The comparison between model and data for the present small diameter thermosyphon is shown in this section. Defining the error as the difference between the theoretical and experimental values, divided by the experimental results, the theoretical predictions of the vapor pressure are compared with data. The pressure transducer is located at the topmost part of the condenser and is compared with theoretical results for vapor pressure at $x = 0.89m$. Table 6.4 shows a quite good comparison, which means that the model was able to predict the data with great precision.

Besides, the overall thermal resistance data, obtained from the ratio of the difference between the average thermocouple readings at the evaporator and condenser regions and the power transferred by the device (see Eq. 4.1), is compared with model predictions. The comparison of the thermal resistance models with data showed that the best combination was achieved with El-Genk and Saber (1988a and b) correlation for the evaporator pool and ESDU (1981) correlation for the condenser. Table 6.5 shows this comparison. As this thermosyphon, with small diameter, operates in confinement conditions not considered in the model, the present comparison still can be considered good.

Table. 6.4 Comparison between theoretical and experimental results for vapor pressure at the topmost position of the small diameter thermsyphon

Power input [W]	Theoretical pressure [Pa]	Measured pressure [Pa]	Error [%]
40.7	4725	4914	−3.85
80.2	5780	6044	−4.37

Table. 6.5 Comparison between the theoretical results and experimental data for thermal resistance of the small diameter thermosyphon

Power Input [W]	Theoretical thermal resist. [K/W]	Measured thermal resist. [K/W]	Error [%]
40.7	0.10	0.07	41
80.2	0.08	0.08	−3

6.3 Water–Copper Heat Pipe

The proposed water-copper heat pipe case do be analyzed consists of a heat pipe operating horizontally, constructed from a tube of *25.4* mm of external diameter, inner diameter of *22.2* and *0.46* m of total length. The evaporator, adiabatic section, and condenser lengths are *0.18*, *0.05* and *0.23* m, respectively. The wick structure, consisting of five ($n = 5$) layers of $N = 160$ [1/inch] mesh is placed against the inner tube surface. It is made of copper wires with a diameter of $D_w = 63\mu m$ and a wire spacing of $95\mu m$. Water is used as a working fluid. The heat to be transported is delivered by a constant flux electrical heater, at a rate of *53.1 W* at $T_0 = 70^o C$. The heat is removed from the condenser by air in forced convection at a velocity of *6* m/s and a temperature of *25*°C. These characteristics are also listed in Table 6.6. A schematic of the present problem heat pipe is presented in Fig. 6.7.

Table. 6.6 Water-copper heat pipe characteristics

Water-copper heat pipe characteristics	
Inner diameter [mm]	25.4
Outer diameter [mm]	22.2
Evaporator length [m]	0.18
Adiabatic section length [m]	0.05
Condenser length [m]	0.23
Total length [m]	0.46
Working fluid	deionized water
Operation position	horizontal
Wall material	copper
Wick	Copper wire screen – 160 mesh
Wire diameter [μm]	63
Wire spacing [μm]	95
Number of layers	5
Crimping factor	1.1
Operation temperature	$T_0 = 70$ °C

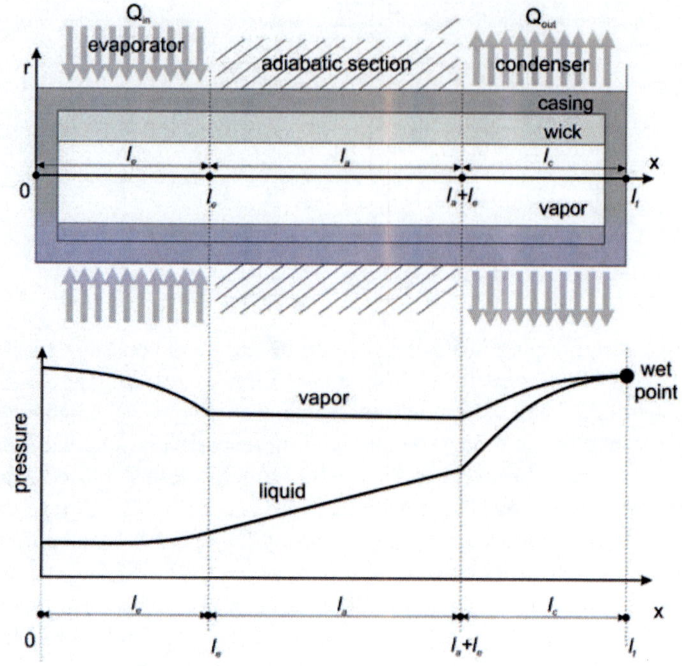

Fig. 6.7 Schematic of the heat pipe and pressure distributions (vapor and liquid)

Vapor and Liquid Pressure Distribution Curves

Vapor Mass Flow Rate – evaporator Section

In the evaporator section ($0 \leq x \leq l_e$), the vapor generation rate, per evaporator length, can be considered constant. All vapor transported through the heat pipe $\dot{m}_t$ is produced in the evaporator. Considering $\dot{m}_v(x = 0) = 0$, one has:

$$\frac{d\dot{m}_{v,e}(x)}{dx} = \frac{\dot{m}_t}{l_e} \tag{6.25}$$

where $\dot{m}_t$ is the total mass flow rate. The integration of this expression results in:

$$\dot{m}_{v,e}(x) = \int_0^x \frac{d\dot{m}_{v,e}}{dx} dx = \frac{\dot{m}_t}{l_e} x \tag{6.26}$$

In steady state conditions, the total mass flow rate can be estimated from the heat input, as:

$$\dot{m}_t = \frac{q}{h_{lv}} \tag{6.27}$$

Then, the space derivative of the vapor mass flow rate in the evaporator is:

$$\frac{d\dot{m}_{v,e}(x)}{dx} = \frac{q}{l_e h_{lv}} \tag{6.28}$$

For the present case, the latent heat of vaporization of water, at $T_o = 70$ °C, is $h_{lv} = 2333 \cdot 10^3\ J/kg$ and the vapor generation rate per evaporator length is:

$$\frac{d\dot{m}_{v,e}(x)}{dx} = 1.264 \cdot 10^{-4} \frac{kg}{m \cdot s}$$

Vapor Mass Flow Rate – adiabatic Section

In the adiabatic section ($l_e \le x \le l_a$), no vapor is generated or condensed. Therefore, all the vapor generated is transported through the adiabatic section, i.e.:

$$\frac{d\dot{m}_{v,a}(x)}{dx} = 0 \quad e \quad \dot{m}_{v,a}(x) = \dot{m}_t \tag{6.29}$$

For the present case, $\dot{m}_t = 2.276 \cdot 10^{-5}\ kg/s$.

Vapor Mass Flow Rate – condenser Section

In the condenser section ($l_a \le x \le l_c$), the vapor condensation rate is also considered constant along the condenser length. The vapor mass flow decreases with the length, according to the expression:

$$\frac{d\dot{m}_{v,c}(x)}{dx} = -\frac{\dot{m}_t}{l_c} \tag{6.30}$$

The undefined integral in x of the vapor flow rate generation is:

$$\dot{m}_{v,c}(x) = -\int \frac{d\dot{m}_{v,c}(x)}{dx} dx = -\frac{\dot{m}_t}{l_c} x + C \tag{6.31}$$

The mass flow at the beginning of the condenser can be considered as the boundary condition to determine the integration constant, i.e., for $x = l_e + l_a$, $\dot{m}_v = \dot{m}_t$, resulting in $C = \dot{m}_t l / l_c$. The substitution of C in the last equation, results in:

$$\dot{m}_{v,c}(x) = -\frac{\dot{m}_t}{l_c}(l - x) \tag{6.32}$$

The mass flow rate space derivative can be related to the vapor generation by the expression:

$$\frac{d\dot{m}_{v,c}(x)}{dx} = \frac{q}{l_c h_{lv}} \tag{6.33}$$

In the present example,

$$\frac{d\dot{m}_{v,c}(x)}{dx} = 9.896 \cdot 10^{-5} \frac{kg}{m \cdot s}$$

The total vapor generated in the evaporator is the same of the total liquid condensed in the condenser, as expected.

Vapor pressure distribution

The vapor pressure distribution at any position x can be calculated by integrating Eq. 3.48, resulting in:

$$\Delta p_{v,e} = \int_0^x \frac{dp_v}{dx} dx = -\int_0^x \left(\frac{(f_v Re_v)\mu_v \dot{m}_v}{2A_v r_v^2 \rho_v} + \beta \frac{2\dot{m}_v}{\rho_v A_v^2} \frac{d\dot{m}_v}{dx} \right) dx \tag{6.34}$$

Substituting Eq. 6.9 and 6.10 (developed for thermosyphons but also valid for heat pipes, see Chap. 4), in the last equation, results:

$$\Delta p_{v,e} = -\int_0^x \left(\frac{(f_v Re_v)\mu_v}{2A_v r_v^2 \rho_v} \frac{\dot{m}_t}{l_e} x + \beta \frac{2}{\rho_v A_v^2} \left(\frac{\dot{m}_t}{l_e} \right)^2 x \right) dx \tag{6.35}$$

Assuming that the vapor velocity distribution in the cross–section does not vary along the pipe, $\beta = 1$. For a laminar flow of an incompressible fluid (water), $f_v Re_v = 16$. Taking the water properties from thermodynamic tables at *70ºC* and considering that, at $x = 0$, $p = p_0$, the last equation results in:

$$p_{v,e}(x) = p_0 - \left[\frac{\dot{m}_t}{l_e} \frac{x^2}{2} \left(\frac{(f_v Re_v)\mu_v}{2A_v r_v^2 \rho_v} + \beta \frac{2}{\rho_v A_v^2} \frac{\dot{m}_t}{l_e} \right) \right] \tag{6.36}$$

The diameter of the central region of the heat pipe, where the vapor flows, can be calculated by the expression:

$$D_v = D_i - 2nSD_w \tag{6.37}$$

S is the crimping factor, which accounts for the accommodation of several screen layers against the wall and is expressed by the ratio of the real to the theoretical thicknesses of the wick structure, which, in the present case, is taken as 1.1, typical value for metal screen in a cylindrical tube. The S value can change with the heat pipe fabrication process adopted.

Besides, considering that the spacing between screens and between screen and tube wall is equivalent to that of one wire diameter, the vapor flow area for the present example is $A_v = \pi r_v^2 = 3.403 \cdot 10^{-4}\, m^2$.

In the present case, p_0 is considered as the saturated vapor pressure of water at the working temperature (70 °C). Therefore, the pressure distribution, after the substitution of the appropriate values, is:

$$p_{v,e}(x) = 31202 - 1.475x^2[Pa]$$

Therefore, the total pressure drop for the evaporator section ($0 \leq x \leq l_e$) is

$$\Delta p_{v,e} = -0.04778[Pa]$$

In the adiabatic section ($l_e \leq x \leq l_a$), no phase change happens and only the first term of the pressure drop equation (Eq. 3.48) is valid. Therefore, the pressure distribution is:

$$\Delta p_{v,a} = \int_{l_e}^{l_e+l_a} \frac{dp_v}{dx} dx = -\int_{l_e}^{l_e+l_a} \frac{(f_v Re_v)\mu_v \dot{m}_v}{2A_v r_v^2 \rho_v} dx \tag{6.38}$$

Besides, the pressure distribution in the adiabatic section is:

$$p_{v,a} = p_0 + \Delta p_e - \int_{l_e}^{x} \frac{(f_v Re_v)\mu_v \dot{m}_t}{2A_v r_v^2 \rho_v} dx = p_0 + \Delta p_e - \frac{(f_v Re_v)\mu_v \dot{m}_t}{2A_v r_v^2 \rho_v}(x - l_e) \tag{6.39}$$

The substitution of the appropriate values results in the following equation:

$$p_{v,a}(x) = 31202.002 - 0.2804x[Pa]$$

and in the following pressure drop:

$$\Delta p_{v,a} = -0.01402[Pa]$$

In the condenser section ($L_a \leq x \leq L_c$), the pressure distribution is also based on Eq. 3.48, changing the sign from minus to plus, as pressure is recovered. Therefore, the vapor pressure drop can be determined from:

$$\Delta p_{v,c} = \int_{l_e+l_a}^{x} \frac{dp_v}{dx} dx = +\int_{l_e+l_a}^{x} \left(\frac{(f_v Re_v)\mu_v \dot{m}_v}{2A_v r_v^2 \rho_v} + \beta \frac{2\dot{m}_v}{\rho_v A_v^2} \frac{d\dot{m}_v}{dx} \right) dx \tag{6.40}$$

where the derivative of vapor mass flow rate is given by Eq. 6.09 and the mass flow rate by Eq. 6.10. Performing the integration, the following expression results:

$$p_{v,c}(x) = p_0 + \Delta p_e + \Delta p_a + \frac{\dot{m}_t}{l_c}\left[\frac{(f_v Re_v)\mu_v}{2A_v r_v^2 \rho_v} - \beta\frac{2}{\rho_v A_v^2}\frac{\dot{m}_t}{l_c}\right]\left[l_t x - \frac{x^2}{2} + \frac{l_c^2 - l_t^2}{2}\right] \tag{6.41}$$

For the present case, using the appropriate thermophysical parameters, the expression is:

$$p_{v,c}(x) = 31202.909 - 0.18325x^2 + 0.16859x\ [Pa]$$

The pressure drop in the condenser is:

$$\Delta p_{v,c} = 0.009695\ [Pa]$$

The calculation of the total pressure drop of the vapor phase along the heat pipe is the summation of the three parcels:

$$\Delta p_v = \Delta p_{v,e} + \Delta p_{v,a} + \Delta p_{v,c} = -0.05211\ [Pa]$$

Figure 6.8 shows the theoretical vapor pressure distribution along the heat pipe. The vapor pressure variation in this case is so small that, for practical purposes, it can be neglected.

Liquid pressure distribution

To determine the pressure distribution in the liquid phase of the working fluid, Darcy's model is employed. Considering laminar flow, the liquid pressure gradient is given by Eq. 3.12. As steady sate conditions are assumed, liquid phase presents the same mass flow equations as the vapor phase, with the opposite direction ($\dot{m}_v = -\dot{m}_l$). The permeability, needed as input parameter for the liquid pressure distributions, must

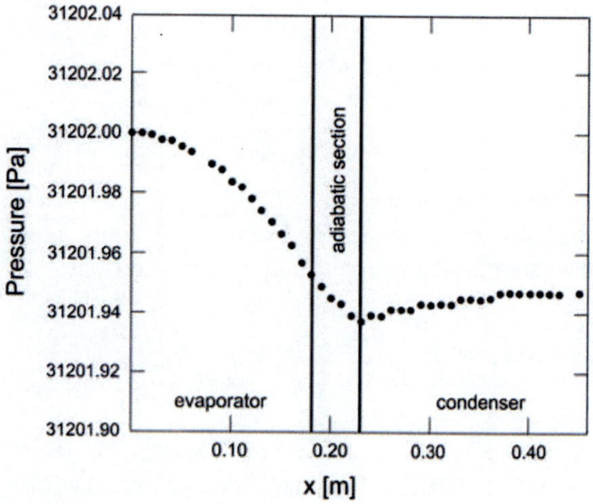

Fig. 6.8 Vapor pressure distribution along the heat pipe

be known. The expression for superposed wire screens is adopted, which requires the determination of the porosity, given by Eq. 3.20, resulting, for the present case, in:

$$\varepsilon = 1 - \frac{1}{4}\pi \mathrm{SN}D_w = 0.6571$$

where *N [1/inch]* is the screen mesh number (the number of metal wires in one inch of screen). Note that the parameter *N* should be converted to the same dimensional system before being applied. The permeability (Table 3.1) is:

$$\mathrm{K} = \frac{\mathrm{D}_w^2 \varepsilon^3}{122(1-\varepsilon)^2} = 7.854 \cdot 10^{-11} m^2$$

The diameter of the cross section area of the vapor region is taken as the inner tube diameter minus the thickness of the mesh formed by superposed screens. The spacing between screen layers is of the same dimension of the screen wire diameter. For a mesh formed by $n = 5$ layers of metal screens, the vapor diameter is:

$$D_v = D_i - 2(2nSD_w) = 0.02081\, m$$

and the wick cross section area is:

$$A_w = \pi\left(D_i^2 - D_v^2\right)\varepsilon = 30.77 \cdot 10^{-6} m^2$$

The effective capillary radius is given by the Table 2.1. From, the expression for wire screens:

$$r_{ef} = \frac{w + D_w}{2} = \frac{\left(95 \cdot 10^{-6}\right) + \left(63 \cdot 10^{-6}\right)}{2} = 7.9 \cdot 10^{-5} m$$

Taking the liquid properties at 70 °C and using Eq. 3.12, one gets, along the evaporator ($0 \leq x \leq L_e$):

$$\Delta p_{l,e}(x) = \int_0^x \frac{dp_l}{dx} dx = \int_0^x \left(\frac{\mu_l}{K A_w h_{lv} \rho_l}\right) q dx = \frac{\mu_l}{\rho_l K A_w} \int_0^x \dot{m}_{l,e} dx \tag{6.42}$$

As $\dot{m}_{l,e}(x) = \dot{m}_l / l_e x$:

$$\Delta p_{l,e}(x) = \frac{\mu_l}{\rho_l K A_w} \frac{\dot{m}_l}{l_e} \frac{x^2}{2} \tag{6.43}$$

For the adiabatic section ($l_e \leq x \leq l_a + l_e$):

$$\Delta p_{l,a}(x) = \int_{l_e}^{x} \frac{dp_l}{dx} dx = \frac{\mu_l}{\rho_l K A_w} \int_{l_e}^{x} \dot{m}_l dx = \frac{\mu_l \dot{m}_l}{\rho_l K A_w}(x - l_e) \quad (6.44)$$

For the condenser section ($l_e + l_a \leq x \leq l_e + l_a + l_c$):

$$\Delta p_{l,c}(x) = \int_{l_e+l_l}^{x} \frac{dp_l}{dx} dx = \frac{\mu_l}{\rho_l K A_w} \int_{l_e}^{x} \dot{m}_l dx = \frac{\mu_l}{\rho_l K A_w} \int_{l_e+l_a}^{x} \frac{\dot{m}_l}{L_c}(l_t - x) dx$$
$$= \frac{\mu_l}{\rho_l K A_w} \frac{\dot{m}_l}{l_c} \left\{ l_t[x - (l_e + l_a)] - \frac{1}{2}[x^2 - (l_e + l_a)^2] \right\} \quad (6.45)$$

To obtain the pressure distribution using the last three equations, the pressures at the positons $x = 0$, $x = l_e$ and $x = l_e + l_a$ need to be determined.

In optimized designed heat pipes, the vapor pressure must be equal than the liquid pressure at the condenser rear position, i.e., in the "wet point" (see Fig. 3.14). In this condition, the heat pipe operate in the lowest capillary pressure, resulting, therefore, in the lowest liquid pressure drop along the device. Considering that, for the present case, the vapor pressure drop is negligible and that the pressure drop at the position $x = l_t$ is $p_0(T_0)$, one has:

$$p_l(x = 0) = p_0 - \Delta p_{l,c}(x = l_t) - \Delta p_{l,a}(x = l_e + l_a) - \Delta p_{l,e}(x = l_e) \quad (6.46)$$

Substituting the appropriate expressions, the following equation results after manipulation:

$$p_l(x = 0) = p_0 - \frac{\mu_l \dot{m}_l}{\rho_l K A_w}\left[\frac{l_e}{2} + l_a + \frac{l_c}{2}\right] = p_0 - \frac{\mu_l \dot{m}_l}{\rho_l K A_w} l_{ef} \quad (6.47)$$

$$p_l(x = l_e) = p_0 - \frac{\mu_l \dot{m}_l}{\rho_l K A_w}\left(\frac{l_e}{2} + l_a\right) \quad (6.48)$$

$$p_l(x = l_e + l_a) = p_0 - \frac{\mu_l \dot{m}_l}{\rho_l K A_w} l_a \quad (6.49)$$

Substituting the appropriate thermophysical properties, the liquid pressure distribution along the wick structure is given by the following function by parts:

$$p_l(x) = \begin{cases} 30210.17 + 10806.45x^2 & 0 < x \leq l_e \\ 29859.84 + 3890.322x & l_e < x \leq l_e + l_a \\ 29412.65 + 7780.65x - 8457.22x^2 & l_e + l_a < x \leq l_t \end{cases} \quad (6.50)$$

Figure 6.9 shows the liquid pressure distribution within the wick structure and a comparison with the vapor distribution, for the present case. It is easily observable

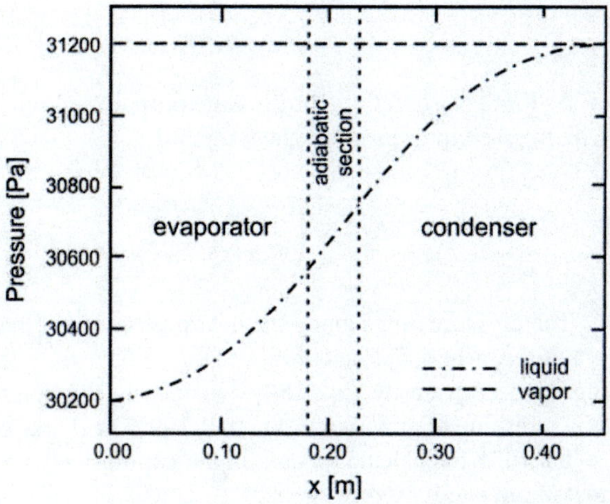

Fig. 6.9 Vapor and liquid distribution

that the liquid pressure drop is much higher than that of the vapor, which, in most applications, is neglected.

Operational Limits

In this section, the operational heat transfer limits of the heat pipe case under analysis are determined, with the objective of verifying whether the heat pipe is able to transfer the required 53.1 W.

To determine the heat transfer limits of the analyzed heat pipe, the important thermophysical properties must be obtained for the device working temperature level of 70 °C.

Capillary Limit

The capillary limit, usually the most restrict one for heat pipes, is determined first. For that, Eq. 3.100, here reproduced, is used:

$$\Delta p_{cap} \geq \int_{l_{ef}} \frac{\partial p_v}{\partial x} dx + \int_{l_{ef}} \frac{\partial p_l}{\partial x} dx + \Delta p_{h,N} + \Delta p_{h,ax}$$

Assuming that the fluid wets perfectly the casing surface and the wick structure ($\theta = 0$), for steady state regime, and considering that the wet point is located at the end of the condenser, the maximum capillary pressure drop is given by Eq. 3.3 $\Delta p_{cap} = 2\sigma / r_{ef}$, where the effective radius, r_{ef}, for wire screens, is given in Table 2.1. For the present case, $\Delta p_{cap} = 1632.40\ Pa$.

The pressure drop of the vapor phase could be considered as the same determined in the last section. However, for the sake of safety, a new value is calculated. Considering laminar flow and total kinetic recuperation, this vapor pressure drop can be given by Eq. 3.42:

$$\Delta p_v = \frac{8\mu_v \dot{m}}{\rho_v \pi r_v^4} l_{ef} = 0.001347 \cdot q$$

As for the last section, the pressure drop in the liquid is determined from Eq. 3.29, which, given in terms of the heat input, is:

$$\Delta p_l = \left(\frac{\mu_l l_{ef}}{K\pi \left(r_w^2 - r_v^2\right) \varepsilon \rho_l h_{lv}} \right) q = 18.68 \cdot q$$

The pressure drops along the heat pipe normal direction, due to the action of the gravity, can be given according to Eq. 3.55, while the pressure drop in the axial direction is given by Eq. 3.54. Considering the horizontal position ($\theta = 0°$), the pressures due to gravity are $\Delta p_{gN} = 199.60\ Pa$ and $\Delta p_{gL} = 0\ Pa$.

Inserting the calculated data in the capillary equation above and isolating $\boldsymbol{q}$, the maximum capillary pressure is:

$$\Delta p_{\max,cap} = 76.69\ W$$

For the calculation of this maximum power that the heat pipe can transfer before the capillary limit is reached, laminar incompressible behavior of the fluid flow is considered. Therefore, these hypothesis must be checked, by observing non-dimensional numbers.

The Reynolds number is:

$$Re = \frac{\rho_v v_v D_v}{\mu_v} == \frac{4 q_{max,cap}}{\pi D_v^2 h_{lv}} = 178.6$$

Which is smaller than 2300, confirming the hypothesis of laminar flow.

The Mach number is determined by:

$$Ma = \frac{v}{c} = \frac{4 q_{max,c} / \rho_v \pi h_{lv} D_v^2}{\sqrt{\frac{cp_v}{cv_v} R T_v}} = 0.00106$$

which is much smaller than 0.3, showing that the flow is really incompressible.

Other Operational Limits

The **viscous limit** is obtained using Eq. 3.62, resulting in $q_{max,v} = 11.59 \cdot 10^6\ W$.

The **sonic limit** is determined using Eq. 3.80, resulting in $q_{max,s} = 29607.1 W$.

The entrainment limit is calculated by Eq. 3.105, considering that the hydraulic radius for a screen wick is $r = 0.5 \cdot w$ (w is the wire spacing, see Fig. 2.12) and is $q_{max,e} = 9212.3 W$.

Table. 6.7 Water-copper heat pipe operational limits

Limit	Value [W]
Capillary	76.69
Viscous	$11.59{\cdot}10^6$
Sonic	29,607
Entrainment	9212
Boiling	8900

The boiling limit is estimated using Eq. 3.113, were the effective thermal conductivity is determined through the expression given in Table 3.2 for superposed wire screens and considering $r_n = 2.54 \cdot 10^{-7}\,m$, resulting in $q_{max,b} = 8900.1\,W$.

Table 6.7 shows these limits. The lowest is the capillary but all of them are superior to the required value of 53.1 W, which means that the device is able to transport the nominated heat power.

Using the expressions obtained above for the prediction of the operational limits, a plot as a function of the operating temperature is presented in Fig. 6.10. It can be observed that, for the curve temperature range, the capillary is still the most restrictive limiting factor and so it establishes the maximum heat capacity of the device.

Thermal Resistances

The circuit similar to the one presented in Fig. 4.3 is adopted in this present case. To determine the individual resistances, the thermophysical properties of the working fluid, wick and casing materials must be determined at the operating temperature (*70 °C*).

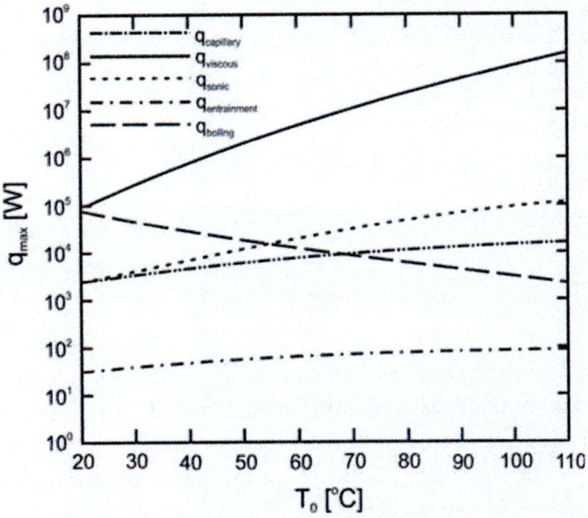

Fig. 6.10 Operational limits as a function of temperature

External Resistances

As the heat flux in the evaporator is known, the external heat source thermal resistance is considered null, i.e.:

$$R_{ex,hs} = 0$$

The cold sink resistance in the condenser region is calculated considering the cross-flow of air over a cylinder, for cooling air properties at *25 °C*. The Reynolds number for the air is:

$$Re_D = \frac{\rho_{air} \cdot V_{air} \cdot D_{ex}}{\mu_{air}} \cong 8594$$

Equation 6.1 is used to determine the Nusselt number for the external flow, resulting in: $\overline{Nu}_D = 49.70$ and, therefore, the external coefficient of heat transfer over the condenser area is $\overline{h}_{ex,c} = 53.16\, W/m^2 \cdot K$. The external cold sink condenser thermal resistance can be calculated by Eq. 4.3, resulting in:

$$R_{ex,cs} = 1.025\, K/W$$

Conduction resistances.

The radial conduction thermal resistance in the evaporator is calculated using Eq. 4.4, resulting in:

$$R_{cond,e} = \frac{\ln\left(\frac{r_{ex}}{r_i}\right)}{2\pi k l_e} = 3.023 \cdot 10^{-4}\, K/W$$

The radial conduction thermal resistance in the condenser is:

$$R_{cond,c} = \frac{\ln\left(\frac{r_{ex}}{r_i}\right)}{2\pi k l_c} = 2.366 \cdot 10^{-4}\, K/W$$

The axial conduction thermal resistance is:

$$R_{cond,ax} = \frac{l_{ef}}{k\pi\left[(r_{ex})^2 - (r_i)^2\right]} = 5.412 \frac{K}{W}$$

As the conduction resistances are associated in parallel and the order of magnitude of the axial resistance is much larger than that of the radial resistance, the axial resistance can be removed from the circuit.

The resistances concerning the wick structure are composed of radial and axial paths. These are calculated by Eq. 4.4 and 4.5, using the effective thermal conductivity of the wick and the wick geometry parameters. Therefore, the effective thermal conductivity, according to Table 3.2 for superposed wire screens, is:

$$k_{ef} = \frac{k_l[(k_l + k_w) - (1 - \varepsilon)(k_l - k_w)]}{(k_l + k_w) + (1 - \varepsilon)(k_l - k_w)} = 1.352\frac{W}{m \cdot K}$$

The wick conduction thermal resistance in the evaporator is

$$R_{w,e} = \frac{\ln\left(\frac{r_i}{r_v}\right)}{2\pi l_e k_{ef}} = 0.04215\frac{K}{W}$$

while the wick thermal resistance in the condenser is:

$$R_{w,c} = \frac{\ln\left(\frac{r_i}{r_v}\right)}{2\pi l_c k_{ef}} = 0.03299\frac{K}{W}$$

Finally, the axial wick resistance is evaluated as:

$$R_{w,ax} = \frac{L_{ef}}{k_{ef}\pi\left[(r_i)^2 - (r_v)^2\right]} = 4017\frac{K}{W}$$

As the axial conduction resistance is associated in parallel with the radial resistances and considering that the order of magnitude of axial is much larger than the radial resistance, the axial resistance can be removed from the circuit. Besides, as the liquid–vapor and vapor–liquid phase change thermal resistances are very small and so are the vapor thermal resistances, these resistances can be neglected, considering that they are associated in series with the conduction resistances (see Fig. 4.3).

Overall thermal resistances.

Once all resistances of the circuit of Fig. 4.3 are determined, the overall thermal resistance in steady state conditions of the heat pipe under analysis can be determined:

$$R_o = R_{\text{ex,hs}} + R_{\text{cond,e}} + R_{\text{w,e}} + R_{\text{w,c}} + R_{\text{cond,c}} + R_{\text{ex,cs}} = 1.10\frac{K}{W}$$

If the external wall temperatures in the evaporator and condenser regions are known, the external thermal resistances can be removed from the thermal circuit, resulting in the following simplified equation for the internal thermal resistance:

$$R_i = R_{\text{cond,e}} + R_{\text{w,e}} + R_{\text{w,c}} + R_{\text{cond,c}} = 0.076\frac{K}{W}$$

These last results show that the external thermal resistance is larger than the combination of all heat pipe internal resistances. This means that, in actual applications, the designer must be more concerned in providing efficient means of the heat to reach and exit the heat pipe than in the thermal performance of the device itself, if the heat pipe transported heat is below the heat transfer limitations.

6.4 Ethanol–Copper Heat Pipe

An ethanol-copper heat pipe, which characteristics are shown in Table 6.8, operates at horizontal position and is constructed from a copper tube of 22.0 mm of external diameter, inner diameter of 20.2 and 0.50 m of total length. The evaporator, adiabatic section, and condenser lengths are 0.25, 0.00 and 0.25 m, respectively. The wick structure consists of ten ($n = 10$) layers of copper screen mesh $N = 160$ [1/inch], inserted against the tube inner surface, which wires have diameter of $D_w = 160\mu m$ and a wire spacing of $263\mu m$. The operating temperature is of $T_0 = 70°C$.. The heat delivery and removal are the same of the previous case.

Thermal Resistances

The same procedure used for the determination of the thermal resistances of the water-copper heat pipe case is also applied in the ethanol-copper case, with the difference that the thermophysical properties of the fluid and the geometry characteristics, of both the heat pipe casing and wick, are different. As the heat input and output mechanisms are the same of the water-copper heat pipe, all the external thermal resistances are the same.

Casing Conduction Resistances

Due to the geometry of the present case heat pipe (equal evaporator and condenser lengths), the case conduction radial thermal resistances, in the evaporator and condenser, are the same. The use of Eq. 4.4, with the appropriate input parameters, results in:

Table. 6.8 Characteristics of ethanol-copper heat pipe

Ethanol-copper heat pipe characteristics	
Inner diameter [mm]	20.2
Outer diameter [mm]	22.0
Evaporator length [m]	0.25
Adiabatic section length [m]	0.00
Condenser length [m]	0.25
Total length [m]	0.50
Working fluid	ethanol
Operation position	horizontal
Wall material	copper
Wick	Copper wire screen – 60 mesh
Wire diameter [μm]	160
Wire spacing [μm]	263
Number of layers	10
Crimping factor	1.1
Operation temperature	$T_0 = 70$ °C

$$R_{\text{cond,e}} = R_{\text{cond},c} = 2.177 \cdot 10^{-4}\, K/W$$

Wick conduction resistance.

The vapor core diameter is obtained from Eq. 6.37, i.e., $D_v = 0.00834m$. The porosity is determined by Eq. 3.20 ($\varepsilon = 0.6735$) and the effective thermal conductivity is calculated according to the expression of Table 3.2, resulting in $k_{ef} = 0.308\,\text{W/m}\cdot\text{K}$.

Therefore, the radial wick thermal resistance, in both evaporator and condenser, is:

$$R_{w,e} = R_{w,c} = 0.7889\, K/W$$

As for the water-copper heat pipe, inserting all the resistances in the thermal circuit and removing the negligible resistances, the overall thermal resistance for the heat pipe working in steady state conditions is:

$$R_{in} = R_{\text{cond,e}} + R_{\text{w,e}} + R_{\text{w,c}} + R_{\text{cond,c}} = 1.578\, K/W$$

Operational Limits

Using the same procedure described in the last case, the plots of the operational limits as a function of operation temperature are shown in Fig. 6.11, where the capillary limit is the most restrictive, for all the temperature range.

It is important to note that the number of layers in this heat pipe is elevated (10 layers) in the attempt to improve the thickness of the wick and so the volume of the working fluid, as the ethanol thermophysical properties are not so good as other fluids' as water, for instance. An interesting discussion of the wick geometry as a function of the fluid properties can be found in Chap. 7, for LHPs or CPLs.

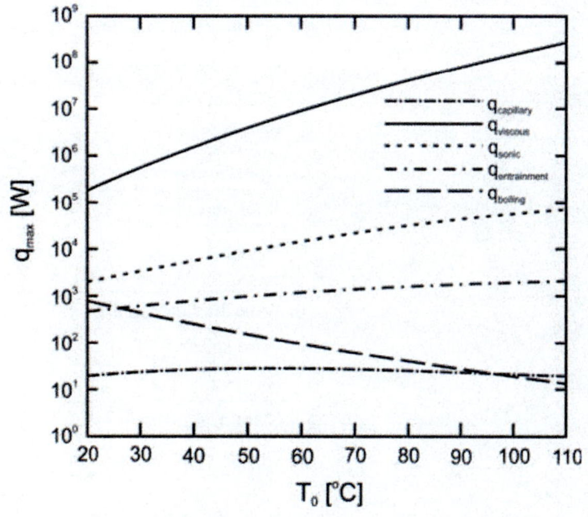

Fig. 6.11 Operational limits for ethanol-copper heat pipe

6.5 Elliptical Cross Section Water-Cooper Heat Pipe

In many applications such as for cooling of electronic components in laptops, elliptical cross section heat pipes are used. They are constructed from cylindrical tubes internally covered with a wick structure, which are partially smashed to reach the desired elliptical geometry.

In this section, an elliptical water-copper heat pipe is considered. The geometry parameters and operation temperatures are depicted in Table 6.9. Figure 6.12 shows a schematic of this heat pipe, which is, actually, very similar to the cylindrical heat pipe considered in Sect. 6.2. At the end of the present section, their performances are compared.

Thermal Resistances

According to Sarmiento et al. (2020), the hydraulic diameter of a tube of irregular cross section area can be assumed as equal to the square root of the cross section area. Using this concept, the "equivalent internal elliptical tube diameter", corresponding to a circular cross section tube, is:

$$D_i = \sqrt{A_{el}} = \sqrt{\pi \left(\frac{D_{i,x}}{2}\right)\left(\frac{D_{i,y}}{2}\right)} \tag{6.51}$$

Table. 6.9 Characteristics of the elliptical cross section heat pipe

Elliptical water-copper heat pipe characteristics	
Major internal cross section axis [mm] $-D_{i,x}$	25.35
Minor cross internal section axis [mm]- $D_{i,y}$	17.9
Wall thickness [mm] $-t_w$	1.59
Evaporator length [m]	0.18
Adiabatic section length [m]	0.09
Condenser length [m]	0.18
Total length [m]	0.45
Working fluid	deionized water
Operation position	horizontal
Wall material	copper
Wick	Copper wire screen – 160 mesh
Wire diameter [μm]	63
Wire spacing [μm]	95
Number of layers	5
Crimping factor	1.1
Operation temperature	$T_0 = 70$ °C

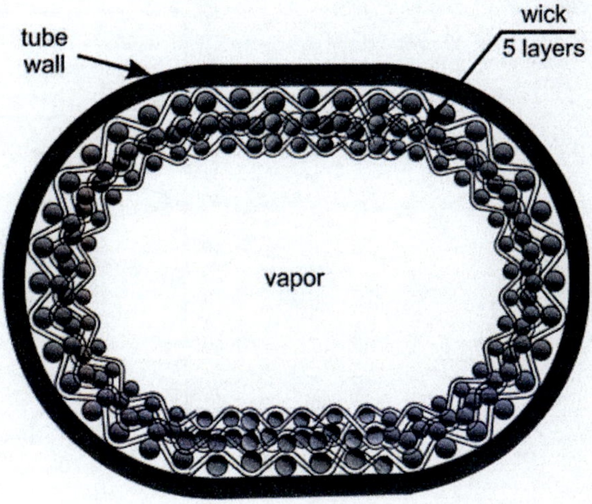

Fig. 6.12 Elliptical cross section water-copper heat pipe

and:

$$D_{ext} = \sqrt{\pi\left(\frac{D_{i,x}+t_w}{2}\right)\left(\frac{D_{i,y}+t_w}{2}\right)} \tag{6.52}$$

The vapor area is determined by:

$$A_v = \pi \cdot \frac{D_{i,x} - 2(2 \cdot S \cdot D_w \cdot n)}{2} \cdot \frac{D_{i,y} - 2(2 \cdot S \cdot D_w \cdot n)}{2} \tag{6.53}$$

Substituting the appropriate values, one has: $D_v = 0.01763\,m$ and $A_w = (A_i - A_v)\varepsilon = 2.999 \cdot 10^{-5} m^2$. The porosity, permeability and effective thermal conductivity are calculated in Sect. 6.3, as the wick characteristics are exactly the same, i.e.: $\varepsilon = 0.6571$, $K = 7.854 \cdot 10^{-11} m^2$ and $k_{ef} = 1.352\,W/m \cdot K$.

These equivalent diameters and data are used for the determination of conduction thermal resistances, resulting in (the evaporator and condenser lengths are equal):

$$R_{\text{cond,e}} = R_{\text{cond},c} = 1.640 \cdot 10^{-4}\,K/W$$

and:

$$R_{w,e} = R_{w,c} = 0.04476\,K/W$$

As for the previous examples, the resistances $R_{pc,lv}$, $R_{pc,vl}$, R_v, $R_{e,ax}$, $R_{c,ax}$, $R_{w,ax}$ can be removed from the thermal circuit. Inserting the remaining important resistances in the equivalent thermal circuit (see Fig. 4.3), the overall thermal resistance is obtained:

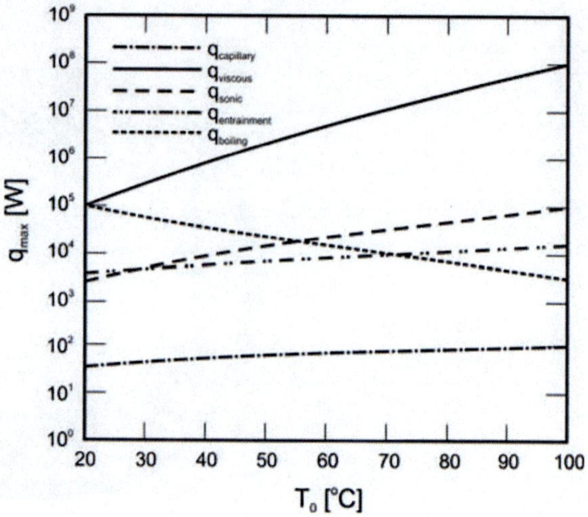

Fig. 6.13 Operational limits for the elliptical cross section ethanol-copper heat pipe

$$R_{in} = R_{\text{cond,e}} + R_{\text{w,e}} + R_{\text{w,c}} + R_{\text{cond,c}} = 0.090\, K/W$$

Operational Limits

The heat transfer limits for this heat pipe are determined following the procedure adopted in the previous examples, resulting in the graphic shown in Fig. 6.13. The capillary limit is again the most restrictive limitation.

Comparing the operational limits of the oval heat pipe with the cylindrical heat pipe (Sect. 6.3), a small reduction of all the limits values is observed, due to the heat pipe geometry. However, this difference is small, showing that the elliptical cross section heat pipe presents a behavior similar to the circular one.

6.6 Water–Copper Grooved Heat Pipe

A water-copper grooved heat pipe, which characteristics are shown in Table 6.10, operates at horizontal position and is constructed from a copper grooved tube of 19.9 mm of external diameter, inner diameter of 12.7 and 0.50 m of total length. The grooves have the geometry shown in Fig. 6.14, i.e., groove width of 0.8 mm, groove diameter of 1.14 mm and pith between grooves of 1.60 mm. The number of grooves is 25 and the opening angle is 74.8°. The operating temperature is of $T_0 = 70°C$.

Operational Limits

Capillary Limit

The capillary limit is determined using Eq. 3.100.

As suggested by Furmetz (2012), in the present case analysis, the working fluid is considered able to wet the grooves differently in the evaporator and condenser

Table. 6.10 Characteristics of water-copper grooved heat pipe

Circular grooved water-copper heat pipe characteristics	
Inner diameter [mm]	12.70
Outer diameter [mm]	19.90
Evaporator length [m]	0.10
Adiabatic section length [m]	0.10
Condenser length [m]	0.30
Total length [m]	0.50
Working fluid	deionized water
Operation position	horizontal
Wall material	copper
Wick	axial grooves
Groove width [mm], w	0.8
Groove diameter, D_g	1.14
Pitch between grooves [mm], s	1.60
Opening angle [°]	74.8
Number of grooves n	25
Operation temperature	$T_0 = 70$ °C

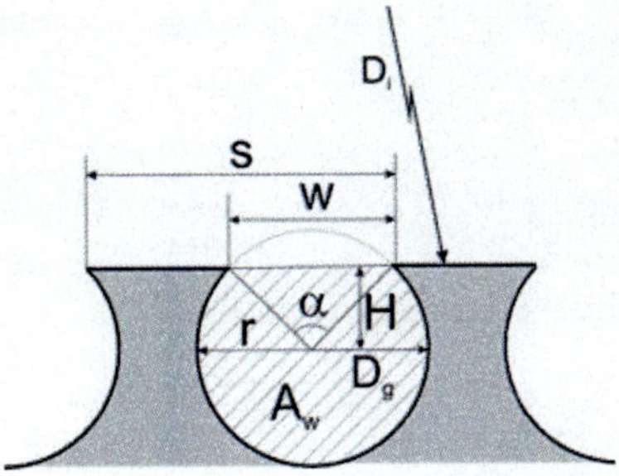

Fig. 6.14 Geometry of the circular grooved water-copper heat pipe

regions. In the evaporator, the liquid intrudes and, in the condenser, it overflows the grooves. The heat transfer mechanisms for liquid–vapor and vapor–liquid phase changes in the evaporator and in condenser, respectively, are deeply affected by the volume and liquid spreading in the region. Figure 6.15 illustrates these hypothesis and the heat transfer paths.

The vapor diameter is an input parameter for the determination of the vapor pressure drops. It varies along the heat pipe length. The liquid height in the groove, in the evaporator region, is at a height in which $H = 0$ and, at the condenser, at a height H (see Fig. 6.15). Therefore, the mean vapor diameter is approximated by the following expression:

$$\overline{D_v} = D_i + \frac{H}{2} = 0.01293\,m \tag{6.54}$$

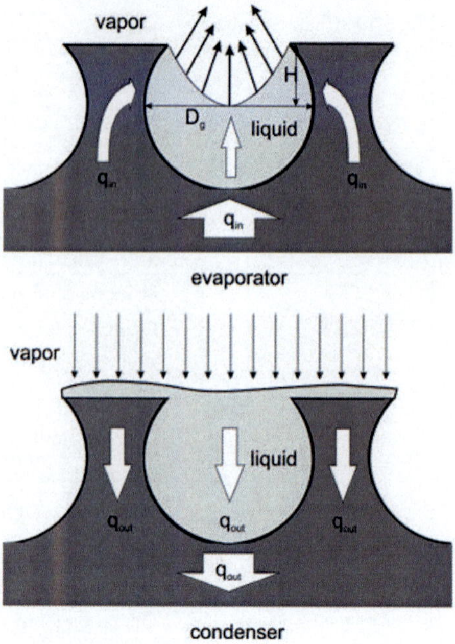

Fig. 6.15 Liquid distribution over the groove in the evaporator and condenser regions

where H is the height of the liquid in the groove:

$$H = \frac{D_g}{2}\cos\left(\frac{\alpha}{2}\right) = 0.0004528\,m \tag{6.55}$$

According to Table 3.2, the porosity of tubes with circular grooves is calculated by the expression:

$$\varepsilon = \frac{D_g[\alpha/2 - 1/2sin(\alpha)]}{2s[1 - cos(\alpha/2)]} = 0.295$$

The permeability is determined using the expression in Table 3.1 for circular grooves heat pipes:

$$K = 0.0221\frac{D_g^{2.2}(2\pi - \alpha + sen(\alpha))^{2.1}}{w^{0.2}(2\pi - \alpha)^2} = 5.25 \cdot 10^{-8}m^2$$

Assuming: steady state regime, fluid wets perfectly the wall ($\theta = 0$) and the "wet point" is located at the end of the condenser, the maximum capillary pressure is given by Eq. 3.3. The effective radius for grooves can be assumed as the groove width, $r_{ef} = w = 8 \cdot 10^{-4}m$, resulting in: $\Delta p_{cap} = 161.2\,Pa$.

The pressure drop of the vapor phase for a total kinetic pressure recuperation and a laminar flow is, according to Eq. 3.42, $\Delta p_v = 0.01065\,q$.

Using Darcy's model, (Eq. 3.29), the pressure drops of the liquid can be calculated. The cross section area of the liquid flow needs to be determined. For the sake of safety, the whole circular area is considered in this calculation, so that:

$$A_w = n\left[\frac{\pi D_g^2}{4} - \frac{\alpha D_g^2}{8} + \frac{D_g^2 \sin\alpha}{8}\right] = 24.13 \cdot 10^{-6} m^2 \tag{6.56}$$

Therefore, the liquid pressure drop in the grooves is $\Delta p_l = 0.0419\,q$.

The pressure drop due to gravity for the device operating in horizontal position is $\Delta p_g = \rho_l d D_v sin\theta = 123.9\ Pa$.

All these pressure drops when included in Eq. 3.100, results in:

$$161.2 = (0.01065 + 0.0419)q + 123.9$$

and, therefore, the maximum power that the heat pipe is able to transfer before the capillary limit is reached is: $q_{max,c} = 708.7W$. This value is obtained for laminar incompressible flow for both vapor and liquid. This hypothesis is quite obvious for the liquid, but, for validating the vapor flow model, the vapor flow Reynolds and Mach number must be checked, resulting in:

$$Re = \frac{\rho_v v_v D_v}{\mu_v} = \frac{4q_{max,c}}{\pi D_v \mu_v h_{lv}} = 2657$$

And

$$Ma = \frac{u}{c} = \frac{4\dot{m}/\rho_v \pi D_v^2}{\sqrt{\gamma R T_v}} = \frac{4q_{max,c}/\rho_v \pi h_{lv} D_v^2}{\sqrt{\frac{c_{p,v}}{cu_v} R T_v}} = 0.025$$

The flow is uncompressible (Ma< <0.3) but the vapor flow regime showed to be turbulent (Re > 2300). Therefore, the pressure drop turbulent model for vapor flow must be used (Eq. 3.53):

$$\Delta p_v = \left(1 - \frac{4}{\pi^2}\right)\frac{\dot{m}_v^2}{8\rho_v r_v^4} + \frac{8\mu_v \dot{m}_v}{\rho_v \pi r_v^4} l_a = 0.00003944\,q^2 + 0.00355\,q$$

With this new expression for the power transferred, Eq. 3.100 can be used for determining a new value for the capillary limit, resulting in: $q_{max,c} = 553.5W$.

Other Limits

The **viscous limit** is calculated using Eq. 3.62, resulting in $q_{max,v} = 1.465 \cdot 10^6\ W$. The **sonic limit** can be determined by Eq. 3.80 as $q_{max,s} = 11.419\ KW$.

The maximum heat power due to the entrainment limit can be determinate by Eq. 3.105, which gives $q_{max,e} = 865.8\ W$.

The **boiling limit** is determined according to Eq. 3.113. For that, the effective thermal conductivity parameter must be calculated. For simplicity, the grooves are considered completely filled. The expression given in Table 3.2, for circular groove wick structure, which is actually an adaptation and combination of the porous media and liquid models in series and in parallel, is used. In this expression, the diameter of the groove is substituted by the height of the liquid in the groove $\left(D_g/2 + H\right)$, resulting in:

$$k_{ef} = \varepsilon k_l + (1-\varepsilon)\left(\frac{k_l k_w \left(D_g/2 + H\right)}{0.185(k_w)(s-w) + k_l\left(D_g/2 + H\right)}\right) = 0.5172\frac{W}{m \cdot K} \tag{6.57}$$

Besides, assuming $r_n = 2.54\ 10^{-7}\,m$, the maximum heat transfer, considering the boiling limitation is $q_{max,b} = 6.918\ KW$.

Figure 6.16 shows the graphic of these operational limits as a function of the temperature. Comparing all the limits, the capillary limit, although high, is again the strictest one, followed by the entrainment limit.

Thermal Resistances

The wall temperatures are considered known and so only the internal thermal resistances of the grooved water-copper heat pipe are determined. Neglecting the presence of circular grooves, the radial thermal resistances are determined considering a bulk solid tube. Therefore, the radial conduction resistances in the evaporator and condenser are $R_{cond,e} = 0.001821\ K/W$ and $R_{cond,c} = 0.000607\ K/W$, respectively.

Due to the different liquid distribution over the grooves (see Fig. 6.15), the thermal conductivity mechanism is different in the evaporator and condenser. Therefore, to

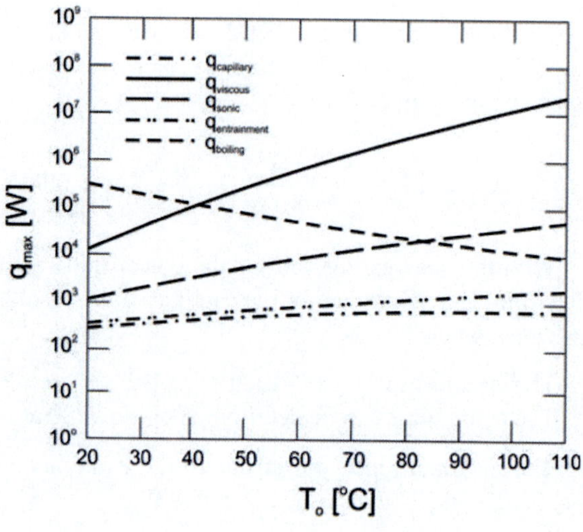

Fig. 6.16 Operational limits for the grooved water-copper heat pipe

determine the wick conduction thermal resistance, two different effective thermal conductivities are calculated.

For the evaporator, the same value, obtained from the expression used for the boiling limit (see Eq. 6.41) is used $\left(k_{ef,e} = 0.5172\,W/m \cdot K\right)$.

The effective thermal resistance for the wick structure in the condenser is determined using the model that considers the porous media and liquid in parallel (see Table 3.2), resulting in:

$$k_{ef,c} = \varepsilon(k_l) + (1 - \varepsilon)(k_w) = 277.9\frac{W}{m \cdot K}$$

Therefore, the wick conduction thermal resistances in the evaporator and condenser are, respectively, $R_{w,e} = 0.05438\,K/W$ and $R_{w,c} = 3.37 \cdot 10^{-5}\,K/W$.

The $R_{pc.lv}$, $R_{pc.vk}$, R_v, $R_{e,ax}$, $R_{c,ax}$, $R_{w,ax}$ are neglected. Thus, the overall internal thermal resistance is:

$$R_{in} = R_{cond,e} + R_{w,e} + R_{w,c} + R_{cond,c} = 0.0568\frac{W}{K}$$

This heat pipe was constructed and tested, in an apparatus similar to the one described in the schematic of Fig. 6.3. Figure 6.17 shows the comparison between data and theoretical results. The vertical bars represent the experimental uncertainty range. Both model and data show that the thermal resistances of the device are quite small and that they do not change much with the power input increase. The model is able to follow the data trends although the relative difference to the data can be large. This difference can be attributed, among other factors, to the fact that the model is one-dimensional while the conduction through the tube material is highly two-dimensional as the groove geometry is very complex.

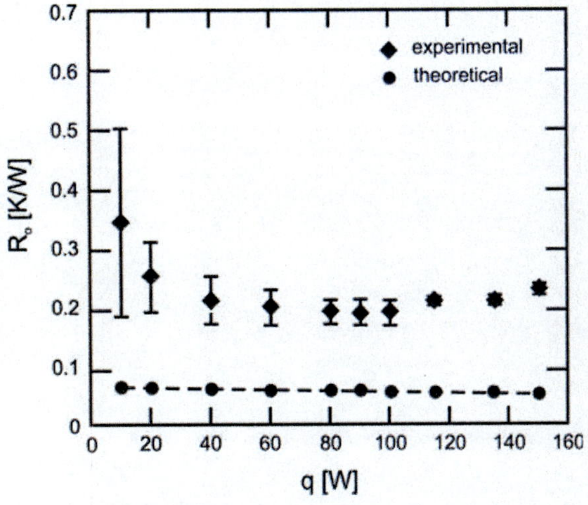

Fig. 6.17 Theoretical and experimental thermal resistances as a function of the power input for the water-copper heat pipe

6.7 Conclusion

The main goal of the present chapter is to offer to the reader actual design examples of thermosyphons and heat pipes, with the use of several models presented along the Chaps. 2 to 4. The use of these models may seem obvious and simple, but several questions may arise to the engineer during designing. Therefore, these examples may work as a guide and offer to the designer confidence in the hypotheses selection and in the correct adoption of procedures in the design of these two-phase devices.

References

ESDU, Heat Pipes - Performance of Two-Phase Closed Thermosyphons. Engineering Sciences Data Unit 81038, London (1981)

El-Genk, M.S., Saber, H.H.: Heat Transfer Correlations for Small, Uniformly Heat Liquid Pools. Int. Journal Heat Mass Transfer **41**(2), 261–274 (1998)

El-Genk, M. S. and Saber, H. H., Thermal Conductance of Evaporator Section of Closed Two-Phase Thermosyphons (CTPTs), 7th AIAA/ASME Joint Thermophysics and Heat Transfer Conference, **3**, 99–106, (1998b)

El-Genk, M. S. and Saber, H. H., Operation Envelope for Closed, Two-Phase Thermosyphons, 10th International Heat Pipe Conference, Stuttgart, Germany (1997)

Furmetz, M., Design, development and verification of the eROSITA thermal control system, Ph.D Theses, Technische Universität München (2012)

Groll, M., Rosler, S.: Operation Principles and Performance of Heat Pipes and Closed Two-Phase Thermosyphons. Journal of Non-Equilibrium Thermodynamic **17**, 91–151 (1992)

Incropera, F. P., DeWitt, D. P. Fundamentos de Transferência de Calor e de Massa. Rio de Janeiro, LTC (2008)

Sarmiento, A.P.C., Soares, V.H.T., Milanese, F.H., Mantelli, M.B.H.: Heat transfer correlation for circular and non-circular ducts in the transition regime. Int. Journal Heat Mass Transfer **149**, 119165 (2020)

Part II
Special Devices

Chapter 7
Classification According to Operational Principles

The description, operation and modelling of some special thermosyphon and heat pipe devices, designed for specific operation ranges and/or applications, is presented. In this chapter, the thermosyphons and heat pipes are classified according to their operational principles (capillary pumped devices, pulsating heat pipes, hybrid devices, among others). Design tools, models and some experimental results are presented.

Two-phase devices depend on the working fluid phase change to operate. The return of liquid from the condenser to the evaporator may be performed due: to the gravity action, such as in thermosyphons, to the capillary pressures provided by porous media, as in heat pipes, to the liquid pumping effect of expanding vapor bubbles in narrow tubes, as in pulsating heat pipes, etc. Besides, the evaporator and condenser may share a common vapor channel, such as in thermosyphons or heat pipes or, in the case of loops, the liquid and vapor may flow through different tubes.

7.1 Loop Thermosyphons—LTS

Loop thermosyphons (LTS), also known as separated thermosyphons, are characterized by independent evaporators and condensers, connected to each other by tubes where only vapor or liquid circulates, as illustrated in Figs. 1.3 and 3.18. Due to this flexible configuration, heat can be transported along large distances passively, without the need of extra energy.

Loop thermosyphons can be large, as in some industrial applications where the heat source and the heat sink are far apart. Loop thermosyphons can also be quite small, as those used for electric/electronic cooling.

One of the greatest advantages of LTS is that the evaporator and condenser can be designed separately. In this section, models for designing the evaporator and condenser are presented in separated subsections.

M. B. H. Mantelli et al., *Thermosyphons and Heat Pipes: Theory and Applications*,
https://doi.org/10.1007/978-3-030-62773-7_7

LTS with Flat Plate Evaporator

The flat plate evaporator for LTS described in this section is designed for temperature management of electric/electronic equipment in avionics, although it can be used in other applications. A schematic of this device, proposed by Simas et al. (2019) is shown in Fig. 3.29. A tridimensional model, able to predict the temperature distribution and the thermal resistance of a LTS with a flat evaporator, which takes into account the concentration and position of the heat source over the external evaporator wall, is presented. A hydraulic model for the pressure drops of the working fluid flowing along the thermosyphon is also shown.

Simas et al. (2020) proposed a LTS which condenser is quite simple, composed by a helical heat exchanger. However, the evaporator has the shape of a flat thin box, designed to be accommodated inside racks containing electronic boxes in airplanes. As discussed in Loop Thermosyphon Section of Chap. 3, these authors observed the presence of the Geyser Boiling in several operating conditions, for either distributed or concentrated thermal loads applied in different positions in the evaporator and for different filling ratios (liquid volume from 40 to 60% of the evaporator capacity). Geyser Boiling was eliminated by the application of a wick structure inside the flat closing surfaces of the evaporator. Considering that for avionics and other applications Geyser Boiling is not desirable, the model described is regarded to the evaporator assisted with the wick structure, i.e., without the presence of this phenomenon.

The evaporator of the loop thermosyphon considered is fabricated from ten square stacked copper plates, with 3.18 mm of thickness, which are stacked and diffusion bonded. Each internal plate is manufactured by water cutting process. The closing external plates do not have any cuts, enclosing the inner volumes. To eliminate Geyser Boiling, the flat evaporator holds a wick structure composed by five sheets of a copper metal screen, spot welded on the inner surface of each flat closing plates. This porous medium is also responsible for the spreading of the working fluid along the evaporator, i,e,, the fluid is able to reach any part of the evaporator, regardless of the heat source position or device inclination. Figure 7.1 shows these plates. The stacked plates form four vertical chambers, to be filled with working fluid, and one horizontal channel, that feeds the vertical chambers. To fill the chambers with approximately the same amount of fluid, the filling channel has different cross section areas along the evaporator width. This filling channel geometry was designed based on a numerical simulation. The fins that separate the chambers also have the objective of improving the mechanical strength of the evaporator. As shown in Fig. 3.29, the liquid line reaches the evaporator from the left down corner and the vapor leaves the evaporator from the upper edge in the central position.

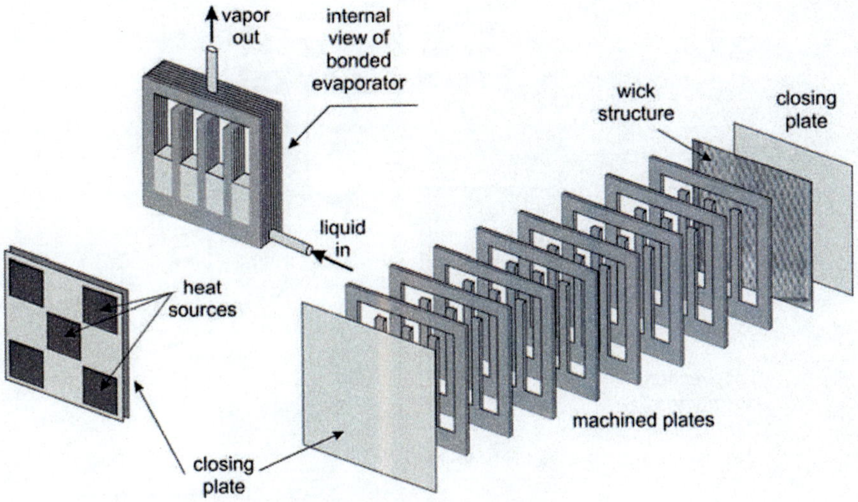

Fig. 7.1 Exploded view LTS flat evaporator with heat sources

Flat Plate Evaporator Thermal Model

For the thermal model, the following hypothesis are adopted:

- Steady-state regime;
- Liquid and vapor laminar incompressible fluid flows;
- The liquid and vapor tube walls are adiabatic;
- Negligible radiation heat transfer along the device;
- Uniform heat over the evaporator wall;
- Working fluid is a pure substance;
- Nucleate pool boiling regime;
- Film condensation;
- Negligible presence of non-condensable gases;
- Negligible contact resistances;
- Adiabatic beans (fin) tips;
- Wick structure spreads the liquid along the internal surface of the closing plate;
- The role of the wick structure in the heat transfer is negligible (zero thermal resistance).

The analogy between electrical and thermal circuits is used for modelling the evaporator of the LTS. The thermal circuit proposed is composed of thermal resistances combined in series, as shown in Fig. 7.2. Each thermal resistance is individually calculated, so that the temperatures of several positions along the device can be estimated. Of especial importance is the determination of the external evaporator surface temperature, which can be a function of the position along the plate. From the thermal circuit of Fig. 7.2, the external evaporator temperature can be estimated by:

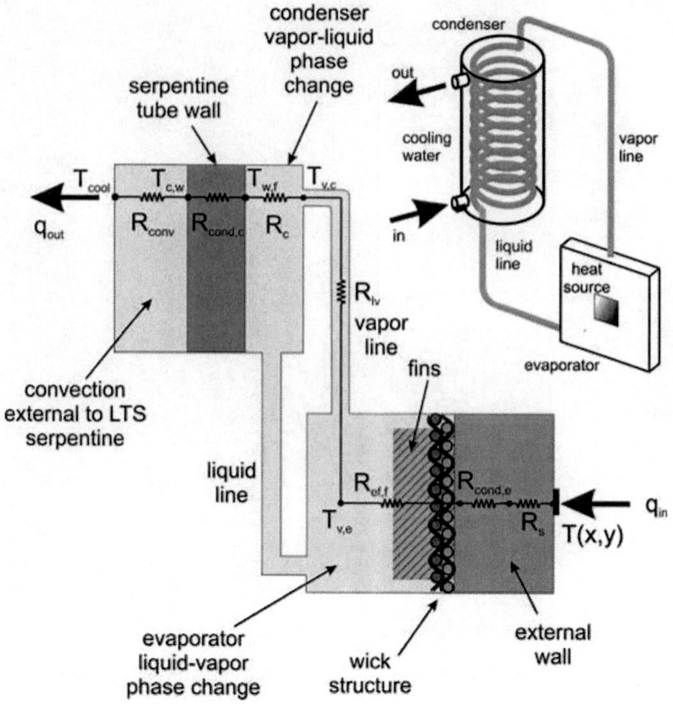

Fig. 7.2 Schematic of the LTS thermal circuit

$$T(x, y) = T_{cs} + q_{in}\left(R_{conv} + R_{cond,c} + R_c + R_{vl} + R_{ef,f} + R_{cond,e} + R_s\right) \quad (7.1)$$

where T_{cs} is the heat sink (cold source) temperature, q_{in} is the input heat power, R represents the thermal resistances, while the sub index "*conv*" stands for convection, "*cond,c*" and "*cond,e*" stand for conduction on the condenser and evaporator walls, respectively, "*c*" for condensation, "*vl*" for vapor line, "*ef*" refers to effective thermal resistance of the set of fins (including wick structure located at its basis) and "*s*" represents the spreading resistance. Besides, in Fig. 7.2, the temperature sub-index "*cw*" is related to the interface between the condenser wall and the cooler (heat exchanger) fluid, "*wf*" represents the interface between the wall and loop thermosyphon working fluid, "*c*" and "*e*" for the working fluid in the condenser and evaporator regions, respectively, and "*w*" to the evaporator wall in the base of the fins. Neglecting any heat losses in vapor and liquid lines and considering steady state conditions, the heat absorbed by the evaporator is assumed to be equal to that delivered by the condenser ($q_{in} = q_{out} = q$).

Following the thermal circuit in the counter heat flow direction, the first thermal resistance is regarded to the heat convection within the condenser heat exchanger, i.e.:

$$R_{conv} = \frac{1}{h_{ex} A_{ex,c}} \tag{7.2}$$

where $A_{ex,e}$ is the tube external area of heat exchange in the condenser region and where h_{ex} is the external convection coefficient of heat transfer. The next resistance, due to the conduction through the condenser wall, is estimated by:

$$R_{cond,c} = \frac{t_c}{k_e A_{ex,c}} \tag{7.3}$$

where t_c is the condenser wall thickness and k_e is thermal conductivity material of the wall. This same expression is used for the determination of $R_{cond,e}$, with the use of the evaporator thickness t_e and of the evaporator base area A_e.

The resistance associated to the working fluid vapor-liquid phase change, which can be estimated by an expression similar to Eq. 7.2, with the use of an appropriated condensation phase change coefficient $h_{c,}$ can be obtained by several literature correlations, such as those of Groll and Rosler (1992) and Kaminaga et al. (1997), depending on the design of the cooler that removes the heat from the condenser.

As discussed by Simas et al. (2020), in order to estimate the thermal resistance in the vapor line, R_{vl}, a simplified thermodynamic analysis is employed. Considering phase change of a fluid in ideal conditions, the Clausius-Clapeyron equation defines the slope of the vapor pressure curve with temperature, i.e.:

$$\left.\frac{dp}{dT}\right|_{T_{sat}} = \frac{h_{lv}}{T_{sat}(\upsilon_v - \upsilon_l)} \tag{7.4}$$

where υ_v and υ_l are specific volumes of vapor and liquid respectively. Using the first order approximation for dp/dT, the following expression can be writen:

$$\Delta T = \left(\left.\frac{dp}{dT}\right|_{T_{sat}}\right)^{-1} \Delta p \tag{7.5}$$

Thus,

$$\Delta T_{vl} = \left(\frac{T_{sat}(\upsilon_v - \upsilon_l)}{h_{lv}}\right) \Delta p \tag{7.6}$$

The calculation of the pressure drop along the vapor line deserves a detailed analysis, to be presented in the next section.

A more sophisticated model is necessary to describe the heat transfer in the evaporator core. As observed in Fig. 7.2, the internal working fluid receives heat directly from the wall by the supporting pillars (beams), which actually work as fins, exchanging the (latent) heat, used to promote the liquid-vapor phase change of the working fluid. A schematic of the internal configuration of the evaporator is shown in Fig. 7.3. As for literature models (Incropera 1998) the temperature of the base is

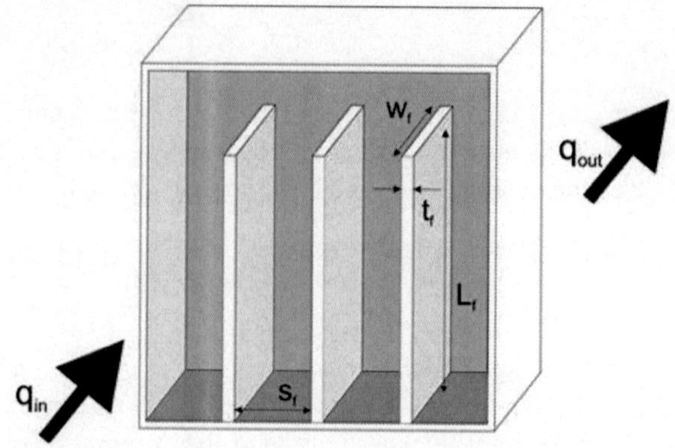

Fig. 7.3 Schematic of the beams (fins)

considered known, while the fin tip is taken as adiabatic. In the present case, a set of four fins (three beams plus the two closing lateral pillars, which, together, correspond to the fourth fin) is considered. The heat transfer coefficient due to the liquid-vapor latent heat can be evaluated using literature correlations for nucleate pool boiling (h_b) such as Foster and Zuber (1955) and Kutatelatze (1952).

According to Incropera and DeWitt [9], the effective thermal resistance of the set of fins can be estimated by:

$$R_{ef} = \frac{1}{\eta_0 h_b A_t} \tag{7.7}$$

where η_o is the efficiency of the set of fins, h_b is the convection heat transfer coefficient and A_t is the total heat exchange area of the complete set, including the fin base surface which is not covered by the fins. The efficiency of the fins can be expressed as:

$$\eta_0 = 1 - \frac{n_f A_f}{A_t}(1 - \eta_f) \tag{7.8}$$

where n_f is the number of fins (four in the present case) and η_f is the efficiency of a single fin, given by:

$$\eta_f = \frac{\sqrt{PA_{tr}kh_b}\tanh\left(\sqrt{\frac{Ph_b}{A_{tr}k}}\right)}{A_f h_b} \tag{7.9}$$

where, as shown in Fig. 7.3, p is the perimeter of the cross section area of a fin $(2t_f + 2w_f)$, A_{trc} is the fin cross sectional (transversal) area $(w_f \cdot t_f)$, A_f is the fin surface $(2w_f \cdot l_f)$, k is the thermal conductivity of the material. From these equations, the effective heat transfer coefficient of the fin set can be given as:

$$h_{ef} = \frac{1}{A_{bp} R_{ef}} \tag{7.10}$$

where A_{bp} is the total area of the base plate, including the area occupied by the fins.

In the present thermal model, the wick layer is thin and porous and so its conductive influence in the heat transfer within the device can be neglected. Actually, the major effect of the wick is to guarantee capillary pumping needed to the working fluid to be spread along the internal surface of the evaporator wall, which receives heat externally, avoiding local dry-outs and providing more nucleation sites, to improve the evaporation of the working fluid.

The next conduction resistance, the spreading resistance R_s, is resulting from the application of concentrated heat over the external evaporator wall. It is included in the thermal circuit, in series with the transversal conduction resistance of the condenser, as shown in Fig. 7.2. Therefore, the overall resistance of the evaporator R_e is composed by the association in series of the spreading (radial) R_s, conduction (transversal) $R_{cond,e}$ and the finned heat sink (R_{ef}) resistances, $R_e = R_{cond,e} + R_{ef} + R_s$. The spreading resistance is tridimensional and can be estimated using the model proposed by Musychka et al. (2003), for the thermal problem which schematic is shown in Fig. 7.4. According to these authors, the temperature difference, due to spreading resistance, for the plane $z = 0$, is given by the expression:

$$T(x, y, 0) - T_{ex} = A_0 + \sum_{m=1}^{\infty} \cos(\lambda_m x) A_m + \sum_{n=1}^{\infty} \cos(\lambda_n y) A_n + \sum_{m=1}^{\infty} \sum_{n=1}^{\infty} \cos(\lambda x) \cos(\lambda y) A_{m,n} \tag{7.11}$$

where T_{ex} is the external temperature. The Fourier coefficients are obtained from the expressions:

$$A_m = \frac{2q\left[\sin\left(\frac{2x_h+c}{2}\lambda_m\right) - \sin\left(\frac{2x_h-c}{2}\lambda_m\right)\right]}{abck\lambda_m^2 \phi(\lambda_m)} \tag{7.12}$$

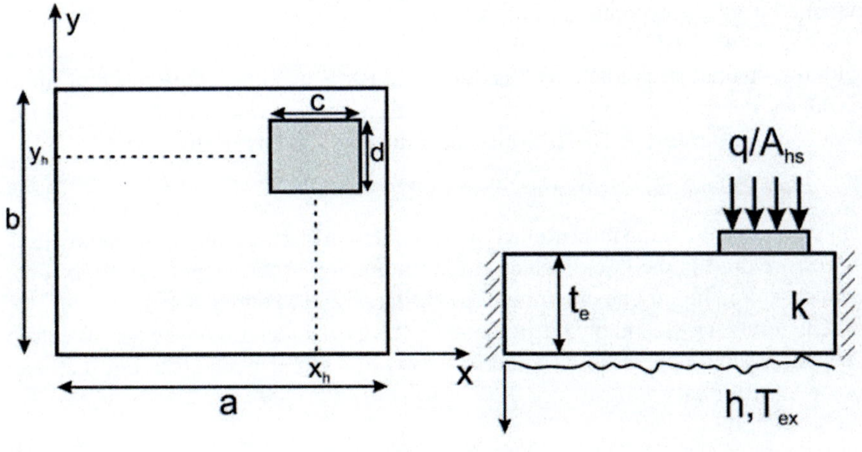

Fig. 7.4 Schematic of the concentrated heat model for the spreading resistance model

$$A_n = \frac{2q\left[\sin\left(\frac{2y_h+d}{2}\zeta_n\right) - \sin\left(\frac{2x_h-c}{2}\zeta_n\right)\right]}{abdk\zeta_n^2\phi(\zeta_n)} \tag{7.13}$$

$$A_{m,n} = \frac{16q\cos(\lambda_m x_h)\sin\left(\frac{1}{2}\lambda_m c\right)\cos(\zeta_n y_h)\sin\left(\frac{1}{2}\zeta_n d\right)}{abcdk\beta_{m,n}\lambda_m\zeta_n\phi\left(\beta_{m,n}\right)} \tag{7.14}$$

and where:

$$\phi(\xi) = \frac{\xi\sinh(\xi t_e) + (h/k)\cosh(\xi t_e)}{\xi\cosh(\xi t_e) + (h/k)\sinh(\xi t_e)} \tag{7.15}$$

where the eigenvalue ξ is accordingly substituted by λ_m, ζ_n or $\beta_{m,n}$, given by the expressions:

$$\lambda_m = \frac{m\pi}{a},\ \zeta_n = \frac{n\pi}{b},\ \beta_{m,n} = \sqrt{\lambda_m^2 + \zeta_n^2} \tag{7.16}$$

Finally, the last coefficient is related to the region of the input heat flux over the heating area, determined by:

$$A_0 = \frac{q}{ab}\left(\frac{t_e}{k} + \frac{1}{h}\right) \tag{7.17}$$

This set of equations, used for the determination of R_s, is quite versatile and can be employed even when a uniformly distributed heat source along the whole plate is imposed.

The thermophysical properties of the working fluid can be obtained from the literature (tables) or they can be estimated by polynomials, where the temperature is the input variable, as proposed by Faghri (2016). Saturation pressure also can be obtained from a polynomial.

Simas et al (2020) compared their data for the experimental thermal resistances with the theoretical predictions, resulting in a RME maximum difference of up to 6.6%, for the heat source positioned in the edge (see Fig. 7.4) and of 4.8%, for the heat source centralized. These results show the good quality of the model.

Flat Plate Evaporator Hydraulic Model

The flow of the working fluid through the several loop thermosyphon elements causes a pressure drop in the liquid, which is the summation of the evaporator, vapor line, condenser and liquid line pressure drops (Milanez and Mantelli 2010).

The following well known expression can be used to determine the pressure drop of laminar single phase flow (liquid or vapor) in a circular pipe (see Eq. 3.29 and Fox et al. 1999):

$$\Delta p = \frac{128\dot{m}_l\mu_l l}{\pi\rho_l D_i^4} \tag{7.18}$$

where l is the length of the vapor or liquid line and D_i its internal diameter. Considering the heat necessary for the liquid-vapor phase change, the mass flow rate is given by: $\dot{m} = q \cdot h_{lv}$.

Actually, a two-phase flow is observed in the evaporator. The liquid pressure drop can be neglected as it is much smaller than the vapor pressure drop and so the pressure drop is estimated from the expression (Singh et al, 2009):

$$\Delta p_e = 32\left(\frac{l_e}{D_h}\right)\left(\frac{\mu_v}{\rho_v h_{lv}}\right)q \tag{7.19}$$

where l_e is the evaporator length and $D_h = 4A/P_w$ is the hydraulic diameter, where P_w is the wet perimeter. This expression was developed for a circular duct. As, in the present case, the evaporator has a prismatic shape, A is the cross section of the prism. The hydraulic losses due to the liquid flow through the present wick structure should be computed and added to ΔP_e. Using the first order approximation of Darcy's law (Eq. 3.12) the following expression results:

$$\Delta p_w = \left(\frac{l_w}{A_w}\right)\left(\frac{\mu_l}{\rho_l h_{lv}}\right)\left(\frac{1}{K}\right)q \tag{7.20}$$

where l_w is the length that the liquid actually travels throughout the wick, A_w is the cross section area of the wick and K is the permeability of the wick formed by metal screens, determined by the use of Eqs. 3.19 and 3.20.

However, the single phase flow simplification (only one phase considered) cannot be applied to the estimation of the condenser pressure drops. Different models can be used to determine two-phase flow pressure drops (see "Pressure Drop Limit in Thermosyhons" in Chap. 3 and "Loop Heat Pipe Modelling" later in this chapter). According to Collier and Thome (1994), the pressure drop in the axial direction, of a two-phase flow, can be estimated by applying the following expression:

$$-\frac{dp}{dz} = \frac{2f^{2\phi}\dot{m}''^2}{\rho_l D_i}\left[1 + X\left(\frac{\upsilon_{lv}}{\upsilon_l}\right)\right] + \frac{g\sin\theta}{(\upsilon_l + X\upsilon_{lv})} + \dot{m}''^2\upsilon_{lv}\frac{dX}{dz} \tag{7.21}$$

where θ is the device inclination angle with the horizontal, $X = \dot{m}_v/\dot{m}$ is the vapor quality and $f^{2\phi}$ is the two phase friction factor, given by:

$$f^{2\phi} = 0.079\left(\frac{\dot{m}'' D_i}{\overline{\mu}}\right)^{-\frac{1}{4}} \tag{7.22}$$

where the equivalent viscosity is $\overline{\mu} = X\mu_v + (1 - X)\mu_l$.

In the determination of the pressure drop, an equivalent length of the tube must be considered, taking into account the elbows and turns of the tube, as proposed by Fox et al. (1999) to a single-phase flow. Due to the lack of more specific models, these equivalent lengths are also applied for two-phase flows.

To determine the pressure drop along the condenser, due to the variation of the vapor quality, its total length must be divided into several regions, in which the pressure drop is determined, considering constant vapor quality for each region.

Flat Loop Thermosyphon (FLTS)

The flat LTS and its cross section are shown in the schematics of Fig. 7.5 (Moreira and Mantelli 2019). The LTS consists basically of three parallel plates that form two vertical cavities: the evaporator (left side) and the condenser (right side). During operation, concentrated or spread heat is applied on the evaporator plate, at the left side, and removed from the condenser plate, at the right side. The wick structure applied over the evaporator internal face has the objective of keeping the whole internal surface covered by a film of liquid, to avoid any dry-out conditions, allowing the heat source to be located at any region of the evaporator. The internal plate is responsible for the organization of the vapor flow within the device. This is basically the same solution used for the flat evaporator, discussed in the last section.

To guarantee mechanical strength for the device, which, depending to the working fluid may be subjected to internal pressures, columns are provided, dividing the vapor volume into channels. Details of the liquid pool in the rear region, the division of the internal volume in parallel channels, the structure columns, the rectangular openings in the upper regions and the resulting vapor paths are also shown in the right side of the figure. Actually, this device can be used as active interface between the interior

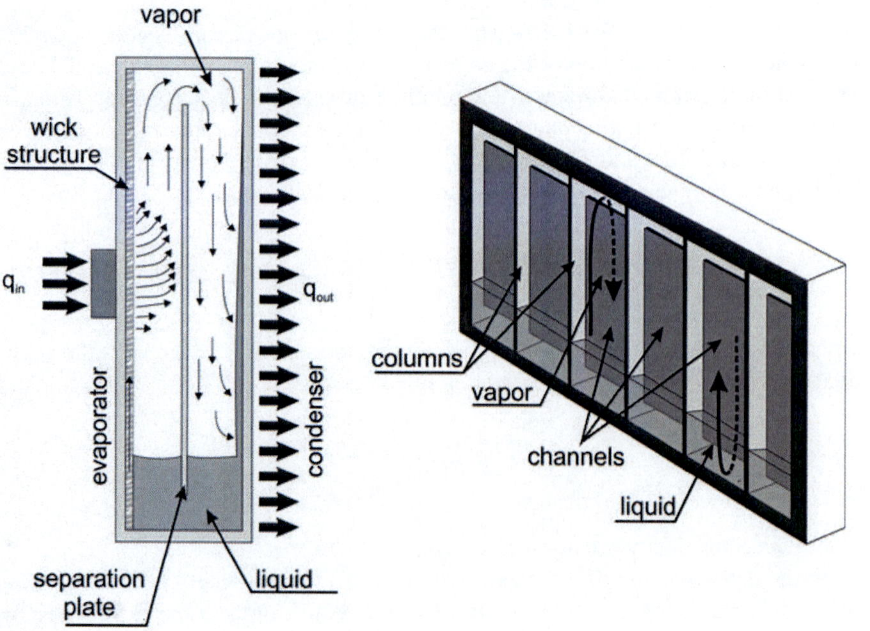

Fig. 7.5 Schematic of the flat LTS, cross section left side and internal view with no evaporator closing plate

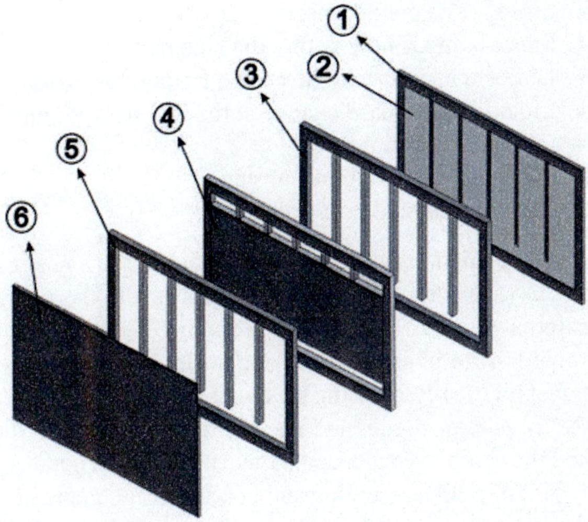

Fig. 7.6 Plates composing the flat LTS

and exterior volumes of electronic cabinet (in avionics, for instance) where the heat generated in the interior region can be easily dissipated to the exterior ambient. Although a wick structure is used, this device is known as thermosyphon as the gravity plays the major role in the heat transfer mechanism.

The flat LTS is a result of piling of nine plates that are individually machined so that the piling generates the internal geometry. The resulting stack is joined by diffusion, as for the last section, which process is described in details in Chap. 5. Figure 7.6 shows these plates, where (1) and (6) are the closing plates; plate (1) has the internal surface covered by a wick structure (2) composed by five layers of a metal screen, united to the plate (1) by spot welding; plates (3) and (5) have the comb shape, generating the empty internal volume occupied by the working fluid in the liquid and vapor phases; the intermediate plate (4) divide the evaporator and condenser sections and has rectangular openings, through which vapor (upper opening) and liquid (rear opening) are able to pass.

For the better performance of the flat LTS, most of the heat must be transferred through working fluid phase change, instead of being transferred by conduction through the solid materials, such as closing plates and columns. Therefore, also to provide fast start-up, the lower the conduction paths, the better is the performance of the flat LTS. The presence of the structure columns must be only to provide the mechanical strength. In this case, low thermal conductance materials, such as stainless steel 304L, can be used for their good mechanical properties and convenient thermophysical properties for the present case.

Flat LTS Thermal Model

According to Moreira and Mantelli (2019), the equivalent thermal resistance circuit is used for the present thermal model. For modelling purposes, the following hypothesis area assumed:

- Steady state conditions;
- Vapor laminar flow within the channels;
- Uniform heat flux in the heated evaporator region;
- Liquid-vapor phase change at the liquid-vapor interface;
- Film condensation;
- Working fluid is a pure substance;
- One dimensional heat transfer.

A schematic of the flat LTS highlighting the heat transfer paths along the device is shown in Fig. 7.7. First, heat crosses the thermosyphon perpendicularly to the external plates of the flat LTS, reaching the liquid stored within the wick structure. Liquid-vapor phase change happens in the meniscus formed in the wick. The vapor travels vertically, crossing the upper window over the separation plate, where it leaves the evaporator region and reaches the condenser. In the condenser, heat is removed and the vapor is condensed. Gravity make the condensate to accumulate at the lower region of both evaporator and condenser, guaranteeing the thermal performance of the device (see Fig. 7.5). As already noted, the flat LTS is divided into channels that actually work as parallel flat LTS. The whole device is composed by parallel cells (six in the present case).

Figure 7.8 shows the overall flat LTS thermal circuit proposed by Moreira and Mantelli (2019), highlighting the association of the unitary cell thermal resistances. In this figure, $R_{ex,e}$ and $R_{ex.c}$ are the external resistances of the evaporator and condenser respectively. The conduction resistance in the evaporator wall has two major components: the transversal one-dimensional conduction thermal resistance $R_{cond,e}$ and the

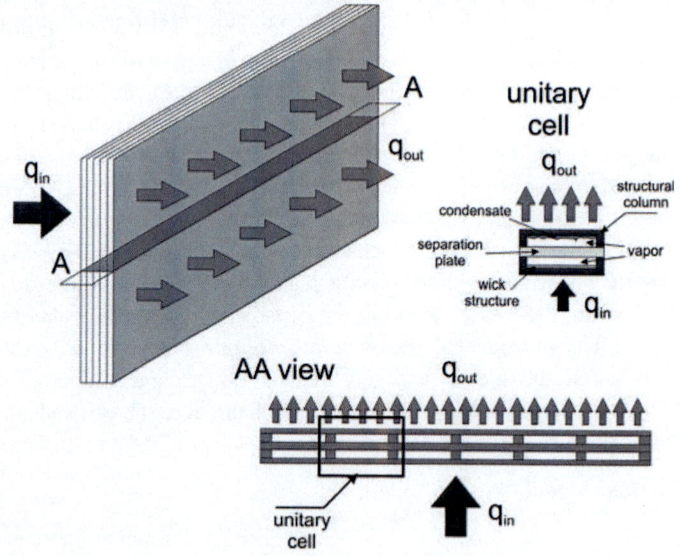

Fig. 7.7 Thermal model schematic of the flat LTS

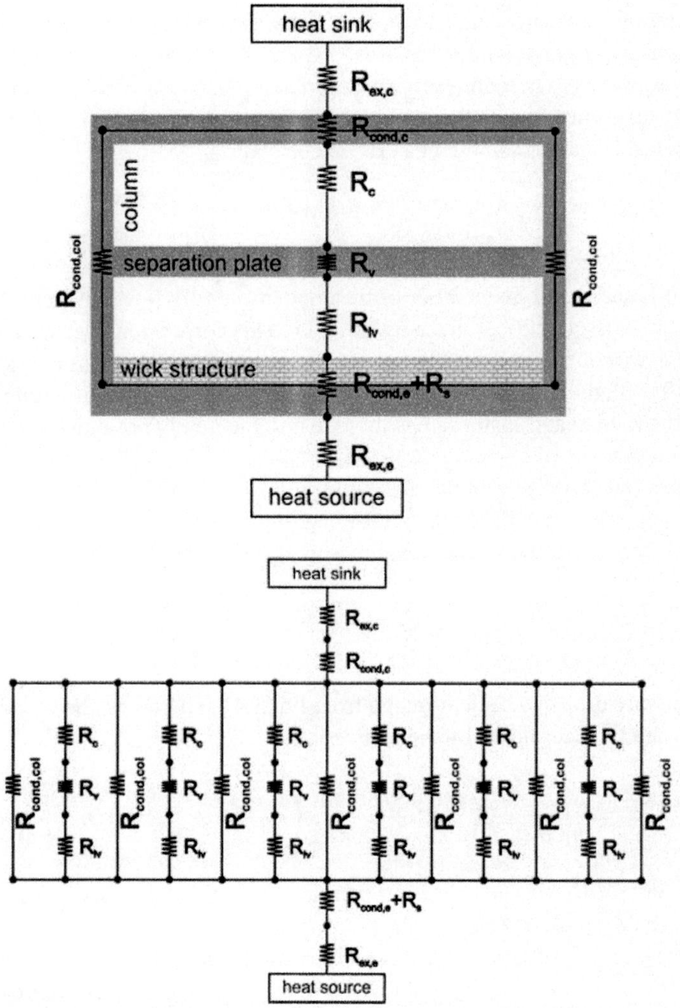

Fig. 7.8 Thermal circuit of the flat LTS

two dimensional spreading resistance along the plate, R_s. The evaporator plate is considered as formed by two plates in perfect contact: the closing plate and the wick structure, soaked with working fluid. The phase change liquid-vapor process is modelled by the R_{lv} resistance. The R_v represents vapor path (up direction in the evaporator section and then down direction in the condenser) thermal resistance. The film condensation in the condenser wall is symbolized by the R_c resistance. After spreading along the evaporator external plate, the heat may also reach the condenser by conduction through the cross section area of the structural columns, represented by the $R_{cond.col}$ path. Finally, the condenser wall thermal resistance is represented by the $R_{cond,e}$.

The external convection and one-dimensional conduction resistances can be determined using expressions similar to Eqs. 7.2 and 7.3, substituting the appropriate physical properties. The condensation thermal resistance is also determined as a convection resistance, where the condensation heat transfer coefficient can be given using the Hashimoto and Kaminaga (2002) correlation:

$$h_c = 0.85\mathrm{Re}_\delta^{0.1\left(0.00067\frac{\rho_l}{\rho_v}-0.6\right)}\bar{h}_L \tag{7.23}$$

where Re_δ is the Reynolds number of the condensate film flow, given by Eq, 2.16, and $\bar{h}_L$ is the Nusselt number given by Eq. 2.11. This correlation, also used by Jafari et al. (2017) was obtained from a linear regression for thermosyphons operating with of water, R113 and ethanol as working fluids, with 20% or error. It is important to note that the total heat transfer rate is divided into the number of unitary cells, six in the present case.

In the present case, the vapor path is not straight and the thermal resistance associated to its vapor pressure drop may not be negligible. According to Peterson (1994), the thermal resistance due to the flow pressure drop can be given by:

$$R_v = \frac{T_v \Delta p_v}{h_{lv}\rho_v q} \tag{7.24}$$

The pressure drop can be determined from Eq. 3.41. Therefore, the vapor thermal resistance can be determined by the expression:

$$R_v = \frac{T_v}{h_{lv}\rho_v q_{cell}}\left(\rho_v u^2 + \frac{8\mu_v \dot{m}_v}{\rho_v \pi r_v^4}\left(\frac{l_e + l_c}{2} + l_a + 50D_h\right)\right) \tag{7.25}$$

where the term $50D_h$ was added to compensate the curve of the vapor (see Fox et al, 1999) where $D_h = 4AA_v/P$.

The phase change liquid-vapor thermal resistance, R_{lv}, is also determined using the conventional convection thermal resistance expression (Eq. 7.2), where the coefficient of heat transfer due to evaporation in the liquid vapor interface is, according to Carey (1992), given by:

$$h_{e,lv} = 0.03\left(\frac{h_{lv}^2}{T_v(\upsilon_v - \upsilon_l)}\right)\left(\frac{M}{2\pi R}\right)^{1/2}\left(1 - \frac{p_v(\upsilon_v - \upsilon_l)}{2h_{lv}}\right) \tag{7.26}$$

where R is the universal gas constant and M is the fluid molar mass.

The last resistance to be determined is the two-dimensional conduction spreading resistance that happens along the evaporator and wick structure plates. For that, the solution of Muzychka et al. (2003) is employed. Figure 7.9 shows a system composed by two-plates and two-sources, which can be representative of the heat transfer process in the evaporator wall of a flat LTS in an actual application condition. This model was already used for the two-dimensional temperature distribution ($z = 0$) of

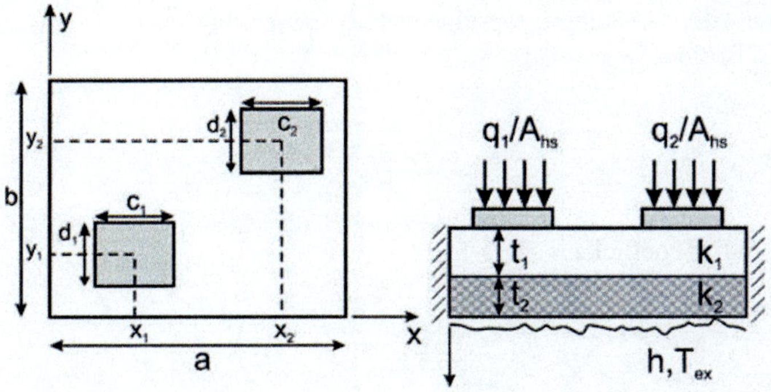

Fig. 7.9 Schematic of the two-source concentrated heat model for the spreading resistance of two parallel plates in perfect contact

the closing flat plate of the evaporator, considering, in that case, only one plate layer. According to Monteiro and Mantelli (2019), the temperature distribution of the one plate subjected to a single source, is given by the expression:

$$\begin{aligned} T(x,y,z) - T_{ex} &= A_0 + B_0 z + \sum_{m=1}^{\infty} \cos(\lambda_m x)[A_m \cosh(\lambda_m z) + B_m \sinh(\lambda_m z)] \\ &+ \sum_{n=1}^{\infty} \cos(\delta_n y)[A_n \cosh(\delta_n z) + B_n \sinh(\delta_n z)] \\ &+ \sum_{m=1}^{\infty}\sum_{n=1}^{\infty} \cos(\lambda_m x)\cos(\delta_n y)\left[A_{m,n}\cosh(\beta_n z) + B_{m,n}\sinh(\beta_{m,n} z)\right] \end{aligned} \tag{7.27}$$

where the eigenvalues are given by Eq. 7.16 and the Fourier coefficients are given by he set of Eqs. (7.12 to 7.14) and:

$$B_i = -\phi(\xi)A_i, \quad i = n;\, m;\; or\;\, m,n \tag{7.28}$$

For two plates in perfect contact, as shown in Fig. 7.9, the expression given by Eq. 7.15 is substituted by:

$$\phi(\zeta) = \frac{\left(\alpha e^{4\zeta t_1} - e^{2\zeta t_1}\right) + \eta\left(e^{2\zeta(2t_1+t_2)} - \alpha e^{2\zeta(2t_1+t_2)}\right)}{\left(\alpha e^{4\zeta t_1} + e^{2\zeta t_1}\right) + \eta\left(e^{2\zeta(2t_1+t_2)} + \alpha e^{2\zeta(2t_1+t_2)}\right)} \tag{7.29}$$

where:

$$\eta = \frac{\zeta + h/k_2}{\zeta - h/k_2} \tag{7.30}$$

and where with ζ assuming the eigenvalues, depending on the coefficient to be calculated. Besides,

$$\alpha = \frac{1 - \frac{k_2}{k_1}}{1 + \frac{k_2}{k_1}} \tag{7.31}$$

with the Fourier coefficients given by:

$$A_0 = \frac{q}{ab}\left(\frac{t_1}{k_1} + \frac{t_2}{k_2} + \frac{1}{h}\right) \tag{7.32}$$

$$B_0 = -\frac{q}{k_1 ab} \tag{7.33}$$

If the heat source was spread along the area of the first plate, the areas: $a \cdot b = c \cdot d$ in Fig. 7.4 and the temperature distribution, using these expressions would result in:

$$T(x, y, 0) - T_{ex} = \frac{q}{ab}\left(\frac{t_1}{k_1} + \frac{t_2}{k_2} + \frac{1}{h}\right) \tag{7.34}$$

where the first terms in the right represents the one-dimensional conduction in the upper plate, the second on the second plate and, the last, the convection thermal resistance.

The overall thermal resistance is determined from the temperature difference between the upper plate mean temperature and the sink temperatures over the heat flux, by the expression:

$$R_s = \frac{\overline{T}_{cd} - T_{ex}}{q} \tag{7.35}$$

For several sources, such as shown in Fig. 7.9, the temperature distribution of the upper surface ($z = 0$) is obtained from the superposition of the solution of the unique source problem.

According to Moreira and Mantelli (2019), this mathematical model compared with data for with an error, in average, less than 25%, for all tested cases (different filling ratios and different heating source positions). This comparison is quite good, considering the complexity of the several physical phenomena involving the operation of the device.

7.2 Capillary Pumped Loops—CPL

As for loop thermosyphons, capillary pumped loops (CPL) and loop heat pipes (LHP) are two-phase devices in which the evaporator and condenser are geometrically independent, being connected by vapor and liquid tubes. This configuration brings geometrical flexibility to the devices: they are easily adaptable to the applications. Besides, this geometry improves the organization of the working fluid streams, avoiding the countercurrent flows of the liquid and vapor phases.

Stenger (1966) was the first to propose the capillary pumped loop concept in the NASA Lewis Research Center, with the intention to improve the thermal performance of heat pipes. However, only in the 1970 decade this device was intensively investigated, especially in USA and Europe, to be used in the thermal control of satellites and spacecrafts. Traditional mechanically pumped circuits were already being considered for space applications, but they have serious disadvantages: they require electrical energy (difficult and expensive to get in space) and cause vibrations. CPLs, on the other hand, are passive devices and cause no vibration. A schematic of a CPL is shown in Fig. 7.10. These devices can be found in several satellites and spacecrafts, including Mars Surveyor and the Hubble Telescope (Swift et al. 2008).

In CPLs, the porous media (wick structure) is located only the evaporator. It has two major tasks: to enable the evaporation of the working fluid and to pump its liquid phase along the whole circuit. The vapor produced in the evaporator reaches the condenser by means of conventional metallic tubes, usually of small diameters. The vapor condenses as it reaches the condenser and loses heat. The resulting condensate returns to the evaporator by a tube, similar to the vapor line.

CPLs also present a liquid reservoir, connected to the evaporator entrance by a tube. The reservoir must be filled with liquid at all times. It is equipped with an electrical heater that controls the internal pressure of the liquid along all the circuit.

When compared to conventional heat pipes, the presence of porous media only in the evaporator can be considered an advantage, as lower friction forces over the vapor flow can be found, resulting in a better heat transfer performance. However, CPL may present start-up problems not common in heat pipes. In addition, the capillary structure in the evaporator of CPLs, known as capillary pumps, can be difficult and expansive to construct.

As already mentioned, the start-up in CPLs may be problematic in some applications. The reservoir must be pre-heated to augment the internal pressure, resulting in the displacement of liquid along the whole circuit, flooding it. When the evaporator temperature reaches the same level of the reservoir temperature, the phase change phenomena starts. As the evaporation proceeds, menisci are formed in the liquid-vapor interface in the wick, resulting in the development of a capillary pressure, responsible for the liquid pumping through the circuit.

It is quite clear that the thermal performance of a CPL depends on the design of the evaporator, also known as capillary pump. There are many different configurations of CPL and Fig. 7.10 presents one of them. In this case, the evaporator is cylindrical and is fed by liquid through a channel located in its center (bayonet). In normal

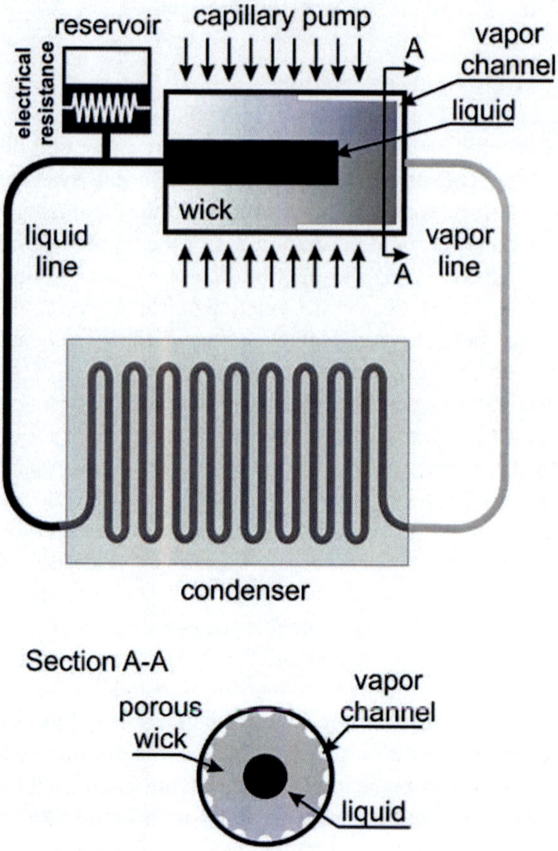

Fig. 7.10 Schematic of a capillary pumped loop

operation, the liquid comes from the condenser, while, in start-up, it comes from the reservoir. The just fed liquid is directed to the evaporator casing by means of a porous medium. The evaporator external walls receive heat and the working fluid changes of phase. Small axial tubes are drilled in the porous media, parallel to the evaporator axis and close to the cylindrical wall, to collect the formed vapor, directing it to the vapor tube.

As the separation between liquid and vapor is supposed to happen in the evaporator, the porous media must be filled with liquid at all times. One of the major problems concerning the operation of CPLs is the retention of vapor within the porous media in the evaporator. If the reservoir does not provide enough liquid to the evaporator in the start-up, vapor may be formed and be stuck within the wick structure. In this case, the capillary forces may not be enough to pump the liquid and the CPL does not start. Also, vapor bubbles located in wrong positions may absorb the capillary pressure variation provided by the wick, hindering the device operation.

CPLs present different operation conditions when in microgravity environment because, in the absence of gravity, vapor bubble tend to grow larger, with more chance to obstruct the working fluid flow passages along the device.

7.3 Loop Heat Pipes—LHP

Conventional heat pipes may not be convenient to transfer high amount of thermal energy, over considerable distances, due to high fluid pressure drops along the wick structures. LHPs were conceived as an alternative to heat pipes. In LHPs, the wick structure is concentrated only in the evaporator. Besides, the liquid and vapor flows in different tubes, resulting in considerable lower pressure drops.

Despite of LHP and CPLs being similar devices, the LHP development started in former Soviet Union in the decade of 1970, independently of the CPL, conceived in USA in this same decade. In 1972, Gerasimov et al., in the Ural Polytechnic Institute, presented the first LHP, which was microgravity tested in 1989 (Dunn and Reay, 2006). Up to the end of 1980 decade, this technology was limited to the Soviet Union, being introduced to Europe and USA in the 1990 decade. Since then, it has been considered a suitable thermal solution for many satellites. As they are easier to construct and have smoother start-ups, LHPs are usually considered as a better succeeded technology than CPLs. However, CPLs present advantages to LHPs such as better temperature control so that it can be a more appropriate thermal control technology for many applications.

Figure 7.11 shows a schematic of a conventional LHP. The major difference between LHP and CPL is in the location of the reservoir: while in CPLs the reservoir is located off the evaporator, in LHPs the reservoir, also known as compensation chamber, is located attached to the evaporator. The compensation chamber and the evaporator are hydraulically connected, through a channel where a capillary structure is inserted (secondary wick). This wick pumps liquid from the compensation chamber to the evaporator, guaranteeing that the primary wick structure is flooded at all times, including during start-ups. Besides, the compensation chamber accumulates any eventual non-condensable gas generated, mainly in the evaporator core, during the device operation.

For a successful LHP start-up, the liquid inventory must be enough to fill the condenser, the vapor and liquid lines and to saturate the wick structure. When heat is applied to the external surface of the evaporator, the liquid evaporates, mainly over the internal evaporator casing wall. Evaporation may be observed in the compensation chamber as well, as some heat may be conducted axially through the wick and casing materials. Therefore, the axial conductive resistance between evaporator and compensation chamber must be high, so that the temperature and pressure levels within the compensation chamber are smaller than within the evaporator. This mechanism allows for the auto-regulation of the liquid inventory in the LHP, with no need of external temperature control. The designer must be careful to design the internal

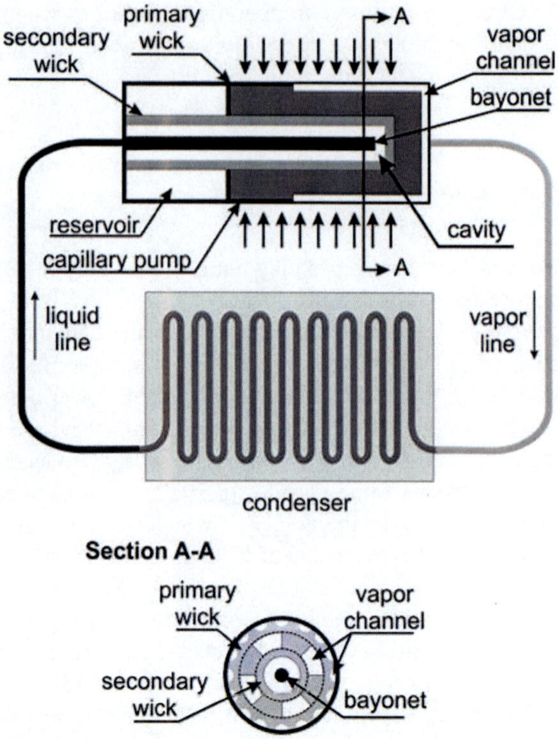

Fig. 7.11 Schematic of the loop heat pipe

geometry of the compensation chamber to avoid the displacement of vapor to the compensation chamber. Details of a typical LHP evaporator can be found in Fig. 7.11.

Modelling

Models are necessary to design efficient LHPs and CPLs. There is not a general model that can be applied to all devices and the designer must be able to develop his/her own model, according to the geometry of the LHP or CPL under consideration. Several LHP and CPL heat transfer models are found in the literature. Launay et al. (2008) proposed an analytical mathematical model based on an equivalent thermal circuit. Singh et al. (2009) presented a theoretical model based on a heat balance (first law of thermodynamics) performed for a LHP with a cylindrical evaporator, similar to the one studied by Ku (1999), who modelled a cylindrical evaporator LHP with a primary and a secondary porous media layers. Kaya et al. (1999) proposed models for the thermal behavior of LHPs based on known geometric and thermophysical properties of materials and working fluids. Siedel et al. (2015) developed an analytical model to determine the thermo-hydraulic behavior of LHPs, where the energy balance equation was combined with two-dimensional (2D) temperature field equations, for a cylindrical evaporator, including the wick structure and casing and where the temperature-sensitive thermophysical parameters were taken for temperatures observed experimentally. Chernysheva and Maydanik (2012) presented a 3D mathematical model for a flat sintered copper porous media evaporator, filled with water, in which the geometries of vapor exit channel, the compensation chamber and the porous structure were considered in their model. Cao and Faghri (1994) developed an analytical model for the liquid pressure and velocity fields and for the temperature distributions of LHPs, where their physical model consisted of a porous media section, heated by solid fins. These researchers used Darcy´s equations to solve the velocity fields.

Numerical solutions are also employed in the design of such devices by several researchers, as Demidov and Yatsenko (1994). Santos (2010) solved numerically thermal and hydraulic heat balance models for ceramic wicks of evaporators of LHPs and CPLs, considering that the wick structure was composed by two layers of different thermal conductivity media. Ji and Peterson (2011) developed a three-dimensional (3D) model to analyze the heat and mass transfer of a flat evaporator of a LHP in which sintered copper powder was the wick structure used in the evaporator.

Determination of the LHP or CPL Evaporator Wick Geometry

To design the geometry (length and diameter) of the wick structure of the evaporators of LHP and CPLs, the designer must consider that enough external area must be provided to allow the heat source to be transferred to the evaporator. In addition, the evaporator must satisfy the capillary pumping requirements of the LHP and this wick must be thin enough to avoid the phase change to happen inside the structure and large enough to provide the working fluid flow and capillary pumping capacity. Figure 7.12, upper drawing shows a schematic of the inside view of a LHP evaporator and the

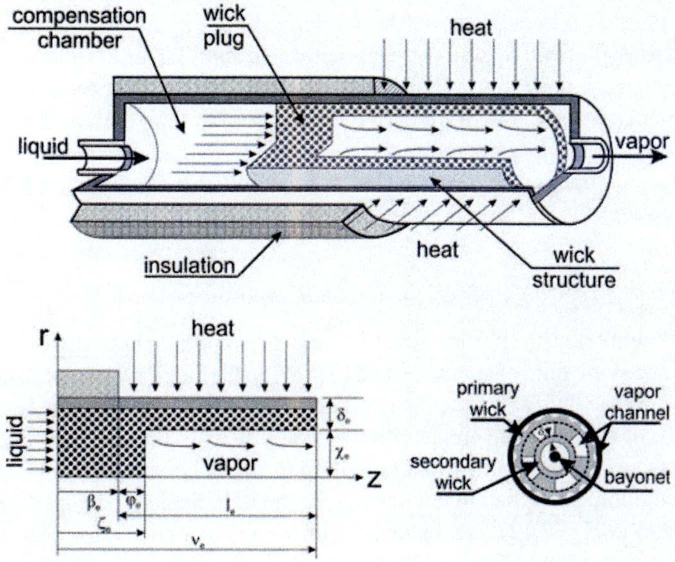

Fig. 7.12 Schematic of a loop heat pipe

lower drawing shows the cross section of a common LHP and/or CPL evaporator, with the wick thickness δ_w highlighted.

From the mass balance principles, applying the Darcy's law and considering that all the input heat that crosses the wick structure in the evaporator is used for the liquid vapor phase change at the vapor-liquid interface, Florez (2016) proposed an non-dimensional expression, which relates the porous medium geometry (ratio of the wick thickness δ_w and its length l_w) to three parameters: wick properties (ratio of the effective thermal conductance and permeability), fluid thermophysical properties and thermodynamic state of the working fluid. They proposed the following expression to be used as a designing tool of evaporator porous media of LHP or CPL evaporators:

$$\underbrace{\left(\frac{\delta_w}{l_w}\right)^2}_{\text{geometry}} \approx \underbrace{\frac{k_{ef}}{K}}_{\substack{\text{wick}\\\text{properties}}} \underbrace{\frac{\mu_l}{\rho_l h_{lv}}}_{\substack{\text{fluid}\\\text{properties}}} \underbrace{\left(\left.\frac{dp}{dT}\right|_T\right)^{-1}}_{\substack{\text{thermodynamic}\\\text{state}}} \tag{7.36}$$

The compensation chamber and the evaporator have slightly different saturated states, due to the temperature differences. These saturation states are thermodynamically linked and must satisfy the relation (Faghri 1995; Ku 1999; Maydanik et al. 2011):

$$\left.\frac{dp}{dT}\right|_T \approx \frac{\Delta p}{\Delta T} = \frac{(p_v - p_{cc})}{(T_v - T_{cc})} \tag{7.37}$$

where p_{cc} and T_{cc} are pressure and temperature of the compensation chamber, respectively. The term *dp/dT* is the slope of the pressure-temperature saturation line at T and this rate depends on the working fluid. High *dp/dT* values result in larger pressure gradients within the evaporator and so in larger evaporator capillary pumping capacity, resulting in larger temperature difference between the compensation chamber and the rest of the LHP. The evaporator temperature is assumed as the reference operation temperature T, for which the derivative is taken. Ku (1999) proposes the compensation chamber as the reference temperature, while Faghri (1995) proposes the average between the evaporator and compensation chamber temperatures.

Case: Copper LHP With Cylindrical Shaped Evaporator:

As already mentioned, the porous media in the LHP evaporator have two objectives: to control the working fluid evaporation and to provide the capillary pumping necessary for conducting the fluid along the device. Therefore, the evaporator must be designed so that most of the heat delivered at the evaporator reaches the working fluid liquid–vapor meniscus. Also, vapor formation within the wick structure must be avoided.

The methodology applied in this section for the LHP depicted in Fig. 7.13 (details are presented in Fig. 7.12), can be used, with some adaptations, for evaporators of different configuration LHPs.

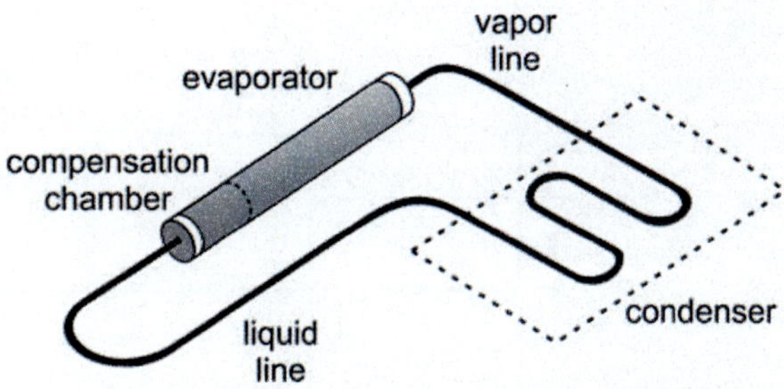

Fig. 7.13 LHP case modelled

Thermal model. In the evaporator under consideration, the condensate enters the evaporator through the liquid line, reaches the compensation chamber (with no wick structure) and crosses the porous media plug that separates the compensation chamber and pumping region. Heat power input is delivered at the condenser external (see Fig. 7.12). The higher the pressure difference between the compensation chamber and the liquid-vapor meniscus interface, the better the system works. Therefore, the objective of the plug is, while allowing the liquid to reach the wick structure in the heated region of the evaporator, to decrease the heat leakage to the compensation chamber, avoiding its temperature increase, which would decrease the pressure difference and so disturb the LHP thermal performance, eventually stopping the system operation.

In other words, the heat loss to the compensation chamber increases the working fluid temperature, decreasing the temperature difference between the compensation chamber and the evaporator meniscus. As a result, the working fluid pressure difference between these same regions of the LHP decreases (see Maydanik et al. (2011b)). Actually, the configuration studied in this work, is similar to the oval evaporator proposed by Maydanik et al. (2011b), which has been successfully tested and considered as an alternative to the conventional evaporators. These authors could use the combination of copper and water with this geometry. Cooper is a material usually avoided due to its high thermal conductivity that causes large heat losses to the compensation chamber. Besides, water is also avoided because, despite of having good thermal properties such as high latent heat of vaporization and surface tension, it presents low pressure sensibility to temperature variations, for low operation temperatures (below 70 °C).

A simplified sketch of the evaporator is shown in Fig. 7.14. The total length of the pumping region of the evaporator ν_e encompasses the plug region (of length β_e) and the length of the evaporator internally covered by a wick structure composed by a layer of length l_e and thickness δ_w, made by sintered copper powder. The total insulated length is ς_e and the insulated length over wick structure is φ_e The radius of the cylinder is ξ_e and the radius of the vapor region is χ_e Two-dimensional cylindrical coordinates were adopted, with the z axis located in the axis of cylindrical wick and the r axis located radial to the wick thickness direction.

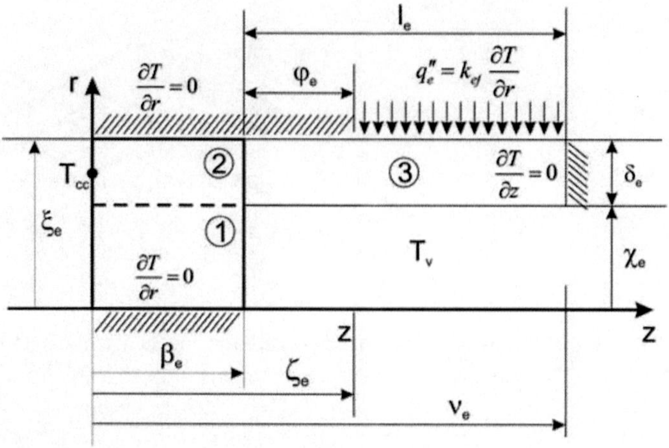

Fig. 7.14 Physical model of the evaporator of Fig. 7.14 LHP

The following hypotheses were adopted: two-dimensional cylindrical coordinate heat transfer, steady state conditions, isotropic wick structure, uniform thermophysical properties of solid and liquid, evaporator (T_v) and compensation chamber (T_{cc}) temperatures prescribed (considered in saturation state, where $T_v > T_{cc}$).

The heat balance results in the Laplace equation:

$$\frac{\partial^2\theta}{\partial r^2} + \frac{1}{r}\frac{\partial\theta}{\partial r} + \frac{\partial^2\theta}{\partial z^2} = 0 \tag{7.38}$$

where $\theta = (T - T_{cc})/(T_v - T_{cc})$, must be solved, subjected to the following boundary conditions:

- Prescribed liquid temperature T_l at $z = 0$: in the compensation chamber, the working fluid, liquid and vapor phases, is in saturated state.
- Thermal insulation at $\partial T/\partial z = 0$ at $z = \upsilon_e$: heat transfer is negligible at end of the evaporator wall.
- Thermal insulation $\partial T/\partial r = 0$ at $r = \xi_e$, for $0 < z < \beta_e$: the whole heat supplied to evaporator is transferred or to the wick meniscus and /or to the compensation chamber. The heat transfer in the interface between the wick plug and vapor interface is negligible.
- Prescribed heat flux $q'' = -k_{ef} \cdot \partial T/\partial r$, for $\varsigma_e < z < \nu_e$.
- Prescribed vapor temperature T_v, at $r = \chi_e$. This temperature is the same of the saturated temperature of the vapor in the meniscus interface.
- Liquid-vapor phase change happens only at the meniscus in the interface between liquid and vapor, at $r = \chi_e$.
- Heat is supplied to the evaporator external wall and reaches the liquid-vapor interface, through the wick structure, which has high thermal conductivity.

- All the heat transferred by the wick structure to the liquid-vapor interface is used to the liquid-vapor phase change of the working fluid.
- The evaporation thermal resistance is much lower than the sintered porous media conduction thermal resistance, which effective thermal conductivity is k_{ef}.
- The surface at $z = \beta_e$ for $0 < e < \chi_e$ is adiabatic, because, as the heat transfer in this face is small, no evaporation is expected at this interface. This is possible if one considers that the whole working fluid flows through of surface at $z = \beta_e$ for $\chi_e < r < \xi_e$, and in this condition, the maximum pressure drop is obtained.

In this 2D model, the heat transfer through the evaporator casing was neglected, in behalf of the design safety, as this hypothesis results in a higher predicted temperature than the actual one. Therefore, only the heat leak to the compensation chamber through the plug wick is considered.

Due to the non-uniformiy of the evaporator geometry, composed of a solid plus a hollow cylinders, the problem is split in three, (sub-dominium 3 for the hollow, and 1 and 2 for the solid cylinder) as shown in Fig. 7.15. The interface at $z = \beta_e$, for $\chi_e < r < \xi_e$, couples the solid and hollow cylinders, sub-dominions 2 and 3. The boundary conditions adopted to this interface is the following: first a heat flux is assumed for the region 2, then the region 2 mathematical model determines the temperature distribution at $z = \beta_e$ and $\chi_e < r < \xi_e$,which is considered as the prescribed temperature distribution boundary condition for region 3. The parity of temperatures couples regions and 1 and 2.

Besides, sub-dominium 2 presents two non-homogeneous boundary conditions. The superposition method is applied and the problem is again divided in two, each one with just one non-homogeneous boundary condition. The same can be said of sub-dominium 3 that has three non-homogeneous boundary conditions and therefore can be divided into three problems.

The well-known method of separation of variables is applied (Arpaci 1966; Ozisik 1993; Carslaw and Jaeger 2004) and the sub-problems, coupled by the boundary conditions, are solved simultaneously resulting in the temperature distributions as

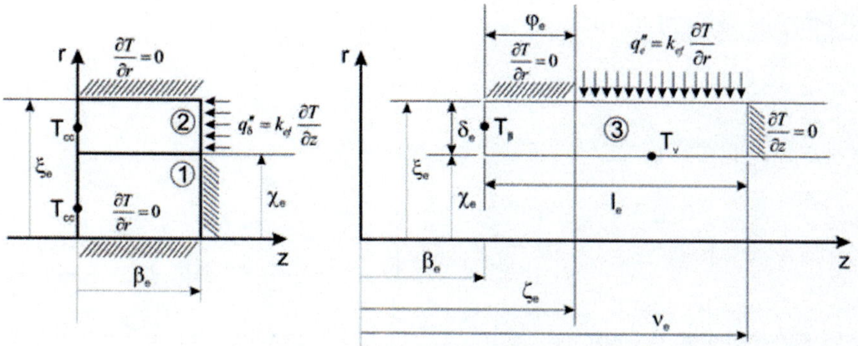

Fig. 7.15 Schematic of the conduction sub-problems boundary conditions (see Fig. 7.14)

given in Table 4.1 for the solid wick plug and Table 4.2 for the hollow wick cylinder In these tables, $\theta_1(r, z)$ refers to the solution of region 1 problem, $\theta_2(r, z) = \theta_{2a}(r, z) + \theta_{2b}(r, z)$ to the solution of region 2 problem and $\theta(r, z) = \theta_{3a}(r, z) + \theta_{3b}(r, z) + \theta_{3c}(r, z)$ to the solution of the region 3 problem. In these solutions, J_0 *and* Y_0, are the Bessel functions of first and second kind and order zero, while of J_1 and Y_1 are the Bessel functions of the first and second kind and order one, respectively. I_o and K_0 are the modified Bessel functions of the first and second kind and order zero, and I_1 and K_1 are the modified Bessel functions of first and second kind and order 1, respectively. λ_m, ζ_m, η_m, υ_n, τ_n and Ξ_m are the eigenvalues and L_m, M_m, N_m, O_n, P_n and Q_n are expressions that associate the several sub-problem outcomes in a unique solution.

Heat Transfer Models

A well-designed evaporator must assure that most of the energy delivered is used to evaporate the working fluid in the liquid-vapor interface in the wick. Heat should not leak to any other LHP component, especially the compensation chamber. Heat leakage increases the liquid temperature and so its pressure, decreasing the pressure drops along the LHP sections, which would eventually prevent LHP operation or its startup. Through the present model, one can predict, from the total energy delivered to the evaporator, the heat lost to the compensation chamber, through the plug solid cylinder, see Fig. 7.14.

In steady state conditions, a simple heat balance within the evaporator determines the total energy delivered to the evaporator. This heat can be divided into the energy transferred to the meniscus interface $\left(q_r|_{r=\chi_e}\right)$(subdomain 3) and the other lost to the compensation chamber $\left(q_z|_{z=\beta_e}\right)$:

$$q_e = q_r|_{r=\chi_e} + q_z|_{z=\beta_e} \tag{7.39}$$

The heat leakage to the compensation chamber, $q_z|_{z=\beta_e}$, can be calculated by two means: considering the heat transferred through the interface between the hollow and the full cylinders or using the Fourier law integrated along the cross-section area of the solid cylinder in $0 < z < \beta_e$ in the subdomains 1 and 2, i.e.:

$$q = q_1 + q_2 \tag{7.40}$$

where q_1 is given by:

$$q_1 = \int_0^{\chi_e} k_{ef} \left.\frac{\partial \theta_1}{\partial z}\right|_{z_i} 2\pi\, r\, dr \tag{7.41}$$

where θ_1 equation is shown in Table 7.1. The heat transferred through subdomain 2 is determined using the temperature distributions θ_{2a} and θ_{2b}, resulted from the solution of the two superposed problems, also given in Table 7.1, resulting in:

Table 7.1 Solutions for the temperature distribution of LHP evaporator wick solid cylinder

Solutions	Eigenvalues
$\theta_1(r,z) = \sum_{m=1}^{\infty} L_m I_0(\lambda_m r)\sin(\lambda_m z)$	$\lambda_m = \frac{(2n-1)}{2}\frac{\pi}{\beta_e}$
$\theta_{2a}(r,z) = \sum_{m=1}^{\infty} M_m\left[\frac{K_1(\zeta_m \xi_e)}{I_1(\zeta_m \xi_e)} I_0(\zeta_n r) + K_0(\zeta_m r)\right]\sin(\zeta_m z)$	$\zeta_m = \frac{(2m-1)}{2}\frac{\pi}{\beta_e}$
$\theta_{2b}(r,z) = \sum_{m=1}^{\infty} N_m\left[-\frac{Y_0(\eta_m \chi_e)}{J_0(\eta_m \chi_e)} J_0(\eta_m r) + Y_0(\eta_m r)\right]\sinh(\eta_m r)$	$\frac{Y_0(\eta_m \chi_e)}{J_0(\eta_m \chi_e)} = \frac{Y_1(\eta_m \xi_e)}{J_1(\eta_m \xi_e)}$
Expressions connecting superposed solutions	
$N_m = \frac{\frac{q''_\delta}{k_{ef}(T_v - T_\infty)}}{\eta_m \cosh(\eta_m \beta_e)} \frac{\int_{\chi_e}^{\xi_e}\left[-\frac{Y_0(\eta_m \chi_e)}{J_0(\eta_m \chi_e)} J_0(\eta_m r) + Y_0(\eta_m r)\right] r\,dr}{\int_{\chi_e}^{\xi_e}\left[-\frac{Y_0(\eta_m \chi_e)}{J_0(\eta_m \chi_e)} J_0(\eta_m r) + Y_0(\eta_m r)\right]^2 r\,dr}$	
$M_m = \frac{I_0(\lambda_m \chi_e)}{\left[\frac{K_1(\zeta_m \xi_e)}{I_1(\zeta_m \xi_e)} I_0(\zeta_m \chi_e) - K_0(\zeta_m \chi_e)\right]} L_m$	
$L_m = \frac{\sum_{n=1}^{\infty} N_n \eta_n \left[\frac{Y_0(\eta_n \chi_e)}{J_0(\eta_n \chi_e)} J_1(\eta_n \chi_e) - Y_1(\eta_n \chi_e)\right]\int_0^{\beta_e}\sinh(\eta_n z)\sin(\lambda_m z)dz}{\lambda_m I_1(\lambda_m \chi_e) - \frac{\lambda_m I_0(\lambda_m \chi_e)\left[\frac{K_1(\zeta_m \xi_e)}{I_1(\zeta_m \xi_e)} I_1(\zeta_m \chi_e) - K_1(\zeta_m \chi_e)\right]}{\left[\frac{K_1(\zeta_m \xi_e)}{I_1(\zeta_m \xi_e)} I_0(\zeta_m \chi_e) - K_0(\zeta_m \chi_e)\right]}\int_0^{\beta_e}\sin^2(\lambda_m z)dz}$	

$$q_2 = \int_{\chi_e}^{\xi_e} k_{ef}(T_v - T_{cc}) \left.\frac{\partial \theta_{2a}}{\partial z}\right|_{z_i} 2\pi\, r\, dr + \int_{\chi_e}^{\xi_e} k_{ef}(T_v - T_{cc}) \left.\frac{\partial \theta_{2b}}{\partial z}\right|_{z_i} 2\pi\, r\, dr \quad (7.42)$$

Another way to determine the heat loss to the wick plug is by the calculation of the heat that leaves the sub-domain 3 at β_e, obtained by the sum of q_{3a}, q_{3b} and q_{3c}, or:

$$q_3|_{z=\beta_e} = \int_{\chi_e}^{\xi_e} k_{ef}(T_v - T_{cc}) \left.\frac{\partial \theta_{3a}}{\partial z}\right|_{z=\beta_e} 2\pi\, r\, dr + \int_{\chi_e}^{\xi_e} k_{ef}(T_v - T_{cc}) \left.\frac{\partial \theta_{3b}}{\partial z}\right|_{z=\beta_e} 2\pi\, r\, dr + \\ + \int_{\chi_e}^{\xi_e} k_{ef}(T_v - T_{cc}) \left.\frac{\partial \theta_{3c}}{\partial z}\right|_{z=\beta_e} 2\pi\, r\, dr \quad (7.43)$$

where θ_{3a}, θ_{3b} and θ_{3c} are given by the expressions given in Table 7.2.

Now, the heat transferred in the liquid vapor meniscus interface, is also determined from the sub-problem solutions 3a, 3b and 3c: (Table 7.3)

$$q_3|_{r=\chi_e} = \int_{\beta_e}^{\nu_e} k_{ef}(T_v - T_{cc}) \left.\frac{\partial \theta_{3a}}{\partial r}\right|_{r=\chi_e} 2\pi\, \chi_e\, dz + \int_{\beta_e}^{\nu_e} k_{ef}(T_v - T_{cc}) \left.\frac{\partial \theta_{3b}}{\partial r}\right|_{r=\chi_e} 2\pi\, \chi_e\, dz \\ + \int_{\beta_e}^{\nu_e} k_{ef}(T_v - T_{cc}) \left.\frac{\partial \theta_{3c}}{\partial r}\right|_{r=\chi_e} 2\pi\, \chi_e\, dz \quad (7.44)$$

Table 7.2 Solutions for the temperature distributions of LHP evaporator wick hollow cylinder

Solutions	Eigenvalues
$\theta_{3a}(r,z)=\sum_{n=1}^{\infty} O_n\left[-\frac{K_0(\vartheta_n\chi_e)}{I_0(\vartheta_n\chi_e)}I_0(\vartheta_n r)+K_0(\vartheta_n r)\right]\cdot[\sin(\vartheta_n z)-\tan(\vartheta_n\beta_e)\cos(\vartheta_n z)]$	$\tan(\vartheta_n\beta_e)=-\cot(\vartheta_n\nu_e)$
$\theta_{3b}(r,z)=\sum_{n=1}^{\infty} P_n\left[\frac{K_1(\tau_n\xi_e)}{I_1(\tau_n\xi_e)}I_0(\tau_n r)+K_0(\tau_n r)\right]\cdot[\sin(\tau_n z)-\tan(\tau_n\beta_e)\cos(\tau_n z)]$	$\tan(\tau_n\beta_e)=-\cot(\tau_n\nu_e)$
$\theta_{3c}(r,z)=\sum_{n=1}^{\infty} Q_n\left[-\frac{Y_1(\Xi_n\xi_e)}{J_1(\Xi_n\xi_e)}J_0(\Xi_n r)+Y_0(\Xi_n r)\right]\cdot\cosh[\Xi_n(z-\nu_e)]$	$\frac{Y_0(\Xi_n\chi_e)}{J_0(\Xi_n\chi_e)}=\frac{Y_1(\Xi_n\xi_e)}{J_1(\Xi_n\xi_e)}$

Expressions connecting superposed solutions

$$O_n=\frac{\frac{q_e''}{k_{ef}(T_v-T_{cc})}}{\vartheta_n\left[-\frac{K_0(\vartheta_n\chi_e)}{I_0(\vartheta_n\chi_e)}I_1(\vartheta_n\xi_e)-K_1(\vartheta_n\xi_e)\right]}\frac{\int_{\varsigma_e}^{\nu_e}[\sin(\vartheta_n z)-\tan(\vartheta_n\beta_e)\cos(\vartheta_n z)]dz}{\int_{\beta_e}^{\nu_e}[\sin(\vartheta_n z)-\tan(\vartheta_n\beta_e)\cos(\vartheta_n z)]^2dz}$$

$$P_n=\frac{1}{\left[\frac{K_1(\tau_n\xi_e)}{I_1(\tau_n\xi_e)}I_0(\tau_n\chi_e)+K_0(\tau_n\chi_e)\right]}\frac{\int_{\beta_e}^{\nu_e}[\sin(\tau_n z)-\tan(\tau_n\beta_e)\cos(\tau_n z)]dz}{\int_{\beta_e}^{\nu_e}[\sin(\tau_n z)-\tan(\tau_n\beta_e)\cos(\tau_n z)]^2dz}$$

$$Q_n=\frac{1}{\cosh[\Xi_n(\nu_e-\beta_e)]}\frac{\int_{\chi_e}^{\xi_e}\theta(r)\left[-\frac{Y_1(\Xi_n\xi_e)}{J_1(\Xi_n\xi_e)}J_0(\Xi_n r)+Y_0(\Xi_n r)\right]r\,dr}{\int_{\chi_e}^{\xi_e}\left[-\frac{Y_1(\Xi_n\xi_e)}{J_1(\Xi_n\xi_e)}J_0(\Xi_n r)+Y_0(\Xi_n r)\right]^2r\,dr}$$

Table 7.3 LHP evaporator heat transfer models

Heat leakage
$q = q_1 + q_2$
$q_1 = 2\pi k_{ef}(T_v - T_{cc}) \sum_{m=1}^{\infty} \lambda_m L_m \cos(\lambda_m z_i) \int_0^{\chi_e} I_0(\lambda_m r) r dr$
$q_2 = q_{2a} + q_{2b}$
$q_{2a} = 2\pi k_{ef}(T_v - T_{cc}) \sum_{m=1}^{\infty} \zeta_m M_m \cos(\zeta_m z_i) \int_{\chi_e}^{\xi_e} \left[\frac{K_1(\zeta_m \xi_e)}{I_1(\zeta_m \xi_e)} I_0(\zeta_m r) + K_0(\zeta_m r) \right] r dr$
$q_{2b} = 2\pi k_{ef}(T_v - T_{cc}) \sum_{m=1}^{\infty} \eta_m N_m \cosh(\eta_m z_i) \int_{\chi_e}^{\xi_e} \left[-\frac{Y_0(\eta_m \xi_e)}{J_0(\eta_m \xi_e)} J_0(\eta_m r) + Y_0(\eta_m r) \right] r dr$
Heat transfer to active zone
$q_3 = q_{3a}\vert_{r=\chi_e} + q_{3b}\vert_{r=\chi_e} + q_{3c}\vert_{r=\chi_e}$
$q_{3a}\vert_{r=\chi_e} = 2\pi \chi_e k_{ef}(T_v - T_{cc}) \sum_{n=1}^{\infty} \vartheta_n O_m \left[-\frac{K_0(\vartheta_n \chi_e)}{I_0(\vartheta_n \chi_e)} I_1(\vartheta_n \chi_e) - K_1(\vartheta_n \chi_e) \right] \cdot \int_{\beta_e}^{\nu_e} [\sin(\vartheta_n z) - \tan(\vartheta_n \beta_e) \cos(\vartheta_n z)] dz$
$q_{3b}\vert_{r=\chi_e} = 2\pi \chi_e k_{ef}(T_v - T_{cc}) \sum_{n=1}^{\infty} \tau_n P_m \left[\frac{K_1(\tau_n \xi_e)}{I_1(\tau_n \xi_e)} I_1(\tau_n \chi_e) - K_1(\tau_n \chi_e) \right] \cdot \int_{\beta_e}^{\nu_e} [\sin(\tau_n z) - \tan(\tau_n \beta_e) \cos(\vartheta_n z)] dz$
$q_{3c}\vert_{r=\chi_e} = 2\pi \chi_e k_{ef}(T_v - T_{cc}) \sum_{n=1}^{\infty} \Xi_n Q_n \left[\frac{Y_1(\Xi_n \xi_e)}{J_1(\Xi_n \xi_e)} J_1(\Xi_n \chi_e) - Y_1(\Xi_n \chi_e) \right] \cdot \int_{\beta_e}^{\nu_e} \cosh[\Xi_n (z - \nu_e)] dz$

Hydraulic Model

The hydraulic behavior of the porous structure is modeled aiming the development of a tool for designing LHP wicks. The following hypothesis are adopted:

- The liquid-vapor phase change happens only at the interface between liquid and vapor, in the wick structure. The solid matrix of the porous media has high thermal conductivity so that the heat supplied to the evaporator is able to reach the liquid-vapor interface.
- All the heat transferred by the wick structure to the liquid-vapor interface is used to working fluid phase change. The evaporation thermal resistance is much smaller than the conductance thermal resistance through the sintered porous media.
- The fluid has uniform velocity within the compensation chamber.
- The working fluid is incompressible.
- Gravity effects are negligible.
- The fluid flow within the wick structure is laminar and its Reynolds number is less than one (Re < 1), therefore Darcy´s equation can be applied. The Reynolds number is based on the average velocity within the pore and on its characteristic dimension is the pore diameter.

Similar to the temperature distributions, the pressure field is also obtained from the solution of the following Laplace's Equation, which results from the mass conservation principle and the application of Darcy's law:

$$\frac{\partial^2 p}{\partial r^2} + \frac{1}{r}\frac{\partial p}{\partial r} + \frac{\partial^2 p}{\partial z^2} = 0 \tag{7.45}$$

As the geometry is non-uniform (solid and hollow cylinders), the problem is split into two, one hollow and one solid cylinder, as shown in Fig. 7.16.

The two-dimensional Laplace equation (Eq. 7.45) requires four boundary conditions. Figure 7.16 shows these boundary conditions. For the solid cylinder, they are: prescribed compensation chamber pressure p_{cc} at $z = 0$, hermetic conditions (no pressure gradient) at the evaporator case wall, $r = \xi_e$ and symmetry condition at $r = 0$. At $z = \beta_e$, two different boundary conditions for the solid cylinder are observed:

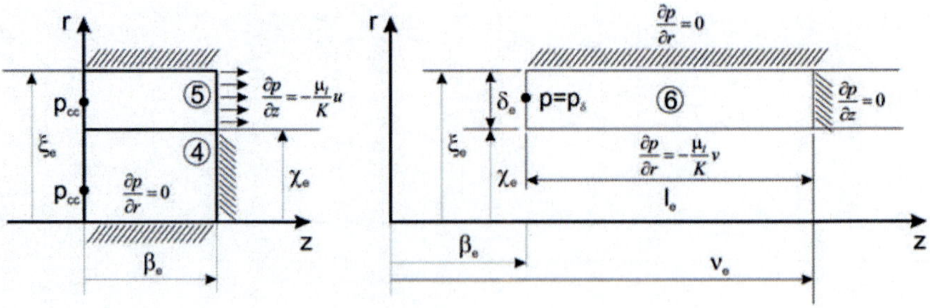

Fig. 7.16 Hydraulic model boundary conditions

for $0 < r < \chi_e$, $\partial p/\partial z = 0$ and for $\chi_e < r < \xi_e$, the pressure gradient is obtained from the Darcy´s expression $\partial p/\partial z = -\mu/K \cdot u$, where K is the permeability of the porous medium.

The solid cylinder is also divided into two regions (4 and 5), to accommodate the non-uniform boundary condition at $z = \beta_e$. The equality temperatures is the boundary condition that couples both regions 4 and 5. However, these regions still have two non-homogeneous boundary conditions, at $z = 0$ and at $z = \beta_e$. To reduce one of these non-homogeneities, the following parameter is defined, where p_{cc} is the compensation chamber pressure:

$$\Phi = p - p_{cc} \tag{7.46}$$

The velocity u, at the non-homogeneous boundary condition on $z = \beta_e$ can be determined by $u = \dot{m}_l/\rho_l A$, in which $\dot{m}_l = q_\delta/h_{lv}$ and q_δ is the heat flux that cross the boundary. Therefore, the non-homogeneous boundary condition at $z = \beta_e$, in terms of Φ variable is:

$$\left.\frac{\partial \Phi}{\partial z}\right|_{z=\beta_e} = 0,\ for\ \ 0 < r < \chi_e \text{ and } \left.\frac{\partial \Phi}{\partial z}\right|_{z=\beta_e} = \frac{\mu_l q_\delta}{K\rho_l A h_{lv}},\ for\ \chi_e < r < \xi_e \tag{7.47}$$

Again, the method of separation of variables is applied (Arpaci 1966; Ozisik 1993; Carslaw and Jaeger 2004) and the sub-problems, coupled by the boundary conditions, are solved simultaneously, resulting in the temperature distribution expressions presented in Table 7.4 for the solid plug, where $\Phi_4(r, z)$ refers to the solution of region 4 problem and $\Phi_5(r, z) = \Phi_{5a}(r, z) + \Phi_{5b}(r, z)$ to the solution of region 5 problem. In these solutions, ς_m, γ_m and φ_m are the eigenvalues and S_m, U_m and R_m are expressions that associate the several sub-problem outcomes in a unique solution.

For the solution of the problem of region 6, the following parameter is used to homogenize the boundary condition in $z = \beta_e$:

$$\Psi = p - p_\delta \tag{7.48}$$

The boundary condition in $r = \chi_e$ can be given substituting the velocity by $v = \dot{m}_l/\rho_l A$ where $\dot{m}_l = q|_{r=\chi_e}/h_{lv}$. Besides, $q|_{r=\chi_e}$ can be calculated by the temperature gradient in $r = \chi_e$, obtained from solution of the temperature field, so that:

$$\left.\frac{\partial \Psi}{\partial r}\right|_{r=\chi_e} = \frac{\mu_l k_{ef}}{K\rho_l h_{lv}} \left.\frac{\partial \theta}{\partial r}\right|_{r=\chi_e},\ for\ \beta_e < r < \nu_e \tag{7.49}$$

Table 7.5 show the pressure distribution solution for the hollow cylinder, region 6 problem.

Table 7.4 LHP evaporator equations for pressure field of LHP evaporator wick solid cylinder

Solutions	Eigenvalues
$\Phi_4(r,z) = \sum\limits_{m=1}^{\infty} R_m I_0(\varsigma_m r)\sin(\varsigma_m z)$	$\varsigma_n = \frac{(2n-1)}{2}\frac{\pi}{\beta_e}$
$\Phi_{5a}(r,z) = \sum\limits_{m=1}^{\infty} S_m\left[\frac{K_1(\gamma_m \xi_e)}{I_1(\gamma_m \xi_e)} I_0(\gamma_n r) + K_0(\gamma_m r)\right]\sin(\gamma_m z)$	$\gamma_m = \frac{(2m-1)}{2}\frac{\pi}{\beta_e}$
$\Phi_{5c}(r,z) = \sum\limits_{m=1}^{\infty} U_m\left[-\frac{Y_0(\varphi_m \chi_e)}{J_0(\varphi_m \chi_e)} J_0(\varphi_m r) + Y_0(\varphi_m r)\right]\sinh(\varphi_m r)$	$\frac{Y_0(\varphi_m \chi_e)}{J_0(\varphi_m \chi_e)} = \frac{Y_1(\varphi_m \xi_e)}{J_1(\varphi_m \xi_e)}$
Expressions connecting superposed solutions	
$U_m = \frac{\frac{q''_\delta \mu_l}{K\rho_l h_{lv}}}{\varphi_m \cosh(\varphi_m \beta_e)} \frac{\int\limits_{\chi_e}^{\xi_e}\left[-\frac{Y_0(\varphi_m \chi_e)}{J_0(\varphi_m \chi_e)} J_0(\varphi_m r)+Y_0(\varphi_m r)\right] r\,dr}{\int\limits_{\chi_e}^{\xi_e}\left[-\frac{Y_0(\varphi_m \chi_e)}{J_0(\varphi_m \chi_e)} J_0(\varphi_m r)+Y_0(\varphi_m r)\right]^2 r\,dr}$	
$S_m = \frac{I_0(\varsigma_m r)}{\left[\frac{K_1(\gamma_m \xi_e)}{I_1(\gamma_m \xi_e)} I_0(\gamma_m r)+K_0(\gamma_m r)\right]} R_m$	
$R_m = \frac{\sum\limits_{m=1}^{\infty} U_m\left[\frac{Y_0(\varphi_m \chi_e)}{J_0(\varphi_m \chi_e)} J_1(\varphi_m \chi_e)-Y_1(\varphi_m \chi_e)\right]\int\limits_0^{\beta_e}\sinh(\varphi_m z)\sin(\varsigma_m z)dz}{\lambda_m I_1(\varsigma_m \chi_e)-\frac{\lambda_m I_0(\varsigma_m r)\left[\frac{K_1(\gamma_m \xi_e)}{I_1(\gamma_m \xi_e)} I_1(\gamma_m \chi_e)-K_1(\gamma_m \chi_e)\right]}{\left[\frac{K_1(\gamma_m \xi_e)}{I_1(\gamma_m \xi_e)} I_0(\gamma_m \chi_e)+K_0(\gamma_m \chi_e)\right]}\int\limits_0^{\beta_e}\sin^2(\varsigma_m z)dz}$	

Evaporator Thermal Circuit Model

Although less precise, thermal circuit simplified models can be useful for the design of LHPs, especially for the prediction of compensation chamber temperatures and of the maximum heat power that the system is able to transfer. Kaya et al. (1999) employed thermal circuit analogy models to determine the operation temperature of cylindrical evaporator loop heat pipes. Later, Singh et al. (2009) also used these models to describe the operation of heat pipe with plane evaporator.

The sketch of the equivalent thermal circuit of the LHP is presented in Fig. 7.17. Vapor and liquid lines are considered adiabatic. The net sensible heat q_l eventually transported by the vapor through the liquid line is removed in the condenser region of the LHP.

In this circuit, the heat q_e, delivered to the evaporator external wall (at temperature T_e) has two different paths to follow. In one path, represented by R_δ, useful heat q_δ is delivered to the wick structure and is transferred up to the wick working fluid meniscus, where phase change takes place, at temperature T_v. In the other path, heat (considered as loss) is conducted to the compensation chamber at temperature T_{cc}, through two paths: q_c through the evaporator casing wall (resistance R_c,) and q_β through the wick plug (resistance R_β). Heat q_s can also be exchanged directly between the compensation chamber and the liquid lines through the solid material resistance R_s. In order to obtain a good device performance, it is expected that this heat leakage is compensated by the working fluid that enters the compensation chamber

Table 7.5 LHP evaporator equations for pressure field of LHP evaporator hollow solid cylinder

Solutions	Eigenvalues
$\Psi_6(r,z) = \sum_{n=1}^{\infty} W_n\left[\frac{K_1(\epsilon_n\xi_e)}{I_1(\epsilon_n\xi_e)} I_0(\epsilon_n r) + K_0(\epsilon_n r)\right][\sin(\epsilon_n z) - \tan(\epsilon_n\beta_e)\cos(\epsilon_n z)]$	$\tan(\epsilon_n\beta_e) = -\cot(\epsilon_n\nu_e)$
$W_n = \frac{-\frac{\mu_l k_{ef}(T_v - T_{cc})}{K\rho_l h_{lv}}}{\epsilon_n\left[\frac{K_1(\epsilon_n\xi_e)}{I_1(\epsilon_n\xi_e)} I_1(\epsilon_n\xi_e) - K_1(\epsilon_n\chi_e)\right]} \frac{\int_{\beta_e}^{\nu_e} \left.\frac{\partial\theta}{\partial r}\right\vert_{r=\chi_e} [\sin(\epsilon_n z) - \tan(\epsilon_n\beta_e)\cos(\epsilon_n z)]dz}{\int_{\beta_e}^{\nu_e} [\sin(\epsilon_n z) - \tan(\epsilon_n\beta_e)\cos(\epsilon_n z)]^2 dz}$	

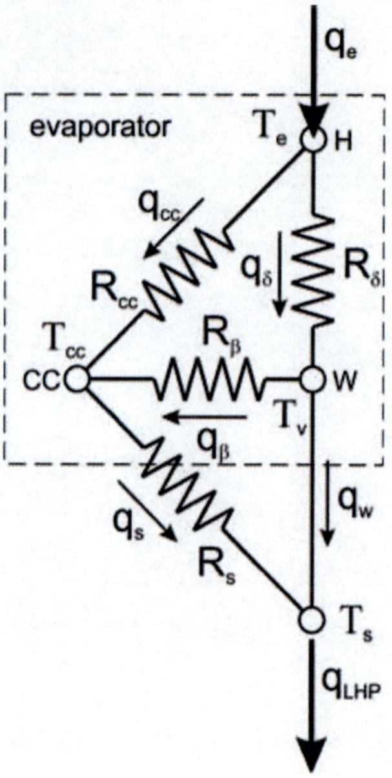

Fig. 7.17 Sketch of the equivalent thermal circuit of the loop heat pipe

in subcooled liquid state. An energy balance at node H (see Fig. 7.17) results in:

$$q_e = q_{cc} + q_\delta \tag{7.50}$$

where q_e is the heat delivered to the evaporator, q_{cc} is the heat transferred through the case to the compensation chamber and q_δ the heat through the wick structure to the liquid vapor meniscus. Using the thermal resistance definition (see Eq. 4.1) this last expression takes the following form:

$$q_e = \frac{T_e - T_{cc}}{R_{cc}} + \frac{T_e - T_v}{R_\delta} \tag{7.51}$$

where R_{cc} represents the conduction thermal resistance through the evaporator casing, i.e., ratio of the total length of the evaporator $l_{t,e}$ to the conduction cross section area, expressed as:

$$R_{cc} = \frac{l_{t,e}}{2\pi\left[(\xi_e + \delta_c)^2 - \xi_e^2\right]k_c} \tag{7.52}$$

where δ_c is the thickness of the casing of the evaporator tube wall and k_c is the thermal conductivity of the casing material. R_δ is the radial conduction thermal resistance of the wick hollow cylinder, given by:

$$R_\delta = \frac{\ln\left(\frac{\xi_e}{\chi_e}\right)}{2\pi l_e k_{ef}} \tag{7.53}$$

Likewise, the heat balance of node W gives:

$$q_\delta = q_w + q_\beta \tag{7.54}$$

where:

$$q_\delta = \frac{(T_v - T_{cc})}{R_\beta} + q_w \tag{7.55}$$

Besides, the thermal resistance of evaporator plug region is expressed as:

$$R_\beta = \frac{\beta_e}{\pi \xi_e^2 k_{ef}} \tag{7.56}$$

Finally, the energy balance for node CC becomes:

$$q_s = q_{cc} + q_\beta \tag{7.57}$$

where q_s is the sensible heat transferred, used for subcooling of the liquid that returns from the condenser, that can be determined from:

$$q_s = \dot{m}_l c_p (T_s - T_{cc}) \tag{7.58}$$

where T_s is the compensation chamber returning fluid temperature. This temperature can be obtained by heat transfer analyses of the condenser region.

Condenser modelling. The condenser is an important component for the performance of LHPs. It should be designed to reject the heat inserted in the LHP in steady state conditions. Besides, the condenser pressure drop and the output temperature of the condensate influence the operation temperature of LHP. The vapor to liquid phase change happens inside the condenser. A typical condenser presents regions with only single phase flows (subcooled liquid, superheated vapor) and others with two phase flows. The pressure drops of these regions must be determined.

In single phase liquid or vapor flow regions, it is considered that only sensible heat is exchanged, while, for two phase flow regions, only latent heat is transferred.

The condenser configuration of the LHP case under study consists of a U shaped bent tube, welded to a flat plate (see Fig. 7.13), from which heat is removed. Figure 7.18 shows a schematic of the cross section of the tube, welded to a cooling plate. The tube is considered straight, with the external face subjected to convection heat transfer to the ambient air. The equivalent thermal resistance network is also shown in this figure.

Fig. 7.18 Interface between the flat plate and the condenser tube (ith position)

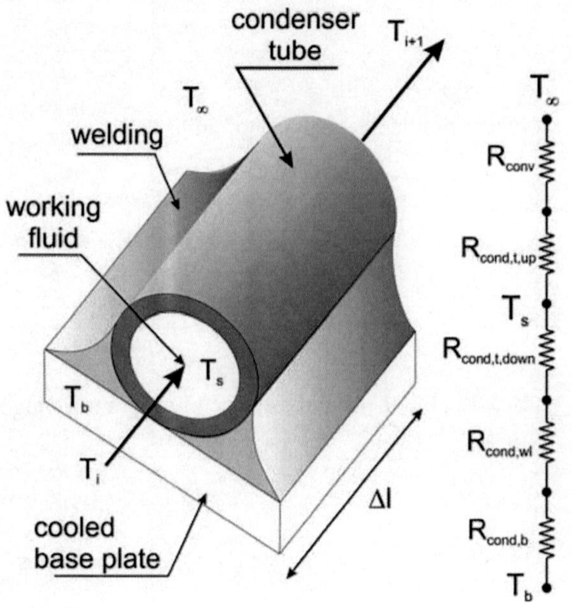

The tube length must be long to guarantee enough area for the working fluid to change of phase from vapor to liquid. The heat exchanged by convection with the environment is:

$$q_{lv} = \bar{h} A^{2\phi} (T_{sat} - T_s) \tag{7.59}$$

where:

$$A^{2\phi} = 2\pi r_i l^{2\phi} \tag{7.60}$$

is the internal area of the condenser in which happens the vapor-liquid phase change and $\bar{h}$ is the equivalent phase change coefficient of a fluid flowing in a cooled horizontal tube which, according to Chato (1962), can be determined from:

$$\bar{h} = 0.555 \left[\frac{g \rho_l (\rho_l - \rho_v) k_l^3 h'_{lv}}{\mu_l (T_{sat} - T_s) D_i} \right]^{\frac{1}{4}} \tag{7.61}$$

where D_i is the internal diameter of tube. The modified latent heat of vaporizations is:

$$h'_{lv} = h_{lv} + \frac{3}{8} c_{p,l} (T_{sat} - T_s) \tag{7.62}$$

Conduction through the cooled flat plate and convection of the tube with environment are the heat transfer mechanisms that provides the energy for the working fluid to change of phase from vapor to liquid:

$$q_{lv} = q_{cond} + q_{conv} \tag{7.63}$$

In terms of thermal resistances, this last equation takes the form:

$$\dot{m}\, h_{lv} = \frac{T_e - T_\infty}{R_{cond,b} + R_{cond,wl} + R_{cond,t,down}} + \frac{T_e - T_b}{R_{cond,t,up} + R_{conv}} \tag{7.64}$$

where $R_{cond,b}$ is the conduction thermal resistance through the base plate close to the welding area, $R_{cond,wl}$ is the conduction through the welding material, $R_{cond,t,down}$ and $R_{cond,t,down}$ are the conduction thermal resistance in the down and up directions, respectively and R_{conv} is the convection heat transfer to the environment. These conduction resistances are obtained from the simple one-dimensional expressions ($R_{cond} = l/kA$ or $R_{cond} = (ln(r_e/r_i))/2\pi kl$), depending on the geometry of the tube plus the base plate.

According to Churchill and Chu (1975), the mean thermal coefficient $\bar{h}$ to be used in the determination of the convection thermal resistance ($R_{conv} = 1/hA$), can be predicted for a fluid flowing within a cooled tube, by the following equation:

$$Nu = \left\{ 0.60 + 0.387 Ra_D^{1/6} \Big/ \left[1 + \left(\frac{0.559}{\text{Pr}} \right)^{9/16} \right]^{8/27} \right\}^2 \quad valid\ for\ Ra_D \leq 10^{12} \tag{7.65}$$

with the Rayleigh number Ra_D, based on external diameter of condenser tubes:

$$Ra_D = \frac{g\beta(T_e - T_\infty)D_c^3}{\nu\alpha} \tag{7.66}$$

For the schematic shown in Fig. 7.18, the area exposed to convection is approximately around ¾ of the external tube area, i.e.:

$$A_c = \frac{3\pi}{2} r_e l^{2\phi} \tag{7.67}$$

where R_{aD} is the Rayleigh number based on external diameter of condenser tube.

The output temperature of the condenser is determined from an energy balance of the second region of the tube, where the already condensed working fluid is subcooled. The heat balance over this condenser region (only liquid phase) in a position I, at temperature T_i (see Fig. 7.18) is:

$$q = hA(T_i - T_s) \tag{7.68}$$

The sensible heat removed to cool down the working fluid between two consecutive positions in the condenser *i* and *i + 1* is:

$$\dot{m}_l c_p (T_i - T_{i+1}) = \frac{\overline{T} - T_\infty}{R_{cond,b} + R_{cond,wl} + R_{cond,t,down}} + \frac{\overline{T} - T_b}{R_{cond,t,up} + R_{conv}} \quad (7.69)$$

where:

$$\overline{T} = \frac{T_i - T_{i+1}}{2} \quad (7.70)$$

Condenser pressure drop modelling. The pressure drop observed in the two-phase flow region represents an important parcel of the total pressure drop observed in LHPs. The length of the condenser within which the working fluid changes phase from vapor to liquid is determined using the thermal model presented in the last section. In this section, models used to calculate the pressure drop are presented.

In Sect. 3.7, a model for the determination of the pressure drop limit (maximum *H*) in thermosyphons was discussed. Although that section was devoted to thermosyphons, usually larger than LHP devices, the two-phase flow pressure drop model can also be used for designing condensers of LHPs. In this section, other classical correlations, which work well with small devices (like the one schematically shown in Fig. 7.13) are presented.

The pressure drop of a fluid flowing within a tube depends on the liquid to tube friction factor f, which can be calculated using the Churchill and Chu (1975) correlation, given by:

$$f = 8\left[\left(\frac{8}{\mathrm{Re}}\right)^{12} + \left\{\left[2.457 \cdot \ln\left(1 \Big/ \left(\left(\frac{7}{\mathrm{Re}}\right)^{0.9} + \left(\frac{0.27 R_a}{D}\right)\right)\right)\right]^{16} + \left(\frac{37530}{\mathrm{Re}}\right)^{16}\right\}^{-1.5}\right]^{1/12} \quad (7.71)$$

where R_a is the tube surface roughness. This expression can be used for the calculation of the "liquid only" friction factor f_l, substituting the Reynolds number:

$$\mathrm{Re}_l = \frac{\dot{m}_l'' D_i}{\mu_l} \quad (7.72)$$

The "liquid only" phase pressure drop is determined by the expression:

$$\left(\frac{dp}{dz}\right)_l = \frac{f_l \left(\dot{m}_l''\right)^2}{2 D_i \rho_l} \quad (7.73)$$

The vapor friction factor f_v can also be determined by Eq. 7.71, substituting the Reynolds number, given by the following expression:

$$\mathrm{Re}_v = \frac{\dot{m}_l'' \, D_i}{\mu_v} \tag{7.74}$$

The "vapor only" phase pressure drop is determined by the expression:

$$\left(\frac{dp}{dz}\right)_v = \frac{f_v \left(\dot{m}_l''\right)^2}{2D_i \rho_v} \tag{7.75}$$

For the two-phase flow region, the friction factor $f_{l2\phi}$ for the liquid is determined by the same Eq. 7.71, substituting the Reynolds number by the expression:

$$\mathrm{Re}_{l2\phi} = \frac{\dot{m}_l'' D_i (1 - X)}{\mu_l} \tag{7.76}$$

where X is the vapor quality. The pressure drop is determined by:

$$\left(\frac{dp}{dz}\right)_l = \frac{f_{l2\phi} \left(\dot{m}_l''\right)^2 (1 - X)^2}{2D_i \rho_l} \tag{7.77}$$

Again for the two-phase flow region, the friction factor $f_{v2\phi}$ for the vapor is determined by Eq. 7.71, substituting the Reynolds number by the expression:

$$\mathrm{Re}_{v2\phi} = \frac{\dot{m}_l'' D_i X}{\mu_v} \tag{7.78}$$

Therefore, the vapor pressure drop in the two-phase region is determined by:

$$\left(\frac{dp}{dz}\right)_l = \frac{f_{v2\phi} \left(\dot{m}_l''\right)^2 X^2}{2D_i \rho_v} \tag{7.79}$$

7.4 Vapor Chamber

The major application of vapor chamber is the substitution with advantages of conventional heat sinks of electronic equipment, spreading concentrated heat over larger areas, where it can be removed by convection.

Vapor chambers can be described as flat heat pipes, where, in one face, there is a wick structure where the working fluid is spread and which external surface is in contact with the concentrated heat source. The liquid spreading along the device is quite important for its efficiency. A single vapor chamber may be able to deal with more than one heat sources. The working fluid in the heated area evaporates and the vapor occupies the void volume over the wick. The opposite surface is subjected to a cooling source, which condenses the vapor of the working fluid that returns to the evaporator through the walls of the devices, as a result of the liquid-metal surface tension. Drops of liquid may fall directly in the wick structure of the evaporator, due to the gravity action.

Figure 7.19 illustrates the working principles of a vapor chamber. Several electronic devices make use of vapor chambers to control the temperature of electronic components. Usually the evaporator of vapor chambers are located in the bottom positions to take advantage of gravity, although wick structures may also take the role of transferring the working fluid to the evaporator regions. Due to the flat geometry, columns are usually provided inside the chamber, as, according to the working fluid and the operation temperature, the chamber pressure can be high or low and so the closing plate can be deformed.

Vapor chambers are subjected to the heat transfer limitations, as for other two-phase devices. A good design of a vapor chamber must take into consideration the wick structure, vapor path and phase change phenomena. Special porous media can be used to allow the liquid spreading along the device, with low pressure drops even when the devices operate in inclined positions (Mizuta et al. 2016).

Michels et al. (2012) constructed and tested a hollow finned vapor chamber prototype, as shown in Fig. 7.20. As the vapor is able to reach up to the top of the fins, each fin actually works as a condenser of a thermosyphon. This device showed an overall thermal resistance 20% smaller than the conventional heat exchanger. The correct amount of working fluid depends on the heat transfer rate.

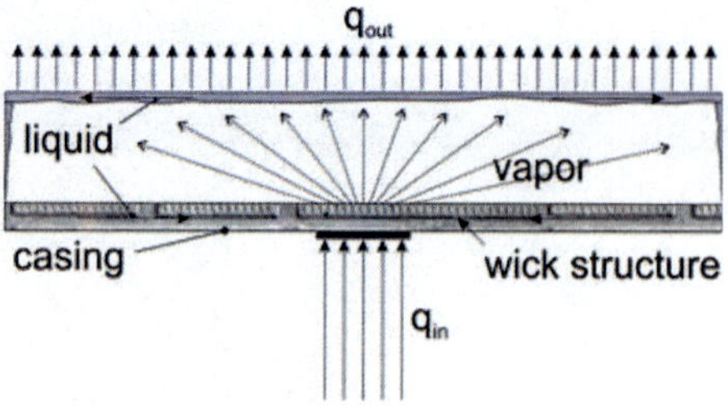

Fig. 7.19 Schematic of vapor chamber

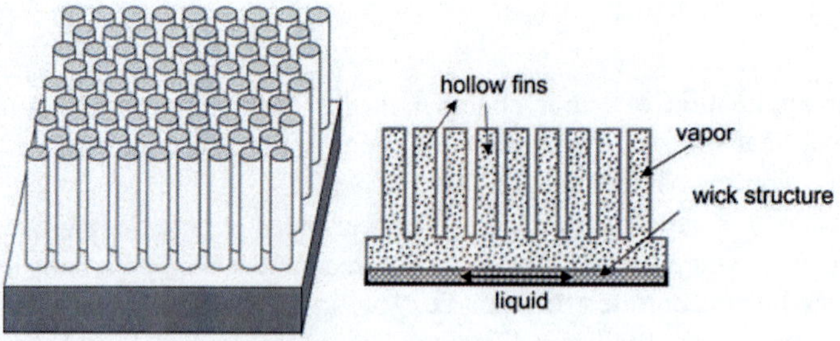

Fig. 7.20 Vapor chamber with hollow fins

Blet et al. (2017), state that gravity may have a negative effect over the vapor chamber performance if they operate with large inclinations with the horizontal position.

Boukhanouf et al. (2006) investigated experimentally the thermal performance of a vapor chamber ($250 \times 200 \times 5$ mm^3) in which the working fluid was water. A thermographic camera was used to observe the temperature distribution in the evaporator and condenser. The concentrated heat source area was around 10 times smaller than the evaporator area. The vapor chamber showed excellent heating spreading conditions, resulting in a thermal resistance 40 times smaller than that of a copper block of same external dimensions.

Chen et al. (2007) compared experimentally vapor chamber with bulk cooper and aluminum blocks of $86 \times 71 \times 5$ mm^3 for water as working fluid. They concluded that the overall thermal resistance of the vapor chamber decrease with the increasing input power. Besides, the vapor chamber mitigated the hot spots. Tsai et al. (2013) also investigated experimentally the influence of the inclination angle on the overall thermal resistance of a copper $90 \times 90 \times 3.5$ mm^3 vapor chamber, with the wick structure made of sintered copper. Water was the working fluid. They concluded that the device worked better at horizontal position (evaporator in the bottom). Ji et al. (2012) evaluated the influence of working fluid, filling ratio and inclination on a vapor chamber that used cooper foam as wick structure. This device operates with heat fluxes of up to 216 W/m^2.

Liu et al. (2018) designed and tested a vapor cooper chamber with solid internal columns, which were used to provide mechanical resistance to the device avoiding padding due to the working fluid vapor pressure. The columns were wrapped up by a copper foam to help in the fluid spreading. They showed that these columns increased the capillary limit of the device, contributing for the liquid circulation.

Zeng et al. (2018) studied the performance of an aluminum vapor chamber for electronics cooling, where the wick structure consisted of micro-grooves, for the thermal control of high power electric devices. Tests were performed under several inclinations. They showed that the vertical position compromised but not prevented the device operation.

Thermal Resistance Model

The thermal resistance model, similar to the one describe in Chap. 4, can be applied to model vapor chambers.

Wei and Sikka (2006) proposed a thermal circuit to model the heat transfer phenomena in a vapor chamber. Figure 7.21 shows the thermal circuit adopted. Six resistances, in series, are considered: conduction through the evaporator wall $\left(R_{cond,e}\right)$, conduction through the wick structure in the evaporator $\left(R_{w,e}\right)$, liquid-vapor phase change resistance $\left(R_{pc,lv}\right)$, vapor-liquid phase change resistance $\left(R_{vl,pc}\right)$, conduction through the wick structure in the condenser $\left(R_{w,c}\right)$ and conduction through the condenser wall $\left(R_{cond,c}\right)$. Spreading resistance can be considered and theoretical models (see Muzychka et al. 2003) can be used to estimate the evaporator conduction resistance, as for in Sect. 7.1.

Fig. 7.21 Thermal circuit for a vapor chamber

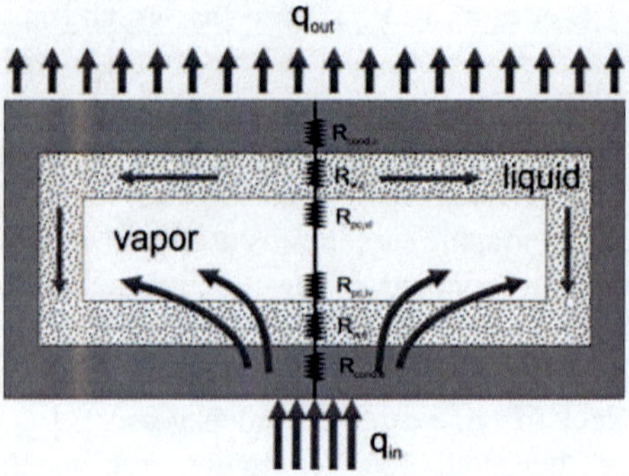

Prasher (2003) also developed a thermal circuit similar to the one presented here. He concluded that the overall thermal resistance is very sensitive to the wick structure characteristics (thickness and effective thermal conductivity) and the wick structure conduction is the largest resistance of the vapor chamber circuit.

7.5 Pulsating Heat Pipes (PHP)

In 1990, Akachi registered a patent regarding new heat transfer devices, the pulsating heat pipes (PHP), which are based on the phase change mechanisms of a fluid, similar to the heat pipes but without the use of a wick structure. Since then, many researchers around the world are devoting efforts in modelling, improving their thermal performances and developing different designs, suitable to real applications.

Basically, a traditional PHP is made from small diameter tubes, meandered in serpentine shape, vacuumed and charged with a controlled amount of working fluid. Figure 7.22 shows schematics of PHPs. As for other two-phase heat transfer devices, a PHP is composed by three sections: evaporator, adiabatic section and condenser.

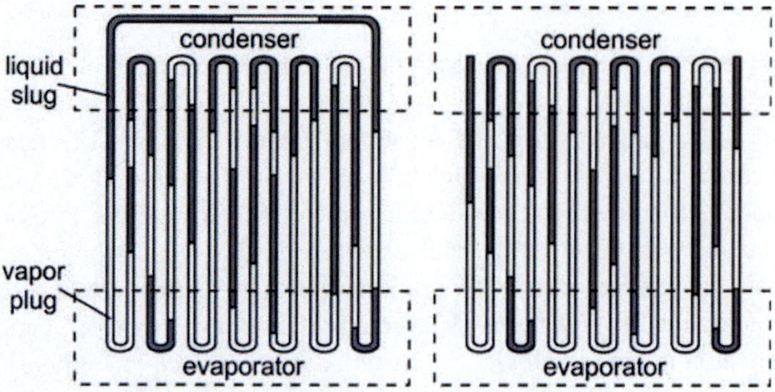

Fig. 7.22 Schematic of pulsating heat pipes: left: close circuit; right open sealed circuit

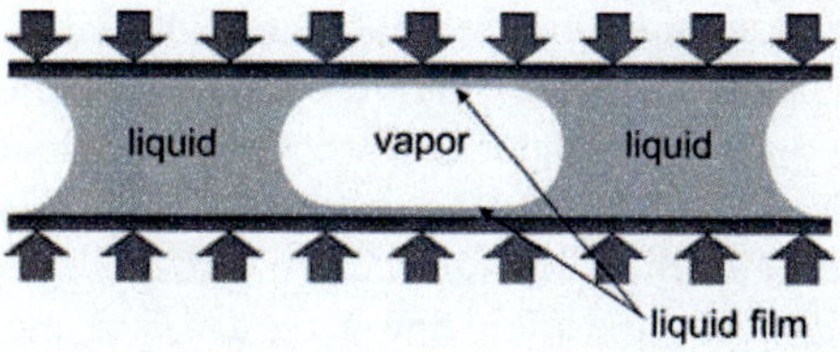

Fig. 7.23 Vapor slug and liquid plug in a pulsating heat pipe in the evaporator section

In the evaporator section, liquid receives heat and evaporates, forming vapor bubbles that grow up to the tube diameter. These bubbles can occupy a considerable volume of the tube. They are intercalated with liquid plugs and the set (vapor slugs and liquid plugs) move together along the tube, as "train wagons" (see Fig. 7.23) from the evaporator to the condenser, where the working fluid lose heat and the vapor is partially condensed. The expansion (evaporator) and contraction (condenser) of the vapor bubbles results in forces over the working fluid able to force its movement along the tubes.

Heat is transferred from the condenser to the evaporator by two major mechanisms: latent (phase change) and sensible (convection). Actually, PHPs tend to be an isochoric system, and, due this characteristic, the sensible heat transfer by the liquid plugs is the major heat transfer mechanism, representing more than 80% of the total heat transferred (Ma 2015; Kandekhar et al. 2003 and Yuan et al. 2010).

The evaporation is aided by the thin liquid films over the evaporator internal surface caused by the vapor bubble dislocation along the tube (see Fig. 7.23). Evaporation is more effective in thin films (see Chap. 2). The high evaporation process causes the bubble internal pressure to increase quickly. When the bubble internal pressures are equal or larger than the pressure drops in the channel, the vapor bubbles are able to push the liquid plugs up to the condenser, which, in turn, push the bubbles. In the condenser, heat is lost and vapor is partially condensed. The "vanishing" of the bubbles in the evaporator section results in a pressure gradient along the tube, which also contributes for the imbalance of forces and, consequently, to the working fluid displacement. The growing and collapsing of bubbles are random phenomena, causing a thermal excited oscillation of the working fluid inside PHPs, being their operation principle (Ma 2015).

As for the two-phase devices, the working fluid needs to be carefully selected. As discussed in Chap. 5, it should be compatible with the casing material (no generation of non-condensable gases). Water is usually a good working fluid due to its high latent heat, high availability, environmental safety and low cost (Reay and Kew 2006).

Usually, PHPs can be arranged as a closed circuit or open sealed tube (see Fig. 7.22, left and right side, respectively). As already observed, they are usually formed by meandering small diameter tubes in serpentine fashion. PHP can also present a flat plate geometry, as discussed in the following section.

Flat plate PHP

In flat plate PHPs, the meandering tubes are substituted by internal channels located inside a solid block. Usually, internal channels can be made by the superposition and union of grooved flat plates, in their machined faces. For circular cross section channels, semicircular grooves are machined in the superposition faces, while, for square cross section, the grooves are machined in one of the plates, while the other one is used for closing the channels. Figure 7.24 shows a drawing of flat plate PHPs and the cross section of the device with these two channel geometries: round and square. Basically, the liquid flow paths can have any geometry. Comparing to the meandered, more channels per overall area can be provided in the flat plate apparatus, as there is no need to respect the minimum radius of the U turns, as for the meandering PHPs. Besides, the internal surfaces can be treated before the joining process is performed. In PHPS, the conduction through the plate material plays important role as a heat transfer mechanism.

According to Betancur et al. (2018), the fabrication of flat PHPs still represents a challenge. Different technologies have been used to create reliable monolithic pieces, such as the additive manufacture and power bed fusion (Ibrahim et al. 2017). A new fabrication process, the diffusion bounding (see Chap. 5), have been considered (Facin et al. 2018 and Soo et al. 2019). Care must be taken so that the PHP fabrication process does not alter the eventually modified surfaces, as diffusion bonding involves high vacuum and high temperatures.

Influencing Parameters

Different authors have reported considerable improvement in the performance of PHP when the pressures inside the device are unbalanced, forcing the liquid to flow in a preferential direction. This can be obtained, for instance by introducing a capillary media or check valves (Chien et al. 2014; Betancur et al. 2018 and Yang et al. 2015).

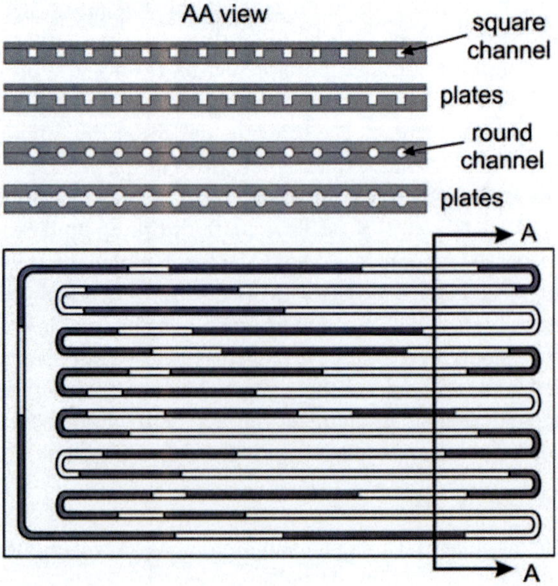

Fig. 7.24 Flat plate pulsating heat pipe: round and square channels

These procedures decrease the temperature oscillations and so increase the thermal performances, decreasing the power input needed for startup and decreasing the gravity dependency. However, these modifications improve the complexity of the PHP manufacture and might increase the fabrication costs. The following parameters, which influence the thermal performance of PHPs, are treated separately in the next sections:

- Geometry of the tubes/channels
- Filling ratio
- Heat flux
- Number of turns
- Inclination angle
- Cross section geometry
- Internal channel surface finishing

Geometry of the Tubes/Channels

The PHP channels or tubes must be dimensioned so that the solid-liquid surface tension must be enough to keep the liquid plug, which are separated by vapor slugs (Hoesing 2014). For meandering PHPs, the internal diameter of the tubes is the most important geometry parameter. Khandekar and Groll (2004) affirms that the pulsation is only observed in determined range of diameters, in which the Bond (Bo) number must satisfy:

$$Bo^2 = \frac{D^2 \cdot g \cdot (\rho_l - \rho_v)}{\sigma} \leq 4 \tag{7.80}$$

Isolating the diameter, results in the following expression:

$$D \leq 2\sqrt{\frac{\sigma}{g \cdot (\rho_l - \rho_v)}} \tag{7.81}$$

Lin et al. (2009) presents a more strict upper boundaries for the diameter:

$$0.7 \cdot \sqrt{\frac{\sigma}{g \cdot (\rho_l - \rho_v)}} \leq D \leq 1.8 \cdot \sqrt{\frac{\sigma}{g \cdot (\rho_l - \rho_v)}} \tag{7.82}$$

As the tube diameter increases, the thermal behavior of the PHP tends to that of two-phase thermosyphons. Besides, the lower is the tube diameter, the higher is the vapor slug capacity of pumping the working fluid, but the higher is the fluid flow friction forces.

Qu e Wang (2013) studied the effect of the internal diameter over the thermal resistance of PHPs in association of the working fluids: water and ethanol. For the same filling ratio (50%), tubes of diameter 1, 2.2 and 2.4 mm were tested. They concluded that the thermal resistance of PHP with water decreased with the increase of the diameter, but the opposite happened with ethanol. Therefore, there is no clear relation between dimeter of the tube and thermal resistance.

Charoensawan et al. (2003) tested the effect of the association of the diameter and number of turns in PHPs, as a function of the power input. They observed that, for a same diameter and evaporator length, the power input that the device is able to transfer increases with the increase of the channel diameter and with the number of turns. Actually, it is expected that the number of U turns in PHPs increases the perturbation level of the working fluid flow and so increases the efficiency of the device (Khandekar et al. 2004). Charoesawan e Terdtoon (2008) tested the effect of this parameter in horizontal PHPs with 50% of filling ratio with water. This device operated at 90 °C at the evaporator section. The PHP was tested with two different evaporator lengths: 50 and 150 mm. They concluded that the influence of the number of U turns is more important for PHPs with larger evaporators.

Filling Ratio

The filling ratio for PHPs is defined as the volume of liquid and the total internal volume of the device. For filling ratios of 0%, the tube has no working fluid and the device transfers heat only by conduction. Heat transfer by natural convection of the working fluid is observed when the PHP is completely filled, because there is no available space for bubbles (Farsula 2009).

Several researchers have developed experimental work to evaluate the ideal filling ratios, with different conclusions. Considering meandering PHPs, Gamit et al. (2015) studied experimentally water PHP for the filling ratios of: 40, 50 and 60%, subjected to low heat fluxes (30 and 40 W), concluding that the thermal resistance tends to increase with increasing filling ratios. Charoensawan and Terdtoon (2008) tested the filling ratios of 30, 50 and 80% of a closed circuit PHP made of 2 mm diameter tube, with 26 turns and operating in vertical position, concluding that the filling ratio affects more the PHP performance when the evaporator length is large. For low evaporator lengths, the heat transfer capacity of the device showed to be small.

The filling ratios for flat plate PHPs were studied by Yang, et al. (2009) for aluminum plates with 40 and 66 channels, with square cross section channels of 2×2 and 1×1 mm^2. Ethanol was used as working fluid. These devices were tested in horizontal and vertical (evaporator down, gravity assisted) positions. They concluded that, for horizontal position, the best filling ratios varied between 50 and 65%, while, in the vertical position, the ideal filling ratio varied according to the power input.

Heat Flux

The heat flux is a very important parameter for the operation of PHPs. In general sense, the higher the power, the best is the PHP performance. For lower powers, pulsation may not occur for operation in vertical positions (evaporator down). On the other hand, high heat fluxes may cause dry-out of the device. Besides, Khandekar et al. (2004) comments that a high heat flux is necessary for the PHP to operate, independently of the conditions, such as the tilt angle.

Gamit et al. (2015) tested PHPs with three different filling ratios, for heat transfer ranging between 10 to 50 W, concluding that the oscillations and so the thermal resistances decrease as the power input increases, especially for PHPs with small filling ratios. The heat flux imposed in the evaporator has a major influence in the flow pattern inside the PHP channels.

Khandekar and Groll (2003) studied the thermal resistance behavior of a PHP subjected to different heat fluxes, illustrated in Fig. 7.25. For low power inputs (Zone I), no pulsating movement of the working fluid is observed: eventually, small amplitude, high frequency oscillation bubbles may be formed. The thermal resistance of the PHP in this region is very high. With the increase of the heat flux, these oscillations gain force (Zone II), their amplitude increase and slug flow is observed. According to Collier and Thome (1994), plug flow regime, in horizontal channels, is similar to the slug pattern in vertical tubes. Also, the bubbles formed tend to be spherical as they move along the device. In Zone II, the flow has random directions and the total heat exchange is composed by sensible and latent heats. At Zone III, the flow tends to follow one preferential direction, decreasing the thermal resistance; oscillations are still observed but with lower amplitude. In Zone IV, as the heat input increases, the flow transition from piston to annular is observed. In Zone V, the dry-out happens and the thermal resistance increases.

Khandekar and Groll (2003) also observed that, in Zone III, at a determined time frame, the fluid flow direction in alternate channels change, so that, in some channels, hot annular/semi-annular liquid-vapor flow moves towards the condenser. The thin liquid film of the annular flow receives heat and vaporizes more intensively than at the nucleate boiling of the piston flow regime. Therefore, a decrease of the thermal resistance can be observed.

Inclination Angle

Although PHPs might operate efficiently no matter the inclination, the tilt angle plays an important role in their thermal performance. These devices present better performance when operate inclined in relation to the horizontal position, if the evaporator is located at the inferior position, as they tend to work as thermosyphons. The larger

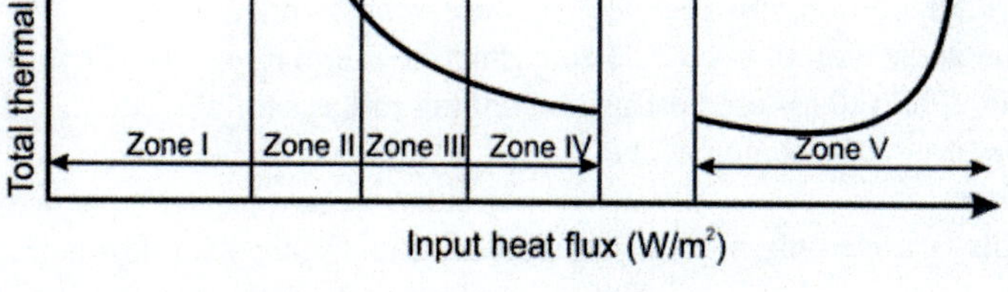

Fig. 7.25 The effect of the input heat flux in the thermal resistance

is the inclination, the larger is the thermal performance of the device, as stated by Khandekar et al. (2003a, b) and Yang et al. (2004), who tested meandered tube PHPs and Xu et al. (2006), who studied flat plate PHP.

Charoensawan and Terdtoon (2008) as well as Khandekar et al. (2003b) demonstrated that there is a minimum number of turns necessary for a PHP be able to start up at horizontal position.

Cross Section Channel Geometry

The effect of the cross section geometry of the channel over the thermal performance of PHPs is the subject of analysis of several researchers. Vassilev et al. (2007), used different cross section channel geometries for the evaporator and condenser sections: squared for evaporator and circular for condenser. They tested smaller than conventional filling ratios for the fluids: water, ethanol and methanol, respectively. They concluded that the channel corners of the squared geometry cause a capillary effect that maintain the evaporator wet along its length, avoiding the occurrence of dry-out. In this case, the devices do not need much working fluid to operate.

Lin et al. (2011) studied the heat transfer characteristics in flat plate PHPs made of aluminum with square cross section areas. Hua et al. (2017) studied the startup of PHPs with rectangular cross section channels, with 4 mm of height. Facin et al. (2018) studied the influence of the channels geometry for flat plate PHPs fabricated by diffusion bonding technique (see Chap. 5). They fabricated and tested PHP with square and round cross sections. Agreeing with other authors, Facin et al. (2018) concluded that squared PHP are more efficient heat transfer devices, presenting lower thermal resistances.

Literature studies concerning mini channel PHPs are also studied in the literature. Lee e Kim (2017) selected two geometries: circular and squared to study the operation limits of micro PHPs of hydraulic diameters of 390, 480 and 570 μm. They concluded that the heat transfer operation limit increases with the increase of the hydraulic diameter for both sections. The square channel was able to stand maximum heat fluxes around 70% larger than the circular one, for the same hydraulic diameter. Khandekar et al. (2003) highlight that about half of the minichannels tested had squared cross section, around 29.5% were circular and the other 20% was divided among trapezoidal, triangular and other geometries. Cai et al. (2006) concluded that microgrooves reduce the evaporator temperature fluctuations, improving the thermal conductivity of the device. Kim and Kim (2018) proposed internal helical microgrooved PHPs, resulting in the enhancement of the heat transport capability, both at horizontal and vertical orientations. The effect of the reentrant cavities was also investigated in a micro PHP and the researchers concluded that such cavities promote the nucleation of bubbles, decreasing the thermal resistance by up to 57%.

Betancur et al. (2020) used diffusion bonding process to fabricate flat plate PHP with special grooved geometry. They tested 10 and 26 parallel channel PHPs, with three different cross section geometries: circular, square and circular with lateral grooves, this last one only in the evaporator section. Figure 7.26 shows drawings of these three cross sections. The diffusion bonding parameters were selected to avoid deformation resulting from the fabrication process. The cross section geometries

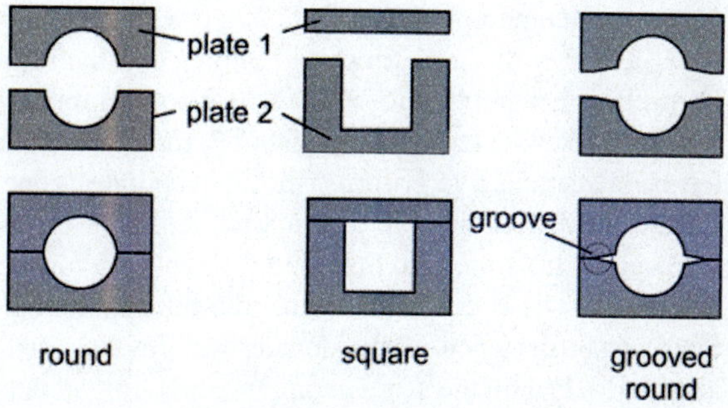

Fig. 7.26 Cross section area of the channels fabricated by diffusion bonding of flat plates

of the bonded PHP were characterized by metallography and microscope analysis. They concluded that the circular cross section channels presented lower deformation resulting from the fabrication process than the square ones. Deformation can be a problem for the squared channel devices, since, due to the diffusion bonding fabrication process (application of high pressure and high temperature), the external plate surfaces tend to lose its flatness, bending slightly, especially towards the channel open spaces. This can be disadvantageous for square and circular cross section geometries, but not for the circular with grooves, as deformation due to fabrication process tends to make the lateral grooves shaper, increasing the capillary effect of the corners over the liquid. The capillary effect is responsible for spreading the liquid along the whole evaporator area, even in the regions occupied by the vapor slug.

Betancur et al. (2020) also observed, in their experimental studies, that the "stop over" phenomenon, which under certain conditions, happens in the circular cross section channels, was vanished for the grooved channel PHP. The "stop over" is a temporary instability of the liquid motion, which suddenly stop to oscillate, with the return of the oscillatory movement after a while. The thermal resistance increases at these moments. This phenomenon was also observed by Qu et al. (2017). Betancur et al. (2020) stated that the amplitude of oscillations with the grooved channel PHP was considerably reduced, as well as the device overall thermal resistance, for high power inputs.

Internal Channel Surface Finishing

According to Betancur et al. (2019), the internal channel surface finishing is an important parameter, as it affects the working fluid boiling process in the evaporator, creating vapor nucleation sites and modifying the fluid wettability behavior, i.e., the liquid contact angle and meniscus radius (Ibrahim et al. 2017). Surface modification process can be chemical, physical, by coating or by the presence of a porous medium.

Literature studies of the effect of the supehydrophobic CuO chemical coatings on the fluid-solid wettability of PHP evaporator channels have been performed. It was concluded that this coating increases the wettability, improving the thermal performance of the device (Ji et al., 2013). Another work (Hao et al. 2014) has shown that hybrid pattern, i.e., the use of CuO (superhydrophobic coating) at condenser and superhydrophilic at evaporator regions reduces the overall thermal resistance.

However, Betancur et al. (2019a) in their work, arrived to different conclusions. They tested a PHP with alternating hydrophilic/superhydrophobic channels at vertical position bottom heat mode and different heat power inputs. The device consists in a copper tube of internal/external diameters of 3.18/4.76 mm, bent into a planar serpentine of ten channels and five U-turns. In the evaporator zone, the tube was functionalized using a superhydrophobic coating, while, in the other regions, the internal surfaces received no treatment, so that the tube kept its hydrophilic bare copper surface characteristics. The startup conditions, the fluid motion along the tubes and the overall thermal performance of the device, was investigated and compared with those of another PHP, with the same geometry, operating under the same working conditions, but with no treated (hydrophilic) surfaces. Distilled water, at 50% filling ratio was used as the working fluid. Power inputs varying from 20 to up to 350 W were applied and the condenser temperature was kept at 20 °C. They concluded that, in general, the alternating PHP presents a worse overall thermal performance. They observed that in some operation conditions, the liquid flow was blocked in the functionalized region, while, for the hydrophilic inserts, more pronounced temperature fluctuations are observed. The superhydrophobic coating actually hindered the liquid film formation, decreasing locally the flow motion. On the other hand, the enhancement of the inner wettability improves the flow motion, as the liquid film that covers the inner surface works as a lubricant.

It should be noted that the modification of the internal surfaces of meandering tubes may not be easy, as the access to these surfaces is difficult. Also, some of these chemical treatments have a very short life, i.e., with the device operation, the hydrophobic properties are lost. Therefore, this technology cannot be considered fully developed for actual applications.

Besides, the chemical surface modification just proposed cannot be applied to flat plate PHPs manufactured by diffusion bonding, because this finishing layer suffer reduction in high temperatures and high vacuum conditions. One alternative is the mechanical treatment of the internal surfaces of the channels before the diffusion bonding process takes place, with the objective of controlling their roughness.

Actually, wall roughness increases the number of nucleation sites and helps PHPs to achieve early start-up conditions (Ibrahim et al. 2017; Qu and Ma 2007). On the other hand, the friction forces observed in highly rough surface channels, cause high pressure drops in liquid flows (Lips et al. 2010). Betancur et al. (2019b) investigated the influence of surface roughness on the thermal performance of two copper flat PHPs, with ten parallel channels of round cross section of 3 mm of hydraulic diameter, fabricated by diffusion bonding technique. In one PHP, the entire internal surface of the channels was polished with standard sandpaper grit N1200, to produce a smooth finishing, while, in the other one, the evaporator region was sanded with standard sandpaper Grit N100. Distillated water was the working fluid. Filling rations of 10, 15, 25, 37.5, 50 and 75% and for tilt angles of 0° and 90°, bottom heat mode were tested. The transient behavior and thermal resistance were analyzed. Start-up conditions, thermal stability and hysteresis were experimentally observed. Depending on the heat input, four regimes were noticed: pure conduction, unstable, stable and dry-out. In general, the hybrid PHP showed lower thermal resistances, except for highest filling ratio of 75%, at vertical position. However, the hybrid device shows stop-over instabilities at power inputs between 140 to 180 W, at vertical position and low filling ratios. These results shows the importance of controlling the channel roughness considering the thermal behavior of PHPs.

PHP Modelling

Several different mathematical approaches have been applied in the modelling of PHPs, as follows (Khandekar and Groll 2003 and Cui et al. 2014):

- Analogy between the oscillatory movement of the working fluid and of a mass-spring system
- Multiple mass-spring systems
- Use of conservation equations applied over a control volume that includes a liquid plug and a vapor slug
- Chaos theory under different conditions
- Neuronal networks
- Semi-empirical methods using dimensional groups.

However, the working principles of PHPs are still not completely understood and the above models can show large deviations when compared to real devices, being of limited applicability. According to Khandekar and Groll (2003) semi-empirical methods using non-dimensional groups are promising if all the possible combinations of applicable force fields and their importance are correctly identified and quantified. Each force on the thermally driven self-excited two-phase flow need to be correctly defined, depending on the boundary conditions to which the PHPs are subjected.

Electrical/Thermal Circuit Analogy

In this section, an electrical/thermal analogy model for flat plate PHPs is presented. This model can be applied for the meandering PHPs, disregarding the thermal resistances associated to the conduction through the plate material . Figure 7.27 illustrates a flat plate PHP composed by a single loop, where heat is transferred, by two parallel

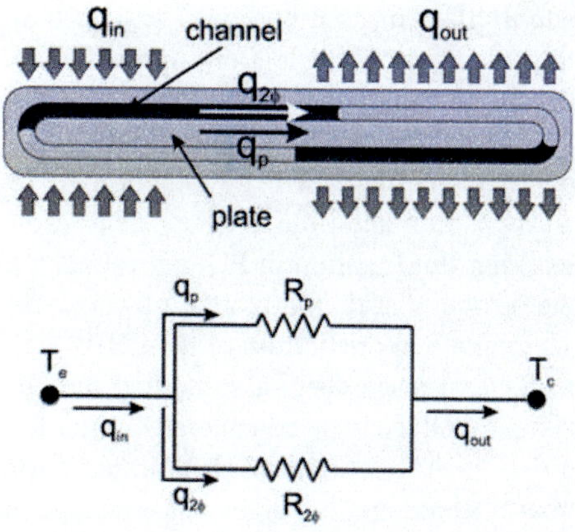

Fig. 7.27 Two channel PHP (up) and the heat transfer thermal resistance network (bottom) for plate and channels

paths: along the channels ($q_{2\Phi}$) and through the solid plate material $\left(q_p\right)$, from the evaporator to the condenser zones. Figure 7.27 also shows the equivalent thermal resistance circuit, where the cross section area of the PHP, in both evaporator and condenser, are considered uniform. The total thermal resistance of this simple loop is:

$$\frac{1}{R_t} = \frac{1}{R_p} + \frac{1}{R_{2\phi}} \tag{7.83}$$

The maximum thermal resistance happens when pure diffusion through the solid material is the only active path $\left(R_{2\phi} \rightarrow \infty\right)$, in such case $R_t = R_p$. This happens when the heat transfer by convection and phase change is negligible.

For a flat plate PHP with N channels in parallel, the resistance given by the last equation must be divided by N. One should note that Betancur (2020) showed that, for copper flat plate PHPs, the transversal heat transfer is able to unify the temperature in the channel transversal direction, so that one-direction heat transfer can be considered.

The two-phase flow thermal resistance through the channel is modelled as a combination of the boiling and convection resistances in parallel. To determine the overall thermal resistance, coefficients of heat transfer in the evaporator and condenser need to be determined. In the next sections, some of the literature correlations regarding the phase-change of a fluid flowing in small diameter open channels are presented and how they can be adapted for the present model is discussed.

Heat Transfer Coefficients in the Evaporator

The convective heat transfer inside the channels is a function of the velocity of the fluid. Both the boiling and condensation within a PHP depend on the working fluid mass flow rate $\left(\dot{m}''\left[\text{kg/m}^2 \cdot \text{s}\right]\right)$. Literature models for mass flow rates are usually applied for open minichannels, which is not exactly the case of PHPs. According to Karayiannis and Mahmoud (2017), different heat transfer mechanisms are present in a boiling of a liquid flowing in small diameter or micro channels. These includes: bubbly flow in pool boiling, nucleate boiling and evaporation at liquid film, thin-film evaporation (nucleation suppression when the liquid thickness is very thin) and, finally, convective evaporation in annular flow.

As stated by Betancur (2020), the major differences between classical microchannel and PHP flow theories is that the heat transfer mechanisms change with position and time in PHPs, as a result of flow direction changes. Actually, the working fluid motion in PHPs presents a highly stochastic nature, especially when the device is at the horizontal position. At the vertical position, the liquid tends to accumulate in the bottom of the PHP.

Several researchers assume that the heat transfer mechanisms at the evaporator zone in PHPs can be considered similar to the convective flow boiling. Chen (1966) proposes the coefficient of heat transfer for boiling of saturated fluids on convective flow as a combination of the phase change (boiling) coefficient h_b, and of a convective coefficient h_{conv}, as:

$$h_e = h_b + h_{conv} \tag{7.84}$$

Carey (1992) affirms that film boiling is observed only for some ranges of film thickness. Some useful expressions, proposed by Ma (2015), Furukawa (2014) and Chen (1966) can be used for the estimation of this coefficient. Betancur (2020) presents a compilation of several literature correlations for open channels that can be used for the estimation of the heat transfer in the evaporator and condenser of PHPs. From these several correlations, Jones et al. (2009) proposes the use of the correlation proposed by Gorenflo, because, differently from the others, it requires the surface roughness as one of the input parameters. It also presents the lowest deviation from the experimental results. Gorenflo's correlation for pool boiling heat transfer coefficient is based on the thermodynamic similarity theory, requiring the use of known reference values, i.e., a heat transfer coefficient h_o for a heat flux q_0'' at a mean average roughness surface $R_{a,o}$. Considering water as the working fluid, data of VDI Heat Atlas, reported by Collier and Thome (1996) can be used as a reference, where $h_o = 5,600 W/m^2K$ for $q_o = 20,000 W/m^2$. Considering q_b as the heat flux necessary to generate vapor in the channel, the heat transfer coefficient ratio h_b/h_o can be given as:

$$\frac{h_b}{h_0} = CF(p_r)\left(\frac{q_b}{q_0}\right)^n \tag{7.85}$$

in which C is a correction factor, and F and n depends on the reduced pressure (p_r). The $F(p_r)$ expression is:

$$F(p_r) = 1.73 p_r^{0.27} + \left(6.1 + \frac{0.68}{1 - p_r}\right) p_r^2 \tag{7.86}$$

while n also depends on the working fluid, which for water is:

$$n = 0.9 - 0.3 \cdot p_r^{0.15} \tag{7.87}$$

Finally, the parameter C is a correction factor that takes into consideration the measured roughness of the internal channel surfaces (R_a) and is determined as a function of a known reference roughness value. The suggested mean roughness presented by Benjamin and Balakrishnan (1997) is adopted in this work, with the value $R_{a,o} = 0.4\mu m$, according to ISO 4287/1984. Thus, factor C is defined by Gorenflo as:

$$C = \left(\frac{R_a}{R_{a,0}}\right)^{0.133} \tag{7.88}$$

For a boiling area A_b, the heat load can be determined as:

$$q_b = h_b A_b \Delta T_{w,sat,e} = \dot{m}_e h_{lv} \tag{7.89}$$

where $\Delta T_{w,sat,e}$ is the difference of temperatures of the wall at evaporator zone and the saturation temperature and $\dot{m}_e$ is the amount of generated vapor in the evaporator.

Heat Transfer Coefficients at the Condenser

As for the evaporator, the heat transfer coefficients at the condenser zone can be determined using correlations for minichannels. Empirical correlations, based on the single-phase forced convection flows, proposed by Dittus and Boelter, can be used (Collier and Thome 1996). For the present case, these correlations needs to be corrected to consider the laminar nature of the flow. Ma (2015) and Furukawa (2014) present some of these correlations. Betancur (2020) propose the use of Akers et al. (1958) correlation to predict h_c, which, as stated by Furukawa (2014), is more precise when applied to low *Re* flows. This expression is:

$$Nu = 5.03(\mathrm{Re}_l*)^{0.33}(Pr_l)^{0.33} \tag{7.90}$$

for:

$$\mathrm{Re}_l* = \mathrm{Re}_l\left(1 + \left(\frac{X}{1-X}\right)\left(\frac{\rho_l}{\rho_v}\right)^{0.5}\right) \tag{7.91}$$

Therefore, the heat transferred in a condenser area A_c is given by:

$$q_c = h_c A_c \Delta T_{w,sat,c} \tag{7.92}$$

where $\Delta T_{w,sat,c}$ is the difference of temperatures of the wall at condenser zone and the saturation temperature.

Mass Flow Rate

The mini-channel correlations used for the prediction of the heat transfer coefficients in the evaporator and condenser depends on the *Re* and *Nu* numbers, which depends of the mass flow rate, which, in turn, depends on the vapor quality, flow velocity, temperature levels and time. In a real PHP, the flow velocity u varies from zero, in relaxation states, to u_{max}, in the oscillation period.

The mass flow rate is a very important parameter, used as input in any model. Khandekar and Groll (2004) stated that the volumetric vapor mass generation under thermodynamic equilibrium $\left(\mathrm{m}''\left[\mathrm{kg/m^3} \cdot \mathrm{s}\right]\right)$ can be expressed as:

$$\dot{m}''' = \frac{4q''}{D \cdot h_{lv}} \tag{7.93}$$

Furukawa (2014) proposed an expression for the liquid flow velocity in the PHP, for two communicating channels:

$$u = \left(\frac{0.08\eta q}{N \cdot F \cdot \rho_l \cdot D \cdot l_t}\right)^{1/3} \tag{7.94}$$

where F is the filling ratio, q the heat input, N the number of channels, D the diameter of the tube and η is the effectiveness parameter (equivalent to Jacob Number), which, for the axial temperature drop in a tube $\Delta T_l (\approx \Delta T_v)$, results in:

$$\eta = \frac{c_{p,l} \cdot \Delta T_l}{h_{lv}} \tag{7.95}$$

Driving Pressure

As already commented, the sensible heat is the major heat transport mechanism, while the latent heat causes the random formations and collapses of vapor slugs, which variable sizes and pressures work as the driving force for the dislocation of the liquid plugs along the PHP (Yuan et al. 2010). According to Ma (2015), the capillary forces must overcome body forces to cause the liquid motion.

Observing a liquid plug in a PHP, a hysteresis in the advancing (θ_a) and receding contact angles (θ_r) can be detected, as shown in Fig. 7.28. Assuming that the radii of curvature in the advancing an receding vapor bubble in motion are the same for all plugs, also supposing laminar regime, rough internal channel wall surface and constant velocity of the slug along the x-axis, the pressure difference necessary to cause the fluid movement within the tube needs to be higher than all the pressure drops acting in the liquid plug, i.e. (Ma 2012):

$$\Delta p_t = \Delta p_r + \Delta p_l + \Delta p_d + \Delta p_a \tag{7.96}$$

where Δp_r and Δp_a are the receding and the advancing pressure differences between vapor and liquid phases, which summation results in the resistance to the liquid flow due to the dynamic contact angle hysteresis effect; Δp_l is the pressure drop due to friction forces and Δp_d is the dynamic pressure difference, which, for an incompressible flow of a liquid plug in a constant cross section area channel, is zero: $\Delta p_D = \rho_l u_1^2/2 - \rho_l u_1^2/2 = 0$. According to Ma (2012), the last equations takes the form:

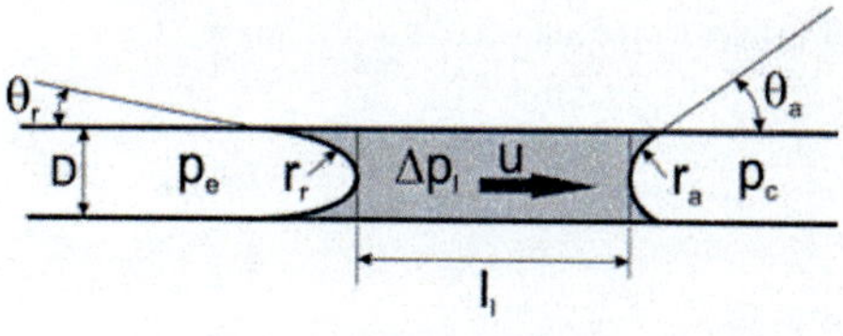

Fig. 7.28 Advancing and rescinding contact angles of liquid tube of a liquid plug in movement

$$\Delta p_t = \frac{32 \cdot \mu_l \cdot l_t \cdot u}{D^2} + N\frac{4\sigma(\cos\theta_r - \cos\theta_a)}{D} \quad (7.97)$$

The first term in the last equation is the friction component of the pressure drop for a fully developed flow in a channel (Hagen-Poiseuille flow, see Chap. 2), while the second term is due to capillary forces, which depend on the dynamic contact angle hysteresis.

Minimum Conditions of the Startup in the Evaporator Region

According to Khandekar et al. (2010), vapor slugs in channels may block the liquid flow if the driving force is not enough to push them. As the phase change is the driving energy of PHPs, the minimum heat load used for liquid-vapor phase change needs to be enough to exert the necessary work in the liquid:

$$q_{b,\min} \geq F_t \cdot u \quad (7.98)$$

where F_t is the total driving force resulting from the total driving pressure drop Δp_t. Thus, the work produced must be greater than all the irreversible losses acting in the PHP and depends on the system ability to increase the vapor pressure at evaporator zone.

The visualization of the bubble in PHPs shows that the PHP can be considered started-up when large bubbles, of the order of magnitude of the diameter of the tube (or of the channel) are formed. These bubbles in movement have the shape of Taylor bubbles. Qu and Ma (2007) proposed the following expression for the determination of the superheat temperature needed to form the Taylor shape bubble:

$$\Delta T_{Taylor} = T_n - T_v = T_v\left(1 \Big/ \left(1 - \frac{RT_v}{h_{lv}}\ln\left(1 + \frac{2\sigma}{p_v}\left(\frac{1}{r_n} - \frac{1}{2(r_{in} - \delta_l)}\right)\right) - 1\right)\right) \quad (7.99)$$

where T_n is the temperature in the nucleation site, T_v is the saturated vapor temperature, R the ideal gas constant, r_{in} the internal radius of the tube or channel and δ_l the liquid film thickness. As the equation shows, ΔT_{Taylor} depends on the activated cavity radius: higher roughness means smaller cavities, resulting in the decrease of the necessary superheating. When the liquid film thickness is of the same dimension order of the nucleation radius, the denominator of the last equation tends to zero and the necessary superheating takes its maximum value, known as the Taylor heat flux. Considering that just before the bubble formation, heat is transferred by conduction through the liquid film of thickness δ_l, the Taylor heat flux can be determined by, for a tube of circular cross section, as:

$$q''_{Taylor} = \frac{\Delta T_{Taylor} k_l}{r_{in}\ln\left(\frac{r_{in}}{r_{in} - \delta_\partial}\right)} \quad (7.100)$$

This maximum Taylor heat can be considered as the minimum heat flux for the startup. For other geometries, the conduction resistance in the tube radial direction, as observed in the denominator of this equations, must be substituted by the appropriate expression.

To obtain the heat flux, the last expression must be multiplied by the external area in which heat is to be applied.

Estimation of the Liquid Film Thickness

The last two expressions depend on the film thickness, which, therefore, must be predicted. According to Carey (1992), in a laminar flow, the heat transfer is dominated by the conduction mechanism and, consequently, the temperature distribution along the liquid layer is linear. In this case, the liquid film can be estimated as:

$$\delta_l = \frac{k_l}{h_e} \tag{7.101}$$

where h_e is the natural convection heat transfer coefficient. In the case where an annular liquid film around the vapor slug is observed, and supposing that $\delta_l \ll D$, Carey (2008) suggests that the liquid film thickness can be determined by the expression:

$$\delta_l = \frac{D \cdot F}{4 \cdot 100} \tag{7.102}$$

However, as reported by different researchers (Hao et al. 2014a, b and Bertossi et al. 2017) the liquid film thickness around a vapor slug is actually not uniform. There is an ideal thickness range, in which the heat transfer is augmented. The increase of the liquid plug motion results in the augmentation of the heat transfer conditions. The oscillation of the liquid plugs causes the formation of a liquid film, up and down in the liquid plug stream. The length and thickness distribution of this film depend on the liquid-solid wettability conditions, as discussed in Chap. 2 (Srinivasan et al. 2015). Figure 7.29 shows a schematics of a vapor slug and liquid plug for a round evaporator channel of a PHP, where the length l_b is associated to the vapor slug tail and l_l to the length occupied by the liquid. In the right side of this figure, details of the film thickness distribution within the length l_b are shown: l_1 represents the length where the liquid film presents a close to constant very thin thickness: in this zone, the strong affinity between liquid and casing material does not allow vapor to be formed. l_2 represents the transition film length, where phase

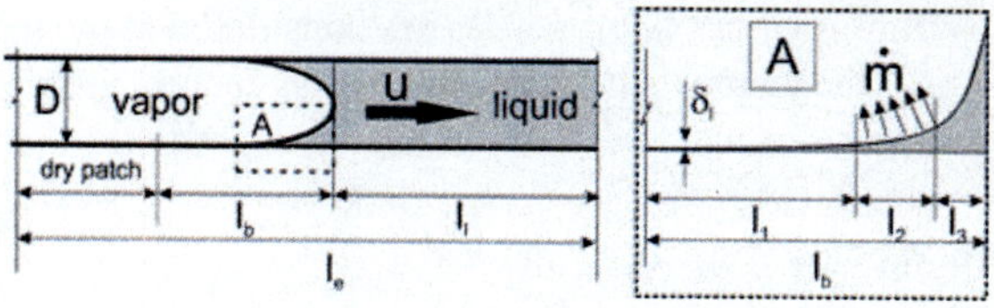

Fig. 7.29 Thickness of the liquid film in the liquid plug tail

change happens, where high heat fluxes are observed, of the order of magnitude of pool boiling nucleation (see Ma 2015). The last length l_3 corresponds to the intrinsic meniscus region. When the velocity is higher than a critical value, the formed liquid film may not be thick enough to allow vapor formation, consequently increasing the elapsed time (relaxation time) for the formation of a new set of vapor slug and liquid plug, which means that the overall fluid motion is slowed down. Dry patches may be formed in the evaporator due to different reasons, which must be avoided, as no working fluid phase change is able to happen in dry regions.

The Thermal Circuit Model

The present model is based on the work of Betancur et al. (2020), for flat plate PHP. In this case and based on this author experimental observations, no heat is considered to be transferred on the channel transversal direction, which is considered of uniform temperature. Therefore, a PHP composed by N interconnected channels can be modelled as an array of N parallel straight minichannels. A thermal circuit model, based on the circuit presented in Fig. 7.27, is proposed for the whole PHP. In this case, the channels are considered to have round cross section geometry, but the present model can be adapted for any other channel geometry. The following hypothesis are adopted:

- The device is subjected to a homogenous heat flux.
- The liquid-vapor phase change at the evaporator zone happens in steady state regime.
- The variation of the pressure inside the vapor bubbles is the PHP driving mechanism that pushes the liquid plugs and vapor slugs along the device.
- The flow is unidirectional.
- A mean working fluid velocity, obtained from the analysis of the thermo-hydraulic behavior of the fluid, is associated to the applied heat load.
- Pressure drops at U-bend turns are neglected.
- Channel length is equal to the device length ($l_t = l_e + l_a + l_c$)
- Vapor quality X is supposed constant and depends mainly on filling ratio Φ. The following expression proposed by Zivi (Carey 2008) is used:

$$X = \frac{1}{\left(1 + \left[\frac{1-\alpha}{\alpha}\right]\left(\frac{\rho_l}{\rho_v}\right)^{2/3}\right)} \tag{7.103}$$

where the void fraction is defined as $\alpha \approx 1 - \text{F}$.

- Thermophysical properties are considered at the saturation state.
- All the channels have constant cross section areas.
- The solid plate conduction heat transfer takes place in a constant cross section area.
- Vapor is considered an ideal gas.

- The bubble slug has two regions: dry, where the vapor is in direct contact with the wall and in which the heat transfer is neglected; and wet, where a liquid film is present and where convective boiling takes place, which is composed by a combination of the sensible and latent heat transfer mechanisms.
- Laminar regime is considered for fluid flow at the condenser zone.
- The lengths occupied by the liquid plugs and vapor slugs are determined as a function of the filling ratio.
- The mean average internal surface roughness R_a is considered uniform along the whole channel.
- Oscillations happen at low frequencies and the total volume of liquid and vapor do not change (steady state conditions).
- Near the liquid plugs, the liquid film has constant thickness around the vapor slugs.
- The gravity effect is neglected

The mass flow rate is determined according to Eqs. 7.93 to 7.95. Betancur (2020) propose to consider, in Eq. 7.95, the minimum superheating necessary to start-up the PHP, at evaporator zone, ΔT_e, while the original expression of Furukawa uses only the temperature drop along the capillary channel. If a mean density $\overline{\rho}$ is defined as $\overline{\rho} \approx \rho_l F$, it is possible to calculate a mass flux $\dot{m}''$, using the most probable velocity u (Eq. 7.94) and the mean density, as:

$$\dot{m}'' = \overline{\rho} \cdot u \tag{7.104}$$

A more detailed thermal circuit, when compared to that of Fig. 7.27, must be mounted. In the present case, the number of liquid plugs per channel is set to be one. According to Betancur (2020), this can be a good hypothesis for flat plate PHPs with 50% of filling ratio operating at the vertical position. The circuit, for a unique channel, is shown in Fig. 7.30. To use this thermal circuit, the lengths of the liquid plug and vapor slug must be determined.

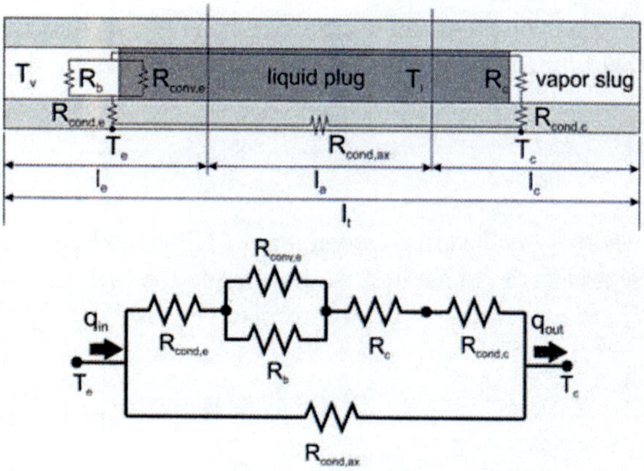

Fig. 7.30 Thermal circuit for unique channel

Considering that the mass of the vapor is negligible, all the volume of the working fluid (established by the filling ratio F(%)), is occupied by the liquid phase. Therefore, if there is no dry patch in the evaporator, the length l_b of the vapor slug in the evaporator is estimated as a function of evaporator length l_e by the expression:

$$l_b = (1 - F/100)l_e \tag{7.105}$$

and the length of the liquid plug is:

$$l_l = (F/100)l_e \tag{7.106}$$

Starting from the evaporator external wall (considering that the external wall temperatures of evaporator and condenser are known, which means that the external thermal resistances are disregarded) the first resistance is the wall conduction thermal resistance in the normal to the channel direction, $R_{cond,e}$, representing the resistance that the heat flux faces to cross the wall to reach the inner region of the channel, where the working fluid, in liquid and vapor states, flow. Once the heat flux achieves the inner channel surface, it faces the evaporator internal resistance $R_{2\phi}$, which is composed by two parallel resistances: R_{conv} concerning the sensible heat transfer mechanism and R_b that represents the phase change mechanism, associated to the interfacial evaporation and pool boiling nucleation. Heat is rejected at the condenser zone, represented by a thermal resistance R_c. Only convection heat transfer (no phase change) is considered in the condenser, because, as for the evaporator, the heat regarding the phase change mechanisms is very low when compared to the sensible heat, being important only as a slug-plug driving mechanism, which was already accounted for in the evaporator model. Finally, to reach the condenser region external surface, the heat needs to cross a conduction thermal resistance in the transversal direction of the wall $R_{cond,c}$, to be rejected to the heat sink.

The diffusive thermal resistances $R_{cond.ax}$, $R_{cond,e}$ and $R_{cond,c}$ depends on the geometry and on the thermal conductivity of the plate material, k_p. It is important to note that, while these resistances can be of less importance for the conventional PHPs with meandering tubes, it is of great importance to flat plate pulsating heat pipes. Before start-up occurs, or when PHP is empty, this is the dominant resistance, determined, for a total constant cross-section area A_t, as

$$R_{cond,ax} = \frac{l_{ef}}{k_p A_t} \tag{7.107}$$

where l_{ef} is the effective length of PHP given by Eq. 3.1. The total transversal cross section area A_t of a PHP with a unique channel, as shown in Fig. 7.31, where W and H are the width and height, respectively, is:

$$A_t = W \cdot H - \left(\pi \frac{D^2}{4}\right) \tag{7.108}$$

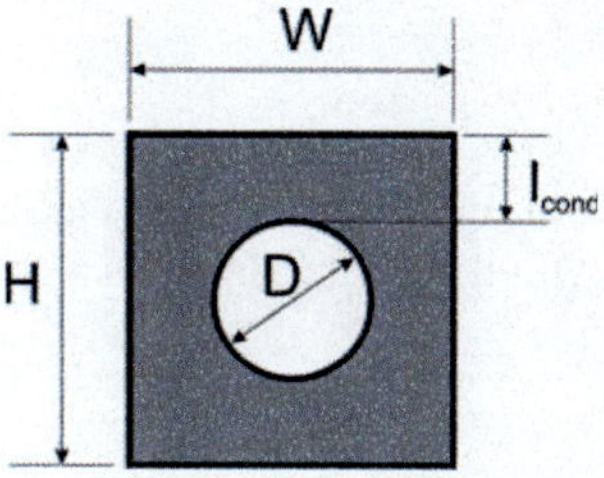

Fig. 7.31 Cross section of a unitary round channel

The axial conductive thermal resistance at evaporator wall $R_{cond,e}$ is determined by:

$$R_{cond,e} = \frac{l_{cond}}{k_p A_e} \tag{7.109}$$

where $A_e = \pi D l_e$ and the equivalent "thickness" l_{cond} is shown in Fig. 7.31.

As shown in Fig. 7.30, part of the evaporator is in contact with vapor and part with the liquid. Besides, as shown in Fig. 7.29, in the vapor slug region, a dry patch and a wet regions are observed. In the dry patch region, the heat transfer is by convection with a gas, resulting in a very high thermal resistance, which means that the heat transfer through this resistance can be neglected. In the wet region, convective boiling heat transfer is observed. The boiling and convection are considered as two resistances in parallel (see Fig. 7.30). Therefore the overall evaporator thermal resistance, for one channel, is given by:

$$\frac{1}{R_e} = \frac{1}{R_b} + \frac{1}{R_{conv,e}} = h_e A_e \tag{7.110}$$

The evaporator thermal resistance can also be expressed in terms of an overall coefficient of heat transfer:

$$h_e A_e = h_b A_b + h_{conv} A_c \tag{7.111}$$

where $A_c = \pi D l_c$.

Remembering that the phase change in the condenser section was neglected, the condenser thermal resistance is given by:

$$R_c = \frac{1}{h_c A_c} \tag{7.112}$$

while the conduction heat transfer through the wall in the condenser is obtained with Eq. 7.109, substituting A_e by A_c.

The channel overall thermal resistance is obtained by the summation in series of the above resistances, resulting in (see Fig. 7.27):

$$R_{2\phi} = \frac{l_{cond}}{k_p A_e} + \frac{1}{h_e A_e} + \frac{1}{h_c A_c} + \frac{l_{cond}}{k_p A_c} \tag{7.113}$$

Considering the conduction path through the plate, which is associated in parallel with this channel resistance, the overall thermal resistance is:

$$\frac{1}{R_t} = \frac{1}{\frac{l_{cond}}{k_p A_e} + \frac{1}{h_e A_e} + \frac{1}{h_c A_c} + \frac{l_{cond}}{k_p A_c}} + \frac{1}{R_{cond,ax}} = \frac{q}{\Delta T} \tag{7.114}$$

Also, considering that the PHP is composed by N parallel channels and that each channel has only one liquid plug that separate two vapor slugs (see Fig. 7.30), the resulting total thermal resistance of the channel is:

$$\frac{1}{R_{2\phi}} = \frac{1}{R_{2\phi,1}} + \frac{1}{R_{2\phi,2}} + \frac{1}{R_{2\phi,3}} + \cdots\cdots + \frac{1}{R_N} \tag{7.115}$$

When all the channels have the same geometry, this expression results in:

$$\frac{1}{R_{2\phi}} = \frac{N}{R_{2\phi,N}} \tag{7.116}$$

If the PHP channels are uniformly spread along the plate, the overall conduction thermal resistance can be given as:

$$\frac{1}{R_{ax,cond}} = \frac{N}{R_{ax,N}} \tag{7.117}$$

This last two expressions are combined in parallel to obtain the overall thermal resistance of flat plate PHP. It should be noted that, for meandering PHPs, the same set of equations as valid, neglecting the plate axial conduction resistance $R_{ax,cond}$.

Energy Balance

To complete the design of PHPs, an energy balance needs to be applied. For steady state conditions, the inlet heat load q_{in} in the evaporator must be equal to heat rejected at condenser zone q_{out}, if all the heat losses are neglected. Besides, the total heat load must be equal the heat transferred by conduction by the plate material q_p, plus the heat transferred by the channel $q_{2\phi}$. In turn, the heat transferred through the channel is also computed as the sum of the phase change (q_b) and sensible (q_{conv}) heats. The heat transferred by diffusion through the metal plate can be determined by the expression:

$$q_p = \frac{k_p A_t \Delta T}{l_{ef}} \tag{7.118}$$

Equating $q_{in} = q_{out}$ and, as $q_{in} = q_b + q_{conv}$ and $q_{out} = q_c$, substituting the appropriate expressions, results in the following equation:

$$\left(\frac{h_b A_b + h_{conv} A_{conv}}{A_e}\right) A_e \left(T_{i,e} - T_{sat}\right) = h_c A_c \left(T_{i,c} - T_{sat}\right) \tag{7.119}$$

where $T_{i,e}$ is the internal temperature of the evaporator wall and $T_{i,c}$ is the internal temperature of the condenser.

The above set of heat equations depend on the temperature level of the PHPs, which, in turn, depend on the heat transferred. Therefore, an iterative procedure is necessary for the prediction of the thermal behavior of PHPs.

Comparison Between Model and Data.

Betancur (2020) compared data with model for four flat plate PHPs, fabricated with copper plates, with overall dimensions of 208 × 150 × 5 mm^3 with round channels of 2.5 mm of diameter and water as the working fluid. They were tested at a bath temperature of 20 °C and subjected to the following inclinations: horizontal, 45°, vertical (bottom heat mode) and vertical (upper heat mode). Details of these PHPs are shown in Table 7.6. The first two PHPs contain 10 round parallel channels, with different roughness at the evaporator zone, while the other two PHPs contain 26 channels, one of them with two lateral grooves, added to the round channel. Experimental results for dynamic contact angles were reported by Betancur et al. (2019).

To take into account the presence of the grooves in the round cross section grooved geometry PHP, (see right side of Fig. 7.26), the following expression, representing an adaptation of the phase change heat transfer coefficient at the evaporator, is proposed:

$$h_b = \frac{(\pi D - NW)h_{b,f} + NWh_{b,g}}{\pi D} \tag{7.120}$$

where W is the largest groove opening and $h_{b,f}$ and $h_{b,g}$ are the boiling coefficients of heat transfer, for the film and groove projection areas respectively. These coefficients are estimated using the same expressions discussed before. It is important to note that, for the thick and continuous film region (see Fig. 7.32 at the center), $h_{b,f} = h_{b,g} = h_b$ and the above equation is not necessary to be applied. However, for the regions where the film is very thin or does not exist and the grooves, due to the capillary forces are filled, this equation is quite useful.

Betancur (2020) shows that, as a general trend, the thermal resistances decrease as the power input increases, for all cases analyzed. Also, the overall thermal resistance model under-estimate the data. The minimum values of the thermal resistances are observed for the vertical position. However, after the device is able to achieve the start-up at the horizontal or vertical upper heat modes, the thermal resistance tends to reach the same order of those observed for the vertical bottom heat model positions. This means that, after activated, the PHP tends to work similarly, whatever the operational inclination is.

Table 7.6 PHP tested experimental

PHP	Ra [μm]	F	N	Inclination	W	D	l_t	l_e	l_a	l_c	l_{cond}	l_{ef}	H	Heat [W]
					[mm]									
Homogeneous	0.03	10	10	Vertical	15	2.5	208	40	88	8	1.25	148	5	20 to 350
Hybrid	3.3	15 25 37.5 50 75		Horizontal 45°										
Round	1.33	30	26	Vertical	5.76									
Round-grooved	1.33	40 50 60		Horizontal Vert. inverted										

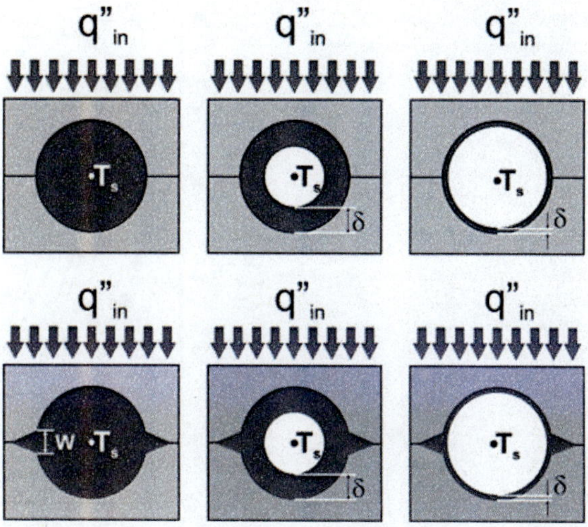

Fig. 7.32 Liquid at the cross section: completely filled left, continuous film in the center and thin film right, for the round and grooves channels

Betancur (2020) also shows that, at vertical position, the theoretical end experimental resistances presented a good agreement, within mean absolute error of 25%, for the cases: 10 channels hybrid finishing (larger roughness in the evaporator) and 10 channels homogeneous. For the 26 channels, the comparison is still good, with the model mostly under-predicting the data with the mean absolute error of about 26%. For the bottom heated mode and the inclination of 45°, the model adjusts very well with data, for both 10 channel devices, when the power inputs were larger than 100 W, with a mean error smaller than 8%. The comparison between data and model is worse at horizontal position, with maximum mean absolute error observed for 10 channel PHP varying between the 50 and 75%, for power inputs larger than 180 W, while, for 26 channel PHPs, this error was of 60% for power inputs larger than 140 W.

For vertical upper heated mode, the error is still worse, as the devices were not able to work at certain conditions. The data also showed that the thermal performance of the PHPs decreased substantially and so the comparison with the model, which presented the higher errors for too small and too large filling ratios (10% and 75%). Therefore, too small and too large filling ratios are not recommended to be used in pulsating heat pipes.

Heat Activation Analysis

It is interesting to note that, for some operational conditions, the overall thermal resistance is equivalent to the conduction thermal resistance through the device. There is a time delay necessary for the PHPs to achieve start-up conditions, especially noticiable for horizontal and vertical upper heat mode positions. In these cases, the PHP transfer heat only by conduction until the working fluid phase change starts. These PHPs need higher activation energy to be able to work without the gravity aid.

Based on a compilation of the data obtained for all the PHPs described in Table 7.6, which filling ratios was of 50% (see Betancur 2020), Fig. 7.33 shows a plot of

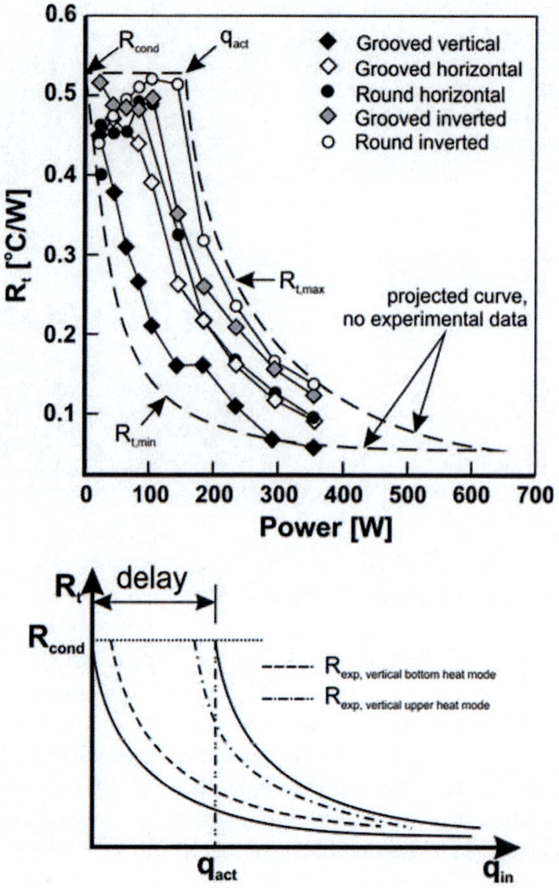

Fig. 7.33 Maximum and minimum thermal resistance envelops for a set of experimental data

the experimental overall thermal resistance as a function of the power input. The horizontal inclination of the thermal resistance curve shows clearly the delay of the startup of the PHP. This figure shows that there is an "activation heat" for each PHP configuration, i.e., a minimum power input necessary to the working fluid to initiate the oscillations. However, the startup, for vertical bottom mode positions is observed at practically any power input. According to Qu and Ma (2007), in vertical bottom mode, the gravity pushes down the liquid plug and helps the liquid film formation around the vapor slug. These two extreme operation conditions can work as envelopes of the operation conditions of a PHP. The lower limit curve of the resistance $R_{t,min}$ of this envelope, can be estimated using Eq. 7.114.

As already observed, the highest differences between model and data are found for the PHP operating at the vertical, upper heat mode condition. Therefore, the proposed model must be corrected to produce reasonable predictions for this operation state, which is equivalent to the maximum total resistance $R_{t,max}$ (upper thermal resistance envelop of the PHP data in Fig. 7.33). The hypothesis adopted here is that, once the

PHP start to work, the transient behaviors of the vertical, bottom and upper heating modes, are the same. In fact, this means that the thermal resistance as a function of power input curve is dislocated to the right position by q_{act}. This activation power is given by Eq. 7.100, considering a boiling area of A_b, i.e.:

$$q_{act} = q''_{Taylor} \cdot A_b \tag{7.121}$$

Therefore, for the maximum total thermal resistance, $R_{t,max}$, the following equation is proposed:

$$\begin{cases} q_{in} - q_{act} \leq 0, \ R_{t,max} = R_{t,cond} \\ q_{in} - q_{act} > 0, \ R_{t,max} = \frac{\Delta T}{q_t - q_{act}} \end{cases} \tag{7.122}$$

where $R_{t,cond}$ is the conduction total thermal resistance of the solid material of the PHP.

7.6 Mini Two-Phase Devices

The literature (Kandlikar 2002) classifies channels according to their hydraulic diameter range, as:

- Conventional: $D_h > 3\,\text{mm}$
- Mini: $200\,\mu\text{m} < D_h < 3\,\text{mm}$
- Micro: $D_h < 200\,\mu\text{m}$.

The fluid flows have different characteristics according to the size and geometry of the channel. In this section, the geometry and performance of small two-phase devices, the mini thermosyphons and mini heat pipes, designed for electronics cooling, are discussed.

Hybrid Loop Thermosyphon/Heat Pipe

A new hybrid loop thermosyphon/heat pipe for electronic cooling application is proposed. Figure 7.34 up shows an exploded view of this device. It basically consists of four special geometry plates, with internal grooves, which are carefully stacked and diffusion bonded at a special furnace (see Chap. 5 for details).

As shown in Fig. 7.35, the union between the first and second plate forms the evaporator and the liquid line, which are in the same level. A layer of a sintered porous wick covers the internal area of the evaporator. The third plate closes the liquid channel and provides the vapor channel path from the evaporator to the condenser, which is in higher height when compared to the liquid channel. Both internal plates have holes that, when piled, compose both the evaporator and condenser. The last plate provides the closing of the evaporator, condenser and vapor line. Details of the device are shown in Fig. 7.35. Two reasons explain the empty space (through hole)

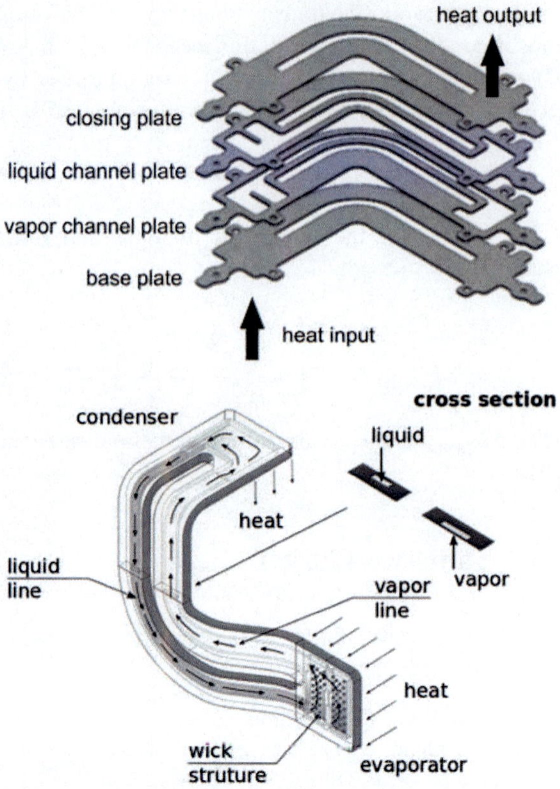

Fig. 7.34 Upper: hybrid mini loop thermosyphon/heat pipe plates, to be piled in the order, bottom to top; lower, actual configuration for cooling of electronic components in printed circuit boards in avionics

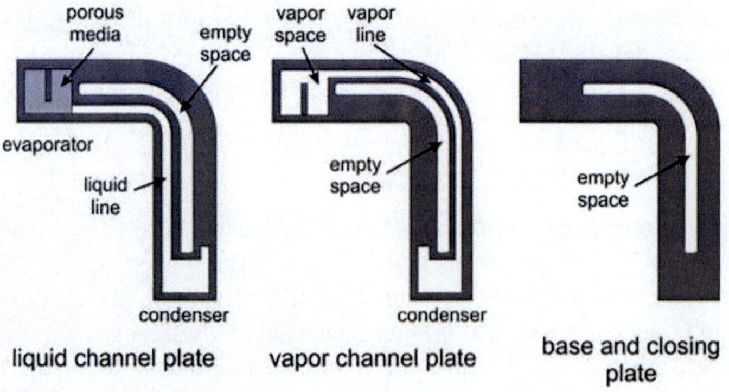

Fig. 7.35 Drawing of the plates: left, liquid channel plate with porous media in the evaporator; center, vapor channel plate; right, base and closing plate

in the middle of the device: physical and thermal separation of the liquid and vapor channels and decrease of the device weight.

Considering the device operating in a perfect horizontal position, the heat, inserted at the rear surface of the evaporator crosses the external wall, reaching the liquid, fed by the liquid channel. The wick structure over the internal surface, by capillary effects, pushes the liquid from the channel. Liquid-vapor phase change happens at the wick and the formed vapor flows through the vapor channel, which is in a superior level of the liquid line, reaching the condenser, where heat is lost and the vapor condenses. Due to the gravity action, the height difference between the liquid and vapor lines separates the working fluid liquid and vapor phases. Therefore, the operation driving forces of this device are the gravity and the porous media liquid pumping effect. For this reason, the device is denominated as hybrid loop thermosyphon.

Figure 7.34-down shows tri-dimensional curves, necessary to accommodate the hybrid loop thermosyphon to the printed circuit board, where the electronic component to be cooled is installed, highlighting the high flexibility of this technology. The heat source (electronic component) is supposed to be thermally attached to one of the faces of the evaporator and the heat sink to one of the faces of the condenser.

Modelling

The hydraulic diameter for the liquid and vapor lines, is determined by the root of the cross section area, as suggested by Sarmiento et al. (2020):

$$D_h = \sqrt{A} \tag{7.123}$$

Steady state conditions are assumed and the analogy between electrical and thermal circuits are used for modelling. The overall thermal resistance is defined as the ratio between the evaporator and condenser temperature difference and the power input. Figure 7.36 shows this thermal circuit. Taking the transversal section of the evaporator, the heat delivered in the evaporator rear surface, has first to cross the conduction thermal resistance imposed by the base plate thickness $\left(R_{cond,e}\right)$ and then cross the wick structure $\left(R_{cond,w}\right)$. Then, at the liquid-vapor interface, the liquid evaporates (R_b) and the vapor, due to pressure differences between the evaporator and condenser section, is able to flow along the vapor channel with a thermal resistance (R_v). At the condenser, the vapor losses heat and changes of phase from vapor to liquid $\left(R_{vl,c}\right)$. Finally, this heat transfer path ends with the transversal conduction thermal resistance at the condenser $\left(R_{cond,c}\right)$. Heat can also be transferred through a parallel conduction path, in the axial direction of the device, composed by the axial

Fig. 7.36 Thermal circuit for the hybrid mini loop heat pipe evaporator, channels (vapor and liquid, also denominated as adiabatic section as they are thermally insulated from the environment) and condenser, given by the thermal resistances: $(R_{cond,ax,e})$, $(R_{cond,ax})$ and $(R_{cond,ax,c})$, respectively.

The next step consists of determining these thermal resistances that compose the circuit. For the present model, the following hypothesis were adopted:

- Steady state
- Incompressible liquid and vapor flows
- Laminar flow
- Uniform heat flux
- Adiabatic liquid and vapor flows in the channels
- Single phase flow in each channel
- Pool boiling at the evaporator
- Wall condensation at the condenser.

The conduction resistances are determined flowing the methodology discussed at Chap. 4 (Design of Thermosyphons and Heat Pipes).

The thermal resistance of the porous medium is a function of its geometry (thickness and area) and of the effective conductivity of the material, including the wick structure and the working fluid. A model for the effective thermal conductivity of the wick structure is presented in Chap. 3.

To calculate the boiling thermal resistance $R_b = 1/(h_b \cdot A_e)$, the coefficient of heat transfer of phase change in the evaporator can be predicted by the Kaminaga expression (see Chap. 4, Thermal Resistance of the Evaporator Pool).

The thermal resistance related to the vapor flow along the vapor channel can be predicted by Eq. 7.24, assuming that the liquid and vapor flows are monophasic. The pressure drop in the vapor flow is predicted by the summation of the distributed and localized head losses. The distributed pressure drop depends on the length of the tube and can be determined using Eq. 2.80, which is rearranged in the form:

$$\Delta p_v = \frac{f \rho_v l_v u_v^2}{2 D_{h,v}} \tag{7.124}$$

where f is the vapor friction factor and l_v is equivalent length of the vapor channel, which takes into consideration the curves that may exist in that region. For laminar flow in a squared cross section area, the friction factor is (Bejan 2004):

$$f = \frac{14.2}{\mathrm{Re}_{D,h}} \tag{7.125}$$

where $Re_{D,h}$ is based on the hydraulic diameter of the channel. As for other heat pipes and thermosyphons which models are presented in this book, the vapor mass flow rate and vapor velocity are obtained from the knowledge of the heat input and of the thermophysical properties of the working fluid.

According to Munson (2009), the localized pressure drops are usually accounted for by means of the following expression:

$$\Delta p_{v,loss} = K_{loss}\frac{u_v^2}{2g} \tag{7.126}$$

where K_{loss} is the loss coefficient, which depends on the tube geometry (elbows, returns and T unions) and on the curve angle. For elbows this coefficient value is $K_{loss} = 1.5$ and, for 45° curves, it is $K_{loss} = 0.4$.

For the vapor-liquid phase change thermal resistance $R_{vl,c} = 1/(h_c \cdot A)$, considering the device operating in vertical or inclined positions, the heat transfer coefficient is determined by Eq. 2.38 with the modified latent heat of vaporization given by Eq. 2.20.

When the condensation happens in horizontal down-facing surfaces, the coefficient of heat transfer can be obtained using the correlation obtained from Gerstmann and Griffith (1967):

$$\overline{Nu} = \begin{cases} 0.69 \cdot Ra^{0.20} & 10^6 < Ra < 10^8 \\ 0.81 \cdot Ra^{0.193} & 10^8 < Ra < 10^{10} \end{cases} \tag{7.127}$$

where the Rayleigh number is obtained by the expression:

$$Ra = \frac{g h_{lv} \rho_l (\rho_l - \rho_v)}{\mu_l k_l (T_v - T_c)} \left(\frac{\sigma}{g(\rho_l - \rho_v)}\right)^{3/2} \tag{7.128}$$

and the Nusselt number, based on the total length of the surface where condensation happens, is:

$$\overline{Nu} = \frac{\bar{h}_c}{k_l}\left(\frac{\sigma}{g(\rho_l - \rho_v)}\right)^{1/2} \tag{7.129}$$

Isolating the coefficient of heat transfer, results in:

$$\bar{h}_c = \overline{Nu} \cdot k_l \left(\frac{g(\rho_l - \rho_v)}{\sigma}\right)^{1/2} \tag{7.130}$$

Almeida (2021) constructed and tested the hybrid mini loop thermosyphon/heat pipe shown in the lower part of Fig. 7.34. The characteristics of the tested device are shown in Table 7.7. Figure 7.37 shows a comparison between the steady state overall thermal resistances between model predictions (based on the thermal circuit of Fig. 7.36) and data. Two conditions where plotted: the tube without working fluid and with working fluid. Without working fluid, the tube works only by conduction and the curve is flat, as the thermal properties of the tube material shows weak dependence of the temperature. The difference between data and model for pure conduction can be attributed to the hypothesis of one dimensional heat flux used in the model while the actual flux is two dimensional. Besides, the data shows a different behavior of the model for low power inputs, when the tube was not working

Table 7.7 Parameters of the testes PHP

Parameters	
Thickness of the base and closing plates (mm)	0.7
Thickness of the vapor and liquid tubes (mm)	1.5
Total thickness (mm)	4.4
Length of the device (mm)	17.5
External width of the device (mm)	20
Porous media thickness (mm)	1.5
Evaporator internal length (mm)	15.0
Evaporator internal width (mm)	15,0
Cross section area of vapor line (mm)	6,0
Vapor line length (mm)	88.15
Cross section area of liquid line (mm)	3.8
Liquid line length (mm)	70.28
External condenser length (mm)	15.0
External condenser width (mm)	20.0
Internal condenser length (mm)	10.0
Internal condenser width (mm)	15.0
Working fluid volume (ml)	0.5

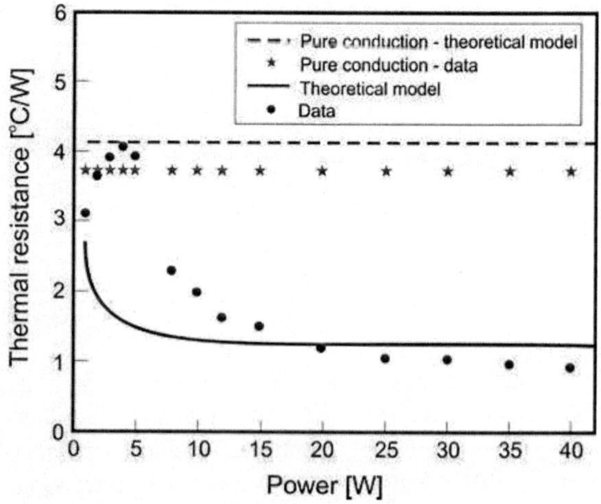

Fig. 7.37 Comparison between theoretical model and data for the mini hybrid loop thermosyphon

properly. For the tube filled with working fluid, the comparison of model and data can be considered good for heat inputs larger than 12 W, where the difference between data and predictions (given by the ratio of the difference between data and prediction over data) is lower than 30%. For low power inputs the experimental data shows that the thermal resistance can increase with increasing power, up to around 5 W, surpassing the pure conduction thermal resistance. This happens because, before the vapor bubble can pump the liquid to the condenser, the trapped liquid, accumulated in the evaporator, is able to storage heat, increasing the temperature difference between evaporator and condenser and so the thermal resistance of the thermosyphon.

Wire-Plate Heat Pipes

Mini heat pipes are among the technologies to be used for dissipation of concentrated heat, such as in electronics. They can be a good alternative for grooved mini heat pipes due to its high quality of the sharp grooves, associated with simple manufacturing.

Actually, wire-plate heat pipes can be classified as grooved mini heat pipes. The capillary pumping of the working fluid is performed by the sharp grooves formed between plates and wires. The vapor moves through the core between wires.

Wire-plate mini heat pipes are fabricated by the sandwich between two flat plates filled with a few wires. They operate as any normal heat pipe, with three regions: evaporator, condenser and between them, an adiabatic section. The sharp edges formed between the plates and the wires form very efficient grooves, able to provide high pumping capacity to the working fluid. Figure 7.38 shows a schematic of such devices. However, usual fabrication processes such as welding with the deposition of material between the plates and wires, block partially the grooves, decreasing its efficiency. The diffusion bonding, described in Chap. 5, is a very convenient fabrication methodology, consisting of subjecting the sandwich (two plates and wires) to controllable high temperature and pressure, under high vacuum and at a controlled time.

Liquid Layer and Meniscus Geometry.

Not many models are available for wire-plate heat pipes. Wang and Peterson (2002) are among the researchers who studied this heat pipe configuration. Launay et al. (2004) studied the theoretical behavior of wire-plates mini heat pipes bonded by diffusion. Paiva et al. (2011) develop a hydrodynamic model for these mini heat pipes, where the contact angles between working fluid and metal were adjusted, so that data and model presented a good comparison. Paiva et al. (2015) developed a thermal model, for both evaporator and condenser, considering that the heat transfer through the liquid film cross section area can be modelled as a thermal resistance circuit.

A unit cell is modelled, comprising a channel, composed by the half volumes of two wires, the closing plates and the liquid and vapor allocated, as shown in Fig. 7.38. Each channel has four grooves. This figure shows the liquid accumulated

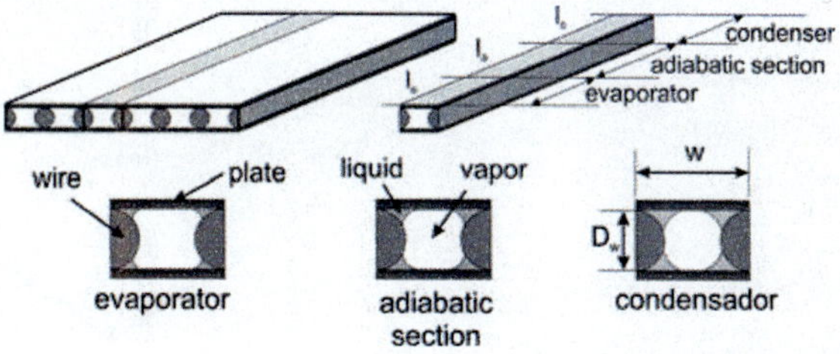

Fig. 7.38 Wire plate mini heat pipe

in the grooves at the evaporator, adiabatic and condenser sections, disregarding the influence of gravity.

The liquid accumulates in the grooves, forming a liquid layer, which geometry is detailed in Fig. 7.39 where the three basic angles are depicted: the liquid-metal contact angle θ, the half-angle of the wire-liquid opening β_1 and the half-angle of the meniscus curvature β_2. The contact angle θ between the liquid and metal depends on the combination of the liquid and the the metal. The liquid surface tension determines the meniscus radius r_m. Although the contact angle does not change due to the variation of the volume of liquid accumulated in the meniscus, its profile actually changes along the length of the heat pipe, as shown in Fig. 7.39. Wang and Peterson (2002) studied the geometry of the liquid profile within the groove, resulting in the following trigonometric relations:

$$\beta_1(x) = \mathrm{atan}\left(\frac{-\sin\theta \cdot r_m(x) + \sqrt{(\sin\theta \cdot r_m(x))^2 + 4r_w \cos\theta \cdot r_m(x)}}{2r_w}\right) \quad (7.131)$$

$$\beta_1(x) + \beta_2(x) + \theta = \frac{\pi}{2} \quad (7.132)$$

Fig. 7.39 Geometry of the meniscus of the wire-plate mini heat pipe

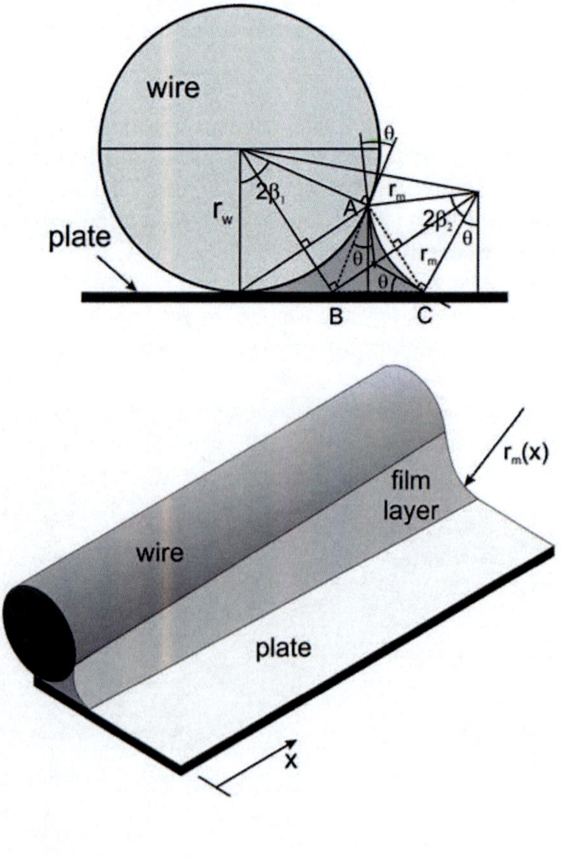

$$\beta_2(x) = \frac{\pi}{2} - \theta - \operatorname{atan}\left(\frac{-\sin\theta \cdot r_m(x) + \sqrt{(\sin\theta \cdot r_m(x))^2 + 4r_w \cos\theta \cdot r_m(x)}}{2r_w}\right) \tag{7.133}$$

where r_w is the wire radius, and $r_m(x)$ is the radius of meniscus, which is a function of the axial distance x. With these angles, the liquid channel cross section area, which also depends x, can be determined as:

$$A_l(x) = 2r_w r_m(x) \sin\beta_1 \sin\beta_2 - r_w^2(\beta_1 - \sin\beta_1 \cos\beta_1) - r_m^2(x)(\beta_2 - \sin\beta_2 \cos\beta_2) \tag{7.134}$$

Considering four menisci at each channel, the cross area of the wire-plate mini heat pipe occupied by vapor is:

$$A_v(x) = r_w(2w - \pi r_w) - 4A_l(x) \tag{7.135}$$

The liquid and vapor cross section areas are used to calculate the fluid velocity along x, for each phase. The vapor occupies the maximum cross section area at the beginning of the evaporator ($x = 0$) and the minimum at the end of the condenser ($x = L$). The opposite is valid for the liquid. Furthermore, the wet and dry areas of the channel are important to calculate the flow stresses applied to vapor and liquid phases. These areas are calculated integrating the wet and dry cross section perimeters (which are functions of x) along the heat pipe's length. Based on the geometry of the liquid layer shown in Fig. 7.39, the following expressions are obtained for the liquid-vapor, solid-liquid and solid-vapor cross section perimeters, respectively:

$$P_{lv}(x) = 8\beta_2(x) r_m(x) \tag{7.136}$$

$$P_{sl}(x) = 8r_w(\tan\beta_1(x) + \beta_1(x)) \tag{7.137}$$

$$P_{sv}(x) = 2(w + \pi r_w) - 8r_w(\tan\beta_1(x) + \beta_1(x)) \tag{7.138}$$

According to Wang and Peterson (2002) the maximum radius of meniscus occurs when all wires inside the channel are wet by the liquid phase, i.e., when $\beta_1 = \pi/4$. On the other hand, the minimum possible meniscus radius is equivalent to the edge of the groove, which, for a perfect groove (such as those resulting from diffusion bonding) tends to zero. Paiva et al. (2015) proposed 100 μm as a practical and reasonable minimum radius of meniscus.

Hydraulic Model

In this sub-section, models to determine the liquid and vapor mass flow rates and pressure drops are proposed. For these models, the following hypothesis are assumed:

- Steady state regime is considered.
- The heat flux is uniform at the evaporator and condenser and so the vapor and liquid mass flow rates vary linearly in these sections.
- The adiabatic section is perfectly insulated.
- The mass flow rate in the adiabatic section is constant.
- The liquid-vapor friction is neglected.
- The flow is uncompressible ($M \leq 0{,}2$).

For a unique cell, at the evaporator section ($0 \leq x \leq l_e$) considering that the vapor generation rate is zero at $x = 0$ and that the maximum vapor mass flow rate leaves the condenser region at $x = l_e$, the mass flow rate at the evaporator is (see also several examples in Chapter 6):

$$\dot{m}_{v,e}(x) = \frac{\dot{m}_t}{l_e}x = \frac{q}{h_{lv}l_e}x \tag{7.139}$$

In terms of heat flux $\left(q''\right)$, for a unit cell and considering that heat is delivered in one of the faces of the heat pipe, of area $A = w \cdot l_e$, the vapor mass flow rate per channel is:

$$\dot{m}_{v,e,ch}(x) = \frac{q''wl_e}{h_{lv}}x \tag{7.140}$$

At the adiabatic section ($l_e \leq x \leq l_e + l_a$), ideally, no phase change occurs and so the mass flow rate is constant. Thus, the vapor mass flow rate is given by:

$$\dot{m}_{v,a,ch}(x) = \frac{q''2wl_e}{h_{lv}} \tag{7.141}$$

In the condenser region ($l_e + l_a \leq x \leq l_e + l_a + l_c$), considering that, at the end of the condenser, the vapor flux is zero, the vapor mass flow is:

$$\dot{m}_{v,c,ch}(x) = \frac{q''wl_e}{h_{lv}}(x - l_t) \tag{7.142}$$

where l_t is the total length of the heat pipe, including evaporator, adiabatic section and condenser.

Vapor pressure distribution. Literature works (Wang and Peterson 2002 and Paiva et al. 2015) predict the vapor pressure distribution solving numerically (Runge-Kutta method) a system of several balance equations. However predictions using Chi (1976) model is much simpler and still precise for low Reynolds number applications. Therefore, the wire-plate mini heat pipe vapor pressure distribution is determined using Eq. 3.48, here reproduced:

$$\frac{dp_v}{dx} = -\frac{(f_v\text{Re}_v)\mu_v\dot{m}_{v,ch}}{2A_{v,vh}(x)r_v^2\rho_v} - \beta\frac{2\dot{m}_{v,ch}}{\rho_v A_{v,ch}^2}\frac{d\dot{m}_{v,ch}(x)}{dx} \tag{7.143}$$

where β is taken as one and r_v is the vapor radius, here assumed as:

$$r_v = \frac{A_{v,ch}}{P} \tag{7.144}$$

The parameter $f_v Re_v$, known as Poiseuille number, is present in almost all models and correlations for pressure drops in laminar fluid flows. It depends on the cross-section fluid-flow area. For the wire plate heat pipe, Paiva et al. (2015) suggests that, for $w = 2D_w$, the vapor cross section area of the evaporator and adiabatic sections can be approximated by a rectangle of 2:1 aspect ratio and a square, respectively, while the condenser can be approximated to a circular channel (see Fig. 7.38). According to Bejan (2004), for these cross section profiles, the Poiseuille number for each section is settled as $f_v Re_v = 17$ at the evaporator, $f_v Re_v = 14.2$ at the adiabatic section, and $f_v Re_v = 16$ at the condenser.

Substituting the appropriate expression for the vapor mass flow rate for each heat pipe section, the vapor pressure expression can be obtained from the integration of Eq. 7.144, resulting in:

$$p_v(x) - p_0 = \int_0^x \frac{dp_v}{dx} dx \tag{7.145}$$

This expression is usually determined numerically, where p_0 is the pressure at the beginning of the evaporator. This pressure is usually unknown, but the pressure at the beginning of the adiabatic section is normally used as a reference and it is considered as equal to the saturated pressure evaluated at the local temperature. An integration of Eq. 7.145 can also be used to determine this pressure, resulting in:

$$p_0 = p_{v,sat}(l_e) - \int_0^{l_e} \left.\frac{dp_v}{dx}\right|_e dx \tag{7.146}$$

Liquid pressure distribution. For the liquid phase, another approach is required to determine the pressure drop. Actually, for the wire-plate heat pipe, the liquid phase profile maintains basically the same shape, but its cross section area varies significantly along the wire-plate heat pipe length.

In steady state conditions, the total mass of working fluid inside the heat pipe is constant. Therefore, at any location, the mass variation (in time) of working fluid in liquid state is converted to vapor, i.e.:

$$\dot{m}_l(x) = -\dot{m}_v(x) \tag{7.147}$$

For the liquid phase, the analogy with capillarity structures can be applied to determine the pressure drops. In this case, the Darcy model is used for the pressure drop predictions. For an ordinary porous channel, Darcy propose the following equation:

$$\frac{dp_l}{dx} = -\frac{\mu_l \cdot \dot{m}_l}{\rho_l K A_{pm}} \tag{7.148}$$

where K is the permeability of the capillarity structure and A_{pm} is the porous media cross section area, while the subscript l stands for the liquid phase.

Permeability, which concept was developed for porous media, is a characteristic of the capillarity structure (see Chap. 3). Shah and Bhatti (see Faghri 2016) applied this idea to isosceles triangular grooves, such as the one shown in Fig. 7.40. Based on the expression for the permeability of rectangular grooves with porosity equal to one, (see Table 3.1), these researchers proposed the following expression:

$$K(x) = \frac{2r_{h,l}^2 \varphi(x)}{f_l \mathrm{Re}_l} \tag{7.149}$$

where $r_{h,l}$ is the meniscus hydraulic radius, which, for a groove with the geometry of Fig. 7.40, is $r_{h,l} = w \cdot \cos\beta$ and $\varphi(x)$ is an equivalent groove porosity, which, for this same groove, is $\varphi = w/2S$, where S is the pitch between grooves.

According to Batista (2021), a close observation of Fig. 7.39 shows that the ABC isosceles triangle can be considered a good approximation for the liquid channel geometry. For this case, Eq. 7.147 can be rewritten, for one channel of the unit cell as:

$$\frac{dp_l}{dx} = -\frac{\mu_l \left[\dot{m}_l(x)/4\right]}{\rho_l K(x) A_{pm}} \tag{7.150}$$

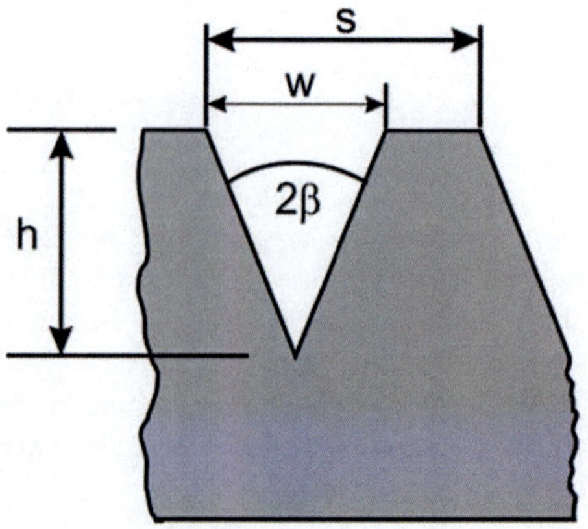

Fig. 7.40 Schematic of the groove e channel modeled in the literature

remembering that each cell has four grooves and that the liquid layer geometry, and so the permeability, varies along the length $K(x)$. Moreover, for the present geometry, the total cross section area of the porous medium for one groove is considered as the sum of one meniscus cross-sectional area and a quarter of wire, yielding:

$$A_{pm} = A_l(x) + \left(\pi r_w^2/4\right) \tag{7.151}$$

For the present case, based on successful results obtained for liquid flows in small channels (Sarmiento 2020), another definition of hydraulic diameter is used:

$$D_h(x) = \sqrt{A_l(x)} \tag{7.152}$$

the permeability (see Eq. 7.148) is written as:

$$K(x) = \frac{\left(D_h^2(x)/8\right)\varphi(x)}{f_l \mathrm{Re}_l} \tag{7.153}$$

The equivalent groove porosity $\varphi(x)$ is also modified for the wire-plate geometry. Based on Faghri (2016) definitions, the relation between groove opening length and groove pitch is adapted, being expressed, instead of $\varphi = w/2S$ (see Fig. 7.40), as:

$$\varphi(x) = \frac{\overline{AC(x)}}{2r_w} = \tan\beta_1(x)\sin\beta_1(x) \tag{7.154}$$

where $(2r_w)$ is defined as pitch of the groove, and $\overline{AC(x)}$ is the groove opening.

The Poisuille number $f_l Re_l$ is determined considering that the liquid phase profile maintains the same shape, but that its cross section area varies significantly along the wire-plate heat pipe. Still reasoning that the ABC triangle is a good approximation for the liquid layer geometry, the correlation of Shah and Bhatti, developed for an isosceles triangular groove (see Faghri, 2016) can be used to calculate $f_l Re_l$ parameter. The modified correlation, adapted for the ABC triangle, is:

$$f_l \mathrm{Re}_l(x) = \frac{12(B(x)+2)\left(1-\tan^2\beta_2(x)\right)}{(B(x)-2)\sqrt{\left[\tan\ \beta_2(x) + \sqrt{1+\tan^2\beta_2(x)}\right]}} \tag{7.155}$$

with:

$$B(x) = \sqrt{4 + \frac{5}{2}\left(\cot^2\beta_2(x) - 1\right)} \tag{7.156}$$

The liquid pressure drop curve is obtained integrating Eq. 7.150 along the mini heat pipe device, i.e.:

$$p_l(x) = p(x = l_t) - \int_{l_t}^{x} \frac{dp_l}{dx} dx \qquad (7.157)$$

where $p(x = l_t)$ is a reference pressure, at the end of the heat pipe, which can be determined considering the pressure drop due to the meniscus curvature, which is maximum at this point.

The curvature radius can be obtained based on the Young-Laplace model (see Chap. 2). According to Wang and Peterson (2002), for a wire-plate groove, the maximum curvature radius of the meniscus is:

$$r_{m,max} = \frac{r_w}{\cos\theta - \sin\theta} \qquad (7.158)$$

so that this pressure is:

$$p(x = l_t) = p_v(x = l_t) - \frac{\sigma}{r_{m,max}} \qquad (7.159)$$

Actually, the radius of the meniscus, used as input parameter for the present model (see Eq. 7.134) is a function of x and needs to be modelled. The Young-Laplace expression can be applied resulting:

$$r_m(x) = \frac{\sigma}{p_v(x) - p_l(x)} \qquad (7.160)$$

It should be observed that the surface tension (σ) vary with temperature, which changes along the device; this variation may be relevant in some cases.

To determine the pressure distribution, an iterative process is necessary. In the first step, an expression for the meniscus radius as a function of the x distance, must be guessed. A linear distribution is suggested:

$$r_m(x) = r_{m,min} + \frac{r_{m,max} - r_{m,min}}{l_t} x \qquad (7.161)$$

where, according to Paiva (2015), the minimum meniscus radius can be considered as 100 μm. After the computation of all necessary parameters, a new meniscus radius is obtained using Eq. 7.160, until convergence.

Thermal Model

The thermal resistance circuit model is adopted. For a unit cell, it is basically composed of four resistances, as shown in Fig. 7.41, where R_e represents the evaporator, R_v the vapor, R_c the condenser and R_{cond} the conduction resistances, this last through the heat pipe solid materials. Using the parallel and series combination of the resistances, the total resistance is:

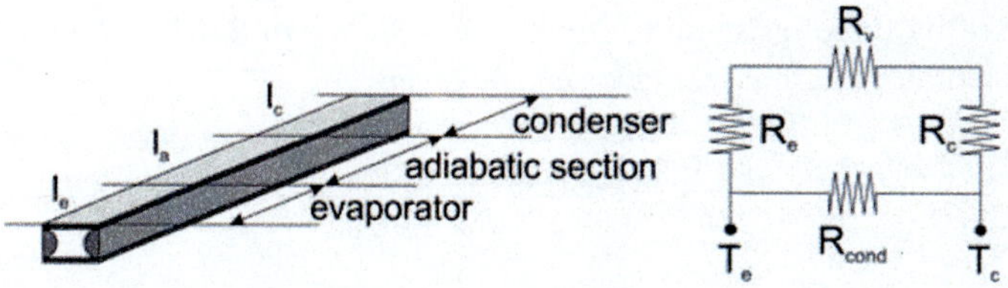

Fig. 7.41 Wire-plate mini heat pipe overall thermal circuit

$$R_t = \frac{R_{cond}(R_e + R_v + R_c)}{R_e + R_v + R_c + R_{cond}} \tag{7.162}$$

Considering that the wire and plates are made of the same material, the conduction thermal resistance of Fig. 7.41 can be determined using the effective length concept (see Eq. 3.1):

$$R_{cond} = \frac{l_{ef}}{kA_t} \tag{7.163}$$

where A_t is the total cross section area, including wire and plate, which, for the present case, is constant along the heat pipe.

The heat transfer phenomena at the condenser and evaporator are very similar, so that the same model can be used, considering that the appropriate geometry and coefficients are used. The wire-plate evaporator is modelled here.

Depending on how the heat is delivered to the evaporator, different models can be used. However, if the solid matrix is considered isothermal, no transversal conduction resistance exists and so the equivalent thermal circuit is composed only by the convection and phase change resistances.

Considering that the solid matrix and the vapor are at the same temperature (no transversal heat conduction resistance) and that the heat is delivered from both upper and lower external surfaces of the evaporator (see Fig. 7.42), the thermal circuit between the interface of the liquid layer and plate (at $T_{l,p}$) and the vapor (at T_v,),

Fig. 7.42 Wire plate evaporator thermal circuit with heat delivered in both plates

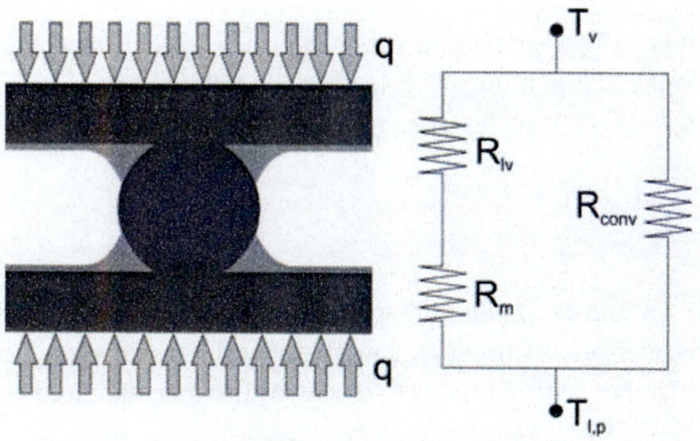

is composed by three resistances, combined as shown in the Fig. 7.42, so that the overall evaporator thermal resistance can be given as:

$$R_e = \frac{R_{conv}(R_{lv} + R_m)}{R_{lv} + R_m + R_{conv}} \tag{7.164}$$

where R_{lv} is the liquid-vapor phase change thermal resistance (boiling at the evaporator and condensation at the condenser), R_{conv} is the convection and R_m is the meniscus thermal resistances. It is important to note that the above model was developed for a unit cell depicted in Fig. 7.41. For a wire heat pipe composed of n parallel channels, the overall thermal resistance is:

$$R_{t,e} = \left[\sum_{j=1}^{n}\left(\frac{1}{R_e}\right)_j\right]^{-1} = \frac{R_e}{n} \tag{7.165}$$

The convection and phase change resistances are determined using the well-known convection thermal resistance expression:

$$R_{lv} = \frac{1}{h_{lv}A_{lv}}, \qquad R_{conv} = \frac{1}{h_{conv}A_{sv}} \tag{7.166}$$

where A_{lv} is the liquid-vapor interface area, A_{sv} is the soli-liquid interface area and h_{lv} and h_{conv} are obtained from suitable models (see Chaps. 3 and 4) for the phase change and convection heat transfer. Batista (2021) used the Kutateladze expression (Eq. 4.13) to predict the phase change coefficient of heat transfer and a literature model for gas flowing in channels for the convection heat transfer coefficients.

The vapor thermal resistance accounts for the pressure and temperature variations that occurs in the core of the heat pipe along its length. This resistance is evaluated using the Clapeyron equation (see Eq. 3.109 here reproduced):

$$\frac{dp}{dT} = \frac{h_{lv}}{T_{lv}\left(\frac{1}{\rho_v} - \frac{1}{\rho_l}\right)} \tag{7.167}$$

where T_{lv} is the liquid vapor interface temperature, which is considered equal to the vapor temperature. Neglecting the vapor term $((1/\rho_v) \ll (1/\rho_l))$ and considering $dp/dT \approx \Delta p/\Delta T$, the vapor resistance can be written as:

$$R_v = \frac{\Delta T}{q_v} = \frac{T_v \Delta p_v}{\rho_v h_{lv} q_v} \tag{7.168}$$

In the design of wire-plate mini heat pipes, the heat transferred through the vapor is, a priori, an unknown parameter and its determination requires an iterative process, together with all the resistances of the thermal circuit.

Meniscus liquid film thermal resistance. The determination of the thermal resistance through the liquid layer (meniscus) can be quite complex, as the heat flux lines are two-dimensional. Several works in the literature attempt to predict this resistance considering a complicated combination of several thermal resistances (Wand and Peterson 2002 and Paiva et al. 2015). A one-dimensional approximation, which shown to be accurate (see Batista 2021) is applied here.

Batista (2021) proposes a 1D model that tries to embody the major physical characteristics of the meniscus, while reducing the geometry complexity. The liquid film, which has an irregular format, is set to be equivalent to the ABC isosceles triangle of Fig. 7.39, as also shown in Fig. 7.43. This approximation considers the meniscus flat and neglects the heat transfer through the volume of liquid accumulated at the corner of the wire and plate region. This is not a bad hypothesis as the liquid film is very thin in the inner region of the groove, so that large liquid metal adhesions forces are expected with no evaporation (see Chap. 2). At this region the liquid temperature is considered the same of the plates.

As already observed, all the cross section solid material is considered at uniform temperature and all the liquid-solid interfaces are considered at uniform temperature, at the solid material level. The liquid accumulated at the deepest region of the groove is also considered at the solid temperature. Therefore, the two edges of the isosceles triangle that represents the liquid film, are at the solid temperature, while the edge representing the liquid vapor interface is at the vapor saturation temperature. In these conditions, the heat flux lines are symmetrical in relation to the $\overline{BA'}$ line, which means that there is an adiabatic surface that divides the isosceles into two right triangles. The "actual" heat flux lines in the isosceles triangle are also shown in Fig. 7.43. The lines tend to be straight, especially close to the symmetry line. Hence, to simplify

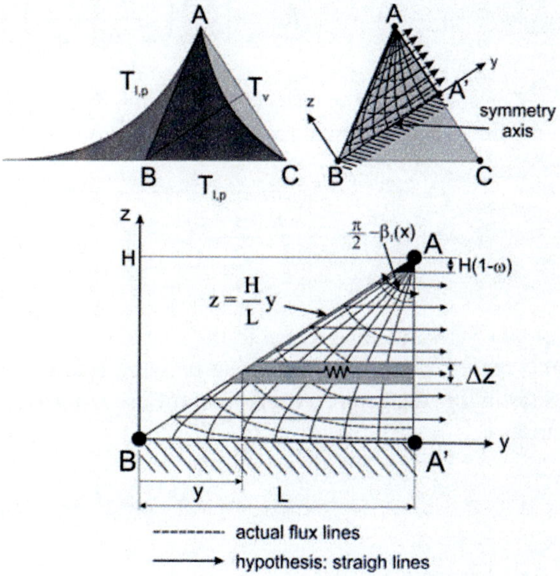

Fig. 7.43 Physical model for the geometry of the liquid layer and heat flux lines. The liquid is at an axial x position

the thermal resistance model, the flux lines are considered straight, as also shown in Fig. 7.43 (see Batista 2021). Actually, if the heat transfer in the deep groove region is to be considered, straighter would be the lines, therefore this hypothesis is reasonable.

From Fig. 7.39, the triangle edge is given by:

$$\overline{AB} = \overline{BC} = r_w \cdot \tan \beta_1(x) \tag{7.169}$$

The height of the isosceles triangle, $\overline{BA'}$ (see Figs. 7.43 and 7.39) is given by:

$$\overline{BA'} = L = r_w \tan \beta_1(x) \cdot \cos \beta_1(x) \tag{7.170}$$

The edge $\overline{A'A}$ is:

$$\overline{A'A} = H = r_w \tan \beta_1(x) \cdot \sin \beta_1(x) \tag{7.171}$$

With these hypotheses, the liquid layer can be divided into several slices and the thermal resistance of the liquid film can be approximated as the sum in parallel of these resistances. As shown in Fig. 7.43, the one dimensional conduction thermal resistance of the liquid layer slice, highlighted in gray is given by:

$$R_{m,\Delta z} = \frac{L - y}{k \cdot \Delta z \cdot \Delta l} \tag{7.172}$$

where Δl is the length of interest, in the perpendicular to the drawing page direction. The overall thermal resistance for a section located in the axial x position is:

$$R_m \simeq \left[\sum \left(\frac{L - y}{k \cdot \Delta y \cdot \Delta z} \right)^{-1} \right]^{-1} \tag{7.173}$$

Considering that $y = L/H \cdot z$ (see Fig. 7.43), in the limit, when $\Delta z \to dz$, the total liquid layer thermal resistance in the grooved layer is:

$$R_m = \frac{L}{k \cdot \Delta l \cdot H} \left[\int_0^{wH} \frac{dz}{H - z} \right]^{-1} = -\frac{L}{k \cdot \Delta l \cdot H} \frac{1}{\ln(1 - w(x))} \tag{7.174}$$

$\omega(x)$ is very small, close to the triangle corner at the $\overline{AA'}$ edge, at point A, At this region, due to the disjoining pressure (see Chap. 3), there is no evaporation. Substituting the geometry relations of Eqs. 7.169 to 7.171, the following expression arises:

$$R_m(x) = -\frac{\cot \beta_1(x)}{k \cdot \Delta l} \frac{1}{\ln(1 - w(x))} \tag{7.175}$$

As expected, the liquid layer thermal resistance varies with x. The last parameter to be determined is $w(x)$. It cannot assume the null value because, if so, the same point A would have two different temperatures, which is physically impossible.

Paiva (2011) observed that the thin liquid layer thickness t_f in wire-plate mini heat pipes, varies from $0.1\,\mu m$ at the evaporator to $10\,\mu m$ at the condenser. From Fig. 7.43 and using triangle resemblance, the thickness of the film t_f can be relate to the effective heat transfer width $H(1-w)$ and with the right triangle edges by the expression:

$$w(x) = 1 - \frac{t_f \cot\left(\frac{\pi}{2} - \beta_1(x)\right)}{r_w \tan \beta_1(x) \cdot \sin \beta_1(x)} \tag{7.176}$$

where $t_f = 0.1\,\mu m$ for the evaporator and $10\,\mu m$ for the condenser.

Integrations along the interest length should be performed to obtain mean thermal resistances of the liquid layer, to be substituted in the thermal circuit.

Temperature distribution. The thermal circuits shown can be used to predict the temperatures of the wire-plate mini heat pipe in some specific points. However, in many applications, the temperature distribution along the external surface of the device may be important. For that purpose, the present temperature profile analysis considers the heat pipe as a solid plate of the same external dimensions, using the effective thermal conductivity and effective length concepts. From the overall thermal resistance R_t of the wire-plate mini heat pipe, the effective thermal conductivity can be estimated as:

$$k_{ef} = \frac{l_{ef}}{A_t R_t} \tag{7.177}$$

where A_t is the total cross section area.

Therefore, considering steady state conditions, the problem to be solved is depicted in Fig. 7.44, consisting on the solution of the following ordinary differential equation:

$$\frac{d^2T}{dx^2} \pm \frac{\dot{q}}{k_{ef}} = 0 \tag{7.178}$$

Fig. 7.44. Physical model for temperature distribution of wire-plate mini heat pipes

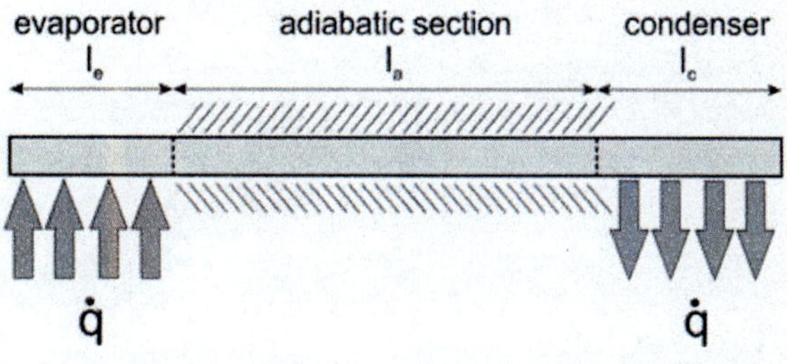

where the heat input and output are considered as generated or absorbed heat $\dot{q}[W/m^3]$ at the evaporator and condenser, respectively. According to the region, the boundary conditions changes. The temperature distribution must be determined backwards, from the end of the condenser. The temperature obtained by the model at the interfaces between regions is considered as the boundary condition for the next region. The resulting temperature distribution, where $l_t = l_e + l_a + l_c$, for the condenser is:

$$T(x) = T_c(x = l_t) + \frac{\dot{q}(l_t)^2}{2k_{ef}}\left[-\left(\frac{x}{l_t}\right)^2 + 2\left(\frac{x}{l_t}\right) - 1\right] \quad (7.179)$$

For the adiabatic section, the temperature distribution is:

$$T(x) = T_c(x = l_t) + \frac{\dot{q}(l_c)^2}{k_{ef}}\left[\frac{1}{2} + \left(\frac{x - l_t}{l_c}\right)\right] \quad (7.180)$$

and for the evaporator is:

$$T(x) = T(x = l_t) + \frac{\dot{q}l_e^2}{2k_{ef}}\left\{-\left(\frac{x}{l_e}\right)^2 + 1 + \frac{2l_c^2}{l_e^2}\left[\frac{1}{2} + \left(\frac{l_e - (l_t)}{l_c}\right)\right]\right\} \quad (7.181)$$

Hybrid Wire-Plate/Sintered Wick Mini Heat Pipes

Special mini heat pipes have been developed and considered as alternatives for the thermal control of small devices, such as electronic gadgets (smart phones, tablets, laptops, etc.). Among these devices, the hybrid technology is gaining importance. Figure 7.45 shows the configuration proposed by Paiva et al. (2015) and Batista (2021). In the left side, it is shown the hybrid configuration while, in the right, the wire-plate heat pipe. Both are composed of wires along the whole length of the tube. The wire provide good permeability (low pressure drop of the liquid flow) but its liquid pumping capacity is limited. This fact is compensated with the presence of metal powder sintered wick at the evaporator. The wick structure improves both the liquid pumping capacity and liquid boiling process, improving the number and quality of vapor nucleation sites. Figure 7.46 shows the cross sections of the evaporator of

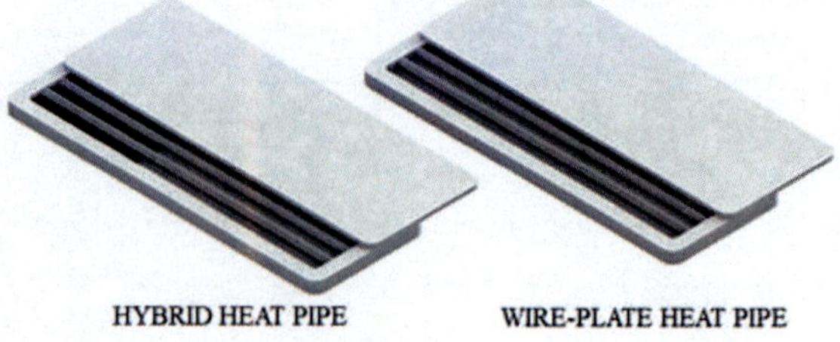

Fig. 7.45 Left: schematic of the hybrid wire-plate and sintered wick mini heat pipe and right schematic of the wire-plate

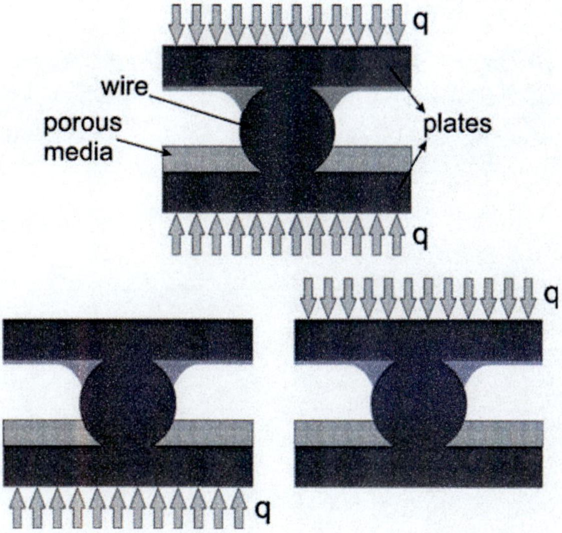

Fig. 7.46 Evaporator cross section area of the hybrid mini wire-plate with sintered porous media

several configurations of hybrid mini heat pipes. It is interesting to note that this technology is very flexible geometrically, being able to adapt to the application, such as the device developed by Batista, which schematic is shown in Fig. 7.47.

The hydraulic models used are the same for the wire-plate mini heat pipes and for sintered wick. However, some adaptations are needed, especially at the evaporator. First, for the evaporator, the cross section area for the vapor flow channel need to be recalculated, taking into consideration the geometry of the wires, plates and porous medium (see Fig. 7.46). Second, the permeability and liquid flow area must be adapted, to consider the presence of the porous medium. Third, a different analysis of the liquid pressure drop is required. Actually, Darcy's model is still applicable, but the permeability must be adjusted (see Chap. 3 models). Forth, the liquid pressure drop calculation is quite different. At the evaporator, there are two different paths for the liquid: through the two top side wire plate channel meniscus (pressure drops are calculated as in the last section) and through the porous media on the bottom side.

The thermal model depends on the hybrid configuration. Also, how heat is delivered to the device can be important on the choice of the thermal circuit to be used. If heat is applied in both surfaces and the plate is made of metal, a good hypothesis is that the device is isothermal in the transversal direction. If the heat is delivered in one of the plates, it is important to know if its internal face is in contact with the porous media or if there is liquid accumulated in wire-plate grooves, as the evaporator thermal resistance circuit can be quite different. Besides, transversal conduction thermal resistances maybe necessary to be included. However, for high thermal conductivity materials such as copper, a good simplifying hypothesis is to consider any transversal section of the evaporator at uniform temperature. For the determination of conductive thermal resistances of the porous medium, model for the effective conductivity must be employed (see Chap. 3).

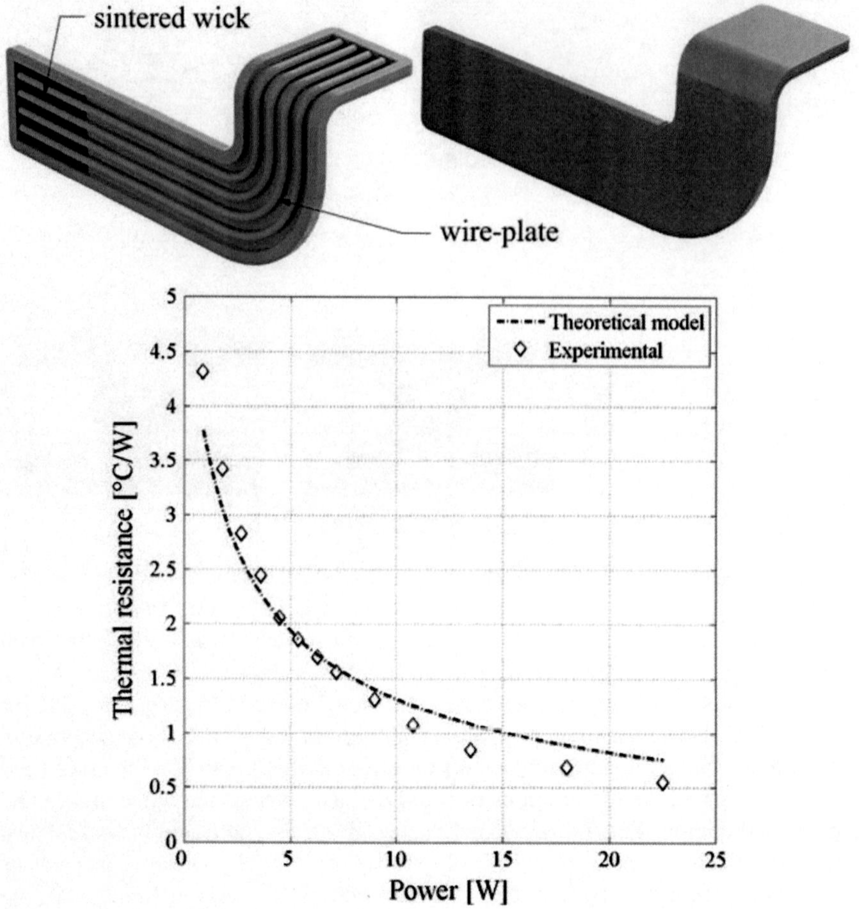

Fig. 7.47 Hybrid wire-plate sintered wick mini heat pipe tested for cooling of electronic devices in avionics

Batista (2021) developed a hybrid wire-plate and sintered wick mini heat pipe to be used to cool electronic devices in avionics. The schematic of the device is shown in Fig. 7.47. The models presented in this chapter for wire-plate mini heat pipes and those presented in Chap. 4 were used by this author for modelling. The major parameters of the constructed and tested device (diffusion bonding technique was used, see Chap. 5) are shown in Table 7.8. Figure 7.47 also shows the comparison of the thermal resistances predicted by the theoretical model and experimental data. A RMS (root mean square) deviation of less than 16% was observed, which is a very good result for this geometrically complex heat pipe, showing the quality of the model.

Table 7.8 Testing parameters of hybrid wire-plate sintered wick mini heat pipe

Parameters	
Material (plates and wire)	Copper
Copper thermal conductivity (W/mK)	401
Working fluid	Water
Working fluid volume (ml)	0.5
Heat pipe thermal conductance (no working fluid) (K/W)	5.5
Condenser temperature (°C)	20
Operation orientation (see Fig. 7.47)	Vertical
Evaporator length (mm)	20
Condenser length (mm)	18.7
Adiabatic section length (mm)	80.8
Wire radius (mm)	1.7
Plate dimensions: width, length, height (mm)	22.0, 124.5, 2.9

7.7 Closure

In this chapter, different thermosyphon and heat pipe technologies, classified according to their operational principle, are described and several models proposed. These models are in many aspects similar to the ones presented in Chap. 4, which methodology was used in the several problems/designs in Chap. 6 and used again here. As much as possible one-dimensional steady state conditions must be adopted, to simplify the expressions and so make feasible the design of equipment based on analytical models. Of course, when these simplified models are not able to take care of some important details, numerical simulations and experimentations might be very useful.

The present chapter clearly shows the breadth, versatility, broadening and high performance of thermosyphon and heat pipe technologies. Finally, it has shown that with good models, there are unlimited possibilities for the designer to use creativity and transform ideas into equipment.

References

Arpaci, V.S.: Conduction Heat Transfer, Addison-Wesley (1966)

Akers, W.W., Deans, H.A., Crosser, O.K.: Condensing Heat Transfer Within Horizontal Tubes. Chem. Eng. Prog. **54**, 87–90 (1958)

Almeida, J.C.P.: Development of a Hybrid Mini Loop Thermosyphon For Electronics Cooling Applications, Master´s dissertation. Federal University of Santa Catarina, Brazil, Mechanical Engineering Department (2021)

Batista, J.V.C.: Development of a Hybrid Mini Heat Pipe for Electronics Cooling Applications, Master´s dissertation. Federal University of Santa Catarina, Brazil, Mechanical Engineering Department (2021)

Bejan, A.: Convection Heat Transfer, 3rd edn. John Wiley & Sons, New York (2004)

Benjamin, R.J., Balakrishnan, A.R.: Nucleation Site Density in Pool Boiling of Saturated Pure Liquids: Effect of Surface Microroughness and Surface and Liquid Physical Properties. Exp. Thermal Fluid Sci. **15**, 32–42 (1997)

Betancur, L.A., Flat Plate Pulsating Heat Pipes: New Fabrication Technologies, Experiments and Modeling, Doctoral Thesis, Federal University of Santa Catarina, Mechanical Engineering Department, Florianópolis, Brazil (2020)

Betancur, L.A., Facin, A., Gonçalves, P., Paiva, K., Mantelli, M.B.H., Nuernberg, G.: Study of Flat Plate Closed Loop Pulsating Heat Pipes With Alternating Porous Media, Joint 19th International Heat Pipe Conference and 13th IHPS International Heat Pipe Symposium. Pisa, Italy (2018)

Betancur, L.A., Flórez, M.J.P., Mantelli, M.B.H.: Experimental Study of Channel Roughness Effect in Diffusion Bonded Pulsating Heat Pipes. Appl. Therm. Eng. **1**, 114734 (2019a)

Betancur, L., Mangini, D., Mantelli, M.B.H., Marengo, M.: Experimental Study of Thermal Performance in a Closed Loop Pulsating Heat Pipe with Alternating Superhydrophobic Channels. Thermal Science and Engineering Progress **1**, 100360 (2019b)

Betancur, L, Hulse, P., Melian, I., Mantelli. M.B.H., Diffusion Bonded Pulsating Heat Pipes: Fabrication Study and New Channel Proposal, Journal of the Brazilian Society of Mechanical Sciences and Engineering, 42, p. 467 (2020)

Bertossi, R., Ayel, V., Mehta, B., Romestant, C., Bertin, Y. and S. Khandekar, Motion of Liquid Plugs Between Vapor Bubbles in Capillary Tubes: A Comparison Between Fluids, Heat and Mass Transfer, **53**, 3315–3327 (2017)

Blet, N., Lips, S., Sartre, V.: Heat Pipes for Temperature Homogenization: A Literature Review. Appl. Therm. Eng. **118**, 490–509 (2017)

Boukhanouf, R., Haddad, A., North, M.T., Buffone, C.: Experimental Investigation of a Flat Plate Heat Pipe Performance Using IR Thermal Imaging Camera. Appl. Therm. Eng. **26**(17), 2148–2156 (2006)

Cai, Q., Chen, C. and Asfia, J., Heat Transfer Enhancement of Planar Pulsating Heat Pipe Device, in Proceedings of the ASME 2006 International Mechanical Engineering Congress and Exposition, Heat Transfer, **2**, 153–158 (2006)

Cao, Y., Faghri, A.: Analytical Solutions of Flow and Heat Transfer in a Porous Structure With Partial Heating and Evaporation on the Upper Surface. Int. Journal of Heat and Mass Transfer **37**, 1525–1533 (1994)

Carey, V.P.: Liquid-vapor Phase-change Phenomena: an Introduction to the Thermophysics of Vaporization and Condensation Process in Heat Transfer Equipment. Series in Chemical and Mechanical Engineering, Taylor and Francis, Hebron (1992)

Carslaw, H.S. and Jaeger, J.C., Conduction of Heat in Solids, 2nd edn, Oxford Science Publications (2004)

Charoensawan, P., Khandekar, S., Groll, M., Terdtoon, P.: Closed Loop Pulsating Heat Pipes: Part A: Parametric Experimental Investigations. Appl. Therm. Eng. **23**, 2009–2020 (2003)

Charoensawan, P., Terdtoon, P., Thermal Performance Correlation of Horizontal Closed-Loop Oscillating Heat Pipes, In: 9TH Electronics Packaging Technology Conference, IEEE (2008)

Chato, J.C.: Laminar Condensation Inside Horizontal and Inclined Tubes. ASHRAE J. **4**, 52–60 (1962)

Chen, J.C.: Correlation for Boiling Heat Transfer to Saturated Fluids in Convective Flow. Industrial and Engineering Chemistry Process Design and Development **5**, 322–329 (1966)

Chernysheva, M.A., Maydanik, Y.: 3D Model for Heat and Mass Transfer Simulation in Flat Evaporator of Copper-Water Loop Heat Pipe. Appl. Therm. Eng. **33**, 124–134 (2012)

Chi, S.W.: Heat Pipe Theory and Practice A Sourcebook, MacGraw-Hill (1976)

Churchill, S.W., Chu, H.S.: Correlating Equations for Laminar and Turbulent Free Convection from a Vertical Plate. Int. Journal of Heat and Mass Transfer **18**, 1323–1329 (1975)

Collier, J.G., Thome, J.R.: Convective Boiling and Condensation, 3rd edn. Science Publications, Oxford, Oxford (1994)

Cui, X., Zhu, Y., Li, Z., Shun, S.: Combination Study of Operation Characteristics and Heat Transfer Mechanism for Pulsating Heat Pipe. Appl. Therm. Eng. **65**, 394–402 (2014)

Demidov, A.S., Yatsenko, E.S.: Investigation of Heat and Mass Transfer in the Evaporator Zone of Heat Pipes Operating by Inverted Meniscus Principle. Int. Journal Heat Mass Transfer **37**, 2155–2163 (1994)

Facin, A., Betancur, L., Mantelli, M., Mera, J.P.F., Coutinho, B.H.: Influence of Channel Geometry on Diffusion Bonded Flat Plat Pulsating Heat Pipes, 19th International Heat Pipe Conference and 13th International Heat Pipe Symposium. Pisa, Italy (2018)

Faghri, A.: Heat Pipe Science and Technology, 2nd edn, Global Digital Press (2016)

Fasula, C., Oscillating Heat Pipes (OHP), A Special Problems Paper, Masters of Science, University of Rhode Island, (2009)

Foster, H.K., Zuber, N.: Dynamics of Vapor Bubbles and Boiling Heat Transfer. AIChE J. **1**, 531–535 (1955)

Fox, R, W., Pritchard, P. J. and McDonald, A. T., Introduction to Fluid Mechanics, 4th Ed., Mc-Graw-Hill (1999)

Furukawa, M., Rationalized Concise Descriptions of Fluid Motions in an Oscillating/Pulsating Heat Pipe, Journal of heat Treansfer, **136**, p. 092901(11) (2014)

Gamit, H., More, V., Mukund, B., Mehta, H.B.: Experimental Investigations on Pulsating Heat Pipe. Energy Procedia **75**, 3186–3191 (2015)

Gerstmann, J. And Griffith, Peter, Laminar Film Condensation on the Underside of Horizontal and Inclined Surfaces, International journal of Heat and Mass Transfer,. 10, 567–580 (1967)

Groll, M., Rosler, S.: Operation Principles and Performance of Heat Pipes and Closed Two-Phase Thermosyphons. Journal of Non-Equilibrium Thermodynamic **17**, 91–151 (1992)

Hao, T., Ma, X., Lan, Z., Li, N., Zhao, Y., Ma, H.: Effects of hydrophilic surface on heat transfer performance and oscillating motion for an oscillating heat pipe. International Journal Heat and Mass Transfer **72**, 50–65 (2014)

Hao, T., Ma, X., Zhong, L., Li, N., Zhao, Y.: Effects of Superhydrophobic and Superhydrophilic Surfaces on Heat Transfer and Oscillating Motion of an Oscillating Heat Pipe, Journal of Heat Transfer, **136**, p. 082001(13) (2014)

Hashimoto, H., Kaminaga, F.: Heat transfer characteristics in a condenser of closed two-phase thermosyphon: Effect of entrainment on heat transfer deterioration. Heat Transfer - Asian Research, **31**(3), 212–25 (2002)

Hoesing, M.: Integration of a Pulsating Heat Pipe in a Flat Plate Heat Sink. The Journal of Underground Research **12**, 21–36 (2014)

Hua, C., Wang, X., Gao, X., Zheng, H., Han, X., Chen, G.: Experimental Research on the Start-Up Characteristics and Heat Transfer Performance of Pulsating Heat Pipes with Rectangular Channels. Appl. Therm. Eng. **126**, 1058–1062 (2017)

Ibrahim, O.T., Monroe, J.G., Thompson, S.M., Shamsaei, N., Bilheux, H., Elwany, A., Bian, L.: An Investigation of a Multi-Layered Oscillating Heat Pipe Additively Manufactured From Ti-6Al-4 V Powder. Int. J. Heat Mass Transf. **108**, 1036–1047 (2017)

Ji, L., Peterson, G.P.: 3D Heat Transfer Analysis In A Loop Heat Pipe Evaporator With A Fully Saturated Wick. Int. Journal of Heat Mass Transfer **54**, 564–574 (2011)

Ji, Y., Chen, H., Kim, Y. J. Yu, Q., Ma, X. and Ma, H., Hydrophobic Surface Effect on Heat Transfer Performance in an Oscillating Heat Pipe, Journal of Heat Transfer, **134**, p. 074502 (2012)

Kaminaga, F., Hashimoto, H., Feroz, C., Goto, K. and Matsumura, K., Heat Transfer Characteristics of Evaporation and Condensation in a Two-Phase Closed Thermosyphon, 10th International Heat Pipe Conference, Stuttgart, Germany (1997)

Kandlikar, S.G.: Fundamental Issues Related To Flow Boiling in Minichannels and Microchannels. Exp. Thermal Fluid Sci. **26**, 389–407 (2002)

Karayiannis, T.G., Mahmoud, M.M.: Flow Boiling in Microchannels: Fundamentals and Applications. Appl. Therm. Eng. **115**, 1372–1397 (2017)

Kaya T, Hoang T, Ku J. and Cheung M., Mathematical Model of Loop Heat Pipes, AIAA 99–47, 37th Aerospace Sciences and Meeting and Exhibit, Reno (1999)
Khandekar, S., Charoensawan, P., Groll, M., Terdtoon, P.: Closed loop pulsating heat pipes - Part B: Visualization and semi-empirical modeling. Appl. Therm. Eng. **23**, 2021–2033 (2003a)
Khandekar, S., Dollinger, N., Groll, M.: Understanding Operational Regimes of Closed Loop Pulsating Heat Pipes: an Experimental Study. Appl. Therm. Eng. **23**, 707–719 (2003b)
Khandekar, S., Groll, M.: An Insight Into Thermo-Hydrodynamic Coupling in Closed Loop Pulsating Heat Pipes. International Thermal Sciences **43**, 13–20 (2004)
Khandekar, S. and Groll M., On the Definition of Pulsating Heat Pipes: An Overview, Proc. 5th Minsk International Seminar (Heat Pipes, Heat Pumps and Refrigerators), 1–12 (2003)
Khandekar, S., Panigrahi, P.K., Lefèvre, F., Bonjour, J.: Local Hydrodynamics of Flow in a Pulsating Heat Pipe: a Review. Frontiers in Heat Pipes **1**, 1–20 (2010)
Kim, W., Kim, S.J.: Effect of Reentrant Cavities on the Thermal Performance of a Pulsating Heat Pipe. Appl. Therm. Eng. **133**, 61–69 (2018)
Ku, J., Operating Characteristics of Loop Heat Pipes (1999), Proceeding of 29th ICES, Denver 12–15 July (1999)
Kutateladze, S. S., Heat Transfer During Condensation and Boiling, State Scientific And Technical Publishing House Of Literature On Machinery. 2nd ed., Russia, (1952)
Launay, S., Sartre, V., Mantelli, M.B.H., Paiva, K.V., Lallemand, M.: Investigation of a Wire Plate Micro Heat Pipe Array. Int. J. Therm. Sci. **43**, 499–507 (2004)
Launay, S., Sartre, V., Bonjour, J.: Analytical Model for Characterization of Loop Heat Pipes. J. Thermophys. Heat Transfer **22**, 623–631 (2008)
Lee, J., Kim, S.J.: Effect of channel geometry on the operating limit of micro pulsating heat pipes. Int. J. Heat Mass Transf. **107**, 204–212 (2017)
Lin, Y.H., Kang, S.W., Wu, T.Y.: Fabrication of Polydimethylsiloxane (PDMS) Pulsating Heat Pipe. Appl. Therm. Eng. **29**, 573–580 (2009)
Lips, S,. Bensalem, A., Bertin, Y., Ayel, V., Romestant, C. and Bonjour, J., Experimental Evidences of Distinct Heat Transfer Regimes in Pulsating Heat Pipes (PHP)," Applied Thermal Engineering, **30**, 900–907 (2010)
Liu, W., Gou, J., Luo, Y., Zhang, M.: The Experimental Investigation of a Vapor Chamber With Compound Columns Under The Influence of Gravity. Appl. Therm. Eng. **140**, 131–138 (2018)
Ma, H. Oscillating heat pipes, Springer (2015)
Ma, Y.D.: Motion Effect on the Dynamic Contact Angles in a Capillary Tube. Microfluid. Nanofluid. **12**, 671–675 (2012)
Maydanik, Y., Vershinin, S., Chernysheva, M., Yushakova, S.: Investigation of a Compact Copper-Water Loop Heat Pipe With a Flat Evaporator. Appl. Therm. Eng. **31**, 3533–3541 (2011)
Michels V., Milanez F. H., Mantelli M. B. H., Vapor Chamber Heat Sink With Hollow Fins, Journal Of The Brazilian Society Of Mechanical Engineering, 24, 3, 233–237 (2012)
Milanez, F.H., Mantelli, M.B.H.: Heat Transfer Limit Due to Pressure Drop of a Two-Phase Loop Thermosyphon. Heat Pipe Science and Technology, An International Journal **3**, 237–250 (2010)
Mizuta, K., Fukunaga, R., Fukuda, K., Nii, S., Asano, T.: Development and Characterization of a Flat Laminate Vapor Chamber. Appl. Therm. Eng. **104**, 461–471 (2016)
Moreira Jr., A.A., Mantelli, M.B.H.: Thermal Performance of a Novel Flat Thermosyphon For Avionics Thermal Management. Energy Convers. Manag. **202**, 112219 (2019)
Muzychka, Y.S., Culhan, J.R., Yovanovich, M.M.: Thermal Spreading Resistance of Eccentric Heat Sources on Rectangular Flux Channels. ASME J. Electronic Packaging **125**(2), 178–185 (2003)
Ozisik, M.N., Heat Conduction, 2nd edn, John Wiley and Sons, (1993)
Paiva, K.V.: Desenvolvimento de Novas Tecnologias para Mintubos de Calor: Análise Teórica e Experimental, Doctoral Thesis 2011. Federal University of Santa Catarina, Florianópolis, Mechanical Engineering Department (2011)
Paiva, K.V., Mantelli, M.B.H., Slongo, L.K.: Experimental Testing Of Mini Heat Pipes Under Microgravity Conditions Aboard A Suborbital Rocket. Aerosp. Sci. Technol. **45**, 367–375 (2015)

Paiva, K.V., Mantelli, M. B. H., Slongo, L. K., Thermal Behavior Analysis of a Wire Mini Heat Pipe, Journal of Heat Transfer, **133,** 121502–1–121502-9 (2011)

Peterson, G.P.: Heat Pipes Modeling Testing and Applications, John Wiley and Sons (1994)

Prasher, R.S.: A Simplified Conduction Based Modeling Scheme for Design Sensitivity Study of Thermal Solution Utilizing Heat Pipe and Vapor Chamber Technology. Journal of Electronic Packaging, ASME **125**(3), 378–385 (2003)

Qu, W., Ma, H.B.: Theoretical Analysis of Startup of a Pulsating Heat Pipe. Int. J. Heat Mass Transf. **50**, 2309–2316 (2007)

Qu, J., Wang, Q.: Experimental Study on The Thermal Performance of Vertical Closed-Loop Oscillating Heat Pipes and Correlation Modeling. Appl. Energy **112**, 1154–1160 (2013)

Reay, D.A. and Kew, P.A.: Heat Pipes Theory, Design and Applications, 5th edn, Butterworth-Heinemann (2006)

Santos, P.H.D., Theoretical and Experimental Study of the Application of Ceramic Porous Wick in LHP and CPL, Doctoral Thesis, POSMEC, Federal University of Santa Catarina, Brazil, Brazil (2010)

Sarmiento, A.P.C., Soares, V.H.T., Milanese, F.H., Mantelli, M.B.H.: Heat transfer correlation for circular and non-circular ducts in the transition regime. Int. Journal Heat Mass Transfer **149**, 119165 (2020)

Siedel, B., Sartre, V., Lefèvre, F.: Complete Analytical Model of a Loop Heat Pipe With a Flat Evaporator. Int. J. Therm. Sci. **89**, 372–386 (2015)

Simas G. V., Mera, J.P.F. and Mantelli, M.B.H., Study of a Loop Thermosyphon Evaporator for Thermal Control of Aircrafts, Journal of Heat Transfer, 141, 091812-1-8 (2019)

Simas, G.V., Flórez, J.P.M., Mantelli, M.B.H.: Improving Heat Transfer and Eliminating Geyser Boiling in Loop Thermosyphons: Model and Experimentation. Int. J. Heat Mass Transf. **156**, 119832 (2020)

Singh, R., Akbarzadeh, A., Dixon, C., Mochizuk, M.: Theoretical Modelling of a Miniature Loop Heat Pipe. Heat Mass Transf. **46**, 209–224 (2009)

Soo, D., Kim, D., Ho, S., Kim, Y.: Comparative Thermal Performance Evaluation Between Ultrathin Flat Plate Pulsating Heat Pipe and Graphite Sheet for Mobile Electronic Devices at Various Operating Conditions. Appl. Therm. Eng. **149**, 1427–1434 (2019)

Stenger F.J., Experimental Feasibility Study of Water-Filled Capillary-Pumped Heat-Transfer Loops, Report No. NASA TM X-1310, National Aeronautics and Space Administration, Lewis Research Center, Cl, Ohio (1966)

Swift, W. L, Dolan, F. X. and Zagarola, M.V., The NICMOS Cooling System—5 Years of Successful On-Orbit Operation, AIP Conference Proceeding, **985**, 799–806 (2008)

Tsai, M., Kang, S., Paiva, K.V.: Experimental Studies of Thermal Resistance In a Vapor Chamber Heat Spreader. Appl. Therm. Eng. **56**, 38–44 (2013)

Vassilev, M., Avenas, Y., Schaeffer, C., Schanen, J.L., Schulz, J.H.: Experimental Study of a Pulsating Heat Pipe with Combined Circular and Square Section Channels, pp. 1419–1425. IEEE Industry Applications Annual Meeting, New Orleans (2007)

Wang, Y.X., Peterson, G.P.: Analysis of Wire-Bonded Micro Heat Pipe Arrays. AIAA Journal of Thermophysics and Heat Transfer **16**(3), 346–355 (2002)

Wei, X. and Sikka, K., Modeling of Vapor Chamber as Heat Spreading Devices, 10th Intersociety Conference on Phenomena in Electronics Systems, ITHERM, San Diego, 578–585 (2006)

Xu, G., Liang, S., Vogel, M. and Katoh, T., Thermal Characterization of Pulsating Heat Pipes, Proceedings of 10th Intersociety Conference on Phenomena in Electronics Systems, ITHERM, San Diego, 556–560 (2006)

Yang, K.S., Cheng, Y.C., Liu, M.C., Shyu, J.C.: Micro Pulsating Heat Pipes With Alternate Microchannel Widths. Appl. Therm. Eng. **83**, 131–138 (2015)

Yang, H.; Khandekar, S. and Groll, M., Operational Characteristics of Flat Plate Closed Loop Pulsating Heat Pipes, In: 13th International Heat Pipe Conference, Shanghai, China (2004)

Yang, H., Khandekar, S., Groll, M.: Performance Characteristics of Pulsating Heat Pipes as Integral Thermal Spreaders. Int. J. Therm. Sci. **48**, 815–824 (2009)

Yuan, D., Qu, W., Ma, T.: Flow and Heat Transfer of Liquid Plug and Neighboring Vapor Slugs in a Pulsating Heat Pipe. Int. J. Heat Mass Transf. **53**, 1260–1268 (2010)

Zeng, J., Zhang, S., Chen, G., Lin, L., Sun, Y., Chuai, L., Yuan, W.: Experimental Investigation on Thermal Performance of Aluminum Vapor Chamber Using Micro-Grooved Wick With Reentrant Cavity Array. Appl. Therm. Eng. **130**, 185–194 (2018)

Chapter 8
Classification According to Operational Temperature

Thermosyphons and heat pipes can operate in basically any temperatures, from cryogenic to very high levels. Although the physical phenomena and models described in the former chapters can be applied to devices operating in many different temperature ranges, in the present chapter, discussions about the thermal behavior, modeling and fabrication aspects, typical for some extreme operational conditions, are discussed.

8.1 Cryogenic Heat Pipes

Operating at narrow ranges of low temperatures (below 200 K), one of the major applications of cryogenic heat pipes is the thermal control of sensors in satellites and spacecraft. This very low temperature level makes this device very sensitive to heat leaks from the surrounding environment (satellite structure, for instance). Therefore, these parasitic heat loads can affect significantly the operational temperature of cryogenic heat pipes, adding thermal loads that can easily reach its maximum transport capacity. Therefore, for designing purposes, the transient response of the heat pipe temperature must be well known. Once the device operates in steady state conditions, modelling can be performed using standard procedures.

Unlike low and medium temperature heat pipes, a cryogenic heat pipe typically starts from a supercritical state (see point F in Fig. 3.2). The entire heat pipe must be cooled below the saturation temperature of the working fluid before nominal operation begins. Cryogenic start-ups have received the attention of many researchers around the world (Colwell 1977; Brennan et al. 1992; Rosenfeld et al. 1995; Yan and Ochterbeck 1999).

Although the present model can be applied for cryogenic heat pipes of different configurations, the device for which a model is presented in this section is based on the one studied by Couto et al. (2006). It is made of stainless steel casing tube of 3/4 in (1.9 cm) of external diameter and 0.8 m of length, with wall thickness of 1.3 mm and the condenser of 0.3 m. Due to the action of the parasitic heat transfer loads, the

M. B. H. Mantelli et al., *Thermosyphons and Heat Pipes: Theory and Applications*,
https://doi.org/10.1007/978-3-030-62773-7_8

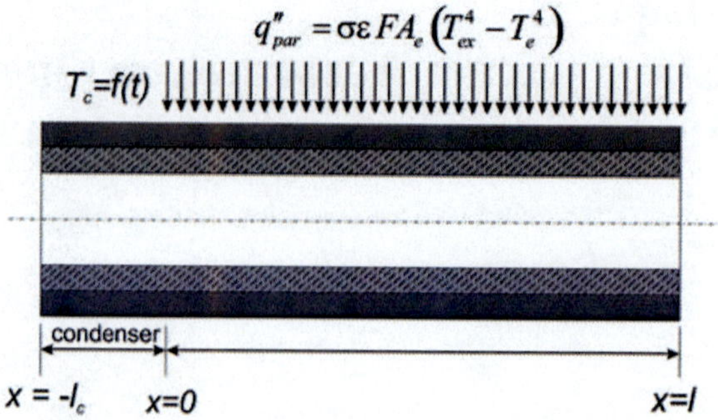

Fig. 8.1 Schematic of the physical model for cryogenic heat pipe

remaining of the heat pipe is considered as the evaporator region. Eight layers of 160 mesh metal screens were used as the wick structures performing an overall thickness of around 1.6 mm, although the model presented also considers the use of grooves.

Figure 8.1 shows a drawing of the axially grooved cryogenic heat pipe for which the supercritical start-up is modelled. In the beginning of operation, the heat pipe temperature T_i is uniform and above the critical temperature. The surface temperature, at the condenser region, is time variable and considered known, $T_c = f(t)$. The remaining length of the heat pipe is considered under the influence of a radiation parasitic heat load. For spacecraft applications, this parasitic heat load represents the radiation heat transfer between the heat pipe and the environment, such as spacecraft structure or heat loads from the space environment, including: direct sun irradiation, earth emission, albedo, etc.

According to Couto et al. (2005), the supercritical start-up process of a cryogenic heat pipe can be represented generically in a pressure-specific volume diagram, similar to the one presented at Fig. 3.2 and reproduced in Fig. 8.2. It is worth noting that, depending on the initial condition of the heat pipe, the working fluid at the condenser can achieve a subcooled condition before reaching the saturated condition, which can prevent the device of working properly.

In the majority of cases, the initial specific volume of the working fluid is approximately the same, or a little smaller than its critical specific volume. At the beginning of the process, the heat pipe is considered to be isothermal at T_i. As the temperature decreases, the condenser reaches the critical temperature at point 2a before the vapor pressure decreases below the critical pressure (point 2b). From this instance, the vapor inside the condenser region changes from supercritical fluid to subcooled liquid, as its temperature is below the critical temperature. On the other hand, the vapor pressure is still greater than the critical pressure. The subcooled liquid fills the wick structure and the vapor space in the condenser region, forming a liquid slug. As the temperature of the condenser decreases, the subcooled liquid slug extends into the heat pipe wick structure until the vapor pressure equals the critical pressure at point 3c. At this point, the leading edge of the subcooled slug is at a critical condition.

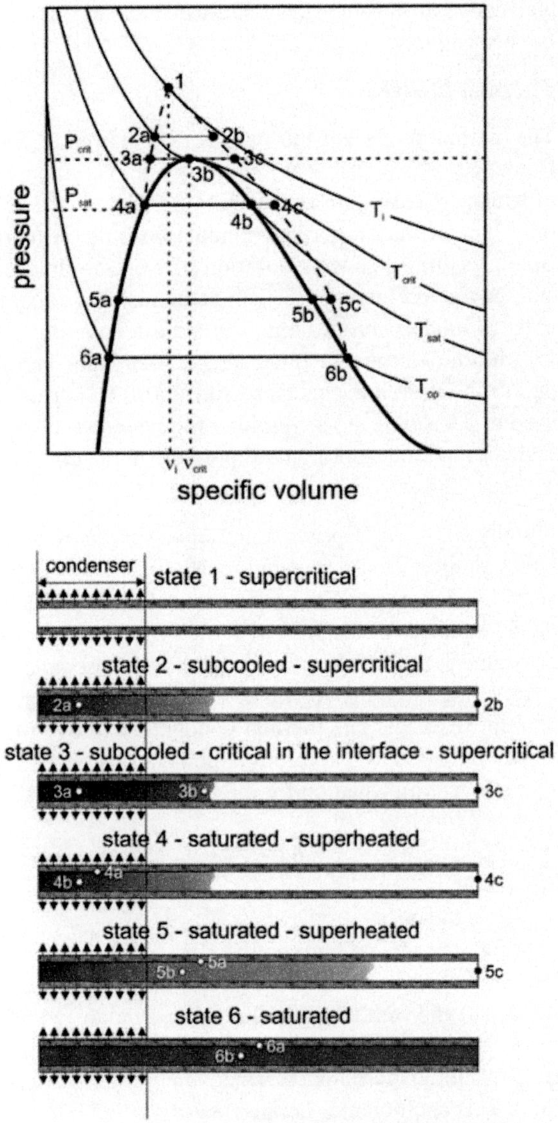

Fig. 8.2 Thermodynamic states for cryogenic heat pipe start-up

From then on, the length of the subcooled liquid slug decreases until the condenser temperature reaches the saturation pressure at point 4a and the condenser region is filled with saturated working fluid: saturated liquid in the wick structure (point 4a) and saturated vapor in the vapor space (point 4b). The remaining length of the heat pipe remains dry, in a superheated condition. As the temperature of the condenser continues to decrease (line 5a – 5b – 5c), more liquid is condensed along the wick structure. The process continues until the heat pipe is completely primed and the thermodynamic state at the wick structure and vapor region is given by points 6a and 6b, respectively.

Thermal Model

The thermal model for the supercritical startup of cryogenic heat pipes include the following submodels: heat pipe wall, liquid column, and internal vapor pressure, all of which are solved for two operation stages. In the first stage, the heat pipe contains only supercritical vapor and conduction is the primary mechanism of heat transfer. In the second stage, condensation of working fluid occurs in the condenser region and the models involves the simultaneous solution of both a conductive and a liquid column models. From them, the axial temperature profile and the liquid column length are determined. Interestingly, the vapor pressure model, usually neglected due to the very low vapor pressure drops in normal applications (see Chap. 4), is used to determine the beginning of the condensation process and is iteratively used to determine the working fluid pressure at the condenser region.

First Stage Heat Transfer Model

Initially, the heat pipe contains only supercritical vapor. According to Fig. 8.1, although the surface temperature is a function of time, the condenser temperature is considered to be uniform at any time t. Figure 8.3 shows a schematic of the physical model adopted for the present modelling, where the origin of the coordinate system is located at the interface between the condenser and the transport sections. The heat conduction at the dry region of the heat pipe wall and wick structure is considered one-dimensional. The thermal conduction through the working fluid is negligible when compared to that through the heat pipe wall. Therefore, the one-dimensional heat conduction equation for the physical model presented in Fig. 8.1 is given by:

$$\left(\rho c_p\right)_{ef}\frac{\partial T}{\partial t} = \frac{\partial}{\partial x}\left(k_s\frac{\partial T}{\partial x}\right) + \frac{\sigma\varepsilon F A_c\left(T_{ex}^4 - T_e^4\right)}{V_s} \tag{8.1}$$

where the second term on the right hand side accounts for the radiative parasitic heat load over the external surface. To account for the heat capacity of the working fluid, Yan and Ochterbeck (1999) suggested the addition of a constant coefficient to the heat capacity of the solid wall, to regard the total change of the internal energy, from the initial to the final thermodynamic states, for both heat pipe wall and working fluid. This coefficient is defined as:

$$\beta = \frac{\Delta E_s + \Delta E_l}{\Delta E_s} \tag{8.2}$$

where ΔE_s and ΔE_l are the internal energy change from initial to the final state. Besides:

$$\left(\rho c_p\right)_{ef} = (1+\beta)\rho_s c_s \tag{8.3}$$

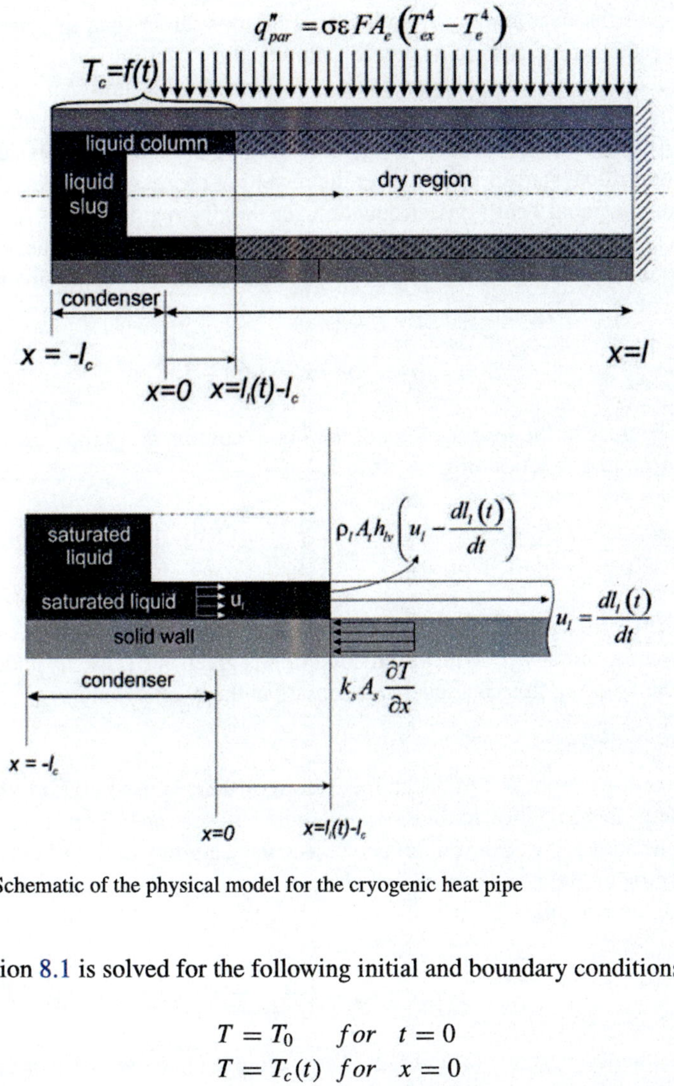

Fig. 8.3 Schematic of the physical model for the cryogenic heat pipe

Equation 8.1 is solved for the following initial and boundary conditions:

$$\begin{array}{lll} T = T_0 & for & t = 0 \\ T = T_c(t) & for & x = 0 \\ \frac{\partial T}{\partial x} = 0 & for & x = l_c \end{array} \tag{8.4}$$

Second Stage Heat Transfer Model

As discussed by Couto et al. (2005), when the vapor pressure inside the heat pipe equals the saturation pressure, at a given condenser temperature, $P_v = P_{sat}(T_c)$, vapor begins to condense at the wick structure, resulting in a liquid column, that flows with an average velocity u. Once the wick structure in the condenser region is filled with saturated liquid, the liquid column advances in the evaporator direction with a rewetting velocity u_r. Liquid is vaporized at the leading edge of the liquid column, as the dry region of the heat pipe is at a higher temperature. If the parasitic heat load plus the heat coming from the dry region is larger than that heat needed to vaporize the entire advancing liquid column, the rewetting process stagnates. The total length $l_l(t)$ of the liquid column is considered at uniform temperature, equal to the temperature of the heat pipe wall. At the condenser, this temperature is considered equal to the cryocooler temperature. Figure 8.3 shows the physical model for the supercritical startup, considering the rewetting process of the liquid column. The one-dimensional heat transfer equation for the dry region $l_l(t) - l_c < x < l$ remains as described by Eq. 8.1. The liquid column moves towards the evaporator and the boundary condition at $x = l_l(t) - l_c$ is considered to be a moving boundary condition:

$$T = T_c(t) \quad for \quad x = l_l(t) - l_c \tag{8.5}$$

A heat balance at the leading edge of the liquid column gives the position of the liquid column as a function of time, as follows:

$$k_s A_s \left.\frac{\partial T}{\partial x}\right|_{x=l_l(t)-l_c} = \rho_l A_t h_{lv} \left(u_l - \frac{dl_l(t)}{dt} \right)_{x=l_l(t)-l_c} \tag{8.6}$$

where $dl_l(t)/dt$ is the advancing velocity of the liquid front, which is the same of the rewetting velocity u_r. The initial condition is given at $l_t(t) = l_c$, at the instant in which the working fluid begins to condensate in the wick structure.

Liquid Column Model

The methodology proposed by Ochterbeck et al. (1995) is used. Taking the whole liquid column as the control volume, a momentum balance can be performed, considering that the advancing liquid column is subjected to capillary driving force, induced by the working fluid surface tension, which must surpass the friction (viscous) forces, resulting in the equation:

$$F_{cap} - F_v = \frac{d(m_l u_l)}{dt} \tag{8.7}$$

For a rectangular groove (see Fig. 2.12 and Table 2.1), the velocity of the liquid, which is related to the liquid column length $u = dl_l(t)/dt$, can be given by Ochterbeck et al. (1995), for the unheated surface:

$$u_l = \left(\frac{\sigma h}{2\mu_l l_l} \right) \left(\frac{w}{2h + w} \right)^2 \tag{8.8}$$

where h and w are the groove geometry parameters (see Fig. 2.12). Due to heat loads (as parasitic), the working fluid in the liquid column evaporates. Performing a mass balance between $x = 0$ and $x = l_l(t) - l_c$ (see Fig. 8.3), yields:

$$\frac{d(m_l)}{dt} = \dot{m}|_{x=0} - \dot{m}_{lv} \tag{8.9}$$

Therefore, in the heated region, the liquid velocity is given by:

$$u_l = \frac{\sigma h}{2\mu_l l_c}\left(\frac{w}{2h+w}\right)^2 - \frac{q''_{par}[2\pi r_v(l_l - l_c)]}{\rho_l A_l h_{lv}} \tag{8.10}$$

where q''_{par} is the parasitic heat load.

For screen metal and sintered porous wick structures, the solution of Eq. 8.7 results in the following expression for the liquid column velocity:

$$u_l = \frac{2\sigma\varepsilon' K^{\frac{1}{2}}}{\mu_l l_l} \tag{8.11}$$

where ε' and K are the volumetric porosity and permeability of the wick (see Chap. 3). To account for the parasitic loads for these wick structures, the liquid velocity can be predicted by the following expression, similar to Eq. 8.10:

$$u_l = \frac{2\sigma\varepsilon' K^{\frac{1}{2}}}{\mu_l l_l} - \frac{q''_{par}[2\pi r_v(l_l - l_c)]}{\rho_l A_l h_{lv}} \tag{8.12}$$

Vapor Pressure Model

As the vapor pressures involving cryogenic heat pipes can be quite small, ideal gas law may not be suitable to describe the vapor behavior. Therefore, the vapor pressures are given by:

$$p_v = \rho_v ZRT \tag{8.13}$$

where Z is the compressibility factor, is more convenient to be applied. Z is determined by (Jacobsen et al. 1997):

$$Z = \frac{p_v}{\rho_v RT} = 1 + \overline{\rho}\left(\frac{\partial \overline{E}}{\partial \overline{\rho}}\right) \tag{8.14}$$

where $\overline{\rho} = \rho/\rho_{crit}$ is the reduced density and $\overline{E}$ is the residual Helmholtz energy, determined by the last square fitting of experimental data. This procedure is fully described by Jacobsen et al. (1997) and Stewart et al. (1991).

As the vapor pressure is considered constant and the temperature of the heat pipe varies axially, the specific volume must also vary axially so that the conservation of the working fluid mass in the heat pipe is satisfied. Thus, the process to determine the vapor pressure is iterative. First, the heat pipe length (including the condenser) is divided into finite volumes and a vapor pressure is estimated at each time step.

Equation 8.13 is then solved for each one of the finite volumes, to determine the vapor densities. Then, the mass of fluid in each volume is obtained and the total mass is calculated by the sum of each volume masses. If the so calculated mass is different from the real fluid mass of the heat pipe, another pressure is estimated, until the calculation of the total mass converges to the actual value. When the temperature of the condenser T_c is below the critical temperature, the vapor pressure is compared with the saturation pressure at this condenser temperature, i.e., $p_{sat}(T_c)$. When $p = p_{sat}(T_c)$, the condensation process begins.

As Eq. 8.1 is highly non-linear, a numerical solution based on the finite volume method is recommended. The heat pipe length can be divided into several volumes and Eq. 8.1 can be integrated inside each volume, resulting in an implicit set equations for the temperature. The resulting system of linear equations can be solved, together with the boundary conditions presented in this section, using, for instance, the finite volume techniques available in the literature (Patankar 1980). In this process, the thermophysical properties of the heat pipe wall, as well as of the saturated working fluid, can be considered as a function of the local temperatures.

For the complete start up solution, iterations of the model are performed until the liquid column reaches the evaporator end, $l_l = l$, or until the temperature at the evaporator end varies less than a previously established value. Once the temperature profile of the heat pipe is determined, the vapor pressure equations can be solved to determine the mass distribution of the working fluid.

This modelling was validated by favorable comparison with experimental data, as shown in Couto et al. (2005 and 2006).

8.2 Intermediate Temperatures

As discussed in Chap. 3, for intermediate temperatures levels (between 300 and 600 °C), organic working fluids have been considered as suitable for thermosyphons and heat pipes. Among them, naphthalene is the most used fluid and has been applied in many heat exchangers (Vasiliev et al. 1988; Zhan et al. 2007; Vasiliev 2005). Naphtalene is an aromatic hydrocarbon (C10H8) with molecular mass of 128.16 g/ml. Detailed characteristics of the naphthalene can be found in the work of Bruce et al. (1988) and Anderson (2007).

Vapor pressure is one of the most important parameters to be considered in the selection of the appropriate working fluid. The main concern about using water for temperatures above 250 °C is the high pressures that the saturated vapor can reach. As stated by Mantelli et al. (2010), for intermediate operation temperatures, the naphthalene vapor pressure is more than one order of magnitude lower than that of the water, being, therefore, more appropriate to be used for two-phase devices at this temperature level.

Naphthalene is in solid state at room temperature. This can be a problem for the start-up of thermosyphons, as discussed later in this chapter. Besides, naphthalene can chemically react with the casing metal, generating non condensable gases (NCG), which accumulate on the extreme condenser regions, reducing the condenser heat transfer area. This effect is more evident at the low temperature levels, as the NCG expand and occupy larger regions of the condenser. At high temperature levels, the NCG are compressed in small regions and do not affect much the heat transfer capacity of the device. To avoid this effect, purging process described in Chap. 5 must be employed. Therefore, for designing thermosyphon equipment that operates with naphthalene, it is advisable to consider the presence of NCGs.

Thermal Resistance Model

As for other thermosyphons and heat pipes, the analogy between electrical and thermal circuits is used to model the heat transfer characteristics of naphthalene devices. However, in this case, thermal resistances, representing the heat transfer in the NCG plug region, must be included in the thermal circuit. The modelling procedure for naphthalene thermosyphons and heat pipes are similar to those described in Chap. 4, and, for the sake of illustration, a naphthalene thermosyphon is considered.

The evaporator is taken as fully flooded by the working fluid, therefore, no film region is considered in the thermal circuit. This can be a good approximation, if the evaporator filling ratio (ratio between the volumes of the working fluid and of the evaporator) is high, as the bubbles formed in the pool increases the effective length of the working fluid in the evaporator.

Figure 8.4 shows the sketch of the physical model adopted for a naphthalene stainless steel thermosyphon, operating in vertical position, considering the presence of NCG at the upper region of the condenser. The present theoretical model is constructed based on the following hypothesis:

- Steady state regime;
- Vapor temperature is the same of the adiabatic section;
- Axial conductance and film liquid–vapor phase change thermal resistances are neglected;
- Evaporation happens in nucleate pool boiling regime, with no film evaporation region in the evaporator;
- Adiabatic section does not exchange any heat.
- Condensation happens below the NCG layer.
- Vapor-NGC front is considered flat, i.e., no mass or thermal diffusion is considered.
- The two-phase flow is annular: the liquid flows from condenser to evaporator over the internal surface of the tube, while the vapor runs in countercurrent flux, in the central region;
- Natural convection heat transfer is considered for the NCG condenser external region.

The application of these hypothesis to the Fig. 4.2 thermal circuit, results in the network shown in Fig. 8.4. The following resistances compose the Fig. 8.4 thermal circuit: $R_{ex,hs}$ heat source external thermal resistance in the evaporator; $R_{ex,cs,NCG}$ and $R_{ex,cs,act}$ heat sink external thermal resistance in the condenser, for the NCG and active regions, respectively; $R_{cond,e}$ tube casing radial conduction in the evaporator;

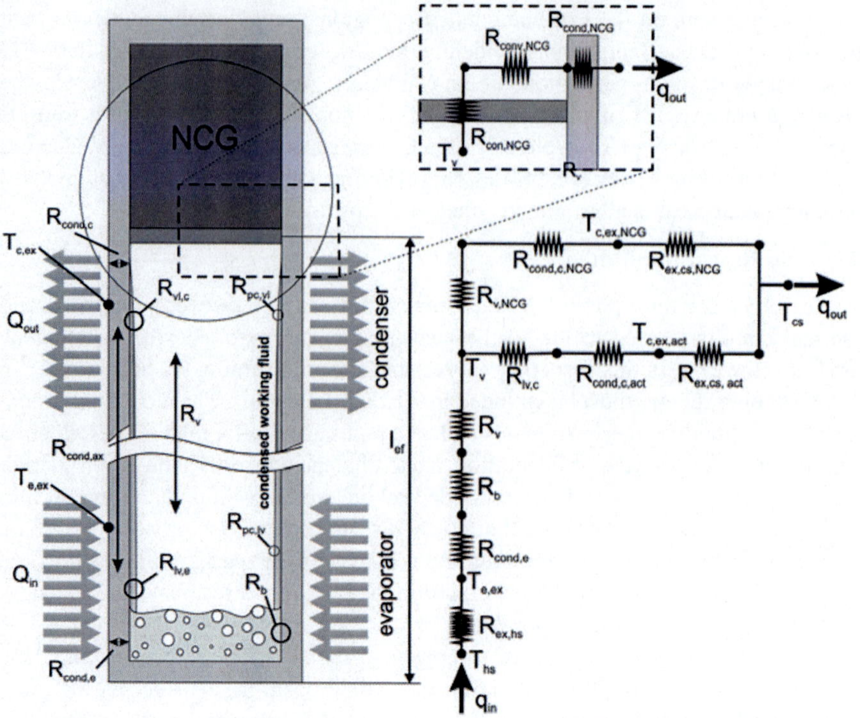

Fig. 8.4 Thermal circuit for naphthalene thermosyphon with non-condensable gases

$R_{cond,c,NCG}$ and $R_{cond,c,act}$ tube casing radial conduction in the condenser for the NCG and active regions, respectively; R_b boiling thermal resistance (bottom of the evaporator region), R_v vapor resistance and $R_{vl,c}$ vapor–liquid phase change in the condenser.

Mantelli et al. (2010) performed an experimental study of the heat transfer rate and the temperature distribution of a naphthalene thermosyphon, in which a controlled amount of NCG (argon) was inserted. The graphics presented in Fig. 4.5 shows the temperature distributions as a function of the thermosyphon length, for several heat loads and different amounts of argon. In these graphics, a large vapor-NCG interface region is observed. Also, it can be observed that the temperature distribution is flat for the thermosyphon without NCG and, as more argon is inserted inside the device, larger is the transition region. Small amounts of NCG are actually allocated at the end of the condenser tube and the temperature distribution on the rest of the condenser region tends to be flat.

Differently from the vapor-NCG flat front hypothesis adopted for the present model, large transition regions are actually observed, especially for thermosyphons with a large amount of NCG. However, Mantelli et al. (2010) shows that the flat front hypothesis is still good for the determination of the overall thermal resistance.

In the next sections, models to obtain the input parameters for the determination of the thermal resistances of the circuit, are presented.

Effective Length of Naphthalene Thermosyphon

As mentioned, NCG accumulate at the upper region of the condenser section. In this region, NCG natural convection is the major heat transfer mechanism, a comparatively ineffective means. Therefore, for modelling purposes, an effective thermosyphon operation length, l_{ef} (see Fig. 8.4), which corresponds to the thermosyphon length where the heat transfer happens indeed, must the determined.

The effective length of the thermosyphon can be determined once the amount of NCG (mass) inside a thermosyphon is known. The NCG pressure is assumed to be the same of saturated vapor pressure, which is a function of the operating temperature. Using the ideal gas law, it is possible to determine the volume of the NCG, and therefore, the length that this gas occupy inside the tube. Thus, the NCG length is given by:

$$l_{NCG} = \frac{1}{A}\frac{N_i R T_{NCG}}{p_{NCG}} \tag{8.15}$$

where N_i is the number of NCG moles (in the working fluid charging conditions), R is the ideal gas constant, A is the tube cross section area, T_{NCG} and p_{NCG} are the temperature and pressure of the NCG. Therefore, to estimate the length occupied by the NCG, the temperature and pressure must be determined. Using the thermal circuit of Fig. 8.4, this temperature is given by:

$$T_{NCG} = 1 - R_{v,NCG}\frac{(T_v - T_{cs})}{R_{v,NCG} + R_{cond,c,act} + R_{ex,cs,act}} \tag{8.16}$$

The NCG temperature is not significantly affected by the vapor temperature if the interface area between vapor and NCG is much smaller than the interface area between NCG and the tube. On the other hand, if the condenser length occupied by the NCG is small, the cross section area is equivalent in the order of magnitude of the NCG tube wall area. Thus, the NCG temperature becomes quite dependent of the heat transfer coefficients between vapor and NCG and between NCG and wall. The larger the NCG plug, the larger is the insulation effect of the NCG in the rear region of the condenser.

Considering that the interface between NCG and vapor is static, the naphthalene and the NCG pressures must be equal. Therefore, knowing the NCG pressures, the naphthalene properties can be obtained from tables.

Vapor Non Condensable Gases Transitions Region

One of the greatest challenges regarding the modelling of naphthalene thermosyphons is to establish a good model for the location and behavior of the transition front, i.e., the region where the vapor of the working fluid and the NCG plug encounter inside the device. Kaiping and Renz (1991) demonstrated both numerical and experimentally that, for a mixture of two gases which present high molecular mass difference, the effect of the thermal diffusion becomes significant. For instance, the molecular mass ratio of naphthalene to air is 4.4, and thus, for thermosyphons with air as NCG, the mass diffusion region is not negligible.

According to Mantelli et al. (2010), many research groups have investigated the thermal behavior of variable conductance heat pipes, usually applied for the thermal control of spacecraft and satellites. In these devices, a controlled amount of NCG is introduced in heat pipes to partially block the working fluid vapor region and so to modify their overall thermal resistance. Edwards and Marcus (1972) developed a one-dimensional model for the prediction of the heat transfer in heat pipes with NCG. It was observed, in steady state conditions, a smooth decrease in the vapor concentration and a corresponding increase in NCG concentration, over a considerable length of the heat pipe. They concluded that the flat-front theory proposed by Marcus and Fleishman (1970), regarding the vapor-NCG border region, does not predict the front behavior very well. Rohani and Tien, in 1973, developed a more precise, steady state, two-dimensional heat and mass transfer numerical analysis in the vapor-NCG region of a gas loaded heat pipe. They applied their model to the following combination of vapor-NCG: water-nitrogen, methanol-nitrogen, ammonia–nitrogen, sodium-argon and water–air.

Later, in the 1980 decade, thermosyphon studies came out. Hijikata et al. (1984) and Peterson et al. (1988) developed a more detailed theoretical and experimental studies for thermosyphons with NCG. Hijikata et al. (1984) considered that, in addition to the axial mass diffusion, NCG also tend to accumulate at the vapor–liquid interface due to the radial diffusion, thus retarding vapor condensation. They developed a two-dimensional model applied for water thermosyphons with air as NCG, where the vapor–gas mixture was assumed to follow the ideal gas law and where the NCG are assumed stationary at the top of the condenser region. The Nusselt model was supposed for the wall condensation and the problem was solved numerically. Peterson and Tien (1989) compared models in which the flat-front and the mass diffusive transport regions were considered for the vapor-NCG interface region, stating that the diffusive two-dimensional results compared favorably with data.

Other models were presented in the literature: Juxiang and Tongming (1992), Maezawa et al. (1994), Zhou and Collins (1995), Hashimoro et al. (1999), in which the flat and transition fronts between vapor and NCG are discussed, using one and two-dimensional models. Obviously, the two-dimensional models describe better the physical behavior of the vapor-NCG front. However, they are also time consuming, demand large computational efforts and so may not be adequate to be used in the design of actual equipment.

Mantelli et al. (2010) concluded that the flat vapor-NCG interface model is precise enough for the determination of the overall thermal resistances of thermosyphons with or without NCG. Therefore, in this chapter, a flat front is considered between the vapor and NCG, which simplifies the designing procedures.

Evaporator and Condenser Phase Change Correlations

Maybe the biggest difficulty in modeling naphthalene thermosyphons using the electrical-thermal circuit analogy, is in obtaining the heat transfer coefficients to be used in the calculation of the evaporator and condenser internal resistances. Actually, most of the correlations are for non-organic fluids, and may not be useful for naphthalene, an organic fluid. It is also important to note that the thermodynamic and heat transport properties of naphthalene are not well known, especially for the saturated conditions.

Mantelli et al. (2010) conducted an experimental study to determine the best correlations for the naphthalene thermosyphon. Four literature correlations: Foster and Zubber and Kutatelaze (see Collier and Thome 1996), Stephan and Abdelsalam (1980) and Cooper (1984), were used to estimate the evaporator coefficient of heat transfer, considering nucleate pool boiling, for the thermosyphon under investigation. It was verified that these correlations predict the heat transfer values with large variations, of up to two orders of magnitude. Therefore, the choice of the appropriated correlation is quite important for the thermosyphon design.

A similar study was performed by Mantelli et al. (2010) regarding the selection of the coefficients of heat transfer associated with the active part of the condenser (not blocked by the NCG). Three condensing film correlations were selected for this study (see Mantelli et al. 1999): Groll and Rosler, Kaminaga and Nusselt. Mantelli et al. (2010) showed that the heat transfer coefficients obtained with the Nusselt correlations can be four times larger than those obtained by Kaminaga, for low temperature levels. The behaviors of the Nusselt and Groll correlations are similar, while the Kaminaga´s correlation does not follow the same trends, as this correlation presents almost an insensitive behavior with temperature, for all temperature levels studied.

Mantelli et al. (2010), through the comparison of the resulting thermal circuit with their own data, concluded that the best model for naphthalene thermosyphons operating without considerable amount of NCG should include the correlation of Foster and Zuber, for the evaporator and of Nusselt for the condenser. On the other hand, for aphthalene thermosyphons operating with considerable amount of NCG, the correlations of Cooper (for evaporator) and Nusselt (for condenser) presented the best results. These authors state that the comparison between data and model improves for high power input levels, so that the difference between model and data can vary from around -150%, for low power inputs (model underpredict data) to down to $+ 15\%$ (model overpredict data).

8.3 High Temperature

Some recent and relevant applications of thermosyphons and heat pipes involve the heat transfer at high temperatures, such as concentrated solar power systems (CSP), photovoltaic solar cells (Boo et al. 2015; Conventry et al. 2015; Tournier and El-Genk

2006 and Zhang and Zhuang 2003) and nuclear applications. In these cases, devices must operate at high temperature levels, above 600 °C. The main characteristics of these two-phase heat transfer devices is the use of liquid metals as working fluids, such as the sodium, which operates at temperatures from 750 to 1500 K. Other working fluids are listed in Chap. 3.

The thermosphysics phenomena that happens in high temperature heat pipes are, in some extend, similar to those described in Chaps. 2 and 3, so that, basically, the same models presented in Chap. 4 can be used to predict the thermal behavior of these devices, once the thermal properties are known. However, modelling high temperature thermosyphons is a complex task, as some thermal behaviors, atypical for other configurations, are observed, such as: frozen start-up, cold tips, high condenser temperature gradients and violent Geyser boiling.

Start-Up From Frozen Working Fluid State

Heat Pipes

The start-up processes (usually at ambient temperatures) from solid state of high temperature working fluids, for two-phase flow devices, may have great practical importance in engineering. As temperatures rise, large gradients can be observed along the length of the device, which can generate significant thermal stresses on the tube walls, eventually affecting its life (Cao and Faghri 1993). Also, when the vapor densities are extremely low, the vapor can reach supersonic speeds and the sonic limit of the vapor phase may be reached.

According to Tournicr and El-Genk (2001), during the start-up from the solid state (frozen working fluid) of liquid metal two-phase devices, three flow regimes can coexist: molecular, transition and continuum.

Figure 8.5 shows a schematic of the startup of a heat pipe based on the description of Cisterna et al. (2021a), El-Genk and Tournier (2011) and Jang et al. (1989). The startup process of a liquid metal heat pipe from the working fluid in solid state can be divided into six stages, based on the working fluid flow regime. In the first stage (Fig. 8.5a), the device is at room temperature and heat starts to be applied to the evaporator section. Heat is transferred only by conduction through the solid material. The temperature of the tube in the evaporator region increases along with the solid sodium contained in this region. Any eventual formed vapor flow happens in molecular regime. In the second stage (Fig. 8.5b), the working fluid in the evaporator changes from solid to the liquid state, but no major evaporation occurs at the liquid–vapor interface. In the third stage (Fig. 8.5c), the temperatures rise with the heat load. The working fluid in liquid and solid states coexist simultaneously in the porous structure and the vaporization of the working fluid occurs at the liquid–vapor interface. At this point, continuum flow is established in both the vapor space and the evaporator region. The continuum vapor front moves towards the end of the cooled region of the heat pipe. In this phase, vapor shock waves may happen as the vapor speed can reach the sonic limit, as a consequence of the low vapor pressures in the condenser section. In fourth stage (Fig. 8.5d), the device can be divided into five regions according to the operation conditions. In region I, flow is continuum and evaporation occurs uniformly along the liquid–vapor interface. In region II, the

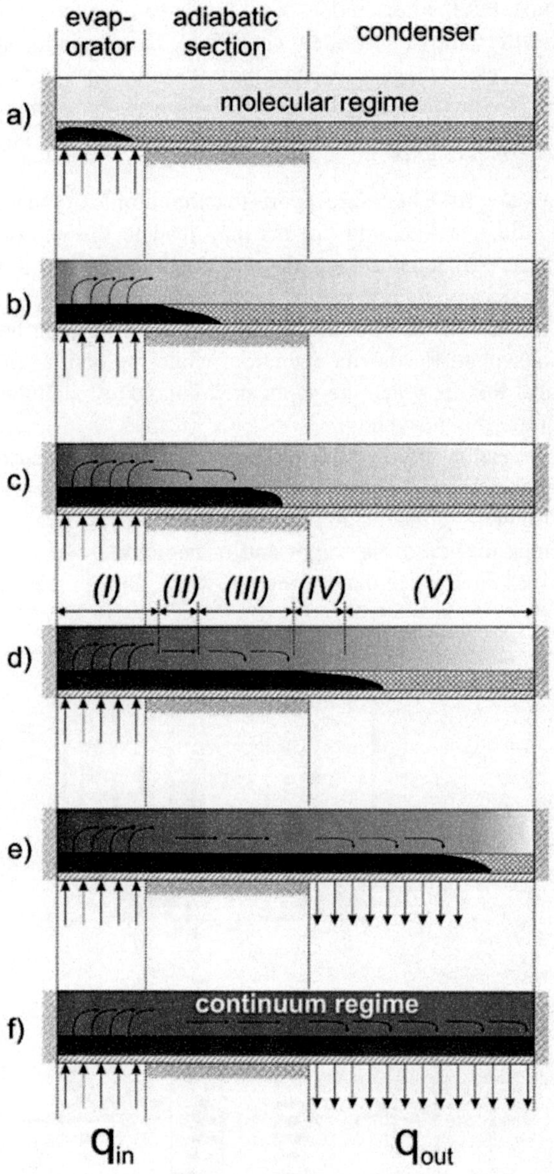

Fig. 8.5 Schematic of the start-up from frozen state of a liquid metal heat pipe

vapor flow is kept in continuum regime with minimal condensation at the liquid–vapor interface. In region III, both the vapor pressure and temperature decrease, while the vapor flow regime is in transition regime. In region IV, the vapor flow reaches supersonic velocities and is choked, with only a minimal amount of vapor being able to reach the condenser, where the flow is in molecular regime (region V). At the fifth stage (Fig. 8.5e), continuum flow exists throughout the entire length of the heat pipe, but the device does not reach steady state conditions (i.e. the evaporator still is affected by intermittent ebullition). Finally, in the sixth stage (Fig. 8.5f), the heat pipe reaches the steady state regime, with vapor in the continuum flow, reaching the end of the condenser and establishing a uniform temperature along the entire device.

Thermosyphons

Although the literature reports that the complexity of the physical processes involving liquid metal thermosyphons may lead to the use of sophisticated and expensive numerical schemes, simple and precise analytical models for the determination of the startup and continuum limits are available.

The startup limit corresponds to the minimum heat flux required for the thermosyphon to start its operation, while the continuum limit predicts the minimum heat flux in which the vapor flow is able to reach the continuum regime along the entire thermosyphon.

Cisterna et al. (2021a) proposed a model based on the analogy between two-phase thermosyphons and the piston-cylinder system (pure substance in a thermally insulated cylinder confined by a piston), as shown in Fig. 8.6. The piston is considered impermeable to the vapor and of negligible mass. The vapor bellow the piston is considered in continuous regime, while, above the piston, it is in free molecular flow.

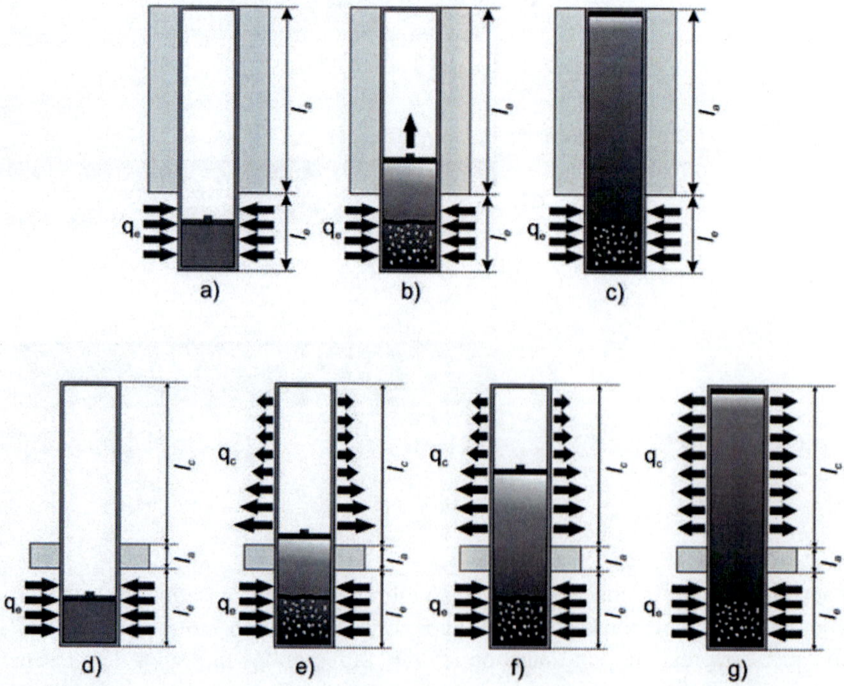

Fig. 8.6 Piston-cylinder thermodynamic schematic of start-up of a liquid metal thermosyphon from frozen state

The piston can be displaced only if there is a pressure difference between the inferior and superior areas of the piston, which forces its movement upwards.

Figure 8.6a shows a piston-cylinder system initially containing solid-state working fluid. The system is divided into two parts: evaporator (where heat is inserted) and adiabatic section, externally isolated, where no heat is exchanged. Initially the cylinder-piston system is in equilibrium (Fig. 8.6a) (the pressures above and below the piston are the same). As already noted, to get the piston to rise, the pressure in the region under the piston must increase. This is accomplished by inserting heat into the evaporator, causing the solid sodium to melt completely. As the substance is considered pure, vapor generation occurs only after all the working fluid has melted completely and its temperature increased to the evaporation level. Therefore, after the melting temperature is achieved and during the melting process, the piston only moves due to the volume expansion of the molten material. Since all working fluid is in the saturated liquid state, any additional heat inserted to the evaporator generates vapor in continuum regime, increasing the pressure and moving the piston up (Fig. 8.6b). From this, any generated amount of vapor is able to move the piston up (see Fig. 8.6c), until it reaches the top of the cylinder, homogenizing the temperature along the entire length of the adiabatic section. As no heat is removed (Figs. 8.6a, c), steady state conditions will never be reached.

Considering that the thermal insulation is removed (Fig. 8.6d–f), a heat rejection section (condenser) is created and the cylinder can now work as a thermosyphon. In this case, it is not possible to affirm that any heat flux imposed in the evaporator is able to move the piston to the top of the cylinder (as in the case of Fig. 8.6c), because, due to the heat removal section (condenser), part of the vapor is condensed and so, the displacement of the piston depends on both the heat inserted in the evaporator and the capacity of the condenser to reject this heat. The larger is the condenser capacity of heat removal (large condensation area or high external heat transfer coefficients), more vapor needs to be generated for the piston to reach the end of the condenser.

As discussed by Cisterna et al. (2021a), up to the startup limit (minimum heat flux required to start the operation of a two-phase thermosyphon), the device behaves like an empty tube, where heat is transferred only by conduction. The start-up stage ends when the vapor front, in the continuum regime, reaches the end of the adiabatic section (condenser base) as shown in Fig. 8.6e. Steady state can be reached at the case depicted in Fig. 8.6.e, only if all evaporator heat flux is transferred to the condenser by axially conduction through the wall, not involving any phase change (the piston is at the end of the adiabatic section). In this condition, the generated vapor volume is negligible and the vapor inside the condenser can be considered in free molecular flow. Consequently, the temperature distribution of the condenser becomes similar to that of a thermal fin (Fig. 8.6e). As more vapor is generated, it starts to condensate in the lower portions of the condenser, which operates with vapor in the continuum regime, while the upper part (volume above the piston) still presents large temperature gradients (typical of fins), characteristic of the region where the vapor is in molecular regime (see Fig. 8.6f)). The continuum limit is reached when the piston reaches the top of the condenser (Fig. 8.6g), i.e., when the vapor is able to be in the continuum regime along the whole thermosyphon. Therefore, the thermosyphon operates in the continuum regime for heat fluxes larger than the continuous limit.

Influencing Parameters for High Temperature Thermosyphons

Filling Ratio

The filling ratio (F), which is defined as the volumetric fraction of the thermosyphon evaporator that is filled by the working fluid, is one of the major factors that influences the operation and performance of two-phase thermosyphons. According to Reay and Kew (2003), two limiting conditions must be considered in the determination of the ideal amount of working fluid in high temperature thermosyphons: too low volumes of fluid can lead to the dry-out of the evaporator and the presence of cold tips, while excessive volumes can cause flooding of the condenser and Geyser boiling. Both effects decrease the device performance. The literature is rich in works dealing with the effect of filling ratios in the operation of low to average temperature thermosyphons. They show that these devices can work with large variations of working fluid volumes (filling ratios typically ranging from 20 to 80%), for several working fluids, such as water and ethanol (see Negishi and Sawada 1983; Noie et al. 2007).

Few works are concentrated in determining the appropriated filling ratios for high temperature thermosyphons. Faghri et al. (1991), performed experiments with different amounts of sodium in two-phase thermosyphons. These authors had problems with the variation of the optical properties of the condenser external wall by oxidation, which changed the radiation heat transfer capacity of the thermosyphon wall, making their results not totally conclusive.

Geyser Boiling

Many literature works, concentrated in intermediate temperature working fluids, associate the filling ratio to the occurrence of Geyser boiling phenomenon (Pabón et al. 2019; Alammar et al. 2018; Liu et al. 2018; Guo et al. 2018; Mantelli et al. 2017; Tecchio et al. 2017; Xia et al. 2017; Jafari et al. 2017; Khazaee et al. 2010; Emami et al. 2009 and Abreu et al. 2003). In general, the results of these studies agree that, by increasing the filling ratio, the frequency of the Geyser boiling decreases, while the amplitude of the oscillations increases.

However, not many works are available concerning high temperature fluids. Mantelli et al. (2017) investigated two filling ratios for sodium thermosyphons: 80 and 100%, concluding that the thermosyphons with 80% of filling ratio transfer more heat than those with 100%. They performed a series of experiments with sodium thermosyphons where they observed violent oscillations caused by the Geyser boiling phenomenon under some specific heat transfer rate levels. Guo et al. (2018) carried out an experimental study with sodium–potassium thermosyphons, also observing the same behavior.

Cold Tip

Another common phenomenon for liquid metal thermosyphons is the cold tip effect. The specialized literature usually relate this phenomenon with low power inputs, typical of startup processes from frozen state (Deveral et al. 1970; Tournier and El-Genk 1996; Cao and Faghri 1993; Tournier and El-Genk 1996). Two hypotheses can be used to explain the cold tip at the end of the condenser: the presence of non-condensable gases (NCG) inside the thermosyphon and the free molecular flow of vapor at the condenser. During startup and for favorable heat removal conditions at the condenser, practically all the generated vapor condenses in the lower regions of the condenser and the tip does not receive enough vapor flow, remaining cold. The low vapor pressures at the end of the condenser can cause the vapor to reach molecular regime. Thus, a temperature profile similar to that of a fin is observed at the condenser.

Internal Heat Transfer Coefficients

For most models, such as those where the analogy between electrical and thermal circuits is applied, the coefficients of heat transfer in the internal regions of the evaporator and condenser need to be known. Generally, thermal resistances associated with phase changes can be determined from the coefficients of heat transfer related to nucleate pool boiling, film evaporation and condensation.

The evaporation of the working fluid in the evaporator can be divided into two mechanisms: nucleate pool boiling in the bottom and film evaporation in the upper region. According to Noie et al. (2003), boiling is the dominant mechanism when the filling ratio is higher than 30%.

However, there are not many correlations involving the heat transfer coefficients for liquid metals in the literature and most of the available correlations are based on experimental observations, since the physical phenomenon is very complex, difficult to model analytically (Jafari et al. 2017). Carey (2018) affirms that nucleate pool boiling in liquid metals is significantly different from that in non-metallic fluids. The high thermal conductivity of the liquid metal transports effectively heat to the liquid–vapor interface, feeding the growing bubbles, which increases rapidly in volume. Despite their faster growth, the period between bubbles in liquid metals is considerably longer than in other fluids. When a bubble departs and the cold liquid metal contacts the wall, the surface temperature decreases locally. This decrease in temperature tends to be more pronounced than for non-metallic fluids under the same circumstances. Due to this fast drop in surface temperature, the waiting period, for the formation of another bubble is longer, resulting in much lower bubble burst frequencies. Besides, the low contact angles between liquid metals and metal surfaces, wet completely normal cavities, which become inactive as nucleation sites, thus requiring higher superheat temperatures for the nucleation to start.

Wilcox and Rohsenow (1969) comment the significant differences observed between film condensation phenomena for metallic and non-metallic fluids. Consequently, correlations based on the Nusselt model may present large discrepancies when compared with experimental data. These authors pointed out that some simplifications applied of the Nusselt model, such as disregarding the effects of the shear stress at the liquid–vapor interface and of the moment and energy transfers by advection in the condensate film, may cause such errors. Therefore, the predicted linear temperature profile is not valid for liquid metals. According to Stephan (1992), the low Prandtl numbers of liquid metals (between 10^{-2} to 10^{-4}), result in a high momentum transfer between the liquid and the vapor phases, making the shear stress in the liquid/vapor interface significant. Chen (1961) presents a correlation for film condensation which includes the effect of shear stress, especially useful for low Prandtl number fluids, as is the case of liquid metals.

It is important to note that most of the literature models to predict the heat transfer coefficients were developed for ideal operating conditions and did not take into account the effect of the temperature gradient generated by the failure of the continuum hypothesis in the condenser. These correlations also did not considered the oscillations due to Geyser boiling phenomenon. The next section aims to fulfill this model gap.

Experimental Evaporator and Condenser Heat Transfer Coefficients

Cisterna et al. (2020a) developed an experimental work on sodium thermosyphons. They constructed four thermosyphons, made of stainless steel (AISI 316L) tubes of diameter of 25.4 mm, with 3 mm of wall thickness, with the filling ratios of 31.5, 69, 96 and 144%. The evaporator, adiabatic section and condenser lengths were 100, 50 and 200 mm, respectively. Sodium with 99.8% of purity was employed and charged in liquid state, following the methodology described in Chap. 5 (Cisterna et al. 2020b). Based on the surface temperature measurements, the heat transfer to the environment was determined by a combination of radiation and natural convection by the expression:

$$\overline{h}_{ex} = \overline{h}_{N,conv} + \varepsilon\sigma\left(\overline{T}_c^2 + \overline{T}_\infty^2\right)\left(\overline{T}_c + \overline{T}_\infty\right) \tag{8.17}$$

where $\overline{h}_{N,conv}$ is the natural convection heat transfer coefficient, calculated by the correlation of Nellis and Klein (2009) for vertical cylinders, $\overline{T}_c$ is the average condenser temperature, T_∞ is the ambient temperature, σ is the Steffan–Boltzmann constant and ε is the condenser external surface emissivity. For high temperature levels, the surface emissivity of stainless steel changes with the temperature: while it presents low emissivities at low temperatures, it can reach quite high levels at high temperatures. Cisterna et al. (2020b) used a thermography camera to measure the emissivity of the stainless steel surface between 500 and 1400 K, resulting in 0.702 ± 0.014. These authors also obtained the evaporator and condenser coefficients of heat transfer, neglecting the conduction radial thermal resistance of the casing walls (which means that the internal and external temperatures of the evaporator and condenser were considered the same). The saturation working fluid temperature was assumed as the same of the adiabatic section wall. The following expressions based on the Newton's law of cooling, were applied for the determination of the evaporator and condenser heat transfer coefficients respectively:

$$h_e = \frac{q}{\pi D_i l_e (T_{w,e} - T_{sat})} \quad and \quad h_c = \frac{q}{\pi D_i l_c (T_{sat} - T_{w,c})} \tag{8.18}$$

Operation Regimes

As already discussed, at the operation beginning, when the working fluid of the high temperature thermosyphon still is in frozen state, the tube is under vacuum conditions. Therefore, any heat applied to the evaporator is transferred to the condenser by conduction through the walls and the thermosyphon behaves like an empty tube fin. The device starts up when effective phase change starts to happen, characterized by a sudden increase of the temperature of the adiabatic and condenser regions, in this order.

Cisterna et al. (2020a) divide the operation of high temperature thermosyphons in three regimes: fin, Geyser boiling and ideal. In the fin regime, the thermosyphons present a large temperature gradients along the condenser, similar to those observed in fins. In the Geyser boiling regime, an intense oscillations of the temperatures are recorded in both the evaporator and the condenser. In the ideal regime, the devices behave like conventional thermosyphons and their temperatures along the device tend to be isothermal. The Geyser boiling phenomenon is treated in details in Sect. 3.8.

Figure 8.7 shows an illustrative plot of the thermal behavior of a high temperature thermosyphon. The graph area is divided, by thick vertical lines, into fin, Geyser boiling and ideal operation regimes. The heat input power increases with increasing

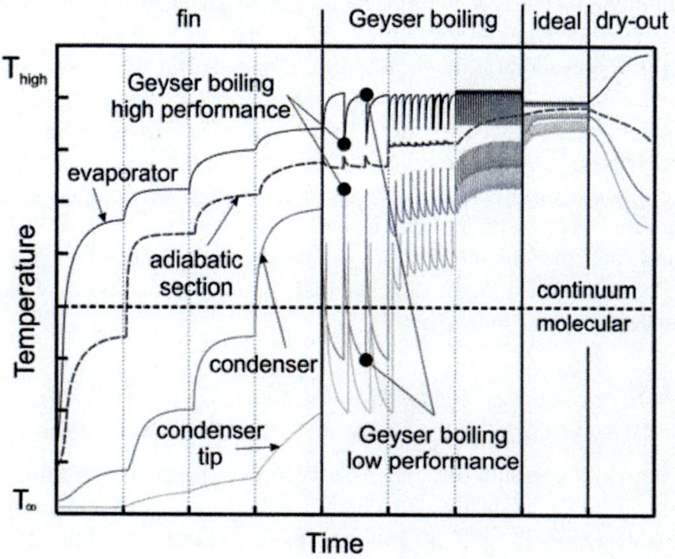

Fig. 8.7 Illustration of the operational conditions of a high temperature thermosyphon

time, but, between vertical lines, the power is kept at constant level. Also, the dry-out regime, when the devices does not operate anymore and, therefore, the temperatures of the evaporator tend to increase while, simultaneously, the temperatures of the condenser tend to decrease, is shown in Fig. 8.7.

Under Geyser boiling regime, the thermosyphon can present two levels of very different heat transfer performances, high and low, which change instantaneously and repeat in each large bubble burst. The low performance Geyser boiling is observed just before the bubble eruption, where there is a quick increase in the temperature difference between the evaporator and condenser. The high performance condition is noted immediately after a large bubble burst, characterized by a very fast temperature homogenization (temperature of the evaporator decreases while the condenser's increases) as shown in Fig. 8.7. This figure also shows that the amplitude of the temperature oscillations decrease and their frequencies increase with the increasing power, until these oscillations are so small and fast that the ideal regime is reached and constant temperatures of evaporator and condenser can be assumed.

In Fig. 8.7, a horizontal dashed axis divides the graphic between continuum and molecular regime regions. It is clear that, for many heat transfer power inputs, the thermosyphon can operate partially in the molecular regime. Only at high temperatures, in the high performance Geyser boiling or ideal regimes, the thermosyphon operates completely outside the molecular regime.

Cisterna et al. (2020a), in agreement with several literature works, observed that, the lower is the filling ratio, the shorter is the startup time and lower is the startup temperature. As for other working fluids, the frequency and the amplitude of the bubble oscillations in the Geyser boiling regime also depends on the filling ratio: as the filling ratio increases, the frequency of the Geyser boiling oscillations decrease, but the amplitude increase. A thermosyphon can be designed with low filling ratios to avoid its operation in Geyser boiling regime. However, dry-out conditions can be created in the evaporator, as the liquid accumulated in the pool is minimal or inexistent. In this case, most of the condensed liquid is located over the evaporator inner walls, as condensed liquid films, in their way back to the evaporator. Actually, any increase of the heat power input may dry-out the working fluid within the descending liquid film. Local dry-out means local high temperature spots, which may compromise the integrity of the device.

To identify whether the temperature oscillation observed can be associated to a Geyser boiling phenomenon, the dimensionless number N, proposed by Pabon et al. (2019) can be used. It is defined as:

$$N = \frac{\Delta p}{\overline{p}} \tag{8.19}$$

where Δp is the pressure difference between two consecutive readings and $\overline{p}$ is the mean pressure under steady state conditions. When $N > 0.1$ the boiling events can be considered Geyser boiling. Pabon et al. (2019) compared this criterion with data from Smith et al. (2018), for thermosyphons filled with several different working fluids, such as water, Ethanol e HFE-700. If pressures are unavailable, this parameter can be approximated by temperature readings, considering working fluid saturated conditions.

Figure 8.8 shows, in the upper plot, the temperature and, in the bottom plot, the N factor, both as a function of time, for one of the sodium thermosyphons tested by Cisterna et al. (2020a) (AISI 316L tubes of diameter 25.4 mm, 350 mm length total, 69% filling ratio—17 g of sodium of 99.8% of purity, see description in the beginning of this section). The shades of gray adopted in the upper plot are related to the temperature reading positions in the thermosyphon: the darker curves represent the evaporator while the lighter the condenser. The number 1 in this plot represents the point in which vapor, in continuum regime, reaches the condenser region. Before that point, there is either vacuum or vapor in molecular regime. In both cases no working fluid phase change happens and heat is transferred only by conduction. Point 2 shows the instant where the first Geyser boiling bubble bursts and vapor reaches the condenser top, causing an almost instantaneous temperature jump in the condenser, and, almost at the same instant, a temperature drop in the evaporator. Point 3 shows the onset of the thermosyphon ideal operation regime, when the heat load is large enough to avoid the Geyser boiling phenomenon. Comparing the upper and lower graphics of Fig. 8.8, it is clear that the parameter N is able to capture the largest bubble bursts and thus, it is good criteria to detect Geyser boiling regime.

In their study, Cisterna et al. (2020a) concluded that the thermosyphons with different filling ratios presented large differences in their thermal performances, especially for the low power inputs (different startup times, onset of Geyser boiling condition, oscillations and frequencies of bubbles). However, once the ideal regime is reached, all configurations presented practically the same behavior. They observed that, while the thermosyphon is operating in the fin regime, the heat transfer resistance in the condenser is much larger (from 4 to 6 times) than that in the evaporator. Therefore, the condenser limits the thermal performance of the thermosyphon. As the vapor enters in the ideal (continuous) regime, the difference in the heat transfer performance of the evaporator and condenser reduces, with both approaching the same order of magnitude.

Evaporator Correlation

For most applications, high temperature thermosyphons must be designed to avoid Geyser boiling regime and cold tip phenomena, as they can strongly affect the performance of the device. For instance, in solar receivers, due to the large variation of the solar radiation during the day, thermosyphons may operate outside the ideal design conditions, due to the large variation of the solar radiation during the day.

It is expected that the comparison between evaporator and condenser data, for high temperature thermosyphons, and common available correlations (some of them presented in Chaps. 4 and 6) is very poor, especially for the fin and Geyser boiling regimes. Furthermore, these correlations do not take the molecular regime into consideration very common in high temperature thermosyphons, as they were not developed for liquid metal working fluids.

Cisterna et al. (2020a) developed a correlation for internal heat transfer coefficients of the evaporator region for high temperature thermosyphons. They used the

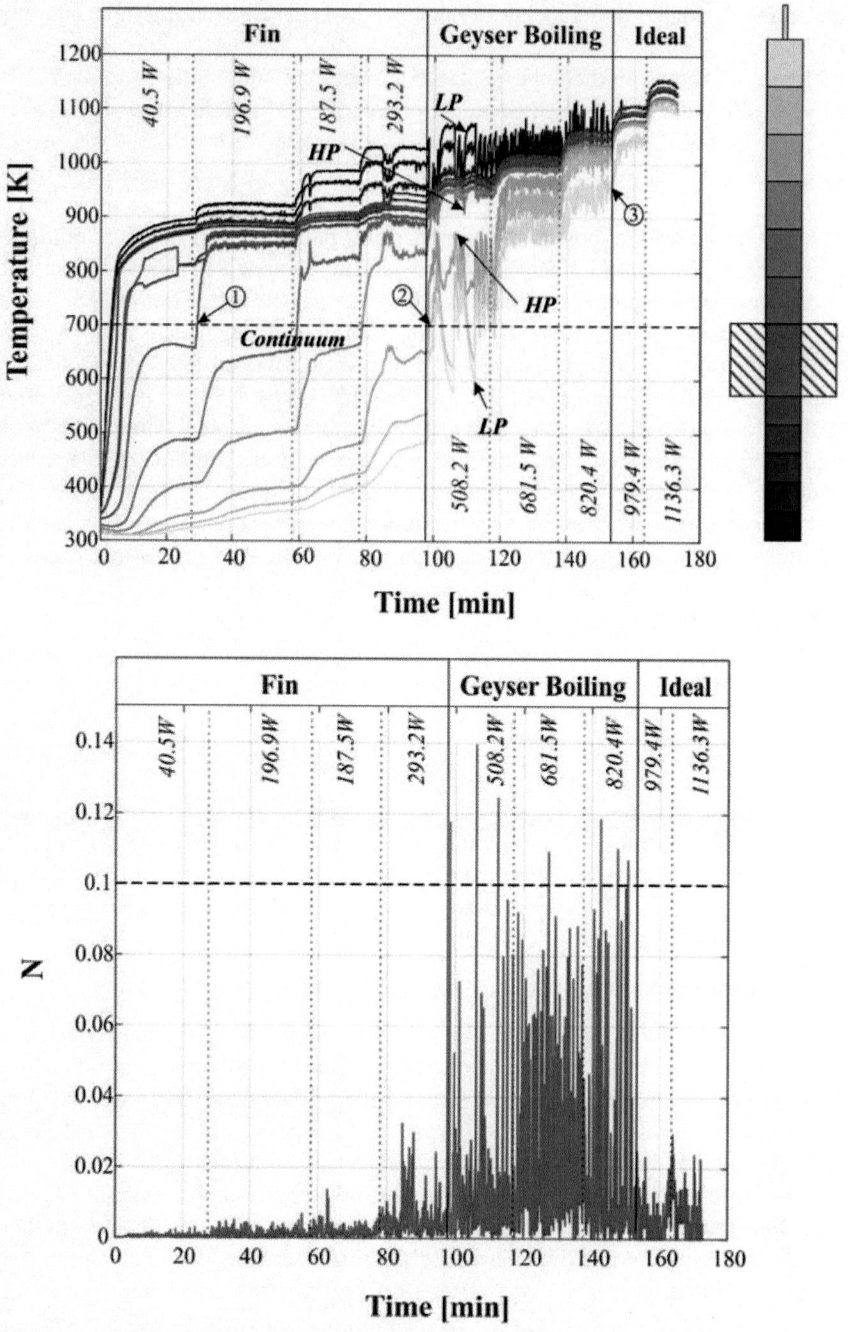

Fig. 8.8 Temperature (upper) and N factor (lower) as a function of time for a 99% pure sodium thermosyphon, with 69% of filling ratio (17.33 g), made of an ANSI 316L tube, of diameter 25.4 mm and of 350 mm of total length (see Cisterna et al. 2020a, for description of the experimental work)

following general format of the pool boiling correlations for thermosyphon (similar to Eq. 4.15, as proposed by Shiraishi et al. 1981, Farsi et al. 2003 and Groll and Rosler 1992):

$$h_e = c\left(\frac{\rho_l^{0.65} k_l^{0.3} c_{p,l}^{0.7} g^{0.2}}{\rho_v^{0.25} h_{lv}^{0.4} \mu_l^{0.1}}\right)\left(\frac{p_{sat}}{p_{atm}}\right)^a \left(q_e''\right)^b \tag{8.20}$$

The literature reports that $a = 0.23, b = 0.4$ are fixed, while c parameter is variable. Cisterna et al. (2020a) adjusted this parameter and stated that the best comparison with data, for the ideal and high performance Geyser boiling regime (see Fig. 8.7) is obtained considering $c = 0.263$.

For both fin and low performance Geyser boiling regimes, the heat transfer coefficient showed a strong dependence on the filling ratio F. Experimental data showed that, the larger the filling ratio, the larger the temperature difference between the wall and the vapor. Thus, for the fin and low performance Geyser boiling regimes, the following structure, which included the parameter F and which is similar to that of Eq. 8.20, is proposed for the correlation:

$$h_e = \frac{c}{F^m}\left(\frac{\rho_l^{0.65} k_l^{0.3} c_{p,l}^{0.7} g^{0.2}}{\rho_v^{0.25} h_{lv}^{0.4} \mu_l^{0.1}}\right)\left(\frac{p_{sat}}{p_{atm}}\right)^a \left(q_e''\right)^b \tag{8.21}$$

After the adjustment of data to model using the least square method, Cisterna et al. (2020a) presented the following correlation for the coefficient of heat transfer in the evaporator:

$$h_e = \begin{cases} \dfrac{4 \cdot 10^{-4}}{F^m}\left(\dfrac{\rho_l^{0.65} k_l^{0.3} c_{p,l}^{0.7} g^{0.2}}{\rho_v^{0.25} h_{lv}^{0.4} \mu_l^{0.1}}\right)\left(\dfrac{p_{sat}}{p_{atm}}\right)^{0.23} \left(q_e''\right)^{0.75}, \; fin\, and\, low\, GB\, regime \\ 0.263\left(\dfrac{\rho_l^{0.65} k_l^{0.3} c_{p,l}^{0.7} g^{0.2}}{\rho_v^{0.25} h_{lv}^{0.4} \mu_l^{0.1}}\right)\left(\dfrac{p_{sat}}{p_{atm}}\right)^{0.23} \left(q_e''\right)^{0.40}, \; ideal\, and\, high\, GB\, regime \end{cases} \tag{8.22}$$

where $m = 1$ for fin and $m = 3$ for low performance Geyser boiling regimes.

Condenser Correlation

Cisterna et al. (2020a) affirm that, although all the condenser correlations for other fluids do not compare well with data for sodium thermosyphons, the correlation of Kaminaga et al. (1997) presented the best fit. Based on this correlation, the following structure is proposed for the high temperature thermosyphons:

$$h_c = c\frac{k_l}{D_i}\mathrm{Pr}_l^a \mathrm{Re}^b, \quad \mathrm{Re} = \frac{4q}{\pi D_i \mu_l h_{lv}} \tag{8.23}$$

In the fin regime, only the bottom regions of the condenser contains vapor in the continuum regime, whereas, in the upper regions, the vapor is in the molecular regime. This causes a large temperature distribution, similar to that of a fin. As it is influenced by the evaporator, the fin regime is strongly dependent on the filling ratio: the larger the filling ratio, the larger is the heat flux for the vapor to reach the upper portions of the condenser. In the low performance Geyser boiling regime, during the period which the liquid pool accommodates before reaching the necessary conditions for a Geyser boiling bubble generation, the vapor mass in the condenser is minimal, as evidenced by the temperature decrease in the condenser and increase in the evaporator. Furthermore, at low pressures, the viscous effects are predominant, which generates a pressure gradient in the vapor region. These effects should be considered in the exponent of the Prandtl number. Thus, for the fin and low performance GB regimes, the correlation for the heat transfer coefficient in the condenser has the following shape, similar to the one presented in the last equation:

$$h_c = \frac{c}{F^m}\frac{k_l}{D_i}\mathrm{Pr}_l^a\mathrm{Re}^b \tag{8.24}$$

In the ideal and high performance Geyser boiling regimes, the entire condenser is in the continuum regime and the temperature is highly homogeneous. Experimental results (Cisterna et al. 2020a) showed that, in these operating regimes, the condensation heat transfer coefficient is smaller than predicted by the Nusselt model, due to the increase in the thickness of the condensate film, caused by the countercurrent flow between the vapor and the liquid. This is especially important for fluids with low Prandtl numbers, such as sodium ($Pr_l \sim 0.043$). Bejan (2004) highlights the importance of the inertia effects (directly proportional to the Prandtl numbers), in the heat transfer coefficients, concluding that the Nusselt analysis is valid only for the cases where this effect is negligible. Bejan (2013) defines the film Prandtl number as a function of the Jakob number as follows:

$$\mathrm{Pr}_f = \frac{\mathrm{Pr}_l(1+Ja)}{Ja} \approx \frac{\mathrm{Pr}_l}{Ja} \quad where \quad Ja = \frac{\rho_l c_{p,l}\left(T_{sat} - \overline{T}_{w,e}\right)}{\rho_v h_{lv}} \tag{8.25}$$

Considering that the phase change from vapor to liquid occurs in saturation state and considering that the temperature of the condenser wall is a unknown parameter that needs to be determined (by mathematical model or by experimentation), the determination of the Prandtl number is iterative, depending on the Jakob number. In order to solve this problem, Cole and Rohsenow (1969) proposed a modified Jakob number, which depends on the nucleate pool boiling critical working fluid temperature. Thus, the condenser heat transfer coefficient for the ideal and high performance Geyser boiling regimes, can be determined by the expression:

$$h_c = c\frac{k_l}{D_i}\mathrm{Pr}_l^a\mathrm{Re}^b\left(\frac{\mathrm{Pr}_l}{Ja*}\right)^d \quad where \quad Ja* = \frac{\rho_l c_{p,l} T_{crit}}{\rho_v h_{lv}} \tag{8.26}$$

Cisterna et al. (2020a) adjusted these coefficients to data, using least squares method, arriving to the following correlation:

$$h_c = \begin{cases} \frac{42.5}{F^m}\frac{k_l}{D_i}\text{Re}\,\text{Pr}^2 & fin\,and\,low\,GB\,regime \\ 2\frac{k_l}{D_i}\text{Re}^2\,\text{Pr}_l^{0.4}\left(\frac{\text{Pr}_l}{Ja*}\right)^{0.5} & ideal\,and\,high\,GB\,regime \end{cases} \tag{8.27}$$

where $m = 1$ for fin and $m = 4$ for low performance Geyser boiling regimes.

Comparison Between Data and Proposed Correlations

Cisterna et al. (2020a) compared their data with the correlations proposed in this section. Figure 8.9 shows the comparison between evaporator model and data obtained for three different filling ratios, F = 0.69, 0.96 and 1.44. However, due to the volume that bubbles occupy in the working fluid pool, both F = 0.96 and 1.44 are considered as $F \geq 1$. As already mentioned, the models for the ideal and high performance Geyser boiling regimes are not sensitive to the filling ratio and the curves are almost the same. The small variations are mostly due to changes of the thermal properties of the fluid with temperature.

On the other hand, the fin and low performance Geyser boiling regimes are sensitive to the filling ratio and two very different theoretical curves can be observed. It is clear that the model is able to capture the data very well, creating a region delimited by the high and low Geyser boiling performance (see hachured area in the graphic of Fig. 8.9) where the values of the thermal resistances can be found. For the ideal and

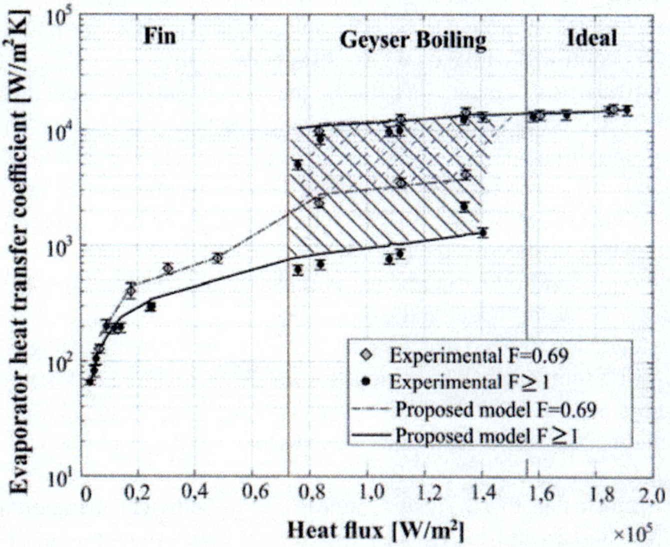

Fig. 8.9 Comparison between evaporator correlation and data for the sodium thermosyphon, for the three regimes (fin, Geyser boiling and ideal), for filling ratios above and under F = 1

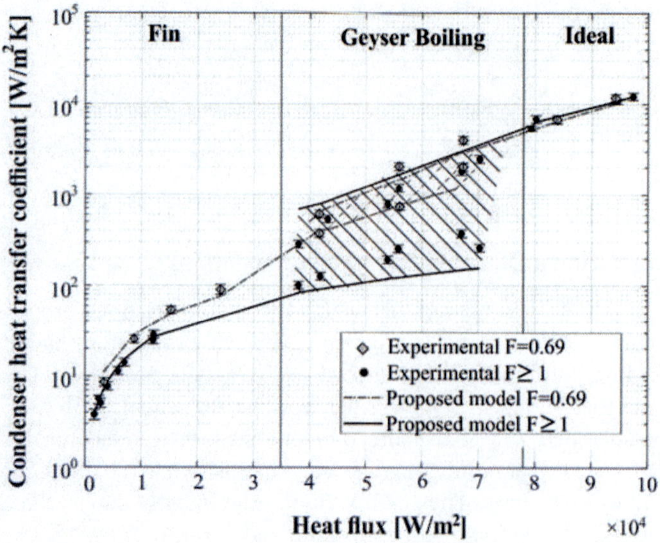

Fig. 8.10 Comparison between condenser correlation and data for the sodium thermosyphon, for the three regimes (fin, Geyser boiling and ideal), for two filling ratios (above and under F = 1)

the high performance Geyser boiling regimes, the experimental data and the model agreed in about 20%. Note that, in this comparison, only the data close to the high performance Geyser boiling curve is considered. If only the ideal region is observed, this difference drops to around 10%. The low performance Geyser boiling and fin regime curves compares with data with a difference of 40%.

Figure 8.10 shows a comparison between the experimental data and the proposed correlation for the condenser heat transfer coefficients. As for the evaporator, the effect of filling ratio is more sensitive to the fin and lower Geyser boiling regimes. The maximum difference between model and data in this region is of 20%. A larger dispersion of experimental data is observed in the Geyser boiling regime, due to the random nature of the phenomenon, which shows larger temperature variations in the condenser than in the evaporator, as it can be observed in Figs. 8.7 and 8.8. The difference between model and data in this region is also larger, around 80%. However, it should be noticed that the model is very precise in proposing boundaries to the Geyser boiling region. The reasons for the Geyser boiling different behaviors in the evaporator and condenser were discussed in Chap. 3. Finally, in the ideal operation regime, the model represents the experimental data well (less than 10% of difference).

It is important to note that the ideal regime is the most efficient operation condition, presenting the highest heat transfer coefficients, in both evaporator and condenser regions and thus, the lowest overall thermal resistance. In any application, the high temperature thermosyphons should be designed to operate at this regime. The good agreement between the proposed model and the data shows that the present model can be used as a design tool for these thermosyphons.

Cold Tip Analysis

In the last section, operation regimes for high temperature thermosyphons were established and correlations were proposed and compared with data. However, the discussion of the cold tip physical phenomena is still missing. The presence of cold tips are observed by sharp temperature distributions along the condenser, which can be an important concern for the applications of high temperature thermosyphons. In this section, the cold tip phenomenon is discussed, in the light of the thermodynamic analysis of the working fluid behavior.

As already discussed, two hypotheses can be used to explain the cold tip phenomenon in liquid metal (high temperature) thermosyphons: the presence of non-condensable gases and the free molecular flow of vapor at the condenser regions. The first guess, for the presence of cold tips, is usually the presence of no condensable gases, as discussed in Sect. 8.2.

However, the literature (Cisterna et al. 2021a, b; El-Genk and Tournier 2011; Tournier and El-Genk 1996, 2001, Cao and Faghri 1993a and Cao and Faghri 1993b) show that cold tips can be associated to the vapor flow in the molecular regime, especially in the extreme regions of the condenser.

According to Cisterna et al. (2021a), free molecular cold tip can be a result of the external large cooling capacity of the condenser, when the vapor generated in the evaporator condenses entirely at its lower region. In this case, at the top of the condenser, the vapor mass and pressure are minimal. This results in a temperature profile along the device similar to a thermal fin, as highlighted in the last section. The continuum hypothesis fails when the average free path length of vapor molecules is larger than the internal diameter of the thermosyphon (Holman 1981).

Tournier and El-Genk (1996) identify the flow regimes in thermosyphons based on the Knudsen number $Kn = \lambda/D$, where λ is the molecular mean free path for vapor and D is the characteristic length, is this case the diameter of the tube:

- Molecular flow: for $Kn > 1.0$
- Transition flow: $0.01 < Kn < 1.0$
- Continuum flow: $Kn < 0.01$.

However, Cao and Faghri (1993) proposed another expression for the limit (transition) temperature T_{tr}, which characterizes the continuum vapor flow regime, also as a function of Kn:

$$\ln\left(\frac{T_{tr}}{p_0}\frac{1.051\cdot\kappa_b}{\sqrt{2}\cdot\pi\cdot\sigma_d^2\cdot D_i\cdot Kn}\right)+\frac{h_{lv}}{R}\left(\frac{1}{T_{tr}}-\frac{1}{T_0}\right)=0 \tag{8.28}$$

where T_0 and $p_0$orrespond to the reference temperature and pressure, R is the universal ideal gas constant, κ_b is the Boltzmann constant and σ_d corresponds to the collision distance between molecules, which, for sodium, can be considered as 3.58 $\times$ 10^{-10} m (El-Genk and Tournier 2011).

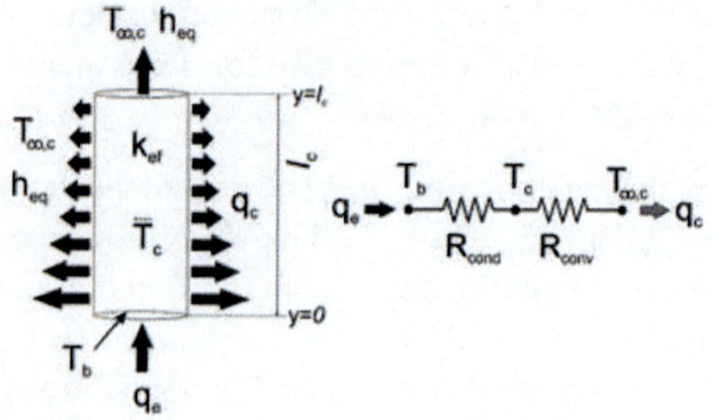

Fig. 8.11 Fin physical model of the tip of thermosyphon condenser

Startup and Continuum Power Limit Models

In order to model the condenser temperature profile for the fin regime, a fin thermal model is used (Cisterna et al. 2021a). For that, the following assumptions are made:

- Steady state regime
- Condenser is considered a cylindrical solid fin, made of a "material" that presents an "effective thermal conductivity", which depends on the vapor generation rate in the evaporator
- Startup is reached when the temperature at the fin base (condenser) is equal to the continuum working fluid temperature
- The fin (including the tip) exchanges heat with the ambient through two mechanisms: natural convection and radiation.

In Fig. 8.11, a schematic of the boundary conditions considered for the development of the startup and continuum limit models for two-phase thermosyphons is presented. Also the resistance network is shown, where the resistance R_{cond} is associated with the heat conduction along the condenser, in this case considered as a cylindrical fin (constant cross section area) and R_{conv} is associated with the external heat transfer to the environment. The total heat transferred by the condenser can be calculated as:

$$q_e = q_c = \frac{(T_b - T_{\infty,c})}{R_{cond} + R_{conv}} = \frac{(T_b - T_c)}{R_{cond}} = \frac{(T_c - T_{\infty,c})}{R_{conv}} \tag{8.29}$$

The resistances of the Fig. 8.11 circuit area calculated as:

$$R_{cond} = \frac{l_c}{k_{ef} A_C}; \quad R_{conv} = \frac{1}{h_{eq} A_c} \tag{8.30}$$

where l_c, the condenser length, is equivalent to the "fin" length, k_{ef} is the effective thermal conductivity of the "fin", A_C is the condenser cross section area, here considered the "solid fin", h_{eq} is the equivalent external convection heat transfer (see Eq. 8.17) and A_c is the total external condenser heat transfer area.

Substituting Eq. 8.30 in Eq. 8.29, it is possible to define a dimensionless number, denominated modified Biot number, Bi_C, which, as the original Biot number (see

Incropera and DeWitt 2008), is ratio between the heat conduction resistance along the condenser, and the external heat transfer resistance. The modified Biot number can also be given as the ratio temperature differences, as follows:

$$Bi_c = \frac{R_{cond}}{R_{conv}} = \frac{\left(T_b - \overline{T}_c\right)}{\left(\overline{T}_c - T_{\infty,c}\right)} = \frac{h_{eq}}{k_{ef}} L \tag{8.31}$$

where L is the characteristic length given by:

$$L = \frac{A_c}{A_C} l_c \tag{8.32}$$

The modified Biot number is a dimensionless parameter used to identify the thermosyphon cold tip regime in free molecular flow. When $Bi_C \gg 1$, the resistance associated with the heat conduction along the condenser R_{cond} is much larger than the resistance associated with the external heat transfer R_{conv}, which means that the capacity of the thermosyphon to rejected heat to the environment is larger than the capacity of the device to conduct heat from the base to the fin tip. Therefore, almost all the vapor generated in the evaporator is condensed in the condenser lower regions and, as a consequence, a large temperature gradient is observed along the thermosyphon, similar to that of a thermal fin. When $Bi_C \ll 1$, the R_{cond} is much lower than R_{conv}, which means that the thermosyphon is able to transfer more heat than the heat rejection capacity of the condenser. In this case, the vapor has sufficient energy to reach the end of the thermosyphon in the continuum regime and the temperature tends to become homogeneous along the entire condenser length, with the thermosyphon assuming its expected behavior, i.e., a super heat conductor. At $Bi_C \approx 1$, the order of magnitude of both resistances is similar ($R_{cond} \approx R_{conv}$), that is, the thermosyphon stops behaving like an empty tube.

Therefore, the condenser temperature distribution for the fin regime, can be predicted using well known thermal fin models (Incropera and DeWitt 2008). Considering convection at the thermosyphon tip area, literature expressions can be used, where the thermal conductivity of the fin is substituted by k_{ef} and the convection heat transfer by h_{eq}. The temperature profile is therefore:

$$\frac{\theta(y)}{\theta_b} = \frac{T(y) - T_{\infty,c}}{T_b - T_{\infty,c}} = \frac{\cosh[m(l_c - y)] + \left(h_{eq}/m \cdot k_{ef}\right)\sinh[m(l_c - y)]}{\cosh(m \cdot l_c) + \left(h_{eq}/m \cdot k_{ef}\right)\sinh(m \cdot l_c)} \tag{8.33}$$

where the parameters M and m are given, respectively by:

$$M = \left(\frac{\pi^2 D_{ex}^3}{4} h_{eq} k_{ef}\right)^{1/2} \theta_b \tag{8.34}$$

$$m = \left(\frac{4h_{eq}}{k_{ef} D_{ex}}\right)^{1/2} \tag{8.35}$$

The heat flux dissipated by the thermosyphon condenser can be determined by:

$$q_c = h_{ex}\left(\overline{T}_c - T_{\infty,c}\right)\varphi \tag{8.36}$$

where φ is a dimensionless condenser fin parameter given by (see Cisterna et al. 2021a):

$$\varphi(Bi_C) = \frac{1}{Bi_C^{1/2}}\left[\frac{\sinh\left(\sqrt{\frac{P_c l_c}{A_c}}Bi_C^{1/2}\right) + \frac{A_C}{A_c}\sqrt{\frac{P_c l_c}{A_c}}Bi_C^{1/2}\cosh\left(\sqrt{\frac{P_c l_c}{A_c}}Bi_C^{1/2}\right)}{\cosh\left(\sqrt{\frac{P_c l_c}{A_c}}Bi_C^{1/2}\right) + \frac{A_C}{A_c}\sqrt{\frac{P_c l_c}{A_c}}Bi_C^{1/2}\sinh\left(\sqrt{\frac{P_c l_c}{A_c}}Bi_C^{1/2}\right)}\right] \tag{8.37}$$

Considering the case where the condenser length is much greater than the condenser diameter ($l_c \gg D_{ex}$), the convection circular area at the top of the condenser is much smaller than the cylindrical area, so that the convection at the top can be neglected ($A_c \gg A_C$). The correction factor expression (Eq. 8.37) takes exactly the same form of the circular pin fin efficiency (see Incropera and DeWitt 2008) and for now on, it is denominated as fin efficiency. It is, therefore, given by:

$$\varphi_C(Bi_C) = \frac{\tanh\left(Bi_C^{1/2}\right)}{Bi_C^{1/2}} \tag{8.38}$$

Using this set of equations, the average condenser temperature $\overline{T}_c$ is determined applying the mean value theorem for integrals, resulting in:

$$\frac{\overline{\theta}_c}{\theta_b} = \frac{1}{l_c}\int_0^{l_c}\left[\frac{\cosh\left[\sqrt{\frac{P_c l_c}{A_c}}Bi_C^{1/2}\left(1-\frac{y}{l_c}\right)\right] + \frac{A_C}{A_c}\sqrt{\frac{P_c l_c}{A_c}}Bi_C^{1/2}\sinh\left[\sqrt{\frac{P_c l_c}{A_c}}Bi_C^{1/2}\left(1-\frac{y}{l_c}\right)\right]}{\cosh\left(\sqrt{\frac{P_c l_c}{A_c}}Bi_C^{1/2}\right) + \frac{A_C}{A_c}\sqrt{\frac{P_c l_c}{A_c}}Bi_C^{1/2}\sinh\left(\sqrt{\frac{P_c l_c}{A_c}}Bi_C^{1/2}\right)}\right]dy \tag{8.39}$$

Still for the case where the top heat exchange area can be neglected, the average temperature in the condenser is:

$$\overline{T}_c = T_{\infty,c} + \left(T_b - T_{\infty,c}\right)\varphi_C \tag{8.40}$$

Cisterna et al. (2021a) observed that most of thermosyphons they tested starts up when $Bi_C \approx 1$. Therefore, these authors suggest the use of.

$Bi_{C,su} = 1$ for the definition of the startup limit. A more strict number that would encompass all dada observed would be $Bi_{C,su} = 4$.

In addition, Cisterna et al. (2021a) perceived that the thermosyphons operated completely in the continuum regime, for all cases where the modified Biot number was less than 0.1. Therefore, they defined $Bi_{C,cont} = 0.1$ as the continuum limit.

Using the equations developed in this section, the following expressions can be used to determine the startup and continuum limits for high temperature thermosyphons, respectively:

$$q''_{su} = h_{eq}(T_b - T_{\infty,c})\varphi(Bi_{C,su}) \tag{8.41}$$

$$q''_{cont} = h_{eq}(T_b - T_{\infty,c})\varphi(Bi_{C,cont}) \tag{8.42}$$

It should be noted that the Biot number Bi was originally conceived for transient conduction problems in fins: when $Bi < 0.1$, the temperature gradients within a fin are negligible compared to the difference of temperatures between the fin and ambient. Likewise, it was observed experimentally that, when $Bi_C < 0.1$, the temperature gradient within the thermosyphon is negligible when compared to the difference between thermosyphon and ambient, i.e., the two-phase thermosyphon operates in continuum regime.

Thermal Resistance Model

The schematic of the piston-cylinder system shown in Fig. 8.11 can be modelled using the thermal circuit methodology. For the sake of generality, the thermosyphon operation mode, for the range between the startup $(Bi_C < Bi_{c,su})$ and the ideal regime $(Bi_C < Bi_{C,cont})$, illustrated by the piston-cylinder system of Fig. 8.6f, was selected to be modelled, because, in this configuration, all operation regimes are observed.

Figure 8.12 shows, in the left, the reproduction of the piston-cylinder system and, at right, the resulting thermal resistance circuit. The high temperature thermosyphon

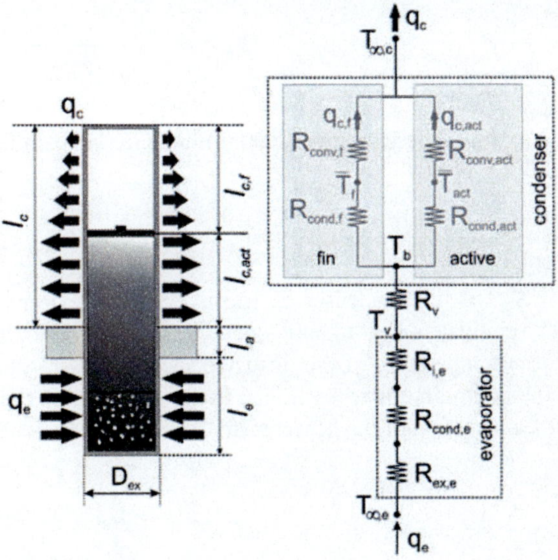

Fig. 8.12 Cold tip thermal resistance model

is divided into four regions: evaporator, adiabatic section, active condenser zone and diffusive condenser zone.

This condenser region comprises the cold tip zone, in which two regimes, separated by the hypothetical piston (continuous under and molecular over the piston) are observed. Neglecting the heat conducted axially through the wall, the resulting thermal resistance network is shown in Fig. 8.12 in the right. In the evaporator, the circuit is composed by the series connection of the resistances: external convection, radial wall conductance and internal phase change. The expressions used for the calculations of these resistances are those depicted in Chap. 4. The condenser resistance is composed by two resistances in parallel. The first path is related to the active length of the condenser when the vapor is in the continuum regime. The second is the fin path resistance, corresponding to the condenser region where vapor is in free molecular flow regime. It is important to highlight that these condenser resistances are in parallel because the fin is a conductive resistance, independent of the vapor action, only dependent of the temperature level at the interface between the active and fin regions. Therefore, the determination of the resistances cannot be made independently as both regions are thermally coupled.

In the active condenser zone, the heat transfer can be calculates as:

$$q''_{c,act} = h_{eq,act} A_{c,act} \left(T_b - T_{\infty,c}\right) \varphi\left(Bi_{C,cont}\right) \tag{8.43}$$

where $\varphi\left(Bi_{C,cont}\right)$ is given by Eq. 8.38, evaluated for the continuum limit, $A_{c,act}$ is the condenser active external area (for the active length $l_{c,act}$) and the equivalent heat transfer coefficient $h_{eq,act}$ is calculated using Eq. 8.17, for the average temperature of the active section of the condenser, calculated using Eq. 8.39, which integration, along the active length, results in:

$$\frac{\overline{\theta}_{c,act}}{\theta_b}(y) = \frac{\cosh\left[Bi_{C,cont}^{1/2}\left(1 - \frac{y}{l_c}\right)\right]}{\cosh\left(Bi_{C,cont}^{1/2}\right)} \tag{8.44}$$

In the diffusive fin path, heat is transferred in parallel to the active condenser zone and is given by:

$$q''_{c,f} = h_{eq,ft} A_{c,f} \left(T_b - T_{\infty,c}\right) \varphi\left(Bi_{C,f}\right) \tag{8.45}$$

$A_{c,f}$ is the condenser fin (diffusion zone) external area, for the fin length $l_{c,f}$. The equivalent heat transfer coefficient $h_{eq,f}$ is calculated as for Eq. 8.43, using the mean temperature of the diffusive section. $\varphi\left(Bi_{C,f}\right)$ is limited by the startup and continuum fin efficiency limits. In this calculation, the fin efficiency is considered a linear function of the heat transfer rate into the condenser, i.e.:

$$\varphi_f\left(Bi_{C,f}\right) = \varphi\left(Bi_{C,su}\right) + \left[\varphi\left(Bi_{C,cont}\right) - \varphi\left(Bi_{C,su}\right)\right]\left(\frac{q_c - q_{su}}{q_{cont} - q_{su}}\right) \tag{7.46}$$

The temperature profile of the fin zone is given by:

$$\frac{\overline{\theta}_{c,f}}{\theta_b}(y) = \frac{\cosh\left[Bi_{C,f}^{1/2}\left(1 - \frac{y}{l_c}\right)\right]}{\cosh\left(Bi_{C,f}^{1/2}\right)} \tag{7.47}$$

It is important to mention that the active and fin lengths of the condenser are, actually functions of the heat transfer rate. The larger the heat transfer rate, the larger the vapor pressure below the piston and, consequently, the larger the piston displacement, the longer the active zone length and the shorter the length of the diffusive zone. It is then necessary to adopt an iterative method to determine the condenser temperature profile, so that the heat transfer rate inserted in the evaporator must be equal to the sum of the heat transfer rate rejected from both the active and the diffusive zones of the condenser:

$$q_e = q_c = q_{act} + q_f \tag{8.48}$$

Validation of the Startup and Continuum Limit Models

Cisterna et al. (2021a) validated the start-up and continuum limit models against data from the experimental work that they performed. Figure 8.13 shows these data in a graphic of the average heat transfer rate as a function of the vapor temperature, where the startup and continuum limit curves are also shown. Experimental points 1 (start-up), 2 (continuum) and 3 (onset of the thermosyphon ideal operation regime), presented in Fig. 8.8, for the F = 69% (17 g) case is also shown in this plot. Point 4, which shows the ideal regime is also depicted. It is possible to observe that the model is able to divide precisely the operation regions of the thermosyphon: bellow the startup limit, the thermosyphons operate as heat conduction devices, showing a linear behavior of the transferred heat as a function of the vapor temperature. Above

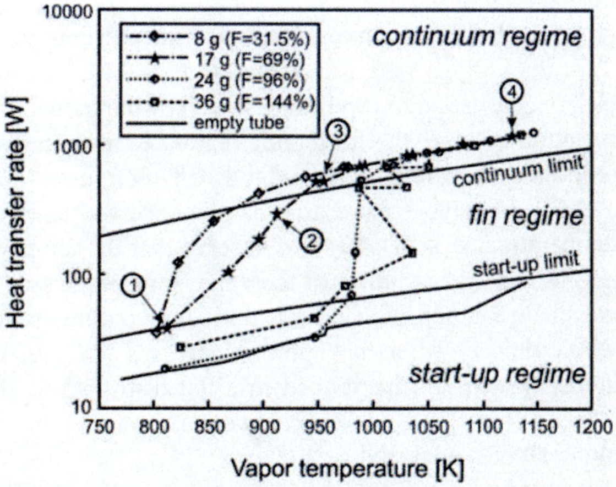

Fig. 8.13 Comparison between continuum and start-up limits and data for thermosyphons with different filling ratios

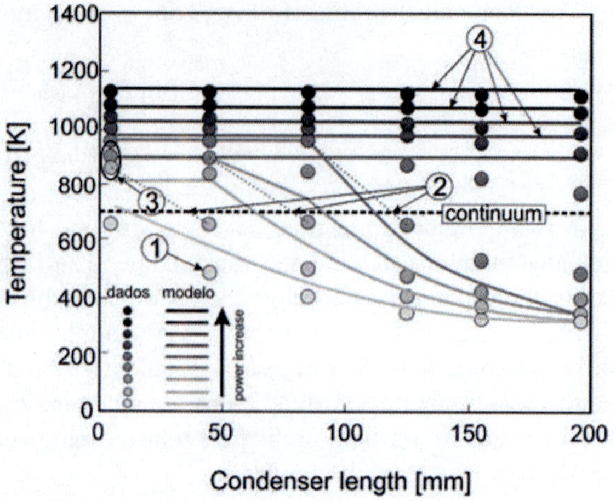

Fig. 8.14 Temperature distribution of 17 g sodium thermosyphon

the continuum limit the devices operate at their higher performance, with high heat transfer rates. Between these limits, in the Geyser boiling regime, the thermosyphons still show a good performance, but present large temperature oscillations (that also lead to mechanical vibrations, see Geyser boiling discussion in Chap. 3) and so this regime is not recommended.

Cisterna et al. (2021a) also reports that the molecular and continuum regime can be visually observed by the reddish colors of high temperature thermosyphons tubes, when subjected to different power inputs: for lower power input levels, the evaporator and the rear part of the condenser are in red bright colors, showing temperatures above 1000 K, while the upper part of the condenser are in darker colors, showing lower temperatures, due to the large temperature gradient along the condenser.

Validation of the Temperature Profile Model

In Fig. 8.14, the theoretical temperature profiles along the condenser length of the thermosyphon are also validated against the same data presented in Fig. 8.8. The full lines represent the model prediction for several heat power input levels in shades of gray: the darker the curves, the higher the power input. The same points (1, 2, 3 and 4) of the Fig. 8.13 are also shown in Fig. 8.14. The temperature distribution along the condenser is quite well represented by the model, including the region where the variation is large, the fin regime region. In these regions (see Fig. 8.6f), the active (continuous) zone presents a flat distribution, until the fin region is reached, where a noticeable temperature drop can be observed tending in the limit to the ambient temperature. It is also easy to observe that the temperatures of the condenser base are, in fact, very similar to the vapor temperature of the adiabatic section and do not change much with power, until the continuous regime is achieved along the whole device. After that, power increases rise the whole thermosyphon temperatures, which are maintained in a flat distribution. Temperature prediction errors $((T_{theoretical} - T_{data})/T_{data})$ between model and data ranged from -20 to $+\,20\%$ along all data analyzed.

Geyser Boiling Limit Model

As already observed, due to its typical vibration, Geyser boiling should be avoided for most applications. This vibration is especially strong for high temperature thermosyphons and may cause mechanical damage to the thermosyphons themselves and/or the device where they are installed. Taking the example of sodium thermosyphons, the working fluid leakage due to the casing rupture can be dangerous, as explosion may happen if the sodium is in contact with the water vapor present in the humid air. Therefore, models to predict the Geyser boiling occurrence are quite important for the design of equipment.

In this section, an analytical model to predict the Geyser boiling limit for high temperature two-phase thermosyphons is presented. As the Geyser boiling is mostly observed in low power inputs, by this limit, it is understood the minimum heat flux required to the thermosyphon to work in the ideal operation regime.

Geyser Boiling Departing Bubble Geometry

One of the input parameters for the Geyser boiling limit model is the diameter of departing bubble, i.e., of the bubble that is able to accumulate and release the energy necessary for the Geyser boiling phenomenon to happen. This diameter results from the balance of forces acting on the bubble, being the most important: the interfacial tension and the buoyancy. The surface tension acts along the contact line between the bubble and the thermosyphon wall and acts in the sense to attach the bubble to the evaporator surface, whereas the buoyancy force pulls the bubble apart from the surface (see Carey 1992). Other forces are the inertia and drag, which are considered minor. The inertia is associated with the liquid flow in the vicinity of a fast growing bubble, helping to pull the bubble off the surface, as observed experimentally by Pabón et al. (2019). The drag forces influence the bubble growth rate and its shape, acting during the bubble release.

The pressure of the liquid column over the bubble, which depends on the filling ratio, has also great influence on the dimension of the bubble formed. Figure 8.15 shows a scheme of a thermosyphon with two different filling ratios, sealed in the

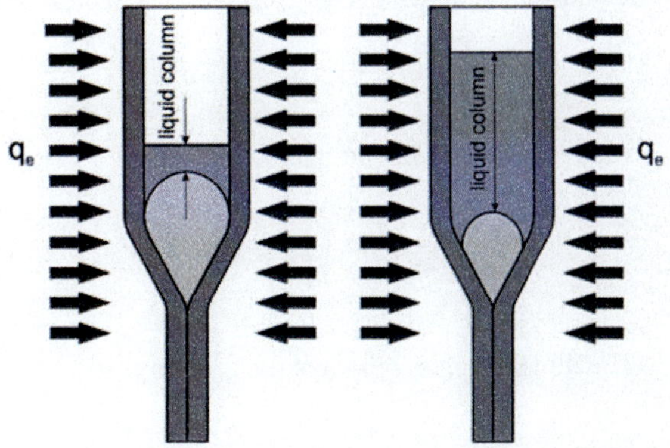

Fig. 8.15 Bubble departure diameter for thermosyphons with different filling ratio values

evaporator bottom side by pinching the tube with a hydraulic press, using a process similar to the one described in Chap. 5 for pinching umbilical tubes (see Cisterna et al. 2020b). Due to the fabrication process, a bubble nucleation site is made available at the bottom of the evaporator, where the Geyser boiling bubble is formed. Considering that, in both cases of Fig. 8.15, the heat transfer rate and the saturation temperature are the same and so is the same the mass of vapor formed, the resulting bubbles have two different dimensions, as they are subjected to different pressures provided by the liquid columns. Therefore, the Geyser boiling bubble in the thermosyphon with larger working fluid volume is smaller than that that with a small filling ratio.

Other parameters that influence the bubble departure diameter are: the wall superheat temperature (or heat flux), the liquid–solid contact angle and the thermophysical properties of the working fluid.

Characterizing Geyser Boiling by Dimensionless Numbers

Bergman et al. (2015) report that the boiling process is quite complex, being governed by the following dimensionless numbers:

- Prandtl (Pr), associated with the thermophysical properties of the working fluid, relates the viscous and thermal diffusion rates of a given fluid,
- Jakob (Ja), relates the maximum sensible energy absorbed by the vapor and the latent energy absorbed by the liquid, to generate the phase change,
- Bond (Bo), relates the buoyant force that acts to detach the bubble from the solid surface and the surface tension that acts to keep the bubble on that surface.
- "Corrected" Grashof number (Gr_c), that has a strong resemblance with the Grashof number (see Cisterna et al. 2020b) and represents the ratio between buoyancy and viscous forces.

It is expected that the model for the minimum heat flux that avoid the thermosyphon operation in Geyser boiling regime contains all these dimensionless parameters. Jensen and Memmel (see Carey 1992) proposed a correlation to determine the bubble departure diameter in pool boiling that has potential to be used for two-phase thermosyphons, as it correlates the bubble departure diameter with the: buoyancy force, surface tension, inertia, thermophysical properties of the working fluid and wall superheat. This correlation is expressed as:

$$Bo^{1/2} = 0.19\left(1.8 + 10^5 K_l\right)^{2/3} \tag{8.49}$$

where the bubble depart parameter K_l is:

$$K_l = \left(\frac{Ja}{\mathrm{Pr}_l}\right)\left\{\left[\frac{g\rho_l(\rho_l - \rho_v)}{\mu_l^2}\right]\left[\frac{\sigma}{g(\rho_l - \rho_v)}\right]^{3/2}\right\}^{-1} \tag{8.50}$$

The dimensionless numbers are given by:

$$Ja = \frac{\rho_l c_{p,l}(T_w - T_{sat})}{\rho_v h_{lv}} \tag{8.51}$$

$$Bo = \frac{g(\rho_l - \rho_v)D_b^2}{\sigma} \tag{8.52}$$

$$\Pr_l = \frac{c_{p,l}\mu_l}{k_l} \tag{8.53}$$

According to Cisterna et al. (2021b), the bubble departure diameter D_b, for a given wall superheating of $\Delta T_w = (T_w - T_{sat})$ and for a working fluid, can be obtained by inserting Eqs. 8.51 to 8.50 in Eq. 8.50 and after manipulation, the following expression is obtained:

$$K_l = \frac{Ja_c}{\Pr_l}\frac{Bo_c^{3/2}}{Gr_c}\frac{\Delta T_w}{T_{sat}} \tag{8.54}$$

which is based on the corrected Jackob, Bond and Grashof numbers, defined respectively as:

$$Ja_c = \frac{\rho_l c_{p,l} T_{sat}}{\rho_v h_{lv}} \tag{8.55}$$

$$Bo_c = \frac{g(\rho_l - \rho_v)D_i^2}{\sigma} \tag{8.56}$$

$$Gr_c = \frac{g\frac{(\rho_l - \rho_v)}{\rho_l}D_i^3}{\nu_l^2} \tag{8.57}$$

Note that both corrected Bond and Grashof numbers are defined as functions of the thermosyphon internal diameter (D_i) and that the corrected Jakob number is defined as a function of the saturation temperature of the working fluid (T_{sat}). Using Eq. 8.50 together with the corrected Bond number, the following expression results:

$$K_l = 10^{-5}\left[12.1 \cdot Bo_c^{3/4} \cdot Co^{3/2} - 1.8\right] \tag{8.58}$$

The dimensionless confinement number Co is defined as:

$$Co = \frac{D_b}{D_i} \tag{8.59}$$

It should be noted that the confinement number contains the departure bubble diameter D_b and is one of the main parameters affecting the Geyser boiling regime, as also stated in the literature (Pabón et al. 2019; Robinson et al. 2019; Smith et al. 2018).

According to Cisterna et al. (2021b), equating Eq. 8.58 with Eq. 8.54, a dimensionless number can be obtained, the "bubble release number", which represents the wall superheat necessary to form a bubble with a pre-determined diameter, able to departure from the surface, for a given liquid pool saturation temperature. This parameter is expressed as:

$$\varphi_b = \frac{\Delta T_w}{T_{sat}} = 10^{-5}\left[12.1 \cdot Bo_c^{3/4} \cdot Co^{3/2} - 1.8\right] Bo_c^{-3/2} \frac{\Pr_l Gr_c}{Ja_c}$$
$$= f(Bo_c, Co)\frac{\Pr_l Gr_c}{Ja_c} \tag{8.60}$$

Therefore, the bubble release dimensionless number involves all the dimensionless numbers used to characterize boiling, as mentioned by Bergman et al. (2015), and also the confinement number, typically used for Geyser boiling.

From Eq. 8.60, the right hand side can be approximated as:

$$f(Bo_c, Co) = 10^{-5}\left[12.1 \cdot Bo_c^{3/4} \cdot Co^{3/2} - 1.8\right] \approx c_1 Bo^{c_2} Co^{c_3} \tag{8.61}$$

Data from Cisterna et al. (2020a, b) for sodium thermosyphons operating between 800 and 1200 K and presenting confinement numbers between 0.2 and 2 were adjusted against the Eq. 8.61, using least square method, resulting in the constants $c_1 = 10^{-3}$, $c_2 = -1.44$ and $c_3 = 1.5$, with a root means squared (RMS) error of 0.74. Therefore, the thermosyphon dimensionless bubble release number can be expressed by this simple correlation:

$$\varphi_b = \frac{\Delta T_w}{T_{sat}} = 10^{-3}\frac{Co^{1.5}}{Bo_c^{1.44}}\frac{\Pr_l Gr_c}{Ja_c} \tag{8.62}$$

The heat flux in the evaporator is determined by the expression:

$$q_b'' = h_p(T_w - T_{sat}) = h_p \Delta T_w \tag{8.63}$$

where h_p is the nucleate pool boiling heat transfer coefficient. Cisterna et al. (2020a) showed that the nucleate boiling correlation based on the Farsi et al. (2003) expression, reproduced here:

$$h_p = 0.263\ K_T\ q_e''^{0.4} \tag{8.64}$$

presents the better agreement with data for sodium thermosyphons operating in the ideal regime, where:

$$K_T = \left(\frac{\rho_l^{0.65} k_l^{0.3} c_{p,l}^{0.7} g^{0.2}}{\rho_v^{0.25} h_{lv}^{0.4} \mu_l^{0.1}}\right)\left(\frac{p_{sat}}{p_{atm}}\right)^{0.23} \tag{8.65}$$

Therefore, the heat flux necessary for releasing a bubble with determined confinement number and saturation temperature can be obtained using a combination of the last four equations, resulting in:

$$q''_b = 0.108(K_T \, \varphi_b \, T_{sat})^{5/3} \tag{8.66}$$

Bubble Release Number Validation

The same data used before are used in this section for the validation of the bubble release number. Figure 8.8 plot is reproduced in Fig. 8.16 together with other two

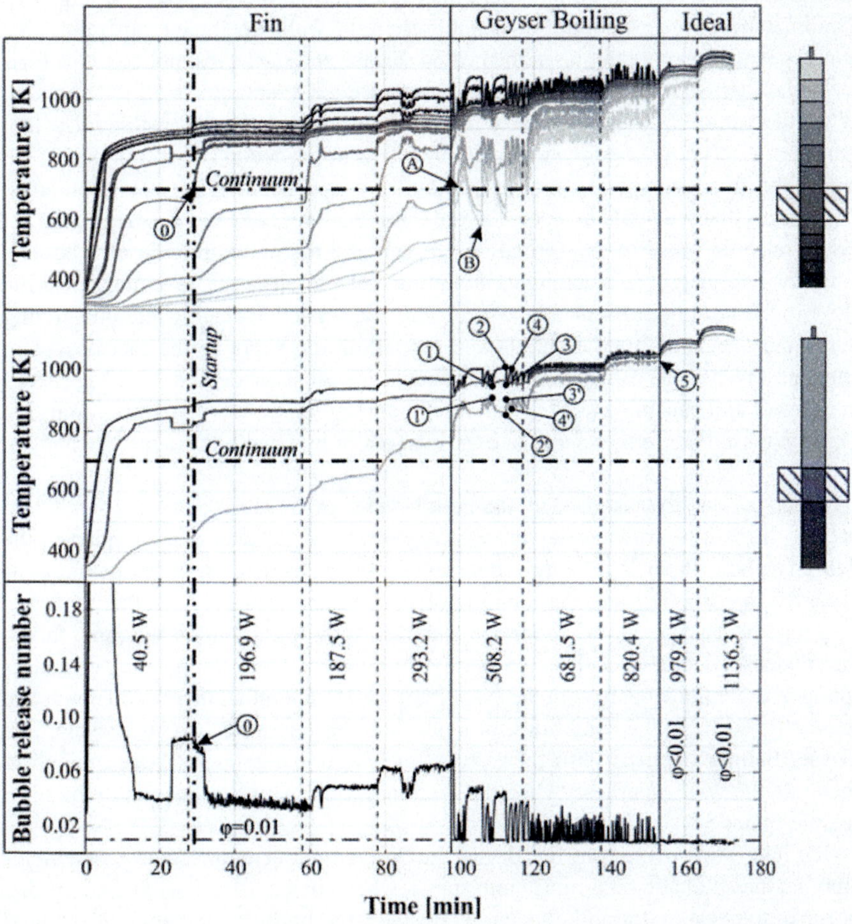

Fig. 8.16 Temperature (upper), average temperature (middle) and bubble release number parameter (φ_b) as a function of time for a 99% pure sodium thermosyphon, with 69% of filling ratio (17.33 g), made of an ANSI 316L tube, of diameter 25.4 mm and of 350 mm of total length (see Cisterna et al. 2020a for description of the experimental work)

graphs: the evaporator, condenser and adiabatic section averages and the bubble release number (φ_b), all against time. As already observed, this thermosyphon operates in the fin, Geyser boiling and ideal regimes, depending on the heat applied to the evaporator.

Among other characteristics, the Geyser boiling can be noted in Fig. 8.16 by the high level of symmetry in the evaporator and condenser temperature oscillations, i.e., when there is a sudden increase or decrease in the evaporator temperature, there is also an almost simultaneous sudden decrease or increase of the condenser temperature, as observed in the set of points: 1 and $1'$, 2 and $2'$, 3 and $3'$ and 4 and $4'$, highlighted in the average temperature plot.

According to Cisterna et al. (2021b), Geyser boiling occurs mainly at low evaporator heat fluxes. At each cycle, before the bubble formation (see point 1 of Fig. 8.16 for illustration), the evaporator does not present the wall superheat temperature that would generate vapor bubbles at the heated surface and the heat supplied to the thermosyphon is used almost exclusively to increase the evaporator wall temperature. When the minimum wall superheat temperature is reached, there is bubble formation (point 2). According to Pabón et al. (2019) and as described in Chapter 3, during the pre-bubble formation period of time, the liquid pool reaches a highly metastable state. This means that the pool has a considerable stored amount of energy and the smallest disturbance can lead to the formation, growth and disruptive release of a bubble. Once the bubble grows, reaching its departure diameter, it raises, reaching a position close to the pool liquid–vapor interface, where it bursts, displacing and pulverizing the liquid that is above the bubble and propelling the vapor to the thermosyphon condenser. When the vapor reaches the condenser wall, it condenses releasing energy to the wall and causing a sudden increase in condenser wall temperature (point $3'$). The condensed liquid then returns to the evaporator by means of gravity, closing the cycle.

When the energy absorbed by the liquid–vapor phase change process (including bubble formation, growth and very fast release) is larger than that supplied to the thermosyphon, there is an abrupt decrease in the evaporator wall temperature, as observed in point 3. Then, the liquid pool wall, again, does not have the minimum wall superheat temperature to generate a bubble. At these conditions, the vapor flux to the condenser decreases and, consequently, the condenser wall temperature decreases (point 4'), simultaneously with the evaporator wall temperature increase. The system is actually building the conditions for a new bubble generation and burst cycle. The Geyser boiling happens until the evaporator heat transfer rate is large enough to maintain the wall superheat, when the thermosyphon begins to operate in the ideal regime (point 5).

It is interesting to observe that, during the Geyser boiling regime, the vapor temperature, at the adiabatic section, is quite stable, when compared to the oscillations that occur in the evaporator and condenser. Besides, the liquid temperature at the pool base in the evaporator can be considered virtually constant during Geyser boiling regime.

Figure 8.16 shows the experimental bubble release number as a function of time for the different heat transfer rates, based on the results presented in Fig. 8.16 upper graphic. Comparing the three graphics of Fig. 8.16, it can be observed that the bubble release number was almost constant and always less than 0.01 for power larger than 980 W, when the thermosyphon operated in the ideal regime. For the fin and Geyser boiling regimes, this parameter presented large variation, especially in the Geyser boiling regime. Therefore, as affirmed by Cisterna et al. (2021b), $\varphi_b = 0.01$ is the upper limit value, where, below it, the thermosyphon operates in the ideal regime. These authors also tested this parameter for other operation conditions, arriving to the same conclusions. They also observed that, before the startup, the bubble release number is very large, as the main heat transfer mechanism is the axial conduction through the thermosyphon wall, requiring large temperature gradients. Besides, their experimental data show that the wall superheat was very affected by the filling ratio, which in turn, affects the Geyser boiling frequency and temperature amplitudes. Actually, the amplitude and frequency of the wall superheat oscillations decrease as the heat flux increase, until the system reaches the ideal regime.

Validation of the Geyser Boiling Limit Model.

Using the dimensionless bubble release number $\varphi_b = 0.01$, the Geyser boiling heat transfer limit can be obtained as:

$$q''_{GB} = 5 \cdot 10^{-5} \left(\frac{\rho_l^{0.65} k_l^{0.3} c_{p,l}^{0.7} g^{0.2}}{\rho_v^{0.25} h_{lv}^{0.4} \mu_l^{0.1}} \right)^{5/3} \left(\frac{p_{sat}}{p_{atm}} \right)^{0.383} T_{sat}^{5/3} \tag{8.67}$$

Figure 8.17 shows a plot of the heat transfer rate as a function of the vapor temperature where data for three sodium thermosyphons of same characteristics of those which data are presented in Fig. 8.8 and 8.16, but with different filling ratios (69, 96 and 144%, i.e., 17.33, 24.16 and 36.25 g of sodium, respectively), obtained by Cisterna et al. (2020a,b, 2021a, b). In this same figure, the Geyser boiling theoretical limit (see Eq. 8.67) and the continuum limit (which depends

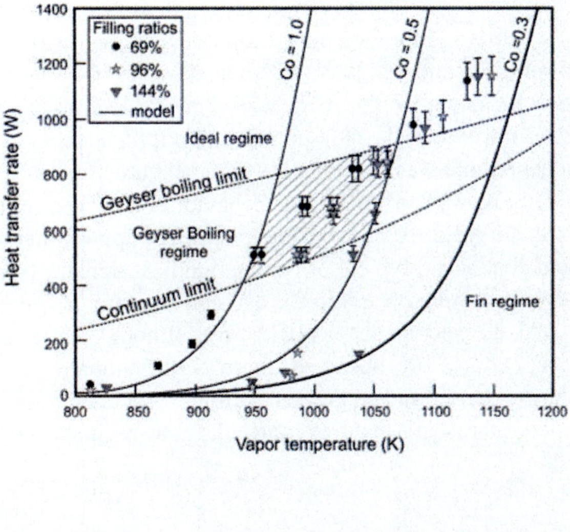

Fig. 8.17 Theoretical map for the thermal behavior sodium thermosyphons and comparison with models

on the vapor operation temperature) curves are plotted. Above the Geyser boiling limit, the thermosyphons operate in ideal regime, bellow the continuum limit, they operate in fin regime and, between these limits, in Geyser boiling regime. Besides, just for reference, theoretical curves (Eq. 8.66) for the heat transfer as a function of vapor temperature for three confinement numbers ($Co = 0.3,\ 0.5\ and\ 1$) are also plotted. All these theoretical curves together form a map that determines the operation conditions of the thermosyphons. Comparing the thermal behavior of the data presented in Fig. 8.16 with that of Fig. 8.17, it can be observed that the map is precise in determining the operation regions of the thermosyphon. Figure 8.16 shows that, between around 950 and 1050 K and from about 400 to 900 W, the thermosyphons operate in Geyser boiling, as predicted in Fig. 8.17 by the shaded area, which shows the Geyser boiling operation region. The thermosyphon behaviors at other operation conditions are also well represented in the map. It is important to note that, for the lower power input levels of the thermosyphons operating in Geyser boiling regime (around 500 W), the departure bubble diameter grows, tending to the tube diameter dimension, as the data get close to the $Co = 1$ curve. On the other hand, for higher power inputs (and higher temperatures), the departure bubbles tend to have half of the tube diameter.

In Fig. 8.17, the several sets of two data points close to each other shows the data obtained from the repeatability study performed by Cisterna et al. (2020a). The vertical bars over the data points represent the uncertainty range. More details of the other data presented in Fig. 8.17 can be found in by Cisterna et al. (2020a,b, 2021a, b).

Avoiding Geyser Boiling Using the Present Models.

The success of the present models can be ultimately checked if a high temperature (sodium) thermosyphon can be designed to operate without going through Geyser boiling, departing directly from the fin to the ideal operation regime. Cisterna et al. (2021b) conducted an experiment using a thermosyphon with the same characteristics of the one which data is shown in Figs. 8.8 and 8.16 but with 144% of filling ratio (the volume of liquid is 44% larger than the volume of the evaporator). The device was subjected to three power input levels: two below the Geyser boiling theoretical limit (35.2 and 104.9 W) and one above (1164.1 W). The set of graphics for these data is shown in Fig. 8.18, where the fin regime can be observed for heat transfer rates lower than the Geyser boiling limit (determined as approximately 200 W). The ideal regime was achieved for high heat transfer rates (larger than 1000 W).

In Fig. 8.18, a very stable behavior is observed for the first power level (35.2 W) with a large difference of temperature, of approximately 400 K, between evaporator and condenser, typical of conduction heat transfer through the wall (fin regime), as also seen more clearly in the average temperature graphic in this same figure.

At the second level (104.9 W), although small oscillations in the evaporator temperatures are observed, there is no symmetry in the temperature variation in the condenser and so Geyser boiling is not characterized. Actually, the vapor in the condenser is still in a free molecular flow regime (under the continuum curve), thus,

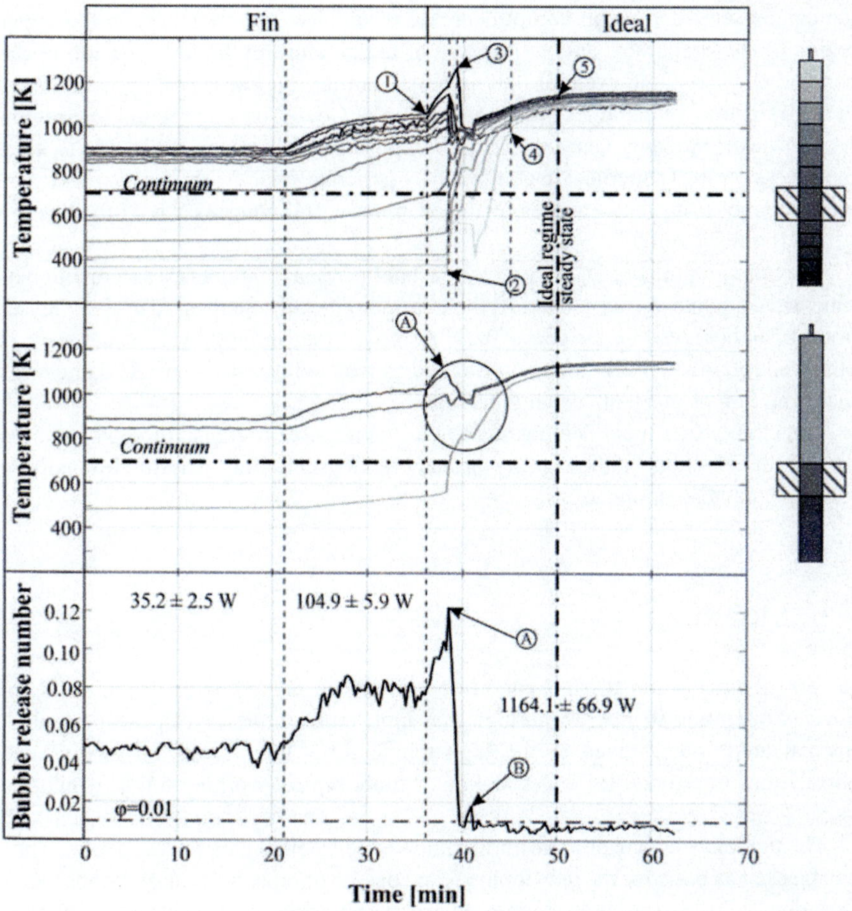

Fig. 8.18 Direct transition between fin and ideal regime. Temperature (upper), average temperature (middle) and bubble release number parameter (φ_b) as a function of time for a 99% pure sodium thermosyphon, with 144% of filling ratio (36.25 g), made of an ANSI 316L tube, of diameter 25.4 mm and of 350 mm of total length (see Cisterna et al. 2020a for description of the experimental work)

the oscillations of temperature in the evaporator do not find a way to propagate to the condenser. Therefore, the thermosyphon is still in the fin regime.

After the steady state was reached at the second power level, the heat transfer rate was increased to 1164.1 W. The transition from the fin to the ideal regime can be clearly observed. Point (1) in Fig. 8.18 marks the end of the fin regime. With the increase of the heat transfer rate, the evaporator temperature rises until it achieves the necessary wall superheat for the bubble's formation, growth and release (Point 2 in Fig. 8.19). Point 3 shows the exact moment when the bubble is released from the evaporator and the working fluid, in vapor and liquid states, being expelled into the condenser. There is an abrupt decrease of temperature especially in the upper region of the evaporator and an increase of temperature in the lower region of the condenser. One single cycle of bubble burst, similar to Geyser boiling is observed in the transition, which does not characterize the Geyser boiling regime (see region A). However, the large heat transfer rate imposed to the evaporator is able to keep the necessary wall superheat to continuously generate small bubbles and enter in the ideal regime. Both temperatures of the evaporator and condenser tend to constant values (points 4 and 5).

The third graphic of Fig. 8.18 shows the bubble release number φ_b as a function of time. In two points (points A and B) there is enough wall superheat to trigger Geyser boiling, as observed by φ_b above 0.01. As no Geyser boiling cyclical behavior is observed, point A and B can be considered as isolated transition events that causes the formation of one isolated large bubble.

Thus, this experiment demonstrates that it is possible to transit directly from the fin to the ideal regime without going through the Geyser boiling, a desirable condition for most practical applications.

8.4 Closure

In this chapter, some thermosyphon and heat pipe technologies and models for devices operating in low, mean/high and high temperature levels, are presented. Special attention is given to the high temperature devices, as the models and correlations developed for water and other more typical working fluids, cannot be used.

The models are confronted with literature experimental data for validation. They can be used as basis for the development and design of other two-phase technologies including loops, pulsating heat pipes, hybrid devices, etc., that operate at cryogenic, moderate/high and very high temperature levels. These technologies involve the use of special working fluids such as liquid gases, organic fluids and liquid metals.

References

Abreu, S.L., Colle, S. Skiavine, J.: Working Characteristics of a Compact Solar Hot Water System With Heat Pipes During Start-Up and Geyser Boiling Periods, Proceedings of the ISES Solar World Congress, Götheborg, Sweden, 14–19 (2003)

Alammar, A.A., Al-Dadah, R.K., Mahmoud, S.M.: Experimental Investigation of the Influence of the Geyser Boiling Phenomenon on the Thermal Performance of a Two-Phase Closed Thermosyphon. Journal of Cleaner Production **172**, 2531–2543 (2018)

Bejan, A.: Convection Heat Transfer, 3rd edn. Wiley, New York (2004)

Boo, J., Kim, S., Kang, Y.: An Experimental Study on a Sodium Loop-Type Heat Pipe for Thermal Transport From a High-Temperature Solar Receiver. Energy Procedia **69**, 608–617 (2015)

Bruce, R.M., Haber, L., McLure, P.: Toxicological Review of Naphthalene. U.S. Environmental Protection Agency, Washington (1998)

Cao, Y. and Faghri, A, A Numerical Analysis of High-Temperature Heat Pipe Startup From the Frozen State, ASME Journal of Heat Transfer, **115**, 247–254 (1993)

Cao, Y., Faghri, A.: Simulation of the Early Startup Period of High-Temperature Heat Pipes From the Frozen State by a Rarefied Vapor Self-Diffusion Model. ASME Journal of Heat Transfer **115**(1), 239–246 (1993)

Carey, V.P.: Liquid-vapor Phase-change Phenomena: an Introduction to the Thermophysics of Vaporization and Condensation Process in Heat Transfer Equipment. Series in Chemical and Mechanical Engineering, Taylor and Francis, Hebron (1992)

Chen, M.M.: An Analytical Study of Laminar Film Condensation: Part 1 – Flat Plates. ASME Journal of Heat Transfer **83**, 48–55 (1961)

Cisterna, L.H.R., Cardoso, M.C.K., Fronza, E.L., Milanese, F.H., Mantelli, M.B.H.: Operation Regimes and Heat Transfer Coefficients in Sodium Two-Phase Thermosyphons. Int. J. Heat Mass Transf. **152**, 119555 (2020a)

Cisterna, L.H.R., Vitto, G., Cardoso, M.C.K., Fronza, E.L., Mantelli, M.B.H., Milanez, F.H.: Charging Procedures: Effects on High Temperature Sodium Thermosyphon Performance. Journal of the Brazilian Society of Mechanical Sciences and Engineering **42**, 416 (2020b)

Cisterna, L. H. R., Fronza, E. L., Cardoso, M. C. K., Milanese, F. H. and Mantelli, M.B.H., Modified Biot Number Models for Startup And Continuum Limits of High Temperature Thermosyphons, International Journal of Heat and Mass Transfer, 165, 120699 (2021a)

Cisterna, L. H. R. Milanez, F. H. and Mantelli, M. B. H., Prediction of Geyser Boiling Limit for High Temperature Two-Phase Thermosyphons, International Journal of Heat and Mass Transfer, **165**, 120656 (2021b)

Cole, R., Rohsenow, W.: M, Correlation of Bubble Departure Diameters for Boiling of Saturated Liquids. Chemical Engineering Progress Symposium Series **65**, 211–213 (1969)

Colwell, G. T., Prediction of Cryogenic Heat Pipe Performance, NASA Final Report NSG-2054 (1977)

Cooper, M.G., Saturation Nucleate Boiling: A Simple Correlation, International Chemical Engineering Symposium Series, 785–793 (1984)

Couto, P., Mantelli, M.B.H., Ochterbeck, J.M.: Experimental Analysis of Supercritical Startup of Nitrogen/Stainless Steel Cryogenic Heat Pipes. J. Thermophys. Heat Transfer **20**(4), 842–849 (2006)

Couto, P., Ochterbeck, J.M., Mantelli, M.B.H.: Analysis of Supercritical Startup of Cryogenic Heat Pipes with Parasitic Heat Loads. J. Thermophys. Heat Transfer **19**(4), 497–508 (2005)

Coventry, J., Andraka, C., Pye, J., Blanco, M., Fisher, J.: A Review of Sodium Receiver Technologies for Central Receiver Solar Power Plants. Sol. Energy **122**, 749–762 (2015)

Deverall, J. E., Kemme, J. E., and Florschuetz, L. W., Sonic Limitations and Startup Problems of Heat Pipes, Contract W-7405, Report LA-4518, 1970.

Edwards, D.K., Marcus, B.D.: Heat and Mass Transfer in The Vicinity of the Vapor-Gas Front in a Gas-Loaded Heat Pipe. ASME Journal of Heat Transfer **94**, 155–162 (1972)

El-Genk, M.S., Tournier, J.M.: Uses of Liquid-Metal and Water Heat Pipes in Space Reactor Power Systems. Frontiers in Heat Pipes FHP **2**, 013002 (2011)

Emami, M.R.S., Noie, S.H., Khoshnoodi, M., Hamed, M.T., Kianifar, A.: Investigation of Geyser Boiling Phenomenon in a Two-Phase Closed Thermosyphon. Heat Transfer Eng. **30**(5), 408–415 (2009)

Farsi, H., Joly, J., Miscevic, M., Platel, V., Mazet, N.: An Experimental And Theoretical Investigation Of The Transient Behavior Of A Two-Phase Closed Thermosyphon. Appl. Therm. Eng. **23**(15), 1895–1912 (2003)

Groll, M., Rosler, S.: Operation Principles and Performance of Heat Pipes and Closed Two-Phase Thermosyphons. Journal of Non-Equilibrium Thermodynamic **17**, 91–151 (1992)

Guo, H., Guo, Q., Yan, X., Ye, F., Ma, C.: Experimental Investigation on Heat Transfer Performance of High-Temperature Thermosyphon Charged With Sodium-Potassium Alloy. Appl. Therm. Eng. **139**, 402–408 (2018)

Hashimoro, H., Kaminaga, F., Matsumura, K.: Study on Condensation Heat Transfer Characteristics in a Thermosyphon With Non-Condensable Gas Effect, 11th International Heat Pipe Conference. Tokio, Japan (1999)

Hijikata, K., Chen, S.J., Tien, C.L.: Non-Condensable Gas Effect on Condensation in a Two-Phase Closed Thermosyphon. Int. Journal Heat Mass Transfer **27**, 1319–1325 (1984)

Holman, J.P.: Heat Transfer, 5th edn. McGraw-Hill, New York (1981)

Incropera, F.P., DeWitt, D.P.: Fundamentos de Transferência de Calor e de Massa. LTC, Rio de Janeiro (2008)

JacobsenR, T., Penoncello, S.G., Lemmon, E.W.: Thermodynamic Properties of Cryogenic Fluids. Plenum, New York (1997)

Jafari, D., Di Marco, P., Filippeschi, S., Franco, A.: An Experimental Investigation on The Evaporation and Condensation Heat Transfer of Two-Phase Closed Thermosyphons. Exp. Thermal and Fluid Science **88**, 11–123 (2017)

Jafari, D., Filippeschi, S., Franco, A., Di Marco, P.: Unsteady Experimental and Numerical Analysis of a Two-Phase Closed Thermosyphon at Different Filling Ratios. Exp. Thermal Fluid Sci. **81**, 164–174 (2017)

Juxing, L., Tongming, X.: A Study on the Property of Heat and Mass Transfer of a Gas-Loaded Thermosyphon, 8th International Heat Pipe Conference. Beijing, China (1992)

Kaiping, P., Renz, U.: Thermal Diffusion Effects in Turbulent Partial Condensation. Int. J. Heat and Mass Transfer **34**, 2629–2639 (1991)

Kaminaga, F., Hashimoto, H., Feroz, C., Goto, K., Matsumura, K.: Heat Transfer Characteristics of Evaporation and Condensation in a Two-Phase Closed Thermosyphon, 10th International Heat Pipe Conference, Stuttgart, Germany (1997)

Khazaee, I., Hosseini, R., Noie, S.: Experimental Investigation of Effective Parameters and Correlation of Geyser Boiling in a Two-Phase Closed Thermosyphon. Appl. Therm. Eng. **30**, 406–412 (2010)

Liu, W., Gou, J., Luo, Y., Zhang, M.: The Experimental Investigation of a Vapor Chamber With Compound Columns Under The Influence of Gravity. Appl. Therm. Eng. **140**, 131–138 (2018)

Mantelli, M.B.H., Ângelo, W.B., Borges, T.: Performance of Naphthalene Thermosyphons with Non-Condensable Gases: Theoretical Study and Comparison with Data. Int. Journal of Heat and Mass Transfer **53**, 3414–3428 (2010)

Mantelli, M.B.H., Carvalho, R.D.M., Colle, S., Moraes, D.U.C.: Study of Closed Two-Phase Thermosyphons for Bakery Oven Applications, 33th National Heat Transfer Conference. Albuquerque, New Mexico (1999)

Mantelli, M. B. H., Uhlmann, T., Cisterna, L. H. R., Experimental Study of a Sodium Two-Phase Thermosyphon, 9th World Conference on Experimental Heat Transfer, Fluid Mechanics and Thermodynamics, Iguazu Falls, Brazil (2017)

Marcus, B. D., Fleischman, G. L. Steady State and Transient Performance of Hot Reservoir Gas Controlled Heat Pipes, Report No. NASA CR-73420, National Aeronautics and Space Administration, Washington, D.C. (1970)

Negishi, K., Sawada, T.: Heat Transfer Performance of an Inclined Two-Phase Closed Thermosyphon. Int. J. Heat Mass Transf. **26**(8), 1207–1213 (1983)

Nellis, G., Klein, S.: Heat transfer. Cambridge University Press, Cambridge (2009)

Noie, S.H., Kalaei, M.H., Khoshnoodi, M.: Experimental Investigation of a Two Phase Closed Thermosyphon, 7th International Heat Pipe Symposium. Jeju, Korea (2003)

Noie, S., Emami, M.S., Khoshnoodi, M.: Effect of Inclination Angle and Filling Ratio on Thermal Performance of a Two-Phase Closed Thermosyphon Under Normal Operating Conditions. Heat Transfer Eng. **28**(4), 365–371 (2007)

Ochterbeck, J.M., Peterson, G.P., Ungar, E.K.: Depriming/Rewetting of Arterial Heat Pipes: Comparison with Share-II Flight Experiment. J. Thermophys. Heat Transfer **9**(1), 101–108 (1995)

Pabón, N.Y.L., Florez, J.P.M., Vieira, G.S.C., Mantelli, M.B.H.: Visualization and Experimental Analysis of Geyser Boiling Phenomena in Two-Phase Thermosyphons. Int. Journal Heat Mass Transfer **141**, 876–890 (2019)

Patankar, S.V.: Numerical Heat Transfer and Fluid Flow. Hemisphere, Washington DC (1980)

Peterson, P.F., Tien, C.L.: Numerical and analytical solutions for two-dimensional gas distribution in gas-loaded heat pipes. ASME Journal of Heat Transfer **111**, 598–604 (1989)

Rosenfeld, J. H., Buchko, M. T., and Brennan, P. J., A Supercritical Start-Up Limit to Cryogenic Heat Pipes in Microgravity, 9th International Heat Pipe Conference (2), 742 – 753 (1995)

Shiraishi, M., Kikuchi, K., Yamanishi, T.: Investigation of Heat Transfer Characteristics of a Two-Phase Closed Thermosyphon. Journal of Heat Recovery System. **1**(4), 287–297 (1981)

Smith, K., Kempers, R., Robinson, A.J.: Confinement and Vapour Production Rate Influences in Closed Two-Phase Reflux Thermosyphons Part A: Flow Regimes. Int. J. Heat Mass Transf. **119**, 907–921 (2018)

Stephan, K.: Heat Transfer in Condensation and Boiling. Springer, Berlin (1992)

Stephan, K., Abdelsalam, M.: Heat Transfer Correlations for Natural Convection Boiling. Int. J. Of Heat and Mass Transfer **23**(1), 73–87 (1980)

Stewart, R.B., Jacobsen, R.T., Wagner, W.: Thermodynamic Properties of Oxygen From the Triple Point to 300K with Pressures to 80 MPa. J. Phys. Chem. Ref. Data **20**(5), 917–1021 (1991)

Tecchio, C., Oliveira, J. L. G, Paiva, K. V., Mantelli, M. B. H., Gandolfi, R. and Ribeiro, L. G. S., Geyser Boiling Phenomenon in a Two-Phase Closed Loop-Thermosyphons, Int. J. Heat Mass Transfer, **111**, 29–40 (2017)

Tournier, J.M.P., El-Genk, M.S.: Liquid Metal Loop and Heat Pipe Radiator for Space Reactor Power Systems. J. Propul. Power **22**(5), 1117–1134 (2006)

Tournier, J.M., El-Genk, M.S.: A Vapor Flow Model for Analysis of Liquid-Metal Heat Pipe Startup From a Frozen State. Int. J. Heat Mass Transf. **39**(18), 3767–3780 (1996)

Tournier, J.M., El-Genk, M.S.: Modeling of the Startup of a Horizontal Lithium Heat Pipe From a Frozem State, 35th Natinal Heat Transfer Conference. Anaheim, Calif. (2001)

Vasiliev, L.L.: Heat Pipes in Modern Heat Exchangers. Appl. Therm. Eng. **25**, 1–19 (2005)

Vasiliev, L.L., Volokhov, G.M., Gigevich, A.S., Rabetskii, M.I.: Heat Pipes Based on Naphthalene. J. Eng. Phys. Thermophys. **54**, 623–626 (1988)

Wilcox, S. J. and Rohsenow, W. M., Film Condensation of Liquid Metals – Precision of Measurement, Report DSR 71475-62, Contract GK 1113 (1969)

Xia, G., Wang, W., Cheng, L., Ma, D.: Visualization Study on the Instabilities of Phase-Change Heat Transfer in a Flat Two-Phase Closed Thermosyphon. Appl. Therm. Eng. **116**, 392–405 (2017)

Yan, Y.H., Ochterbeck, J.M.: Analysis of Supercritical Start-Up Behavior for Cryogenic Heat Pipes. AIAA Journal of Thermophysics and Heat Transfer **13**(1), 140–145 (1999)

Zhan, D., Zhang, H., Liu, Y., Li, S., Zhuang, J.: Investigation of Medium Temperature Heat Pipe Receiver Used in Parabolic Through Solar Collector. ISES World Congress, Beijing, China **1**, 1823–1827 (2007)

Zhang, H., Zhuang, J.: Research, development and industrial applications of heat pipe technology in China. Appl. Therm. Eng. **23**, 1067–1083 (2003)

Zhou, X., Collins, R.E.: Condensation in a Gas-Loaded Thermosyphon. Int. J. Heat and Mass Transfer **38**, 1605–1617 (1995)

Part III
Applications

In Part III of this book, some examples of equipment assisted by thermosyphons and heat pipes, designed to promote the management of heat in real applications are presented. First, thermosyphon heat exchangers are described and design models are discussed. In the sequence, several heat pipe and thermosyphon technology solutions for electronics cooling are presented. Finally, industrial applications of thermosyphons in cooling towers, heating of houses, industrial ovens and dryers, heating systems for oil tanks and for petroleum gas distribution stations are described.

Chapter 9
Thermosyphon Heat Exchangers

9.1 Geometry

One of the most common use of a thermosyphon assisted equipment in industry is in heat exchangers (Vasiliev 2005). Although its use is more recent in some countries, China has successfully and largely employed these devices since 1980 decade (Zhang and Zhuang 2003). Thermosyphon heat exchangers can be used in substitution of more traditional technologies, such as "shell and tubes" or "plates" heat exchangers. One typical application is in the recovery of waste heat in industries (Ramos et al. 2016); in this case, the process air is pre-heated using hot flue gases from industrial chimneys, which would be discarded to atmosphere, causing thermal pollution. Figure 9.1 shows a typical configuration of the thermosyphon heat exchanger designed for this application. Usually, it consists of a bank of thermosyphons in vertical or inclined positions. The inclination can be as small as 7° to the horizon. The evaporators of the thermosyphon are in contact with the hot stream, while the condensers, with the cold stream (usually air) to be heated. Heat is conducted from the thermosyphon evaporators to the condensers. The adiabatic sections, between evaporator and condensers, are usually small or even inexistent. The hot and cold streams can be gas or liquid flows. The equipment is usually isolated from the environment.

If the application requires the tubes to be located in the horizontal position, heat pipes (with internal wick structure) must be employed instead of thermosyphons, to provide the liquid return from the condenser to the evaporator by capillary action.

Thermosyphon heat exchangers present several advantages when compared to traditional heat exchanger technologies. First, as the hot and cold streams are external to the tubes, the maintenance (cleaning) is much easier. Second, if one of the tubes is perforated for any reason, there is no contamination between the streams. The contamination would probably happen in any other heat exchanger technologies, such as shell and tube or plates, for instance. Third, for a typical industrial thermosyphon heat exchanger (with more than 1000 tubes, for instance), the overall heat transfer capacity of the device would be negligibly affected if a few tubes are lost. Forth, in the case of losing some tubes, in normal operation conditions, the neighbor tubes

M. B. H. Mantelli et al., *Thermosyphons and Heat Pipes: Theory and Applications*,
https://doi.org/10.1007/978-3-030-62773-7_9

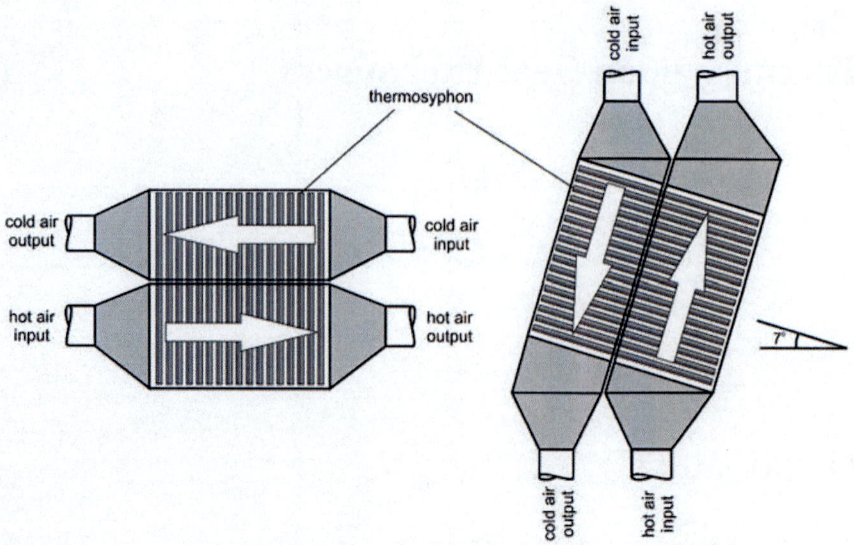

Fig. 9.1 Schematic of the thermosyphon heat exchanger, vertical tubes (left) inclined tubes (right)

are able to compensate the loss of the heat transfer capacity, as thermosyphons are very adjustable to the variations of power inputs. Fifth, thermosyphon geometries are very flexible, so that heat exchangers with many different configurations can be easily adapted to diverse applications. In loops, for example, the evaporator and condenser can be designed as independent heat exchangers, and, if necessary, they can be located apart (see schematic in Fig. 1.3). As drawbacks, it can be quoted that, comparing to other technologies, the thermosyphon heat exchanger fabrication costs may be slightly higher. Note that costs can be considerable higher for heat pipes heat exchangers. Besides, the devices cannot be very compact, as tube external areas must be provided to both hot and cold streams, while, for other heat exchanger technologies, the flows can be separated by a common wall.

Usually the same tube arrangement of the conventional heat exchange equipment (see Zukauskas et al. 2008; Zukauskas et al. 1985; Gnielinski 2013) is adopted for the tubes in the thermosyphon heat exchangers: 30, 45, 60 or 90°, as shown in Fig. 9.2.

Fig. 9.2 Arrangements of thermosyphon tube bank

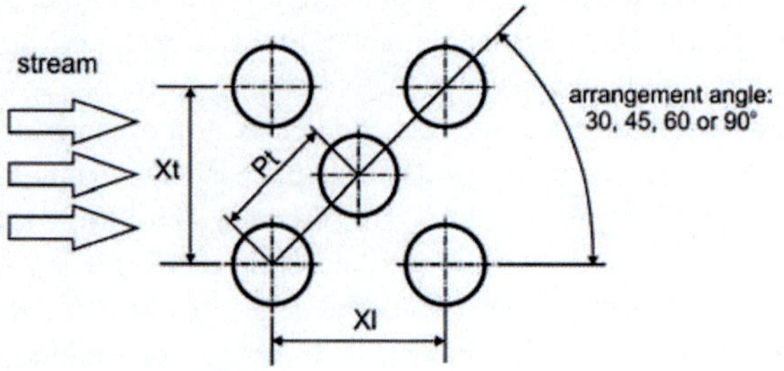

These angles are important because they control the percolation of the streams among the tubes, ultimately determining the flow regime, which can be laminar, turbulent or transitional. It also controls the compactness of the resulting equipment. Another advantage of adopting these arrangements is that the available literature correlations can be used for the prediction of the coefficients of heat transfer, very important parameters for the design of such equipment. Besides, the flow pressure drops are very affected by the tube arrangements and this information is critical to the selection of the liquid pumps and/or gas fans, to be used to supply the required pressure drops in both fluxes of the heat exchanger. Therefore, the tube arrangement affects both the thermal performance and the cost of the equipment.

9.2 Thermosyphon Heat Exchanger Design Methodology

Design procedures are presented in this chapter. The simplified thermal circuit model discussed in Chap. 4 is used for the determination of the overall thermal resistance of thermosyphons, in steady state conditions. Well know methodologies for designing of shell and tube heat exchangers are also used, where coefficients of heat transfer are obtained from literature correlations. Also, the heat exchanger methodology for determination of the pressure drops in tube bank external flows is used, for both smooth and finned tubes.

The method of Effectiveness—Number of Transfer Units (ε-NUT) (Roetzel and Spang 2013; Kays and London 1984) is normally used for designing the tube banks geometry. Effectiveness is defined as the ratio between heat actually transferred and maximum possible heat transfer rate that the heat exchanger could transfer:

$$\varepsilon = \frac{q}{q_{max}} \tag{9.1}$$

where the heat transferred is predicted by the expression:

$$q = UA\Delta T_{lm} \tag{9.2}$$

and the logarithmic mean temperature difference is determined by:

$$\Delta T_{lm} = \frac{\Delta T_h - \Delta T_c}{\ln(\Delta T_h / \Delta T_c)} = \frac{\Delta T_c - \Delta T_h}{\ln(\Delta T_c / \Delta T_h)} \tag{9.3}$$

where ΔT_h is the temperature difference between the hot flux entrance and exit and ΔT_c is the temperature difference between the cold flux entrance and exit, for counter-flux heat exchangers. The global coefficient of heat transfer U is defined based on the total thermal resistance of the equipment, for convection heat transfer in both fluxes.

As a first guess, the thermal resistance of the thermosyphons is considered null. This hypothesis is equivalent to consider that the heat exchange is similar to an one-pass concentric heat exchanger. If the global coefficient of heat transfer is constant, the thermal effectiveness can be determined by:

$$\varepsilon = \frac{\dot{m}_h c_{p,h}(T_{h,in} - T_{h,out})}{C_{min}(T_{h,in} - T_{c,in})} \quad (9.4)$$

where $T_{h,in}$ and $T_{h,out}$ are the input and output temperatures of the hot stream and $T_{c,in}$ and $T_{h,out}$ are those for the cold stream. Moreover, and C_{min} and C_{max} is the minimum and maximum conductance parameter between C_h e C_c, where $C = \dot{m} \cdot c_p[W/K]$. Defining $C_r = C_{min}/C_{max}$, the Number of Transfer Units (NTU) can be determined from:

$$NUT = \frac{UA}{C_{\min}} = \frac{1}{1 - C_r} \ln\left(\frac{1 - \varepsilon C_r}{1 - \varepsilon}\right) \quad (9.5)$$

where *U* is the heat exchanger global coefficient of heat transfer. From this expression, *UA* can be determined and so the total heat power, from Eq. 9.2. On the other hand, the total heat power transferred can also be given by $q = \varepsilon \cdot C_{min} \cdot (T_{h,in} - T_{c,in})$. From these equations, it is possible to determine the total area needed for the heat exchange and so the number of thermosyphons and the energy to be transported by tube. The thermosyphon, on the other hand, are designed to fulfill thc heat transfer requirements.

9.3 Thermosyphon Heat Transfer

The analogy between electrical and thermal circuits, as discussed in Chap. 4, is used to predict the overall thermal resistance of thermosyphons and their heat transfer capacity, needed for the heat exchanger design.

Considering heat exchanger applications, the most important resistances of the thermosyphon thermal circuit are associated with the convection heat transfer in the external surfaces of evaporators and condensers. These external resistances are, actually, around two order of magnitude smaller than the internal ones. This means that the tricky selection of the appropriate correlations for the inside evaporation and condensation processes is not so important in this case, as errors of up to one order of magnitude in the prediction of these coefficients usually cause insignificant effect in the determination of the overall thermal resistance. Actually, in a first approach, for the thermosyphon heat exchanger design, only the external resistances are considered.

This section is devoted to the prediction of the external thermal resistances of the thermosyphon thermal circuit ($R_{ex,hs}$, $R_{ex,cs}$), which schematics is shown in Fig. 4.2. All other resistances are determined exactly as proposed in Chap. 4.

To guarantee that the thermosyphon designed is able to transfer the necessary heat, the thermosyphon heat transfer limits must be determined, as discussed in Chap. 3. Special attention is given to entrainment, sonic and viscous limits.

9.4 External Convection

External Convection Across a Smooth Tube Bank

The coefficients of heat transfer used for external fluid flow across a bank of thermosyphons are the same employed for the conventional shell and tube heat exchangers (Hewitt 2008). For a bank with more than 20 rows of tubes in staggered layout with angles of 30, 45 and 60° (see Fig. 9.2), subjected to a forced convection flow, the following $Nu_D = hD/k$ correlation of Zukauskas (1972), for Prandtl range $0.7 < Pr_D = \nu/\alpha < 500$ and for $1000 < Re_{D,max} = \rho u_{max} D/\mu < 2 \cdot 10^6$ can be used:

$$Nu_D = C\mathrm{Re}_{D,max}^{m}\mathrm{Pr}_D^{0.36}\left(\frac{\mathrm{Pr}}{\mathrm{Pr}_s}\right)^{\frac{1}{4}} \tag{9.6}$$

where u_{max} is the maximum between the following velocities, based on the parameters depicted in Fig. 9.2:

$$u_{max} = \frac{Xt}{Xt - D}u \quad or \quad u_{max} = \frac{Xt}{2(Pt - D)}u \tag{9.7}$$

where D is the diameter of the tube. The properties of the fluid are obtained for the average fluid temperature, between the entrance and exit. The Pr_s parameter, however, is obtained for the tube surface temperature.

The values of the constants C and m in Eq. 9.6 are selected according to the tube arrangement and are given in Table 9.1, based on Zukauskas (1972) work. According to this author, if $Xt/Xl < 0.7$, the heat transfer is inefficient for aligned tube array.

External Convection Over a Finned Tube Bank

For a fixed heat transfer rate, aiming the reduction of the thermosyphon heat exchanger volume, the number of tubes must the reduced. The number of tubes is defined based on the necessary external heat transfer area (and not on the capacity of the device of transferring heat). Finned tubes are usually applied to improve this area and so decrease the volume of the equipment. The correlations used to obtain the heat transfer coefficient of finned tubes are different from those for smooth tubes. In this section, correlations for external convection heat transfer over banks of finned tubes are presented.

Table 9.1 Constants for Eq. 9.6 for an array of tubes for cross surrent flow

Configuration	$Re_{D,max}$	C	m
Aligned	$10 - 10^2$	0.80	0.40
Staggered	$10 - 10^2$	0.90	0.40
Aligned	$10^2 - 10^3$	Consider a single cylinder	
Staggered	$10^2 - 10^3$		
Aligned—$Xt/Xl > 0.7$	$10^3 - 2 \cdot 10^5$	0.27	0.63
Staggered—$Xt/Xl < 2$	$10^3 - 2 \cdot 10^5$	$0.35 \cdot (Xt/Xl)^{1/5}$	0.60
Staggered—$Xt/Xl > 2$	$10^3 - 2 \cdot 10^5$	0.40	0.60
Aligned	$2 \cdot 10^5 - 2 \cdot 10^6$	0.021	0.84
Staggered	$2 \cdot 10^5 - 2 \cdot 10^6$	0.022	0.84

According to Hewitt et al. (1994), the annular fins, usually employed in heat exchangers, can be classified between tall and short, where H is the height and D is the tube diameter, using the criteria:

- Tall: $0.05 < H/D < 0.33$
- Short: $0.2 < H/D < 0.70$.

Schematic of the cross sections of typical finned tubes are shown in Fig. 9.3. The designer must select the finned tube to be used in the heat exchanger. One of the criteria is the efficiency of the fin set. The heat exchange between a bank of thermosyphon and the liquid flow, is given by:

$$q = h(\eta_f A_f + A_b)(T_b - T_\infty) \quad (9.8)$$

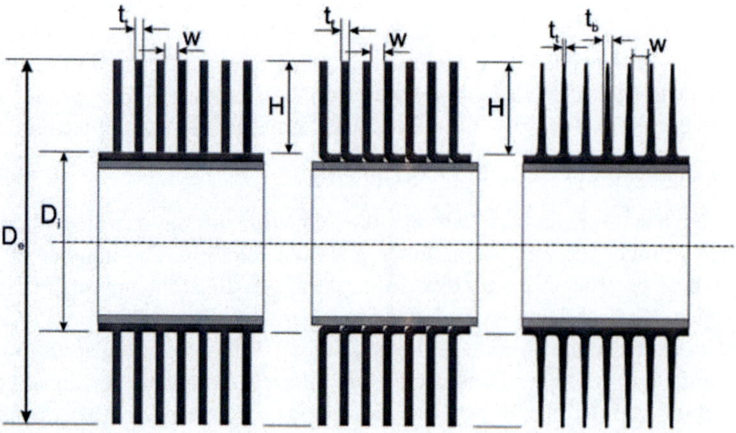

Fig. 9.3 Fin geometries

where the fin heat exchange area is given by:

$$A_f = \left[\frac{1}{2}\pi\left(D_e^2 - D_i^2\right)\right] + \pi D_i w \tag{9.9}$$

and where D_e and D_i are the external an internal diameters of the fin and W is the distance between fins (see Fig. 9.3). A_b in Eq. 9.9 is the area of the base, given by:

$$A_b = \pi D_i (t_f + w) \tag{9.10}$$

where t_f is the fin thickness. In Eq. 9.8, η_f is the fin efficiency, which, for straight circular fins has the form:

$$\eta_f = \frac{\tanh\left(\sqrt{\frac{2h}{t_f \cdot k_f}} \cdot \Gamma\right)}{\sqrt{\frac{2h}{t_f \cdot k_f}} \cdot \Gamma} \tag{9.11}$$

where Γ is the fin parameter given by:

$$\Gamma = \frac{D_i}{2}\left(\frac{D_e}{D_i} - 1\right)\left(1 + 0.35 \ln \frac{D_e}{D_i}\right) \tag{9.12}$$

In these expressions, an average value for the coefficient of heat transfer h is used, taking into consideration the fin effect over the fluid flowing close to it. For fins with variable thickness, such as the one illustrated in Fig. 9.3, the following correction factor is proposed to be applied to the fin efficiency:

$$\xi = 1 + 0.125\left(1 - \sqrt{\frac{t_t}{t_b}}\right)\left(\sqrt{\frac{2h}{t_f \cdot k_f}}\right)H \tag{9.13}$$

where t_t and t_b are the thickness at the top and base, while t_f is the average thickness of the fin.

Tall Fins

Usually, the fins used in thermoyphon tubes are annular and tall. For this case, Hewitt et al. (1994) propose the following correlation for Nusselt number $\left(Nu_{D_i} = hD_i/k\right)$, for finned tubes in staggered arrangement, valid for $2 \cdot 10^3 \leq Re_{D,max} \leq 4 \cdot 10^4$, $0.13 \leq w/H \leq 0.57$ and for the configuration $1.15 \leq Xt/Xl \leq 1.72$:

$$Nu_{D_i} = 0.242 \cdot \mathrm{Re}_{D,max}^{0.658}\left(\frac{w}{H}\right)^{0.297}\left(\frac{Xt}{Xl}\right)^{-0.091} \mathrm{Pr}^{\frac{1}{3}} F_1 \cdot F_2 \tag{9.14}$$

The factor F_1 corrects the correlation with respect to the variation of the physical properties and is especially important for high temperatures. For air heat exchangers, this factor can assume the unitary value ($F_1 = 1$). The factor F_2 corrects for cases where the heat exchanger has few rows and assume the values: 1 for four or more tube rows, 0.92 for three rows, 0.84 for two rows and 0.76 for one row.

Short Fins

For short fins, the correlation proposed by Hewitt et al. (1994), valid for $103 \leq Re_{D,max} \leq 8 \cdot 10^5$, $0.19 \leq w/H \leq 0.66$ and for the configuration: $1.10 \leq Xt/D_e \leq 4.92$ and $0.058 \leq H/D_e \leq 0.201$, is proposed, being given by:

$$Nu_{D_i} = 0.183 \cdot \mathrm{Re}_{D,max}^{0.7} \left(\frac{w}{H}\right)^{0.36} \left(\frac{Xt}{D_e}\right)^{0.06} \left(\frac{H}{D_e}\right)^{0.11} \mathrm{Pr}^{0.36} F_1' \cdot F_2' \cdot F_3' \quad (9.15)$$

The parameter F_1' corrects for the variation of the fluid properties due to temperature variations, being given by:

$$F_1' = \left(\frac{\mathrm{Pr}}{\mathrm{Pr}_s}\right)^{0.26} \quad (9.16)$$

The parameter F_2' corrects for the number of rows and assumes the values: 0.85 for three rows, 0.9 for four rows, 0.95 for six rows, 0.07 for eight rows and 1 for nine and more rows. The parameter F_3' takes into consideration the angular arrangement of the tubes, being considered 1 for 30, 45 and 60° tube arrangement angles (see Fig. 9.2), while, for aligned tubes (90°), this factor is:

$$F_3' = \frac{Nu(for\ aligned\ smooth\ tubes)}{Nu\ (for\ 30^o\ triangular\ alignment)} \quad (9.17)$$

9.5 Flow Pressure Drops

One of the major factors to be observed in the design of any heat exchanger, including the thermosyphon's, is the flow pressure drops, which are, actually, the prime parameter in the selection of the equipment responsible for the flow of the cold and hot streams: liquid pump or the gas fan. Obviously, the pressure drops depends of the tube configuration: smooth or finned.

Smooth Tube Bank

The total pressure drops of a stream percolating a tube bank is composed of two major components: acceleration, associated to entrance and exit in the array (and so it does not depend on the number of tube rows), and a friction factor, associated to the friction and moment pressure drops, observed when the flow go through the sequential tube rows. Hewitt et al. (1994) propose the following expression, which is an adaptation of the kinetic energy pressure drop, for n tube rows:

$$\Delta p = \left(K_a + nK_f\right)\frac{1}{2}\rho u_{max}^2 \tag{9.18}$$

The constant K_a is:

$$K_a = 1 + \lambda^2 \tag{9.19}$$

where:

$$\lambda = \frac{\text{minimum free flow cross section area}}{\text{total front area}} \tag{9.20}$$

This constant may assume the value of approximately 1.5, when the minimum cross section area is perpendicular to the flow direction.

The constant K_f depends on the $Re_{D,max}$ and on the tube arrangement. According to Hewitt et al. (1994), for aligned tube configuration (see Fig. 9.2):

$$Xt/D = 2.5 \quad \begin{cases} 8 \cdot 10^2 < Re_{D,max} < 2 \cdot 10^4, & K_f = 0.075 \\ 4 \cdot 10^4 < Re_{D,max} < 2 \cdot 10^6, & K_f = 0.050 \end{cases}$$

$$Xt/D = 2.0 \quad \begin{cases} 8 \cdot 10^2 < Re_{D,max} < 4 \cdot 10^4, & K_f = 0.115 \\ 8 \cdot 10^4 < Re_{D,max} < 2 \cdot 10^6, & K_f = 0.090 \end{cases}$$

$$Xt/D = 1.5, \quad 10^5 < Re_{D,max} < 2 \cdot 10^6, \quad K_f = 0.120$$

For staggered configuration, in equilateral triangle arrangement ($Pt = Xt = Xl$, see Fig. 9.2), this parameter is more variable and typical values are given in Table 9.2.

The values of K_f for other Reynolds number ranges and tube arrangements can be found in Hewitt et al. (1994).

Finned Tube Bank

For the determination of the pressure drop of the bank of finned tubes, Eq. 9.18 and 9.19 are used. The λ parameter is given by (Hewitt et al. 1994):

Table 9.2 K_f parameter for pressure drop of a bank of smooth tubes (Zukauskas and Ulinskas, 2008)

$Re_{D,max}$	Xt/D			
	1.25	1.5	2.0	2.5
10^2	3.48	1.90	1.40	1.16
10^3	0.90	0.74	0.65	0.45
10^4	0.44	0.45	0.32	0.30
10^5	0.23	0.19	0.15	0.14
10^6	0.20	0.19	0.17	0.11

$$\lambda = \frac{Xt - D_i - 2H \cdot t_f / (t_f + w)}{Xt} \tag{9.21}$$

and K_f is a factor that depends of the geometry of the tubes and of the fins and, for short fins, staggered arrangement and for $10^3 < Re_{D,max} < 10^5$, the following expression is proposed:

$$K_f = 4.71 \cdot \mathrm{Re}_{D,max}^{-0.286} \left(\frac{H}{w}\right)^{0.51} \left(\frac{Xt - D_i}{Xl - D_i}\right)^{0.536} \left(\frac{D_i}{Xt - D_i}\right)^{0.36} \tag{9.22}$$

For long fins, staggered arrangement, $5 \cdot 10^2 < Re_{D,max} < 5 \cdot 10^4$, the following correlation can be used:

$$K_f = 4.567 \cdot \mathrm{Re}_{D,max}^{-0.242} \left(\frac{A}{A_t}\right)^{0.504} \left(\frac{Xt}{D_i}\right)^{-0.376} \left(\frac{Xl}{D_i}\right)^{-0.546} \tag{9.23}$$

where the area ratio is:

$$\frac{A}{A_t} = \frac{\left[\frac{1}{2}\left(D_e^2 - D_i^2\right)\right] + D_i w + D_e t_f}{D_i (w + t_f)} \tag{9.24}$$

9.6 Selection of Tube Geometries and Materials

Several parameters are important for the selection of the materials and geometry of the thermosyphons to be used in the thermosyphon heat exchanger. First, the tube material must to be able to stand for the pressures. Second, the working fluid must operate in the desired temperature level (see Sect. 9.3.1). Third, the tube material and fluid must be chemically compatible, to avoid the formation of non-condensable gases. Forth, the tube material must enable the necessary heat exchange between fluids. Fifth, aspects such as maintenance, cost and availability in the market, must also be considered.

As the use of thermosyphon in heat exchangers can be considered a recent technology, a specific standard for designing such equipment is not available. As thermosyphons are sealed tubes and, especially at high temperatures, their saturated vapor may reach high pressures, one could think that the devices are considered pressure vessels and so its design should follow these specific regulations. However, some standards such as ASME Section VIII, which applies for pressure vessels, cannot be used in this case, as it states: "….the following classes of vessels are not included: (i) vessels having an inside diameter, width, height, or cross section diagonal not exceeding 6 in. (152 mm), with no limitation on length of vessel or pressure." For diameters smaller than 150 mm, the standard is applied only with flammable working fluids. As most the thermosyphons used in heat exchangers have internal diameters smaller than 152 mm the pressure vessel regulations do not apply.

Although not developed to thermosyphons, it is recommended the use of Standard UG-27 (ASME Section VIII, Division I) for the determination of the thickness of the wall of the thermosyphon tube. According to the methodology described in this standard, the thickness of an internally pressurized tube can be determined by the following expression:

$$t = \frac{p \cdot r_i}{\sigma \cdot E_f - 0.6 \cdot p} \tag{9.25}$$

where p is the internal pressure, σ is the maximum stress that the tube material is able to stand and E_f is the mechanical efficiency of the tube, considering the fabrication process (for seamless tube, $E = 1$). This expression is valid when the thickness of the tube does not exceeds half of the internal radius, or when the internal pressure p does not exceeds $0.385\sigma E$.

The tube material admissible stress can be obtained by the Table UCS-23, of ASME, Section VIII, Division 1. Considering possible corrosion effects, an over-thickness of 1 mm for smooth tubes must be supplied. Actually, even thicker tubes are commonly used, due to fabrication concerns (facility of welding the lids to the tubes, for instance).

It is important to keep in mind that the cost of the heat exchanger is directly associated to its mass. Therefore, for costs reasons, it is advisable to select, from the commercial tubes, the thinner and cheaper ones, among those that attend to the Standards.

9.7 Design Methodology

Several parameters are required for the design of a thermosyphon heat exchanger, such as the thermophysical and hydrodynamic characteristics of the hot and cold streams and the volume and geometry of the space where the device is to be installed. For illustration purposes, Table 9.3 presents a list of the data required for designing a typical waste recovery heat exchanger. The designing methodology is described in the sequence of this section.

Internal and external heat transfer aspects of thermosyphons, for the hot and cold streams, need to be considered separately for the design of a heat exchanger.

As already observed, the methodology discussed in Chap. 4 is recommended for the design of the thermosyphons of the heat exchanger.

As also already treated, the total heat transferred, the pressure drops and the total volume of the equipment depend strongly of the tube arrangements inside the heat exchanger. Therefore, usual arrangements such as aligned (90°) and staggered (30, 45 and 60°) must be considered as possibilities. Besides, the designer must decide if the tubes are to be finned to not. While finned tubes have larger heat transfer area

Table 9.3 Necessary data for the desing of a thermosyphon heat exchanger

Parameter description
External tube diameter
Arrangement geometry of the thermosyphon tubes (aligned or staggered)
Overall dimensions of the heat exchanger
Evaporator length
Adiabatic section length
Tube wall thickness
Input temperature of the hot stream
Exit temperature of the hot stream
Input temperature of the cold stream
Mass flow rate of the hot stream
Mass flow rate of the cold stream
Thermophysics characteristics of the hot stream (fouling, corrosion possibility, etc.)
Thermophysics characteristics of the cold stream (fouling, corrosion possibility, etc.)
Local atmospheric pressure
Geometry of fins at the evaporator (if they are to be used)
Geometry of fins at the condenser (if they are to be used)
Thermal conductivity of tube in the evaporator
Thermal conductivity of tube in the condenser
Filling ratio
Working fluid
Additional thermal resistance (fouling, corrosion, etc.)
Estimation of the cold tip length, due to the presence of non-condensable gases

than the smooth ones and the equipment volume is smaller, the overall cost increases significantly.

Isoppo et al. (2009) propose the following methodology for designing thermosyphon heat exchangers. Having the data in hands, the preliminary thermosyphon design and the pre-selected tube arrangement, the method ε-NUT is used for the heat exchange design. Taking the example of the waste recovery heat exchanger, the entrance and exit temperatures of the hot stream, that deliver heat of the thermosyphon evaporator array, are considered known, while for the cold stream, that removes heat from the condenser array, only the entrance temperature is considered known. Therefore, a first guess of the exit temperature is necessary. Knowing these temperatures, the average temperature of the heat exchange can be estimated. The thermal properties are obtained for this average temperature. An estimate of the total heat transferred is made and the exit temperature of the cold stream is determined and compared with the initial guess. If these temperatures are not equal within a pre-defined limit, this temperature is recalculated until convergence.

For the estimation of the total heat transfer, the following procedure is adopted. With all temperatures known, the global coefficient of heat transfer of the heat exchanger, obtained from the ε-NUT method, can be evaluated.

Knowing the overall heat transferred, the overall thermal resistance of the thermosyphon array can be determined, and so the thermosyphon design can be re-evaluated. If important, thermal resistances due to the fouling must also be considered. In practice, as the thermal resistances related to the tube material are not very sensible to the heat exchange operation conditions, they are calculated just once. The same happens to the fouling thermal resistances.

The global coefficient of heat transfer (parameter U) determined by the ε-NUT method is compared with the overall coefficient of heat transfer of the thermosyphon array and the overall area necessary for the heat removal from the hot stream and to the heat delivery to the cold stream, is checked. With the area, the number and arrangement of the thermosyphons can be re-evaluated. With the mass flow rate and the velocity of the streams (design input parameter, see Table 9.3), the Reynolds and Prandtl numbers can be obtained, which also needs thermophysical properties, which, in turn, are evaluated at the heat exchanger average temperature, as already mentioned. These are the input parameters of the heat transfer correlations, considering the selected tube arrangement (aligned, staggered, etc.). With the coefficients of heat transfer, the thermosyphon external heat transferred to the streams can be determined and these values are checked with the expected heat transferred. If this calculated heat transfer does not agree with the previous one within a pre-determined tolerance, a new number of tubes and/or another arrangement can be tested. This procedure is repeated until convergence. Table 9.4 shows all the parameters determined during the design of a waste recovery heat exchanger, while Fig. 9.4 show a design flow chart of such equipment.

The methodology used to design heat exchangers with inclined thermosyphons is very similar to the one just described, as the correlations for the convection heat exchanged with the external streams are the same. Only some corrections about the thermal performance of the inclined thermosyphons must be applied (see Chap. 4).

The application of the methodology described in this chapter may results in the proposal of several different thermosyphon heat exchangers, for the same input parameters. Borges et al. (2007) developed thermo-economic designing tools, using mathematical programming for the selection of the appropriate heat exchanger configuration.

Table 9.4 Parameters determined during the design of a thermosyphon heat exchanger

Determined parameters
Diagonal distance between the tube centers
Longitudinal distance between the tube centers
Ratio of the smallest distance between the tube centers and diameter of the tubes
Spacing between tubes of different rows (smallest slot between tubes)
Fluid flow minimum area in the evaporator
Fluid flow minimum area in the condenser
Reynolds number of the fluid in the evaporator
Reynolds number of the fluid in the condenser
Maximum velocity of fluid in the evaporator
Maximum velocity of fluid in the condenser
External coefficient of convection in the evaporator
External coefficient of convection in the condenser
Heat exchanged in each thermosyphon
Evaporator external convection resistance
Condenser external convection resistance
Evaporator internal convection resistance
Condenser internal convection resistance
Overall thermal resistance of the thermosyphon
Average working fluid temperature, equivalent to adiabatic section temperature
Average heat transferred by the thermosyphons
Number of tubes for the heat exchanger
Number of rows, which depends on the selected arrangement and the number of tubes
Total length of the heat exchanger, which depends on the number of rows
Pressure drop in the evaporator
Pressure drop in the condenser
Corrected number of rows, to complete the array
Number of tubes of the first and odd rows
Number of tubes of the second and even rows
Hot flow exit temperature, in the evaporator region
Cold flow exit temperature, in the condenser region

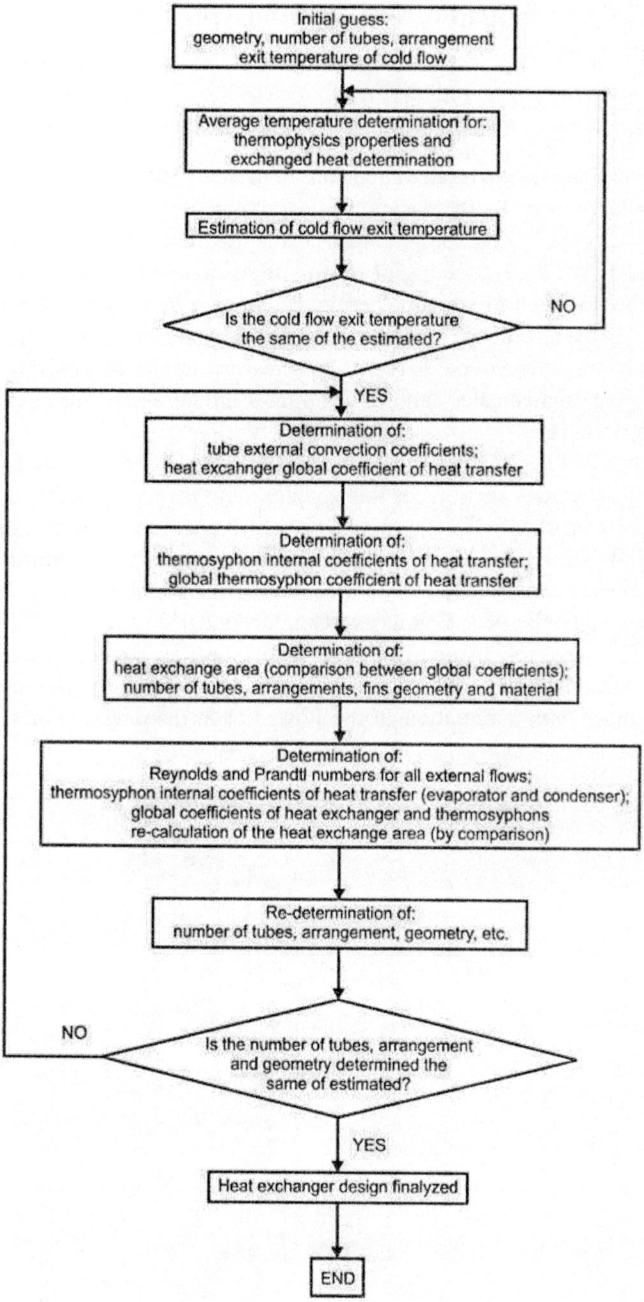

Fig. 9.4 Flowchart for the design of thermosyphon heat exchanger

9.8 Special Configurations of Thermosyphon Heat Exchangers

Use of Baffles

As already mentioned, the overall dimensions of thermosyphons heat exchangers can be considerable reduced by increasing the heat transfer area of each thermosyphon, by using fins. Another well explored means is the use of baffles in both hot and cold streams. Baffles are flow-directing or obstructing vanes or panels. In this case the baffle must be designed to support thermosyphon tube bundle and direct the fluid flow, improving the turbulence of the flow and the heat transfer capacity of the device. On the other hand, the use of baffles raises the pressure drops along the device, which means that more sophisticated fans and/or pumps are necessary for circulating the hot and cold streams.

The design of baffled thermosyphon heat exchangers follows the same procedure describe in the previous sections of this chapter, with the exception of the thermal resistances associated with the external flows, which coefficients of heat transfer are obtained by the same available correlations, subjected to a correction to take into account the presence of baffles. The correction factors are obtained by the method of Bell-Delaware (Bell 1963; Gaddis and Gnielinski 2013).

Figure 9.5 shows a schematic of a thermosyphon heat exchanger where one baffle is applied to the condenser bundle and finned tubes to the evaporator. However, baffles may cause undesirable circulation of the flows inside the device. For verification

Fig. 9.5 Thermosyphon heat exchanger with one baffle on the condenser and finned evaporator

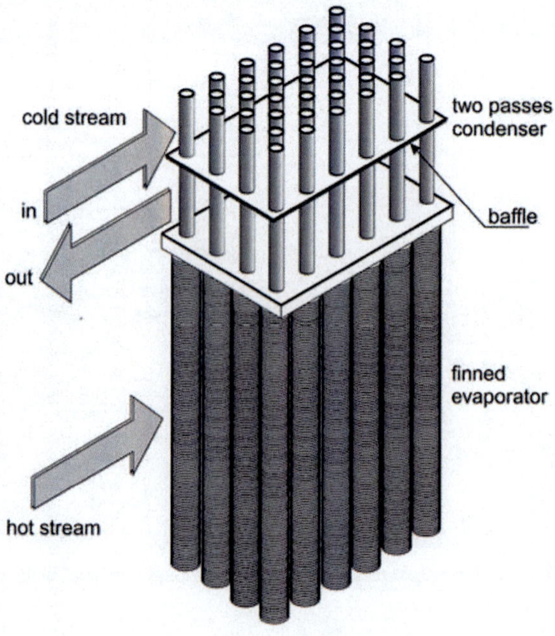

of this possibility, the use of any of the commercial fluid-dynamic software for numerical simulation of the flow around the tubes and inside the equipment shell, is highly recommended.

Modular Heat Exchangers

Besides, the facility of maintenance of thermosyphon heat exchangers can be further explored with the appropriate design of the equipment. While the maintenance of traditional heat exchangers such as shell and tube requires the stop of the industrial process, the maintenance of thermosyphon heat exchangers constructed in modules can be much easier. In this configuration, the thermosyphons are divided into structurally independent compartments, as shown in Fig. 9.6, which, for reference, represents an equipment of around 6 m of height and 10 m of length. Considering that, at least, one module of the heat exchanger is available, when one or more thermosyphons of a specific module are damaged (usually several thermosyphons need to damaged, for the drop of the performance to be noted), only the damaged module is removed and replaced with the spare one. If a replacement module is not available,

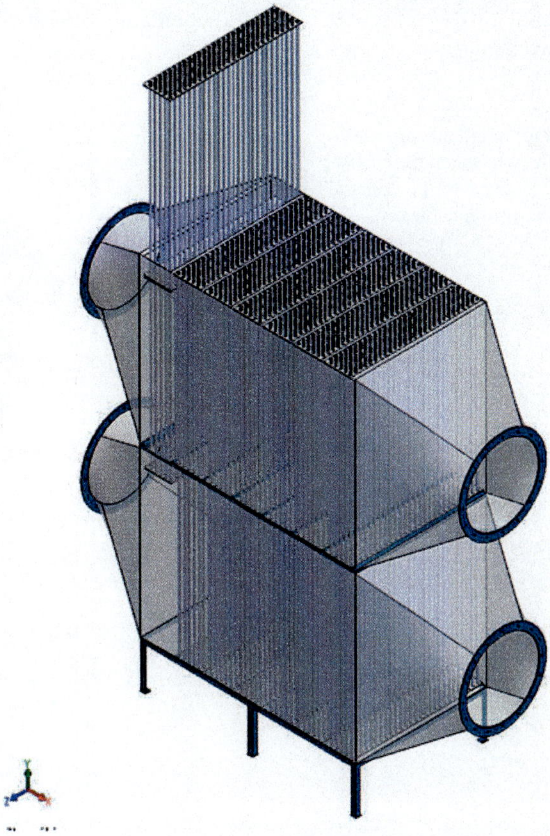

Fig. 9.6 Thermosyphon heat exchanger in modules, typical for use in waste heat recovery in industries

the module can just be removed and a plate located in the separation plane between hot and cold streams to avoid the flow mixing. After that, the device is ready to operate, obviously with lower heat transfer capacity.

Another interesting feature of this configuration is that the overall thermal conductance of the heat exchanger may be adjustable. This means that, if the operation conditions are different from those for which the equipment was designed, one or more modules can be removed from (or added to) the device, adjusting the equipment to the new operation conditions.

Shell and Shell Heat Exchanger

In the petroleum and gas industry, for instance, there are several applications where at least one of the fluid streams is a pressurized liquid, like water and oil. Shell and tube or shell and plate heat exchangers are traditionally used because of the cylindrical geometry of the shell, which can withstand high fluid pressures. For the shell and tube case, one fluid flows inside the tubes and the other one flows across the space between the tube bank and the cylindrical shell. In the present shell and shell thermosyphon heat exchanger, both streams flow within cylindrical casings, able to hold large static pressures. This technology was recently proposed by Sarmiento et al. (2018). Figure 9.7 shows a schematic of the shell and shell thermosyphon heat exchanger.

Fig. 9.7 Shell and shell thermosyphon heat exchanger for pressurized fluid, with baffles

One of the main advantages of shell and shell thermosyphon heat exchanger is that it can be considered a free from contamination technology. At least two leakage points on the same pipe, one inside each shell, would be necessary for the contamination between the two streams, a very unlikely condition. In conventional heat exchangers, a leakage at a plate or at one of the pipes, would result in the mixing of both streams, causing contamination. Furthermore, the design of shell and shell thermosyphon heat exchangers allows for easy replacement of a damaged pipe in the tube bank.

A shell and shell thermosyphon heat exchanger basically consists of two cylindrical shells, separated through a partition plate, to which the thermosyphons are attached.

This geometry facilitates the mounting and replacement of a thermosyphon in case of maintenance of the equipment. As for more conventional thermosyphon heat exchangers, the shell and shell type may work under several inclinations, from vertical to up to 7° with the horizon. The hot stream flows inside the lower shell while the cold stream flows inside the upper shell. Baffles may be employed do organize the flow across the tube bank and increase the turbulence, as for in regular shell and tube heat exchangers.

The design methodology of shell and shell thermosyphon heat exchangers is similar to that described in the previous sections of this chapter. The thermosyphon overall thermal resistance is obtained through the overall thermal resistance circuit model, as described in Chap. 4. The resistances associated with the external single-phase convection are the dominant ones, once the heat transfer coefficients for the single-phase fluid around a bunch of cylindrical tubes, are much smaller when compared with those associated with the internal heat transfer mechanisms (Shabgard et al. 2015).

Both shells can be considered as conventional shell and tube heat exchangers. The hot and cold flow heat transfer coefficients across the evaporator and condenser tube arrays can be estimated using the usual correlations (see Sects. 9.3 to 9.5 and Sect. 9.7), applying the necessary corrections, according to the Bell-Delaware method (Bell 1963; Gaddis and Gnielinski 2013).

9.9 Conclusions

The design of thermosyphon heat exchangers involve the use of several conventional methodologies that need to be combined and organized. The heat transfer areas and geometry of the thermosyphon external flows are determined through the use of shell and tube design methodology (ε-NUT), associated with the use of convection correlations for the hot and cold flows. Once known the heat power that each thermosyphon must transfer and the operation temperature, the thermosyphons can be designed, with the selection of the suitable working fluid and casing material. In the sequence, the tube arrangement must be chosen. The tube material and geometry (finned or smooth tubes) must also be selected carefully, also considering the increase in the heat transfer performance, in the pressure drops and the associated costs.

References

Bell, K., Final Report of the Cooperative Research Program on Shell and Tube Heat Exchanger, University of Delaware, USA (1963)

Borges, T. P.F., Mantelli, M.B.H., Persson, L. G., Techno-Economic Optimization of Thermossyphon Heat Exchangers Design Using Mathematical Programming, 14th International Heat Pipe Conference, Florianópolis, 63–68 (2007)

Gaddis, E., Gnielinski, V., Shell-Side Heat Transfer in Baffled Shell-and-Tube Heat Exchangers, VDI-Gesellschaft Verfahrenstechnik und Chemieingenieurwesen (eds): VDI Heat Atlas, 1st ed., Germany (2013)

Gnielinski, V., Heat Transfer in Cross-Flow Around Single Rows of Tubes and Through Tube Bundles, in: VDI Heat Atlas (eds), VDI-Gesellschaft Verfahrenstechnik und Chemieingenieurwesen, 1st ed., Germany, 725–729 (2013)

Hewitt, G.F. (ed.): in: Heat Exchange Design Handbook. Begell House Inc, New York (2008)

Hewitt, G. F., Shires, G. L. and Bott, T. B., Process Heat Transfer, Begell House Inc. (1994)

Isoppo, D. F., Borges, T. P.F., Mantelli, M. B. H., Development of a Detailed Thermal Model for Designing Heat Pipe Heat Exchangers, 22nd International Conference on Efficiency, Cost, Optimization and Environmental Impact of Energy Systems, 2009, Foz do Iguaçu, 243–252 (2009)

Kays, W.M., London, A.L.: Compact Heat Exchangers, 3rd edn. Krieger Publishing Company, USA (1984)

Ramos, J., Chong, A., Jouhara, H.: Experimental and numerical investigation of a cross flow air-to-water heat pipe-based heat exchanger used in the waste heat recovery. International Journal od Heat and Mass Transfer **102**, 1267–1281 (2016)

Roetzel, W. and Spang, B., Thermal Design of Heat Exchangers, in: VDI Heat Atlas, (eds) VDI-Gesellschaft Verfahrenstechnik und Chemieingenieurwesen, 1st ed., Germany, 33–36 (2013)

Sarmiento, A. P. C., Cisterna, L. H. R., Milanez, F. H., Mantelli, M. B. H., A Numerical Method for Shell and Thermosyphon Heat Exchanger Analysis, X Minsk International Seminar Heat Pipes, Heat Pumps, Refrigerators, Power Sources" (2018)

Shabgard, H., Allen, M.J., Sharifi, N., Benn, S.P., Faghri, A., Bergman, T.L.: Heat Pipe Heat Exchangers and Heat Sinks: Opportunities, Challenges, Applications, Analysis, and State of the Art. Int. Journal of Heat and Mass Transfer **89**, 138–158 (2015)

Vasiliev, L.L.: Heat Pipes in Modern Heat Exchangers. Appl. Therm. Eng. *25*, 1–19 (2005)

Zhang, H., Zhuang, J.: Research, development and industrial applications of heat pipe technology in China. Appl. Therm. Eng. **23**, 1067–1083 (2003)

Žukauskas, A.: Heat Transfer from Tubes in Cross Flow. In: Hartnett, P., Irvine, T.F. (eds.) Advances in Heat Transfer, **8**. Academic Press, New York (1972)

Žukauskas, A., Ulinskas, R.: Efficiency Parameters for Heat Transfer in Tube Banks. Heat Transfer Eng. **6**, 19–25 (1985)

Žukauskas, A. and Ulinskas, R., Banks of plain and finned tubes, in: Hewitt. G.F., Heat Exchange Design Handbook, Part 2, Begell House Inc., New Hork (2008)

Žukauskas, A., Skrinska, A., Žiugžda, J. and Gnielinski, V., Banks of plain and finned tubes, in: Hewitt, G.F. (ed.), Heat Exchanger Design Handbook, 1st ed., New York, pp 3.5.3.1–2.5.3.30 (2008)

Chapter 10
Electronics Cooling

In this chapter, thermosyphon and heat pipe technologies for electronics cooling are discussed. Special attention is given for cooling of avionics.

Along the last decades, microelectronic, telecommunication and power electronic industries have invested financial resources to improve the reliability of equipment, basically by decreasing the operation temperature of the electronic components. At the same time, the size of these equipment reduced, while the power dissipation increased. This combination of parameters resulted in the overheating of electronic components.

A rule of tomb proposed by Cengel and Ghajar (2015) is that, for each 10 °C of reduction in the temperature of electronic component operation, the failure rate of the equipment reduces by half. As stated by Blet et al. (2017), conventional technologies, such as air forced convection, are not able to face the growing heat fluxes in the electronics industry. Consequently, sophisticated thermal control devices, including the use of special fans, coolers, thermoelectricity and two-phase devices (heat pipes and thermosyphons) have been applied. The heat dissipation problem is even more challenging when the electronic components are installed inside hermetic cabinets. In this case, the thermal control is usually performed by conduction, i.e., the heat must travel by conduction through the boards to reach the cabinet wall and then be removed by convection to the surround air. Figure 10.1 shows the schematic of a cabinet containing electronic equipment. This is obviously an inefficient heat transfer means, as among other reasons, a typical board is made of fiber glass, a heat insulating material.

Nowadays, especially for the temperature control of electronic equipment inside cabinets, the interest in using thermosyphon and heat pipe technologies has grown, due to the fact that these devices are cheap, efficient and passive.

M. B. H. Mantelli et al., *Thermosyphons and Heat Pipes: Theory and Applications*,
https://doi.org/10.1007/978-3-030-62773-7_10

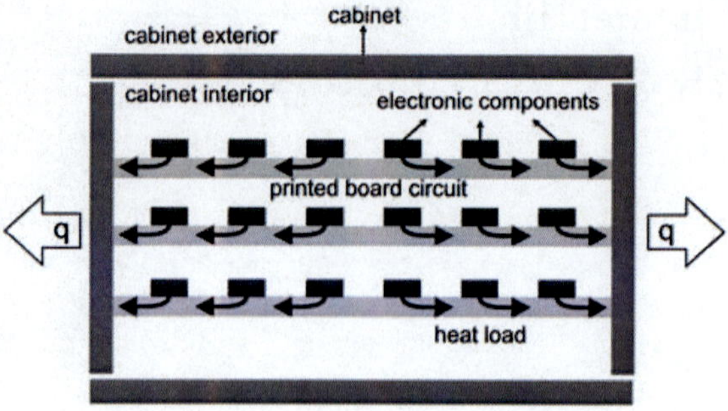

Fig. 10.1 Cooling by conduction of electronic components installed in printed circuit boards inside cabinets

10.1 Electronic Cabinets

Electronic cabinets are actually casings that hold several boards, usually composed of a robust structure, able to support and protect the electronic components, typically located in printed circuit boards. Cabinets may have a great variety of sizes and forms. They are regularly made of aluminum or steel and their material and geometry selection takes into consideration mechanical constraints, such as the ability of the cabinet plates to hold forces and chocks to which the electronic systems can be subjected.

Printed circuit boards usually are made of plastic materials (fiber glass, phenolite, polyester film, fiber, etc.), in which conductor material fillets are fixed. Several electronic components, such as diodes, transistors, resistors, capacitors and, mainly, integrated circuits, are mounted over these boards, which work as rigid basis for these components. The conductor material acts as electrical current paths, substituting the wires.

The integrated circuits or monolithic integrated circuit (also referred to as a chip, or a microchip) is a set of electronic circuits on a small flat piece of semiconductor material, normally silicon. The material of the flat pack is usually ceramic, plastic or polymer. The major function of these packs is to protect the circuit from the damaging effect of the ambient, such as humidity and corrosion, so that the electronic component can be safely manipulated.

10.2 Electronics Cooling

Details of a conventional chip is shown in Fig. 10.2 as well as the heat transfer paths. The chip is glued to the copper base by a glue of high thermal conductivity, creating a path of low thermal resistance. The heat generated spreads through the copper base and then crosses the casing, made of plastic or ceramic material. Then the heat is transferred out of the integrated circuit through the metallic terminals. Figure 10.3 illustrates details of the packing of electronic equipment. The void space of the pack is filled with gas (usually air), which, as for the pack material, has low

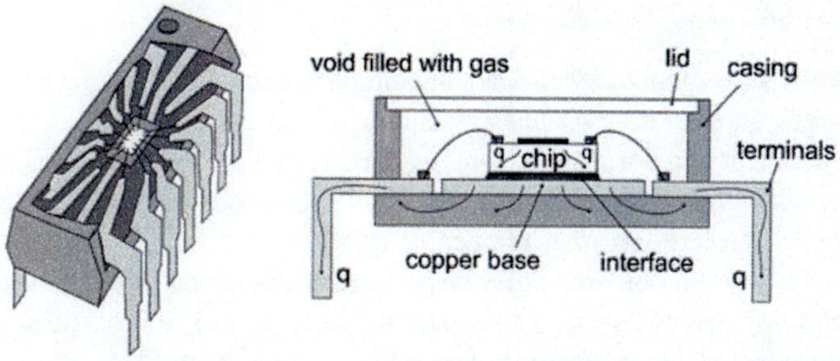

Fig. 10.2 Details of a flat pack of a chip

Fig. 10.3 Schematic of the electronic cabinet, printed circuit board and electronic circuit

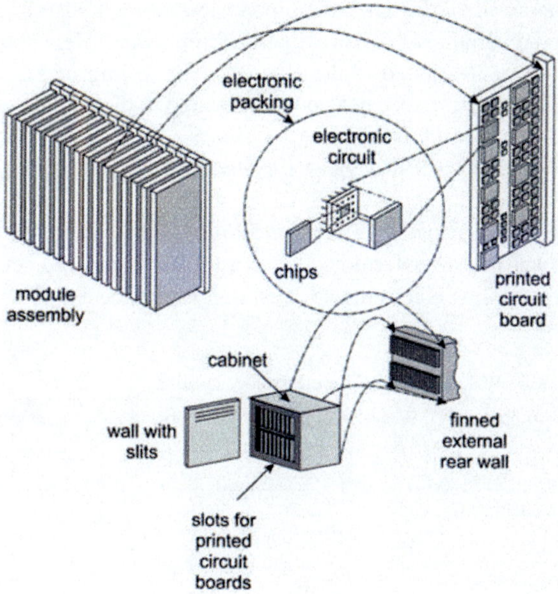

thermal conductivity. The gas represents a high thermal resistance route between the chip and the packing, so that most of the heat is transferred to the printed circuit boards by the metallic terminals. Then, the heat flows to the boards and, in the sequence, to the cabinet. In the cabinet, it reaches the external environment to where it is lost mainly by convection (in some cases, radiation may also be important). The external cabinet walls are ordinarily assisted by fins to increase the heat transfer area by convection to the environment (see Fig. 10.3). The electric/electronic equipment must have an appropriate thermal design, to keep the integrated circuits and their components under controlled temperature levels.

Heat Pipes and Loop Heat Pipes

The thermal management of electronics in laptops and notebooks represents the most common application of the two-phase technologies in the world, with conventional heat pipes with sintered porous media being usually employed. More recently, other configurations of wick structured heat pipes have been considered for use in smaller devices, such as tablets and smartphones.

Basic configurations of heat pipes were already discussed in Chap. 4 and some special technologies in Chap. 7. Table 10.1 shows a compilation of the several heat pipe technologies used for electronic devices. Heat pipes are used to collect heat from the electronic components in printed circuit boards and conduct it to the cabinet walls, to be delivered to the ambient or to a forced convection cooling system. They can be integrated to the printed circuit boards: in this case, the heat pipe evaporator is located under the chip. The heat pipe evaporator can also be installed over the chip in the void volumes between printed circuit boards. Heat pipes (conventional or pulsating) can be integrated to the cabinet walls to improve their heat transfer performance. Vapor chambers can also be integrated to the cabinet walls. Figure 10.4 shows these design configurations.

Some typical devices for electronics applications are briefly discussed in this section. It is known (see Chap. 4) that suitable combinations of capillarity (liquid pumping capacity) and permeability (small pressure drops) are required for an efficient heat transfer device. Paiva (2007) compared several grooved heat pipes used for cooling electronic devices, suggesting the use of wire-plates (Wang and Peterson

Table 10.1 Heat pipes for electronics cooling

Author	Power (W)	Fluid	Evaporator área (cm^2)	Condenser area (cm^2)	Operation temperature (°C)	Observations
Paiva (2007)	55 12 12	Water Acetone Methanol	6,0	9	70 65 55	Mini heat pipe wire plates, cooper
Nikolaenko et al. (2018)	10	R600a	37.7	467.2	19–21	Aluminum 6063, grooved evaporator. Natural convection
Paiva et al (2015)	12.1 8 10	Methanol	6 9 6	9 15 9	32 45 50	Hybrid – S shaped wires, sintered, copper
Cai et al. (2017)	160	Water	Total of 30.2	30.2	110	Two evaporators
Zhou et al. (2017)	5	Water	3.8	3.8	53.7	Ultrathin, copper, easy fabrication

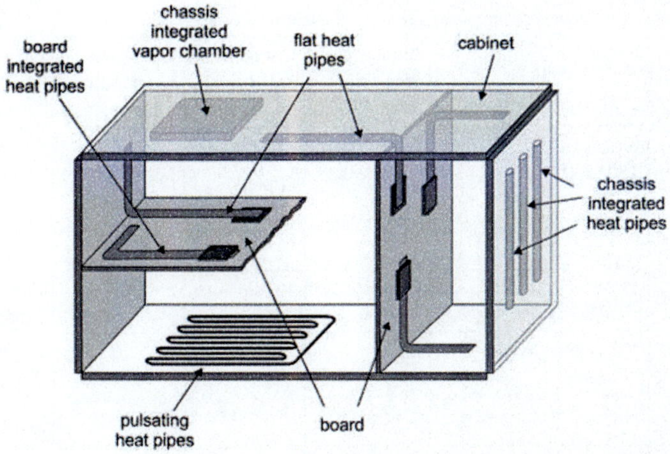

Fig. 10.4 Illustration of two-phase technology assisted electronic cabinet with printed circuit boards

2002). Later, Paiva (2011) developed a hybrid wick structure, where the heat pipe is composed by a sintered porous medium at the evaporator and grooves at both the condenser and adiabatic sections. These special heat pipe technologies are discussed in details in Chap. 7. On the other hand, Florez (2011) studied heat pipes with two layers of porous structures. The first layer was made of copper powder, with average powder size of approximately 50 μm, resulting in structure with high permeability, of $2.89 \cdot 10^{-12} m^2$, that facilitates the working fluid flow to the evaporation zone (low liquid pressure drops). The second layer has higher pumping capacity as it is made of copper powder of 20 μm, resulting in a permeability of $3.71 \cdot 10^{-13} m^2$. This wick arrangement with two layers increased the heat transfer capacity of the device in about 20%, when compared to a unique layer porous media.

As mentioned, the new generation of electronic gadgets, such as tablets and smartphones, get thinner, lighter and more powerful. Therefore, the heat pipes used for their thermal control need to improve their efficiency, while decreasing the size and volume. Zhou et al. (2017) proposed a very thin heat pipe that combines pumping capacity with low permeability. They used a special configuration, in which the wick is installed only at a limited width of the heat pipe, allowing space for vapor flow (see Fig. 10.5).

Loop heat pipes (LHPs) are also very common for cooling of electronic devices. Table 10.2 shows a list of some recent developments of LHPs for electronic cooling.

Fig. 10.5 Cross section of the ultra-thin heat pipe, according to Zhou et al. (2017)

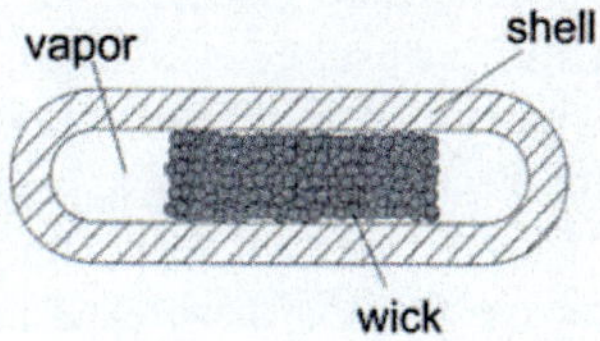

Table. 10.2 Recent developments of LHPs for electronics cooling

Author	Power (W)	Fluid	Evaporator área (cm²)	Condenser area (cm²)	Operation temperature (°C)	Observations
Gabsi et al. (2018)	0–150 150–500	Water	6,0	110.5	50–60 60–80	Operate in variable and fixed conductance regime
Song et al. (2018)	160	R245fa	54.8	467.2	55	Flat evaporator
Shioga and Mizuno (2015)	5	Water	2.3	12.3	50.5	Copper, very thin (0.6 mm) for smartphones
Zhou et al. (2016)	100	Ethanol	63.9	65.4	90	Thermal control of CPUs. Evaporator with three layers of copper or nickel, condenser made of aluminum
Zilio et al. (2018)	60	Water	9.0	15.9	100	Conical coupling with a cooling system
Ramasamy et al. (2018)	225	Ammonia	25.4	36.4	70	Stainless steel, use of fins in the condenser
Maydanik et al. (2018)	400 320	Ammonia	12.6	97.4	86.5 91	Cylindrical and flat evaporator
Pastukhov and Maydanik (2018)	120	Ammonia	56.5	42.9	55	Several heat sources Stainless steel

As already discussed in Chap. 7 (see Figs. 7.10 and 7.11) the evaporators of LHPs are usually cylindrical. This shape may not be convenient for the attachment with electronics heat sources. LHPs with flat thin evaporators have been developed for electronics cooling applications (see Chap. 7). Table 10.3 shows a compilations of these works. It worth noting that new loop technologies (LHP and loop thermosyphon) that have external geometries similar to the flat heat pipes, has been

Table. 10.3 LHPs with flat evaporator

Author	Power (W)	Fluid	Evaporator área (cm^2)	Material	Operation temperature (°C)	Evaporator shape
Liu et al. (2011)	160	Methanol	10.7	Copper	55	Flat disk
Wang et al. (2014)	230	Methanol	12.6	Copper	65	Flat disk
Song et al. (2016a; b)	240	Methanol	14.7	Brass	79	Flat disk
Song, H. et al. (2016a; b)	110	Ammonia	10.2	Copper	35	Flat disk
Maydanik e Pastukhov (1999)	30	Ammonia	8.6	Stainless Steel	–	Flat disk
Maydanik e Pastukhov (1999)	400	Water	19.6	Stainless Steel	–	Flat disk
Maydanik et al. (2017)	300	Ammonia	12.6	Stainless Steel	35	Flat disk
Liu et al. (2011)	60	Acetone	12.0	Copper	47.5	Flat Rectangular
Singh et al. (2007)	50	Water	15.8	Copper	84	Flat Rectangular
Joung et al. (2008)	78	Methanol	41.1	Copper	114	Flat Rectangular
Li et al. (2011)	628	Water	9.0	Copper	95	Flat Rectangular
Li et al. (2013)	300	Water	9.0	Copper	110	Flat Rectangular
Becker et al. (2011)	100	Water	0.5	Copper	40	Flat Oval

considered for thermal management of electronics. These devices are presented in Sect. 7.6.

Thermosyphons

Thermosyphons can usually transfer more heat than heat pipes, being less expensive to produce. Due to these characteristics, their use for the thermal control of electronic/electric devices is increasing. However, as gravity is the driving force in the transfer of liquid from condenser to evaporator, the use of thermosyphon and correlated technologies is restricted to the cases where the evaporator can be located below the condenser. In this sense, Nikolaenko et al. (2018) made a comparison of the thermal performances of a heat pipe and a thermosyphon with similar external geometries, with the same aluminum case and isobutene as working fluid, for thermal control of large electric equipment in field applications, concluding that the thermosyphon was able to transfer 25% more heat.

The use of energy in data centers is increasing in a very fast speed in the present century. For instance, in 2010 the energy consumption was three times larger than in 2000 and represented 0.3% of the world total use of electricity (Delforge and Whitney 2015). The need for cooling of electronic/electric components in the data center increase at the same pace. Thermosyphon cooling technologies, which involves phase change of a working fluid, are passive and exchange heat by natural convection with environment, are very attractive for this application (Pawlish and Varde 2010 and Zhang et al. 2003, 2015a, b, 2016, 2017a, b, 2018).

Zhang et al. (2017) affirm that the majority of the thermosyphons used for cooling data centers still use as working fluids HCFCs and HFCs, forbidden in the Montreal protocol. Ecological fluids for thermosyphons are being tested around the world, including: acetone, isobutene, CO_2 (Tong et al. 2015 and 2016), R410a (Okazaki 2005 and Ding et al. 2017), n-pentane (Chehade 2015), among others. Most of these fluids are inflammable and so not adequate for data centers. From these, CO_2 has shown to be promising, with a heat transfer capacity larger than classical fluids (R22, see Tong et al. 2015) and R410 (Okazaki 2005).

Loop thermosyphons have also being considered for cooling of electronics. Lamaison et al. (2017), modeled and validated experimentally a cooling system for CPU (Central Processing Unit), involving mini loop thermosyphons with two evaporators in different positions and heights. The illustrations shown in Fig. 10.6 is based on the device designed by these researchers to remove a total of 7 W/cm^2 of heat from two different locations at different heights in the electronic equipment. Actually, this heat transfer rate is not so high, as, according to Costa-Patry et al. (2011), thermosyphons are able to remove efficiently up to 100 W/cm^2.

Some researchers have used nano-fluids (working fluid with the addition of nano-metallic particles) in thermosyphons and loop thermosyphons, with the objective of

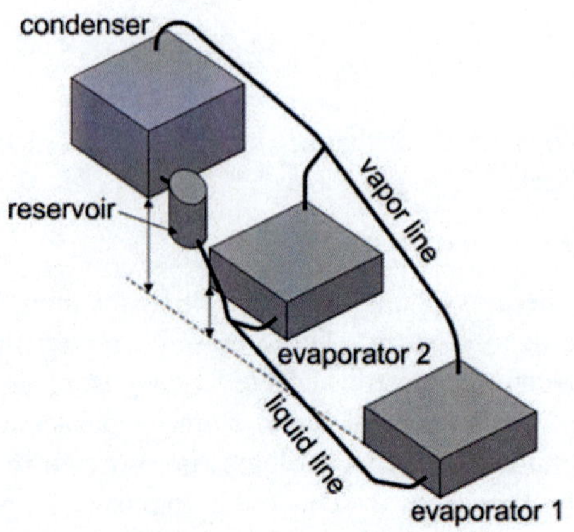

Fig. 10.6 Schematic of a mini loop thermosyphons with two condensers

Table. 10.4 Thermosyphon and loop thermosyphon for electric equipment cooling

Author	Technology	Working Fluid	Method
Samba et al. (2013)	Loop thermosyphon	n-pentane	Experimental
Ling et al. (2016)	Loop thermosyphon	R22	Experimental
Zhang et al. (2014a, 2015a, c)	Loop thermosyphon	R124a, R22	Experimental simulation
Weber and Wyatt (2011)	Thermosyphon	Not Available	Not Available
Ma et al. (2013); Zhou. et al. (2011a, 2012, 2013)	Thermosyphon	Not Available	Experimental simulation
Zhang et al. (2016a, b, 2017)	Thermosyphon	R134a	Experimental simulation
Zhu et al. (2013)	Loop thermosyphon	Not Available	Simulation
Tong et al. (2015, 2016)	Loop thermosyphon	CO_2	Experimental
Zhang, H. et al. (2017b)	Loop thermosyphon	CO_2	Experimental simulation
Okazaki (2005)	Loop thermosyphon	CO_2 R410A	Experimental simulation
Ding et al. (2017)	Loop hermosyphon	R410A	Experimental
Khodabandeh (2004)	Loop thermosyphon	Isobutene	Experimental
Chehade et al. (2015)	Loop thermosyphon	n-pentane	Experimental simulation

increasing the heat transfer capacity of the devices, as shown in the review presented by Ramezanizadeh et al. (2018).

Table 10.4 shows recent literature works concerning the use of thermosyphons for thermal control of data centers.

10.3 Vapor Chambers

The use of vapor chambers for the thermal control of electronic components is under investigation by several researchers.

Naphon et al. (2012a) presented a study of the results of the thermal control of a hard disk of a computer using vapor chamber. They concluded that the vapor chamber was able to reduce the temperature of the hard disk in about 15%. They showed that the amount of liquid (filling ratio) is a very important parameter. These same authors, in another work (Naphon et al. 2012b) investigated experimentally, with great success, the cooling of a computer processing unit (CPU) using a vapor chamber. Water was used as the working fluid.

Chang et al. (2013) constructed and tested a copper vapor chamber prototype to cool electronic components with copper columns (structural). The wick structure is also concentrated in columns, so that the device can operate in unfavorable conditions, i.e., the heat source in the upper position. These authors tested heat transfer rates varying from 5 to 50 W. The spreading resistance showed to be the most important one.

Bose et al. (2017) analyzed the thermal performance of a cylindrical vapor chamber for cooling electronic components. The power input analyzed varied from 15 to 100 W. They concluded that the maximum temperature of the electronic device decreased around 26% with the use of the vapor chamber.

Mizuta et al. (2016) developed a flat vapor chamber, which they denominated of Fine Grid Heat Pipe, composed of several copper plate layers with texturized surfaces to promote the radial displacement of working fluid. They tested a device, which dimensions were 50 × 50 x 52 mm^3. For comparison purposes, a heat exchanger with the same dimensions were also constructed. An infrared camera was used to record the temperature distribution data. The device always presented a thermal resistance inferior to 0.08 K/W. These authors affirm that the proposed geometry results in vapor chambers with lower thermal resistances.

Zeng et al. (2018) studied the performance of an aluminum vapor chamber for electronics cooling where the wick structure consisted of micro-grooves. Tests were performed under several inclinations. They showed that the vertical position compromised but not prevented the device operation.

Chen et al. (2019) developed a vapor chamber for the temperature control of a module of power transistors (insulated gate bipolar transistor), which is cooled by a water heat exchanger, bolted directly to the power transistor. They simulated the temperature distribution of this device, concluding that it is possible to obtain an important reduction of the temperature of the module.

10.4 Avionics Cooling

The number of electronic components in aircrafts has increased since the introduction of digital fly-by-wire flight control systems by the National Aeronautics and Space Administration (NASA) in 1972. Fly-by-wire systems include several on-board electronic equipment designed to replace conventional pneumatic, hydraulic and mechanical powered systems, reducing the fuel consumption and increasing the aircraft autonomy. However, the electric power consumed also increased and so the need for heat dissipation. Size and weight are also important requirements for airplane devices. As the electronic gadgets get smaller and powerful, the power dissipation density increases (Sarno et al. 2009; Savino et al. 2008; Cai and Chen 2007 and Ellis et al. 2014).

One of the major aspects considered for the design of cabinets for the aeronautic industry is the easy access to the electronic boards, simplifying the reparation, in failure events. This aspect is very important because failures may cause the airplane to stay on ground, resulting in great economic losses.

Most avionics installed on aircraft cockpits must work in temperatures up to 100 °C, being 70 °C the most common restriction. Effective cooling systems must ensure that temperatures do not surpass these limits, as excessive temperatures may compromise the electronics.

Cabinet Cooling

According to Sarno et al. (2012), the cooling method involving pure conduction between the dissipating electronic element and the wall is popular within the aeronautic industry, mainly when the cabinets are hermetic. In this case, the heat generated by electronic components has to flow by pure conduction through the components, boards and finally to the external cabinet wall, where it is removed by natural or forced convection. As stated by Jones et al. (2016), under certain circumstances, even high conductivity metal plates (aluminum, copper) cannot spread efficiently the heat and hot spots are observed over the chassis walls.

Usually heat problems are solved by increasing the mass flow rate of the cooling air. This solution has drawbacks: increase of cooler size, periodic maintenance requirement, noise generation and increase of energy consumptions (drained from the airplane turbine) for the compressed air (see Del Valle et al. 2014).

Basically, there are two methods of cooling avionics (cabinets) using air flows: direct, by the impingement of the cooling flow directly in the hot components and indirect (cold walls) where the air flow removes heat from the cabinet that, in turn, received heat from the electronics. The indirect cooling method can be by forced or natural convection, as illustrate in Fig. 10.7. Contamination by dirtiness or condensation of water over electronics (with possible generation of electric dischargers), can be a problem related to the direct cooling solutions. This method is more viable when associated with refrigeration closed cycles, in which the fluid does not interact with the component to be cooled, but flows through tubes within the boards or the cabinet walls.

In avionics, the indirect cooling technique is more employed, as discussed by Sarno et al. (2013). The thermal circuit analogy can be used as design tool, however, several thermal contact resistances need to be estimated and this is not a simple task. Roughly speaking, these thermal circuits must contain the thermal resistances displayed in series: electronic component to the board (the heat is usually conducted through the terminals), board to the cabinet walls and cabinet walls to the cooling system or external convections.

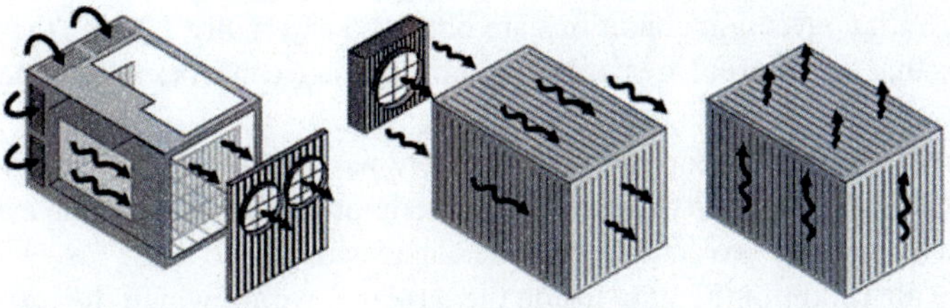

Fig. 10.7 Illustration of cabinet cooling methods: internal forced convection (left) external forced convection (center) and external natural convection

Heat Pipes and Thermosyphons

In this context, heat pipes and/or thermosyphons are good reliable alternatives for the more conventional thermal control processes (Cai et al. 2005, Sarno 2013, Lohse and Schmitz 2010).

Campo et al. (2014) proposed the use of heat pipes integrated to the printed circuit boards. They also suggested to dispose the most dissipating components close to the chassis, where heat is ultimately removed by the external walls. Figure 10.4 illustrates this concept. This authors affirm that the effective thermal conductances of these heat pipe assisted modules vary between 500 and 1200 W/m·K. Their experiments show that the maximum temperature of the hottest electronic component decreased in 28 °C.

Slippey et al. (2014) studied the viability of using carbon fiber polymer in the fabrication of avionic cabinets, as this polymer has mechanical resistance similar to the steel and 75% lower density. However, the thermal conductivity of the carbon fiber polymer is 40 times lower than that of aluminum. This negative characteristic could be mitigated by the use of copper heat pipes in the cabinet walls.

Campo et al. (2014) also proposed the use of heat pipes to improve the heat removal from avionics aluminum cabinet walls. They performed numerical simulations to evaluate this concept, concluding that, for this case (aluminum), the maximum temperature within the cabinet decreased by 24 °C.

Moreira Jr. and Mantelli (2019) proposed the use of vertical flat thermosyphons to be used as walls of cabinets for avionics. The flat loop thermosyphon (FLTS) is described in details in Chapter 7.

In many applications, more than one two-phase heat dissipation technology in series may be necessary to connect the heat source (electronic component) to the heat sink. These devices can be thermally coupled (usually the evaporator of a heat pipe or thermosyphon is connected to the condenser of another device), by means of conical connectors (plug) (Zilio et al. 2018).

Double Condenser Thermosyphon

Most of the avionics cooling systems include the use of electrical energy drained from the airplane engines, for air conditioning, pumps or fans. On the other hand, taking the example of conventional commercial flights, the airplane in cruise flies at 850 km/h at low external temperatures, typically ranging from -62 to -20 °C. Therefore, high convection conditions are observed externally, outside the airplane. This means that the external area of the aircraft fuselage can work as an efficient heat sink.

Oliveira et al. (2015) proposed a completely passive method for the cooling of avionics consisting of a loop thermosyphon. It comprises of one common evaporator, for the heat source and two condensers, one connected to the outside fuselage and the other to the air conditioning duct inside the airplane, within which the refrigeration air conditioning fluid flows by forced convection. Figure 10.8 shows this equipment inside the airplane cabin. The air conditioning is still important for low external convection conditions, such as when the airplane is parked on the ground and so

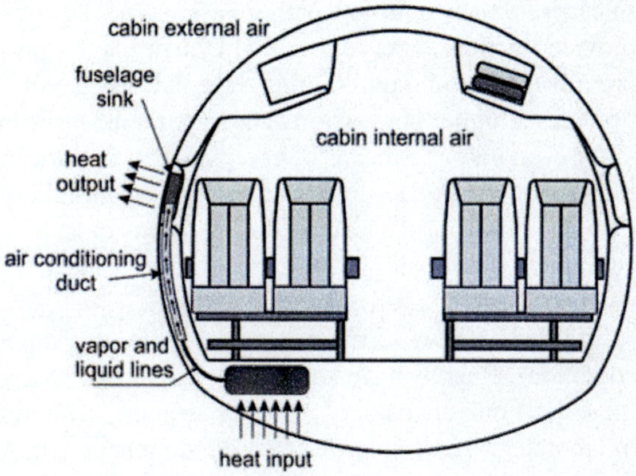

Fig. 10.8 Double-condenser thermosyphon for avionics

subjected externally to natural convection. However, the need for electronic cooling when the airplane is parked is drastically reduced, as most of the automatic controlling systems are down.

If convenient, the heat from several electric device sources (avionics) in the airplane can be directed to a common evaporator of the double-condenser loop thermosyphon. The heat transfer from these sources to the evaporator can be performed by smaller thermosyphons and/or heat pipes, in both simple or loop configurations, which are connected in series. Cylindrical or conical shaped plugs (males) are proposed as the external geometry of the smaller thermosyphons or heat pipes to be mechanically coupled to cylindrical or conical holes (females) that matches the plugs. The mechanical couplings between the double-condenser loop thermosyphon and the thermosyphon and/or heat pipe devices were designed to assure practical fittings and low thermal contact resistances.

These configurations are discussed by Tecchio et al. (2017a, b). Water was selected as the working fluid. In addition to the fact that water presents the best figure of merit number (see Sect. 4.3), water is a non-toxic and fireproof working fluid, very convenient characteristics for use inside cooling systems in aircraft applications. However, at subzero flight temperatures, the water can freeze.

Oliveira et al. (2016) developed a prototype that was successfully tested both in laboratory and in flying conditions, aboard an airplane. In this experiment, one airplane window was allowed to be used as the fuselage external heat sink. The heat source was mimicked by an electrical resistance. Later, Tecchio et al. (2017a, b) flew in an airplane another more complex experiment, involving the use of more than one two-phase technologies coupled in series, to deal with the waste heat from different avionics sources.

It worth noting that, in the system proposed by Oliveira (2015,2016) and Tecchio (2017a, b) no valves are used. It works as follows. The heat delivered to the evaporator causes the working fluid, in liquid state, to evaporate. The resulting vapor has to ways to flow: to the condenser located the exterior surface of the fuselage or to the air conditioning duct, usually located close to the internal surface of the airplane. This selection of the vapor flow path is passive and depends of the convection coefficients provided by the condensers: in flight the fuselage convection is much greater than that provided by the air conditioning duct, but, when parked, the air conditioning usually provides a higher convection. The vapor condenses in the condenser(s) and returns to the evaporator by means of gravity. Therefore, during any airplane maneuvers, the evaporator must be located in an inferior position in relation to the condenser and so the inclinations of the aircraft during flights must be considered in the design of the equipment. The analogy between thermal and electrical resistance circuits, discussed in Chap. 4 can be used as modelling tool.

Figure 10.9 shows a representation of the experimental setup tested in actual flying conditions. Results have shown that the device was able to operate under even the most drastic maneuvers of a carrier airplane, with almost the same performance as in laboratory. It also showed that the double condenser loop thermosyphon, during flight conditions, presented a smooth start up from a freezing water state, which represents a situation where the device starts to transfer heat with the airplane already in flight. In this case the temperatures of the heated devices did not increase much, during the working fluid thawing period. Therefore, water showed to be a convenient working fluid for this device. These experiments also showed that the vapor was able to re-direct itself passively to the air conditioning duct, as the external convection conditions decreased.

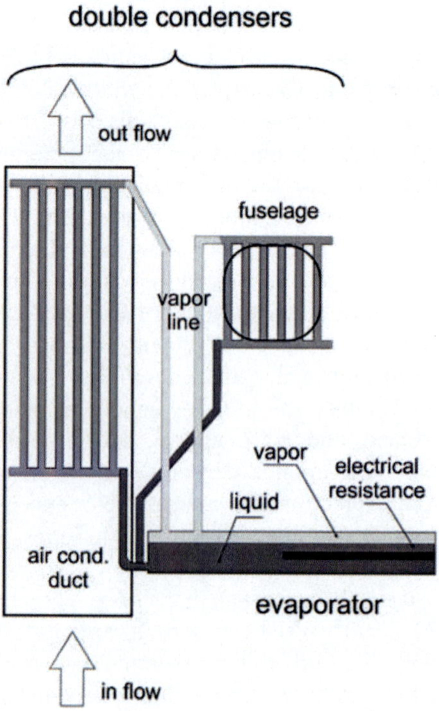

Fig. 10.9 Experimental set up tested in real flight conditions by Oliveira et al. (2015)

As stated in Simas et al. (2019), this research line was continued, with the development of an evaporator with a more convenient geometry, i.e., a flat narrow hollow plate, to be installed between boards or in cabinet walls of electronic equipment. This device is meant to transfer larger amounts of heat power in hermetic cabinets. The evaporator, which was described in Chap. 7, was constructed by diffusion bounding process.

Vapor Chambers and Pulsating Heat Pipes

Vapor chambers, presented in Chap. 7, are frequently used to monitor the temperature of electronic components in avionics.

Jones et al. (2016) studied the use of vapor chambers to optimize the efficiency of heat removal of an avionic cabinet. These authors inserted a commercial vapor chamber of $122 \times 87x3mm^3$ within one of the chassis walls to unify the temperature distribution, removing hot spots and increasing the heat transfer area. In this thermal solution, the electronic components transfer heat to the boards and ultimately to the chassis, which is turn, is cooled by forced convection in a heat exchanger coupled to the chassis. A wick structure assured the transference of the working fluid to the cooled wall, even against gravity. In their study, the authors also compared the performance of the device with a solid aluminum plate installed in the chassis, showing that the thermal resistance of the chassis decreased by 40 times with the presence of the vapor chamber. Figure 10.10 shows a schematic of the vapor chamber integrated to a chassis of a cabinet of avionics.

Reyes et al. (2011) studied the thermal behavior of a vapor chamber of $190 \times 140x15\ mm^3$ for avionics thermal control. The vapor chamber was installed in vertical orientation inside a cabinet wall, to dissipate the heat of a specific PCI (peripheral component interconnect). A conventional heat sink of the same external dimensions of the vapor chamber was also tested for comparison purposes. The vapor chamber

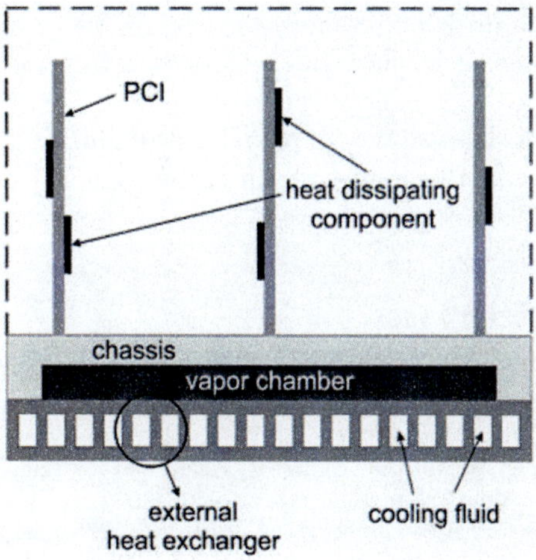

Fig. 10.10 Vapor chamber inserted in the cabinet walls with external heat exchanger

showed to be 30% more efficient than the metallic heat sink, when natural convection was applied, allowing the increase of the heat dissipation by 25 W.

Sarno et al. (2013) tested a passive cooling system for the electronic components of entertainment devices aboard airplanes (audio, video, internet, etc.) in commercial airplanes. These electronic devices are installed inside small boxes, usually positioned bellow the passenger seats. The cooling system is composed by a heat pipe, which connected thermally the electronic components to the box walls and by a loop heat pipe that connected the box walls to the seat structure, made of aluminum, where finally the heat is dissipated by convection to the airplane environment. This passive heat transportation system allowed the duplication of the power dissipation of electronic devices.

Pulsating heat pipes (PHPs) have also been applied for the thermal control of electric equipment cabinet in avionics. Cai et al. (2007) inserted PHPs in the cabinet walls of airplane electronic equipment, creating a low thermal resistance paths between printed circuit boards and the external environment. The PHPs were able to reduce significantly the maximum temperature difference found in the cabinet walls, reducing them by half for some power inputs, in comparison with those found with traditional aluminum walls.

10.5 Closure

In the present chapter, thermosyphon and heat pipe technologies, most of them described in Chaps. 6 and 7, are suggested as thermal control solutions of electronic equipment. A large range of applications, from smartphones, tables and computers to avionics, are discussed. The necessary adjustments of the devices, for making the solution feasible, are shown. The models needed for the designing of such two-phase devices are shown along the book, mainly in Chaps. 2, 3, 4, 6 and 7, therefore no modelling is presented here. In the case of avionics applications, two-phase heat transfer technologies are proposed, from the level of the electronic components inside boxes (mini devices) to the whole refrigeration system (large equipment), opening a great possibility of partial substitution of the air conditioning system, which would certainly improve the energetic efficiency of airplanes.

References

Becker, S., Vershinin, S., Sartre, V., Laurien, E., Bonjour, J., Maydanik, Y.F.: Steady State Operation of a Copper-Water LHP With a Flat-Oval Evaporator. Appl. Therm. Eng. **31**(5), 686–695 (2011)

Blet, N., Lips, S., Sartre, V.: Heat Pipes for Temperature Homogenization: A Literature Review. Appl. Therm. Eng. **118**, 490–509 (2017)

Bose, J.R., Ahammed, N., Asirvatham, L.: Thermal Performance of a Vapor Chamber For Electronic Cooling Applications. J. Mech. Sci. Technol. **31**(4), 1995–2003 (2017)

Cai, Q., Chen, C.: Experimental Investigation of an Avionics Cooling System for Aerospace Vehicle. Journal of Spacecraft and Rockets **44**(2), 439–444 (2007)

Cai, Q., Chen, C., Asfia, J. F., Applications of Advanced Cooling Technology in Avionics System for Aerospace Vehicle, 43rd AIAA Aerospace Science Meeting and Exhibit, Reno (2005)

Campo, D., Weyant, J. and Bryan, M., Enhancing Thermal Performance in Embedded Computing for Ruggedized Military and Avionics, 14th The Intersociety Conference on Thermal and Thermomechanical Phenomena in Electronic Systems · ITHERM (2014)

Cengel, Y.A., Ghajar, A.J.: Heat and Mass Transfer, Fundamentals and Application, 5th ed. McGraw Hill (2015)

Chehade, A., Louahlia-Gualous, H., Masson, S., Lepinasse, E.: Experimental Investigations and Modeling of a Loop Thermosyphon For Cooling With Zero Electrical Consumption. Appl. Therm. Eng. **87**, 559–573 (2015)

Chen, Y., Li, B., Wang, X., Yan, Y., Wang, Y., Qi, F.: Investigation of Heat Transfer and Thermal Stresses of Novel Thermal Management System Integrated with Vapour Chamber for IGBT Power Module. Thermal Science and Engineering Progress **10**, 73–81 (2019)

Costa-Patry, E., Olivier, J., Michel, B., Thome, J.R.: Two-Phase Flow of Refrigerants in 85µm Wide Multimicrochannels: Part II: Heat Transfer With 35 Local Heaters. Int. J. Heat Fluid Flow **32**(2), 464–476 (2011)

Del Valle, P., Munoz, B. P., Advantages of the Dynamic Simulation for the Thermal Management Systems Design. SAE 2014 Aerospace System Technology Conference, SAE International (2014)

Ding, T., Cao, H., He, Z.G., Li, Z.: Experiment Research on Influence Factors of The Separated Heat Pipe System. Especially the Filling Ratio and Freon Types, Applied Thermal Engineering **118**, 357–364 (2017)

Ellis, M., Anderson, W., and Montgomery, J., Passive Thermal Management for Avionics in High Temperature Environments, SAE Technical Paper 2014–01–2190 (2014)

Gabsi, I., Maalej, S., Zaghdoudi, M.C.: Thermal Performance Modeling of Loop Heat Pipes With Flat Evaporator for Electronics Cooling. Microelectron. Reliab. **84**, 37–47 (2018)

Jones, A., Chen, R., Angus, M.: Assessment of Vapour Chamber Heat Spreader Implementation for Avionic Module Thermal Management. Int. Journal of Thermal Technologies **6**(2), 103–111 (2016)

Joung, W., Yu, T., Lee, J.: Experimental Study on The Loop Heat Pipe With a Planar Bifacial Wick Structure. Int. Journal of Heat Mass Transfer **51**, 1573–1581 (2008)

Khodabandeh, R.: Thermal Performance of a Closed Advanced Two-Phase Thermosyphon Loop For Cooling of Radio Base Stations at Different Operating Conditions. Appl. Therm. Eng. **24**, 2643–2655 (2004)

Lamaison, N., Ong, C.L., Marcinichen, J.B., Thome, J.R.: Two-Phase Mini-Thermosyphon Electronics Cooling: Dynamic Modeling, Experimental Validation and Application To 2U Servers. Appl. Therm. Eng. **110**, 481–494 (2017)

Li, J., Lin, F., Wang, D., Tian, W.: A Loop-Heat-Pipe Heat Sink With Parallel Condensers for High-Power Integrated LED Chips. Appl. Therm. Eng. **56**, 18–26 (2013)

Li, J., Wang, D., Peterson, G.P.: A Compact Loop Heat Pipe With Flat Square Evaporator for High Power Chip Cooling. IEEE Trans. Component Packaging Manufacturing Technology **1**(4), 519–527 (2011)

Ling, L., Zhang, Q., Yu, Y., Liao, S., Sha, Z.: Experimental Study On The Thermal Characteristics of Micro Channel Separate Heat Pipe Respect to Different Filling Ratio. Appl. Therm. Eng. **102**, 375–382 (2016)

Liu, Z., Gai, D., Li, H., Liu, W., Yang, J., Liu, M.: Investigation of Impact of Different Working Fluids on the Operational Characteristics of Miniature LHP With Flat Evaporator. Appl. Therm. Eng. **31**(16), 3387–3392 (2011)

Maydanik, Y.F., Vershinin, S.V., Chernysheva, M.A.: Experimental Study of an Ammonia Loop Heat Pipe With a Flat Disk-Shaped Evaporator Using a Bimetal Wall. Appl. Therm. Eng. **126**, 643–652 (2017)

Maydanik, Y.F., Vershinin, S.V., Chernysheva, M.A.: The Results of Comparative Analysis and Tests of Ammonia Loop Heat Pipes With Cylindrical and Flat Evaporators. Appl. Therm. Eng. **144**, 479–487 (2018)

Maydanik Y. F. and Pastukhov V.G., Loop Heat Pipes – Recent Developments, Test Results And Applications, 34th Intersociety Energy Conversion Engineering Conference Vancouver, British Columbia (1999)

Mizuta, K., Fukunaga, R., Fukuda, K., Nii, S., Asano, T.: Development and Characterization of a Flat Laminate Vapor Chamber. Appl. Therm. Eng. **104**, 461–471 (2016)

Moreira, A.A., Jr., Mantelli, M.B.H.: Thermal Performance of a Novel Flat Thermosyphon For Avionics Thermal Management. Energy Convers. Manage. **202**, 112219 (2019)

Naphon, P., Wongwises, S., Wiriyasart, S.: Application of Two-Phase Vapor Chamber Technique for Hard Disk Drive Cooling of PCs. Int. Commun. Heat Mass Transfer **40**, 32–35 (2012a)

Naphon, P., Wongwises, S., Wiriyasart, S.: On The Thermal Cooling of Central Processing Unit of the PCs With Vapor Chamber. Int. Commun. Heat Mass Transfer **39**(8), 1165–1168 (2012b)

Nikolaenko, Y.E., Alekseik, Y.S., Kozak, D.V., Nikolaienko, T.Y.: Research on Two-Phase Heat Removal Devices For Power Electronics. Thermal Science and Engineering Progress **8**, 418–425 (2018)

Okazaki, T.: Performance Analysis of a Cooling System With Natural-Circulation Loop Using CO_2. Trans JSRAE **22**(1), 73–81 (2005)

Oliveira, J.L.G., Tecchio, C., Paiva, K.V., Mantelli, M.B.H., Gandolfi, R., Ribeiro, L.G.S.: Passive Aircraft Cooling Systems for Variable Thermal Conditions. Appl. Therm. Eng. **79**, 88–97 (2015)

Oliveira, J.L.G., Tecchio, C., Paiva, K.V., Mantelli, M.B.H., Gandolfi, R., Ribeiro, L.G.S.: In-flight Testing of Loop Thermosyphons for Aircraft Cooling. Appl. Therm. Eng. **98**, 144–156 (2016)

Paiva, K.V.: Comportamento Térmico em Gravidade e Microgravidade de Mini Tubos de Calor do Tipo Fios-Placas, Master's Dissertation: Mechanical Engineering Department. Federal University of Santa Catarina, Florianópolis (2007)

Paiva, K.V., Mantelli, M.B.H., Slongo, L.K.: Experimental Testing Of Mini Heat Pipes Under Microgravity Conditions Aboard A Suborbital Rocket. Aerosp. Sci. Technol. **45**, 367–375 (2015)

Paiva, K. V., Desenvolvimento de Novas Tecnologias para Mintubos de Calor: Análise Teórica e Experimental, Doctoral Thesis 2011.Mechanical Engineering Department, Federal University of Santa Catarina, Florianópolis (2011)

Pastukhov, V.G., Maydanik, Y.F.: Development and Tests of a Loop Heat Pipe With Several Separate Heat Sources. Appl. Therm. Eng. **144**, 165–169 (2018)

Pawlish, M., Varde, A. S., Free Cooling: A Paradigm Shift in Datacenters, IEEE 5th International Conference on Information and Automation for Sustainability, Colombo, Sri Lanka, 348–354 (2010)

Ramasamy, N. S., Kumar, P., Wangaskar, B., Khandekar and S., Maydanik, Y. F., Miniature Ammonia Loop Heat Pipe for Terrestrial Applications: Experiments and Modeling, International Journal of Thermal Sciences, **124**, 263–278 (2018)

Ramezanizadeh, M., Nazari, M.A., Ahmadi, M.H., Açıkkalp, E.: Application of Nanofluids in Thermosyphons: A Review. J. Mol. Liq. **272**, 395–402 (2018)

Samba, A., Louahlia-Gualous, H., Masson, S., Nörterhäuser, D.: Two-Phase Thermosyphon Loop for Cooling Outdoor Telecommunication Equipments. Appl. Therm. Eng. **50**(1), 1351–1360 (2013)

Sarno, C., Tantolin, C., Hodot, R., Maydanik, Y., Vershinin, S.: Loop Thermosyphon Thermal Management of The Avionics Of An In-Flight Entertainment System. Appl. Therm. Eng. **51**, 764–769 (2013)

Sarno, C., Tantolin, C. and Parbaud, S., Cooling of Seat Electronics Boxes With Loop Heat Pipe, La Rochelle: IMAPS., France (2009)

Savino, R., Abe, Y., Fortezza, R.: Comparative Study of Heat Pipes With Different Working Fluids Under Normal Gravity and Microgravity Conditions. Acta Astronaut. **63**, 24–34 (2008)

Shioga, T. and Mizuno, Y., Micro Loop Heat Pipe for Mobile Electronics Applications, 31st SEMI-THERM Symposium. IEEE, 978–1–4799–8600–2/15 (2015)

Simas G. V., Mera, J.P.F. and Mantelli, M.B.H., Study of a Loop Thermosyphon Evaporator for Thermal Control of Aircrafts, Journal of Heat Transfer, 141, 091812–1–8 (2019)

Singh, R., Akbarzadeh, A., Dixon, C. and Mochizuk, Novel Design Of A Miniature Loop Heat Pipe Evaporator For Electronic Cooling, Journal of Heat Transfer, **129**, 10, 1445–1452 (2007)

Slippey, A., Ellis, M., Conway, B. and Yun, H.C., Heat Pipe Embedded Carbon Fiber Reinforced Polymer Composite Enclosures for Avionics Thermal Management. SAE 2014 Aerospace Systems and Technology Conference Technical Paper (2014)

Song, H., Liu, Z., Wang, D., Zhao, J., Liu, W., Yang, J.: Investigation Of The Flat Disk-Shaped LHP With A Shared Compensation Chamber. Appl. Therm. Eng. **104**, 139–145 (2016)

Song, H., Zhao, J., Zhi-Chun, L., Wei, T., Jin-Guo, Y., Wei, L.: Experimental Investigation of Loop Heat Pipe With a Large Squared Evaporator for Cooling Electronics. Appl. Therm. Eng. **144**, 383–391 (2018)

Song, H., Zhi-Chun, L., Jing, Z., Chi, J., Jin-Guo, Y., Wei, L.: Experimental Study of an Ammonia Loop Heat Pipe With a Flat Plate Evaporator. Int. Journal of Heat Mass Transfer **102**, 1050–1055 (2016)

Tecchio, C., Oliveira, J. L. G, Paiva, K. V., Mantelli, M. B. H., Gandolfi, R. and Ribeiro, L. G. S., Geyser Boiling Phenomenon in a Two-Phase Closed Loop-Thermosyphons, Int. J. Heat Mass Transfer, **111**, 29–40 (2017)

Tecchio, C., Oliveira, J .L. G, Paiva, K.V., Mantelli, M. B. H., Gandolfi, R. and Ribeiro, L. G. S., Thermal Performance of Thermosyphons In Series Connected By Thermal Plugs, Experimental Thermal and Fluid Science, **88**, 409–422 (2017)

Tong, Z., Ding, T., Li, Z., Liu, X.: An Experimental Investigation of an R744 Two-Phase Thermosyphon Loop Used To Cool a Data Center. Appl. Therm. Eng. **90**, 362–365 (2015)

Tong, Z., Liu, X., Li, Z., Jiang, Y.: Experimental study on the effect of fill ratio on an R744 two-phase thermosyphon loop. Appl. Therm. Eng. **99**, 302–312 (2016)

Wang, D., Liu, Z., Shen, J., Jiang, C., Chen, B., Yang, J., Tu, Z., Liu, W.: Experimental Study of The Loop Heat Pipe With a Flat Disk-Shaped Evaporator. Exp. Thermal Fluid Sci. **57**, 157–164 (2014)

Wang, Y.X., Peterson, G.P.: Analysis of Wire-Bonded Micro Heat Pipe Arrays. AIAA Journal of Thermophysics and Heat Transfer **16**(3), 346–355 (2002)

Weber, R. M., Wyatt, W. G., Heat Removal System for Computer Rooms. Patent US 7907395 A1 (2011)

Zeng, J., Zhang, S., Chen, G., Lin, L., Sun, Y., Chuai, L., Yuan, W.: Experimental Investigation on Thermal Performance of Aluminum Vapor Chamber Using Micro-Grooved Wick With Reentrant Cavity Array. Appl. Therm. Eng. **130**, 185–194 (2018)

Zhang, L.Y., Liu, Y.Y., Guo, X., Meng, X.Z., Jin, L.W., Zhang, Q.L., Hu, W.J.: Experimental Investigation and Economic analysis of Gravity Heat Pipe Exchanger Applied in Communication Base Station. Appl Energy **194**, 499–507 (2017)

Zhang, L.Y., Liu, Y.Y., Jin, L.W., Liu, X., Meng, X.Z., Zhang, Q.L.: Economic Analysis of Gravity Heat Pipe Exchanger Applied in Communication Base Station. Energy Procedia **88**, 518–525 (2016)

Zhang, H., Shao, S., Tian, C., Zhang, K.: A Review on Thermosyphon and Its Integrated System With Vapor Compression for Free Cooling of Data Centers. Renew. Sustain. Energy Rev. **81**, 789–798 (2018)

Zhang, H., Shao, S., Xu, H., Zou, H., Tang, M., Tian, C.: Numerical investigation on integrated system of mechanical refrigeration and thermosyphon for free cooling of data centers. Int. J. Refrig **60**, 9–18 (2015)

Zhang, H., Shao, S., Xu, H., Zou, H., Tian, C.: Integrated System of Mechanical Refrigeration and Thermosyphon for Free Cooling of Data Centers. Appl. Therm. Eng. **75**, 185–192 (2015)

Zhang, P., Wang, B., Shi, W., Han, L., Li, X.: Modeling and Performance Analysis of a Two-Phase Thermosyphon Loop With Partially/Fully Liquid-Filled Downcomer. Int. J. Refrig **58**, 172–185 (2015)

Zhang, H., Shao, S., Jin, T. and Tian, C., Numerical Investigation of a CO2 Loop Thermosyphon in an Integrated Air Conditioning System for Free Cooling of Data Centers, Applied Thermal Engineering 2016.12.135 (2017a)

Zhang, H., Shao, S., Xu, H., Zou, H., Tang, M., and Tian C., Simulation on the Performance and Free Cooling Potential of The Thermosyphon Mode in an Integrated System of Mechanical Refrigeration and Thermosyphon, Applied Energy, **185** (2017b)

Zhang, P., Li, X., Shang, S., Shi, W. and Wang B., Effect of Height Difference on the Performance of Two-Phase Thermosyphon Loop Used in Air-Conditioning System, International Refrigeration and Air Conditioning Conference at Purdue, Purdue University, (2014a)

Zhang, H., Shao, S., Xu, H., Zou, H., Tang, M. and Tian C., Numerical Investigation on Fin-tube Three-Fluid Heat Exchanger For Hybrid Source HVAC & R Systems, Applied Thermal Engineering **95** (2016b)

Zhou, F., Chen, J., Ma, G., Liu, Z.: Energy-Saving Analysis Of Telecommunication Base Station With Thermosyphon Heat Exchanger. Energy Build **66**, 537–544 (2013)

Zhou, W., Ling, W., Duan, L., Hui, K.S.: Development and Tests of Loop Heat Pipe With Multi-Layer Metal Foams as Wick Structure. Appl. Therm. Eng. **94**, 324–330 (2016)

Zhou, F., Tian, X., Ma, G.: Investigation Into the Energy Consumption of a Data Center With a Thermosyphon Heat Exchanger. Chin. Sci. Bull. **20**, 2185–2190 (2011)

Zhou, W., Xie, P., Li, Y., Yan, Y., Li, B.: Thermal Performance of Ultra-Thin Flattened Heat Pipes. Appl. Therm. Eng. **117**, 773–781 (2017)

Zhou, F., Chen, J., Ma, G. and Liu, Z., Energy-Saving Potential Study on Telecommunication Base Station Free Cooling With a Thermosyphon Heat Exchanger in China, International, Refrigeration and Air Conditioning Conference at Purdue, Purdue University, USA (2012)

Zhu, D., Yan, D., Li, Z.: Modelling and Applications of Annual Energy-Using Simulation Module of Separated Heat Pipe Heat Exchanger. Energy Build **57**, 26–33 (2013)

Zilio, C., Righetti, G., Mancin, S., Hodot, R., Sarno, C., Pomme, V., Truffart, B.: Active and Passive Cooling Technologies for Thermal Management of Avionics in Helicopters: Loop Heat Pipes and Mini-Vapor Cycle System. Thermal Science and Engineering Progress **5**, 107–116 (2018)

Chapter 11
Other Applications

Several different applications of thermosyphon assisted equipment are described in this chapter. First, the use of thermosyphpns to recover part of the water lost to the ambient in cooling towers is discussed. Following, the use of thermosyphons for the solar heating of house internal ambient is presented. In the sequence, the use of thermosyphons in ovens, to promote efficient consumption of energy and the uniform temperature and humidity distributions is described. Finally, the use of thermosyphons in heating of oil storage tanks and in gas pumping and distributing stations, is discussed.

11.1 Cooling Towers

Water is the main cooling fluid in the industry. Heat exchangers are used to promote heat transfer between fluxes with high efficiency when the temperature difference of fluxes is large. However, at the end of the industrial cycle, the process water is at a too low temperature level to be cooled in heat exchangers, but too high to be reintroduced into the industrial streams. Cooling towers are used to discharge this residual heat to the environment.

Cooling towers can be classified as of direct and indirect contact between air and heated streams. In the indirect contact, the water to be cooled flows through tubes and exchange heat with external air flows. These equipment are similar to conventional heat exchangers. Due to the small temperature difference between the streams, these heat exchangers can reach quite high volumes. However, no water is wasted in this process, which is convenient for regions where water is scarce. In the direct contact cooling towers (also known as wet cooling towers), the water to be cooled is in intimate contact with the cooling air, taken from the environment. Wet cooling towers are the focus of this section.

Actually, wet cooling towers are conceived as a low-cost high efficiency heat removal equipment for large mas flow rates of water streams, at low temperature

M. B. H. Mantelli et al., *Thermosyphons and Heat Pipes: Theory and Applications*,
https://doi.org/10.1007/978-3-030-62773-7_11

levels. However, during the cooling process, water is lost to the environment. This water waste can happen by drift, blow-down or evaporation, being this last the most significant mechanism, representing around 80% of the total water loss to the ambient (Hensley 2009). As the ambient air stream pass through the tower, it becomes hot and humid. Depending on the weather conditions, a water plume is formed at the cooling tower discharge. This "water loss" has to be replenished, usually from natural resources (rivers and lakes, for instance).

The order of magnitude of the cooling capacity of this equipment can vary from $10^{-2}[\mathrm{ton/h}]$, for equipment used for small air conditioning systems, for example, to $10^{4}[\mathrm{ton/h}]$, for large industry applications, such as power plants. In extreme conditions, the evaporated water flow in a single equipment can reach up to $5 \times 10^{2}\left[\mathrm{m}^{3}/\mathrm{h}\right]$ in the fan discharge nozzle. Cooling towers of large industries are usually installed in concrete masonry buildings reaching the height of four to five floors (more than 40 m). They can be constructed as independent units in parallel.

The air discharge in cooling towers is usually at the saturated condition, which means that any heat withdrawal causes condensation of vapor. As this air is usually warmer than the surrounding atmosphere, the outside ambient acts as heat sink, promoting condensation. Special technologies can be used to recover part of the water lost. In an ideal prospect, depending on the temperature and humidity conditions, up to 50% of all the liquid water converted to vapor could be recovered (Mantelli 2016a).

Working Principles of Cooling Towers

Wet cooling towers are devices designed to cool down industrial water by means of heat and mass transfer to the environment air. The equipment provides means to increase the contact of the hot humid air with the ambient air. For this purpose, sprinklers, showers and sprays are used.

According to Hensley (2009), based on the airflow circulation mechanism employed, cooling towers can be classified as natural convection (also known as atmospheric towers) or mechanical draft towers (with single or multiple fans). Cooling towers can still be classified according to the airflow: counter-flow, if the air moves vertically upward trough the fillings structures, in opposite direction to the downward movement of the water to be cooled; and crossflow, if the air flows horizontally through the fillings while the hot water flows in downward direction.

They can also the characterized by the method of heat transfer: evaporative, when the primary cooling effect is the heat absorbed during the partial evaporation of the water to be cooled, which happens when air is brought into direct contact with of the hot water; and dry, when no direct contact happens between air and water.

In evaporative towers, the process water is cooled down by two mechanisms: convection and liquid–vapor latent heat. The liquid–vapor phase change is an endothermic phenomenon that absorbs energy from the water and air, causing their refrigeration. According to Mantelli (2016), the air, pushed by a fan located in the cooling tower exit nozzle, crosses the tower and captures the formed vapor. The resulting humid stream is expelled to the atmosphere through the fan nozzle. The cooling tower thermal performance depends strongly of the humidity and temperature of the ambient air that enters the equipment. Typically, for each 7 K of environment temperature variation, 1% of the cooled water is evaporated (ASHARE 2001).

The working principles of the several configurations of cooling towers are very similar. In this section these principles are presented, taking the wet cross current equipment as illustration, as shown in Fig. 1.1. Typically, cooling towers are designed to cool down process water from temperatures of around 42 °C to about 30 °C. Ambient air enters the equipment through shutters located at its lateral walls. The water to be cooled is sprinkled by showers located over the fillings, close to the lateral walls. The fillings are formed by metal gutters that are designed to retain water, forming a kind of "artificial water falls" that favors the formation of droplets and the water evaporation. Besides the fillings, there are two drift eliminators (one in each side), consisting of an assembly of baffles (or labyrinth passages), through which the air passes prior reaching the central region of the equipment, i.e., the "plenum", from which it exits the tower. The purpose of the drift eliminators is to remove entrained water droplets from the exhaust air, forming large drops that are pushed to the collecting tray by gravity (Fig. 11.1).

Over the years, the cooling tower heat exchange capacity have enhanced and components have been improved in the sense of mitigating the problem of water losses to the environment.

Technologies to Reduce the Waste of Water in Cooling Towers

As stated by Mantelli (2016), although evaporative cooling tower was originally designed as a water conservation device, Hensley (2009) affirms that 3 to 5% of the

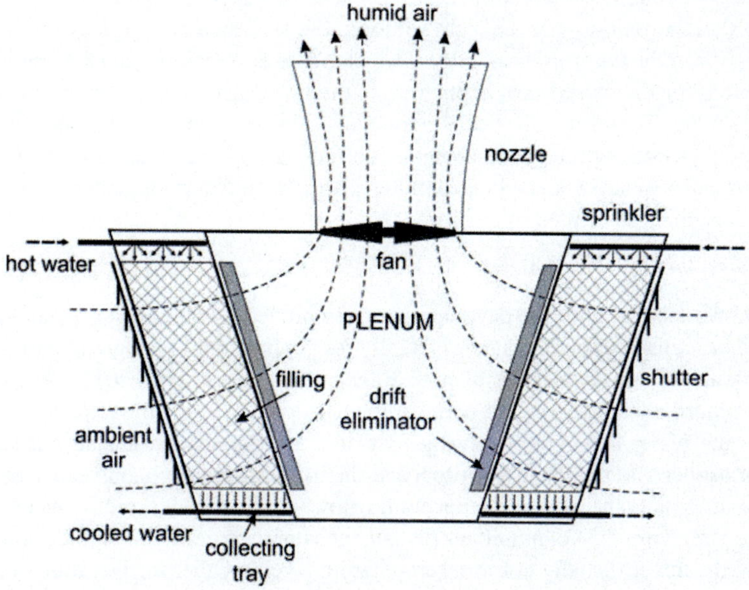

Fig. 11.1 Schematic of a wet cross current water cooling tower

circulating water is lost by evaporation, drift and blowdown. This is not acceptable for regions where water is scarce. In this section, technologies developed for plume abatement and reduction of the water waste are presented.

Plume Abatement

Depending on the weather conditions, plumes over cooling towers can be visible (Lindhal and Jamesom 1993; Tyagi et al. 2012 and Lidhal and Mortensen 2010). Plumes are formed in cooling towers when the air is not able to sustain any more vapor and, consequently, water liquid droplets are spelled to the environment. There are regulations in some countries in which visible plumes must be avoided, due to aesthetics, community relation and safety. Actually, plumes do not pollute the air, as they are essentially pure water vapor made visible, when the heated moist air leaving the tower is cooled below its dew point.

Some new technologies have been proposed in an attempt to reduce the plume in cooling towers, involving condensation of vapor in cold surfaces, in direct contact with the wet flow at the tower discharge (Lindahl 2009, Park et al. 2013, Ghosh et al. 2015 and Czubinski et al. 2013). However, such solutions generally reduce the equipment efficiency, are costly to operate and are not always effective.

Other procedures to prevent plumes may involve the heating of the humid air using natural gas burners, hot water serpentines, chemical reactors sprayed in the tower exit, to name a few. These solutions are expensive, not always operational and may not recover wasted water.

Hybrid towers represents a good alternative for the plume abatement (Asvapoositkul et al. 2014). This technology involves the simultaneous application of direct (wet cooling) and indirect (closed) heat exchangers. Figure 11.2 illustrates two of these solutions. For the right solution, the saturated hot air, leaving the wet cooling tower, mixes with the hot dry air, which heaves a closed heat exchanger. The hot source for the closed heat exchanger is the hot water (to be cooled) that flows through tubes. Therefore, this heated moisture air leaves the tower in a sub-saturated condition, promoting the plume abatement effect, but not the condensation of vapor (no savings of wasted water). In the left, basically the same idea is sketched, but with the heat exchanger located in a different position.

Thermosyphon Heat Exchanger

Several works deal with the development of improved cooling tower technologies, aiming the conservation of water, especially for applications in geographic locations where water is rare (Panjaeshashi et al. 2009 and Al-Bassan et al. 2011). According to Pozzobom et al. (2015), cold permeable walls located in the tower plenum could reduce the water loss by condensing part of the vapor that would be released to the atmosphere. Mantelli (2016) proposed the use of thermosyphons as a means to connect thermally the inner and outer cooling tower environments, providing efficient heat transfer from the hot humid air (usually at saturation conditions) to the ambient outside the tower (usually at lower temperature levels). Other refrigeration systems could be used, but they showed themselves unfeasible, as they require additional equipment and extra energy.

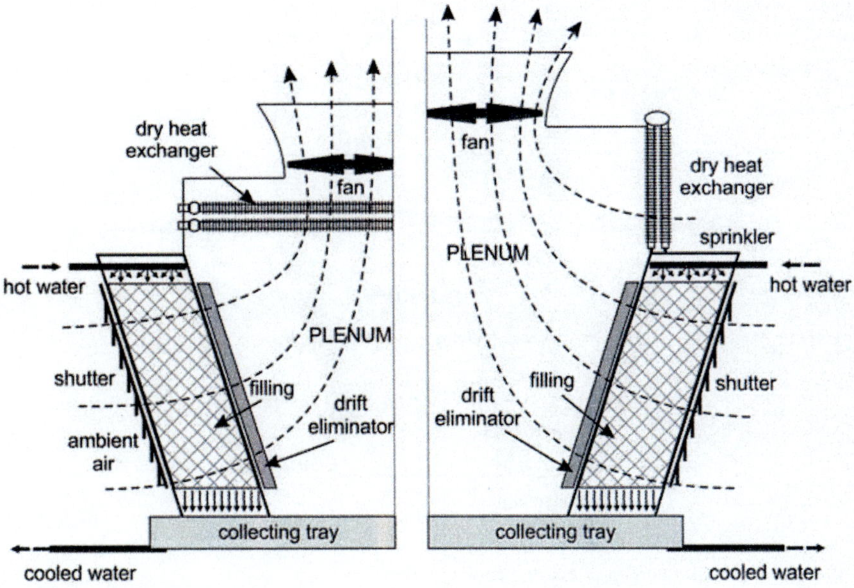

Fig. 11.2 Hybrid cooling tower technologies

Mantelli (2016) proposes the use of a heat exchanger, composed of bunch of thermosyphons, which evaporators are in contact with the hot humid air in the cooling tower plenum and the condensers with the ambient outside the cooling tower, close to the exit fan nozzle. The cooling tower fan induces the humid air flow against the arrangement of thermosyphon evaporators and also promotes some convection outside the tower. The drift eliminator of the cooling tower can be designed to be incorporated as fins to the external surfaces of the thermosyphon evaporators, to improve the contact of the humid air stream with the cooled surfaces. Also, porous media can be used instead of fins, as proposed by Pozzobon et al, (2016), Czubinsk et al. (2013) and Costa et al. (2014).

Thermosyphons are efficient heat transfer devices which, when well designed, can transfer heat even at small temperature differences between evaporator and condenser. To increase the heat transfer area, finned tubes can be used in both condenser and evaporator regions. The colder is the ambient air, the highest is the ability of the device to remove heat from the plenum and the larger is its capacity in condensing vapor. On the other hand, the dryer (less humid) is the ambient air, the largest is its capacity of absorbing vapor during the water cooling and so more water is to be recovered. Actually, the temperature and humidity of the ambient air are very important parameters for the performance of a cooling tower and of the thermosyphon heat exchanger. Figure 11.3 shows a wet cross flow cooling tower assisted by a thermosyphon heat exchanger.

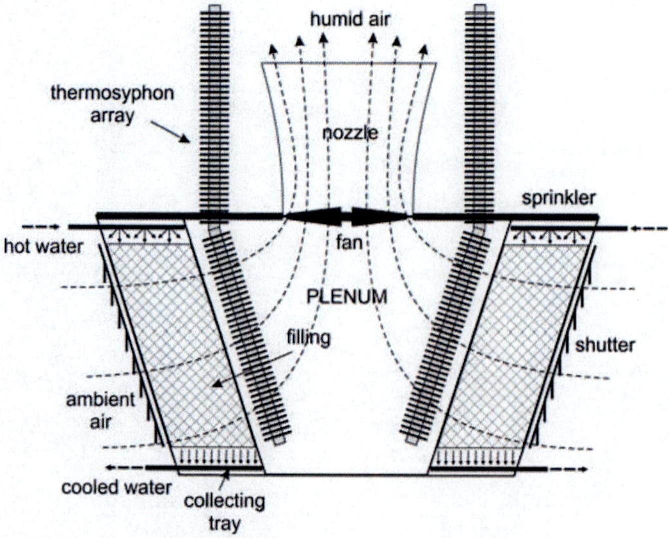

Fig. 11.3 Thermosyphon array assisted cooling tower

Testing Apparatus

As described by Castro et al. (2021), the thermosyphon assisted heat exchanger shown in Fig. 11.3 was tested in a small scale (1:20) experimental bench. As for the real equipment, a fan was located at the tower discharge to induce the air through the splash-type fill and through the drift eliminator. The hot water is discharged at the upper distribution basins, located in the top of the tower, from where it is distributed by sprinklers (composed of orifices located in those basins) to the filling, by gravity action. In its way to the cooled water collecting tray (bottom of the tower), large drops are dismantled into small droplets, resulting in larger areas of direct contact between the air and the water, promoting high heat and mass transfer conditions between the fluxes.

Analytical Model

The major idea behind the use of a thermosyphon heat exchanger in a cooling tower is to reduce the temperature of the saturated humid hot air that leaves the tower, and so, to condensate part of the vapor. This procedure saves wasted water and avoids plume formation.

It is important to note that the presence of the heat exchanger in the cooling tower plenum causes low pressure drops in the humid air streams (see Mantelli, 2016). This means that most of the dehumidification happens due to the temperature decrease by convection and phase change (vapor–liquid) and not by decompression of the humid air.

As Castro et al. (2021) stated, each section of the thermosyphon in the cooling tower, is subject to distinct operational conditions. The evaporator, positioned inside the tower, experiences the forced convection regime, provided by the flow of hot and humid air dragged by the fan. In contrast, the heat transfer in the condenser depends on the external air velocity, which can vary from forced convection, in a windy day, to natural convection, at a calm atmosphere.

As shown in Chap. 4, the thermosyphon overall thermal resistance is defined as the ratio of the difference between the mean temperatures of the evaporator and condenser to the power transferred.

Heat Transferred by the Thermosyphon Condenser Array

In order to quantify the heat transferred outside the cooling tower by the thermosyphons, consolidated convection literature correlations can be used, depending on the conditions to which the thermosyphon is subjected.

The forced convection heat flux over the external surface of the condenser thermosyphon arrays, can be determined from the following expression:

$$q_c = h_{conv} A_{c,t} \left(\overline{T}_c - T_{amb}\right) \quad (11.1)$$

where $\left(\overline{T}_c - T_{amb}\right)$ is the difference between the condenser section mean temperature and the ambient temperature, A_{cd} represents the external condenser surface area, including the fins (with their respective efficiencies, see Incropera and DeWitt 2008) and h_{conv} is the forced convection coefficient of heat transfer. Castro et al. (2021) propose the use of Churchill and Bernstein (1977) correlation, valid for $Re_D Pr > 0.2$, which has the form:

$$Nu_D = \frac{h_{conv} D}{k_l} = 0.3 + \frac{0.62 \mathrm{Re}_D \mathrm{Pr}^{1/3}}{\left[1 + (0.4/\mathrm{Pr})^{2/3}\right]^{1/4}} \left[1 + \left(\frac{\mathrm{Re}_D}{2.82 \cdot 10^5}\right)^{5/8}\right]^{4/5} \quad (11.2)$$

where $R_D = u_l D / \nu_l$. One should note that, in practical applications, this equation is able to predict the heat transfer coefficient with no better than 20% of precision (Bergan et al. 2016).

Heat Transferred by Thermosyphon Evaporator Array

The psychrometric state of the air can be determined from the parameters: pressure, absolute humidity and relative humidity. In the cooling tower interior ambient, close to the evaporator surfaces and due to the cooling effect over the mixture of vapor and air (humid air), temperature and mass gradients are observed. These gradients force both the heat and mass flows from the gas mixture towards the cooled wall. Therefore, gas mixture effects must be considered in the determination of the total heat absorbed by the thermosyphon from the humid air.

A similar to Eq. 11.1 is used for the determination of the heat flux removed from the internal cooling tower environment by the thermosyphons, using the temperature

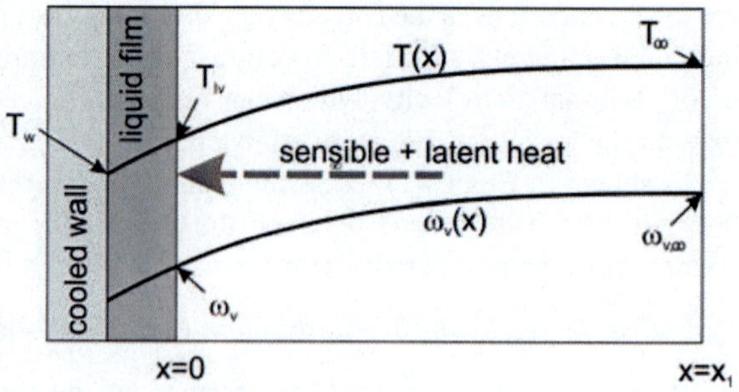

Fig. 11.4 Physical principles of water vapor condensation in the presence of air

difference $\left(\bar{T}_e - T_{air}\right)$, where $\bar{T}_e$ is the evaporator mean temperature and T_{air} is the humid air temperature, for the total evaporator area $A_{e,t}$ (including the fins) and for the global coefficient of heat transfer, which needs to be determined, considering the mixture of gases: air and vapor (see Castro et al. 2021, for details).

Figure 11.4 shows a representation of a vertical cooled surface with a condensate film and binary mixture of air and vapor that flows in the perpendicular to the surface direction. Once the wall temperature is below the dew point for the flow conditions, condensation is observed in the plate. Considering thermodynamic equilibrium in the liquid film surface, the vapor mass fraction in the gas mixture varies between ω_v at $x = 0$ to $\omega_{v,\infty}$ at $x = x_1$. As the wall is colder than the fluid, heat is diffused along the liquid and the gas temperature decreases. Besides, the phase change (condensation) also uses energy, contributing for the cooling of the gas layer (Cremasco 2015).

For the calculation of the heat transferred by convection and phase change, inside the cooling tower plenum, Cremasco (2015) propose the following equation:

$$q = A_e\left[\underbrace{h_{conv,e}\left(\bar{T}_{lv} - T_{db}\right)\frac{\gamma}{1-e^{\gamma}}}_{sensible} + \underbrace{\Omega_m\left(\frac{\omega_v - \omega_{v,\infty}}{1-\omega_v}\right)\rho_{lv}h_{lv}}_{latent}\right] \tag{11.3}$$

where A_e is the total area of the thermosyphon's evaporator, $\bar{T}_{lv}$ is the temperature at the liquid–vapor interface, T_{db} is the dry bulb temperature of the mixture far from the interface, Ω_m is the forced convection coefficient of mass transfer inside the cooling tower, $\omega_v$and $\omega_{v,\infty}$ are the vapor mass fraction at the liquid–vapor interface and at the environment, respectively and ρ_{lv} is the mass concentration at the liquid–vapor interface. $h_{conv,e}$ is the convective heat transfer coefficient, which, as suggested by Castro et al. (2021), can be given by Churchill and Bernstein (1977) expression, shown in Eq. 11.2. γ, known as the Arckmann factor (see Cremasco 2015), represents the influence of the mass flow in the convection heat transfer (sensible heat) and is calculated by:

$$\gamma = \frac{\omega_{lv} c_{p,v}}{h_{conv}} \tag{11.4}$$

where $c_{p,v}$ is the average of the heat capacity of the mixture at the liquid–vapor interface and ω_{lv} is the mass fraction at the liquid vapor interface, determined by the expression:

$$\omega_{lv} = \Omega_m \left(\omega_v - \omega_{v,\infty}\right) \rho_{lv} \tag{11.5}$$

The mass fraction can be determined as:

$$\omega_v = y_v \frac{M_v}{M_{mix}}, \quad \omega_{v,\infty} = y_{v,\infty} \frac{M_{v,\infty}}{M_{mix,\infty}} \tag{11.6}$$

where M_v is the molar mass of the vapor, M_{mix} is the molar mass of the mixture of air and vapor (the sub index ∞ stands for environment) and y_v and y_∞ is the molar fraction of the vapor in the interface liquid–vapor, given by:

$$y_v = \frac{p_{p,v}}{p}, \quad y_{v,\infty} = \frac{p_{p,v,\infty}}{p} \tag{11.7}$$

Here, $p_{p,v}$ and $p_{p,v,\infty}$ are the partial pressures of vapor at the interface and at the environment and p is the total pressure.

The term ρ_{lv} in Eq. 11.5 represents the mixture mass concentration at the gas–liquid interface, which was considered an ideal gas mixture,,i.e.:

$$\rho_{lv} = \frac{p M_{mix}}{R T_{mix}} \tag{11.8}$$

where R is the universal constant of the gases.

The forced convection coefficient of mass transfer Ω_m can be calculated from the Sherwood number (ratio of the convective mass transfer to the rate of diffusive mass transport):

$$Sh = \frac{\Omega_m D}{\Gamma_{v,air}} \tag{11.9}$$

where $\Gamma_{v,air}$ is the diffusion coefficient of vapor in air. The Sherwood number, in turn, can be obtained as a function of the Chilton and Colburn J factor analogy, which establishes the relationship between heat and mass transfer through an expression that, for the flow around rotund bodies, has the form (Sundararajan and Ayyaswamy 1984):

$$J = \frac{Sh_D}{\mathrm{Re}_D Sc^{1/3}} \tag{11.10}$$

where $Sc = \nu/\Gamma$ is the Schmidt number (ratio of the momentum and mass diffusivities). Bedingfield and Drew (1950) propose the following expression for a flow perpendicular to a cylinder:

$$J = \frac{0.281 \cdot Sc^{0.107}}{\mathrm{Re}_D^{1/2}} \tag{11.11}$$

$\Gamma_{v,air}$ can be determined by the Fuller et al. (1966) correlation, that uses a comparison to a reference (known) temperature and pressure, by the expression:

$$\frac{\Gamma_{v,air/T,p}}{\Gamma_{v,air/T_{ref},p_{ref}}} = \left(\frac{p_{ref}}{p}\right)\left(\frac{T_{ref}}{T}\right)^{1.75} \tag{11.12}$$

In the present case, p_{ref} is the atmospheric pressure at the sea level and T_{ref} is adopted as 25 °C, therefore $\Gamma_{v,air/T_{ref},,p_{ref}}$ is the coefficient of diffusion of vapor in air at this pressure and temperature.

Air Mass flow Rate Through the Cooling Tower

The mass flow of air inside a cooling tower plenum is controlled by the fan velocity and can be estimated from a mass balance. Considering the input and output conditions of the air and the vapor mass fractions that enters and leaves the tower to the ambient, the following equation can be obtained:

$$\dot{m}_{air} = \frac{m_{evap}}{t\left(\omega_v - \omega_{v,\infty}\right)} \tag{11.13}$$

where m_{evap} is the total mass of water evaporated, $(w_{int} - w_{amb})$ is the difference between the absolute air humidity inside the tower (before passing through the thermosyphon array) and at the external ambient and t is the time frame considered.

Thermosyphon Heat Exchanger Design

To design a thermosyphon heat exchanger, the methodology described in Chap. 9 can be used. The designer needs to know, a priori, all the characteristics of the ambient air and of the hot flow to be cooled. The maximum efficiency of the cooling tower is a useful parameter, as it relates the input and output water temperatures and the humid bulb temperature of the external ambient air ($T_{hb,amb}$), by the expression:

$$\eta_{ct} = \frac{T_{w,in} - T_{w,out}}{T_{w,in} - T_{hb,amb}} \tag{11.14}$$

Having the ambient air and water flow characterized and using the expressions above, it is possible to calculate the amount of heat to be removed from the hot humid stream. With this heat, the designer is able to determine the necessary external areas of the evaporator and condenser, and so to select the appropriate number of tubes and geometry of the heat exchange areas. The space that the heat exchanger occupies in the cooling tower is also a very important design constraint.

Fins can be used to improve the heat exchange areas and to make the equipment more compact. The geometry of the fins can be determined using the well-known fin design procedures described in Incropera and DeWitt (2008) and Bejan (2000). The designer must also remember that vapor condensation happens at the cooled surface of the fins inside the cooling tower and that enough area must be provided for the liquid drops to form and move to the collecting tray, by the gravity effect.

In a typical problem, the relative humidity of the air inside the tower is around 90% or larger. The temperature of the water stream to be cooled is around 45 °C and the ambient temperature is around 22 °C. It is important to note that the ambient conditions may change during the day and with wind velocities. Equation 11.2, proposed for the convection coefficient of heat transfer, takes into account the temperature and velocity of the ambient air stream. Typically, for calm (no windy days), the internal overall coefficients of heat transfer can be around ten times larger than the external coefficients, while, at windy days (wind velocity of around 5 m/s), these coefficients tend to be of the same order of magnitude. Therefore, usually, the condenser is the limiting heat transfer factor. However, while the external conditions depends on the weather, the internal cooling tower environment is much more predictable as it depends of the fan velocity and of the cooling water conditions.

Theoretical Recovered Water

It is important to note that the maximum heat that the thermosyphon exchanger can transfer is limited by the minimum theoretical heat transfer capacity of the evaporator or of the condenser. This must be taken into consideration in the prediction of the amount of water to be recovered.

As discussed by Castro et al. (2021), when the heat to be absorbed in evaporator is lower than that rejected by the condenser ($q_e < q_c$), the latent heat parcel of Eq. 11.3 is used to estimate the vapor condensed:

$$\dot{m}_{cd} = A_{e,t}\left[\Omega_m\left(\frac{\omega_v - \omega_{v,\infty}}{1-\omega_v}\right)\rho_{lv}\right] \tag{11.15}$$

In the case in which the heat rejected in the condenser is lower than that absorbed in the evaporator, ($q_c < q_e$), the total heat in the evaporation section is determined considering both parcels, sensible and latent separately. With this data, the following parameter, which determines the percentage of the evaporator heat used to change the phase of the water, can be obtained:

$$\Psi_m = \frac{q_{lat}}{q_e} \tag{11.16}$$

This parameter is then applied to the total heat rejected by the condenser array. Therefore, the estimate of the total mass of vapor which is condensed can be given by:

$$\dot{m}_{cd} = \frac{\Psi_m q_c}{h_{lv}} = \frac{q_{lat} q_c}{q_e h_{lv}} \tag{11.17}$$

Maximum Condensation Potential of a Cooling Tower

The maximum theoretical condensation performance of a cooling tower is an important parameter to be determined. Frequently, the heat exchanger, designed to condensate the theoretical maximum vapor from the hot humid air flow that leaves the cooling tower, is not feasible to be constructed. In some cases, for instance, the theoretical spacing between fins can be very small and such finned tubes may not be available in the market and/or are difficult/expensive to fabricate. However, the maximum condensation potential is a parameter that shows the limiting vapor condensation capacity of the equipment and helps in the decision of whether the technology is viable.

The maximum capacity of condensing vapor of a cooling tower assisted by the thermosyphons is achieved when the air that leaves the equipment is at the same temperature and humidity of the ambient air. In this case, the total volume of recovered vapor water would be that of the water absorbed by the air stream, when crossing the tower, until its complete saturation. Therefore, the maximum condensed mass flow rate value can be determined from the expression:

$$\dot{m}_{\text{cd,max}} = \dot{m}_{\text{air}} \left(\omega_v - \omega_{v,sat} \right) \tag{11.18}$$

11.2 House Solar Heating System

Due to the ecological and economical concerns, energy savings in buildings is a very important issue nowadays. In cold weather countries, the efficiency of the heating systems in buildings can be achieved based on passive methods, such as Trombe wall, solar collectors, green walls, wall-thermosyphons, heat pipes, etc.

For the specific case of the Trombe wall, the internal building ambient thermal control is achieved by the use of a great mass (stones, bricks, etc.) wall that receives solar heat during the day and works as heat storages. To avoid heat losses to the environment, a glass wall is constructed parallel to the vertical wall. The glass must be transparent to most of the solar radiation wavelength range and opaque to the infrared range (as most glasses are). A void space filled with air is provided between the heated wall and the glass that thermally insulates the wall from the external environment, avoiding heat losses. Natural convection is used to spread the heat through the environment. A schematic of a Trombe wall can be seen in Fig. 11.5.

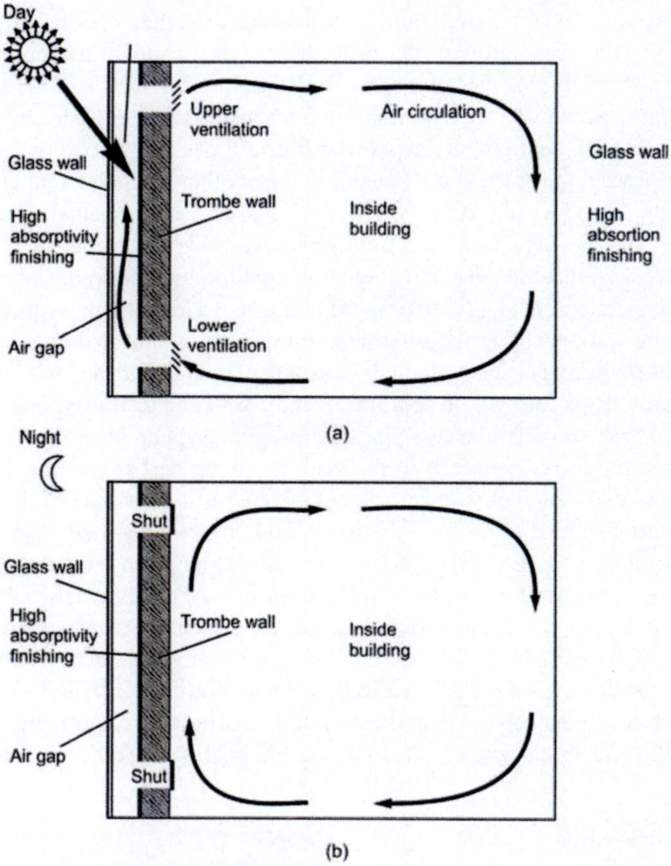

Fig. 11.5 Working principles of a Trombe wall

This technology has been applied since the end of 1960 decade, as commented by Hu et al. (2017). The classic Trombe wall presents inferior and superior openings that allows for the circulation of the hot air located between the solar heated and glass walls, in the direction of the internal housing environment, as discussed by Saadatian et al. (2012) (see Fig. 11.5a). These opening must be closed during the night, to avoid the heat to be lost to the external ambient, as illustrated in Fig. 11.5b. To improve the performance of the system, the surface that receives solar heat must be covered by a high solar absorptance finishing.

The use of heat pipes in solar energy heating systems for building heating purposes has been considered for a while. Susheela and Sharp (2001) presented a detailed analysis concerning the use of heat pipes, including different working fluids, case materials and solar collector configurations.

Zhang et al. (2014) proposed the use of heat pipes integrated to the external walls, to heat housing ambient, by solar energy. Considering north countries in the winter, the building south wall receives higher solar heat fluxes than the others. This heat is directed to the evaporators of the heat pipes, which conduct this energy to the internal environment. Valves "turns off" the heat pipes installed in other walls that could promote eventually the heat transfer from inside to the outside environment. In the summer, the south thermosyphon wall heat pipes are "turned on" and those of the north walls are "turned off", enabling the cooling of the internal ambient of the building. Zhang et al. (2014) made numerical and experimental analyses and concluded that, in a typical winter, this system was able to save about 14% of the heat that would be lost to the external environment through the wall.

Recently, Fantozzi et al. (2016) proposed the use of a loop thermosyphon coupled to a building wall, enabling the solar heat transport from a heated external wall to the internal environment. They studied the saving of energy with and without the use of this device. Their idea was to combine Trombe wall and thermosyphon technologies. The device consists in a two-phase loop-thermosyphon where the evaporator is located in the external surface of the wall, receiving heat from the sun, and the condenser is in its internal surface, where ambient air is to be heated. Figure 11.6 shows the arrangement of the wall-thermosyphon proposed by these researchers. A vertical air circulation gap is formed by the installation of an indoor vertical wall with two venting openings, in rear and top positions. In the external wall, rear and top ventilation openings are also provided. The condenser of the loop thermosyphon is located in the gap within these walls. In the winter sunny days, the outside air vents are closed, while those of the internal wall are open. The condenser heats the air that circulates through the gap by natural convection. During the winter night, when the internal wall temperatures are higher than the external ones, no heat is transported,

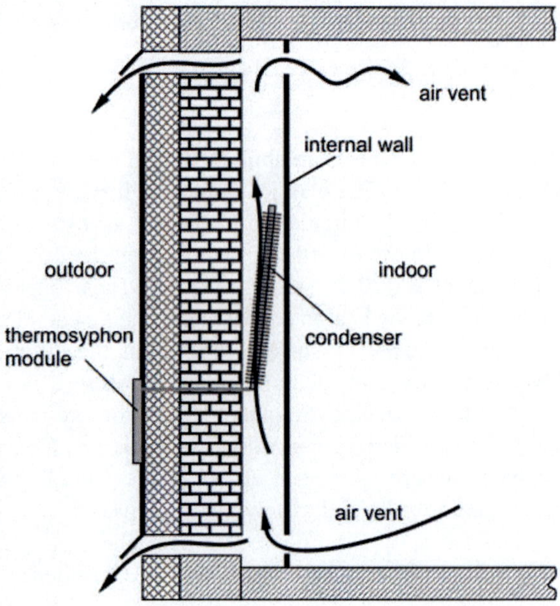

Fig. 11.6 Wall-loop thermosypon

as the thermosyphon works as a thermal diode. In the summer, when the external ambient presents higher temperatures internal ones, all vents are closed and the gap, which is filled with air, works as a thermal insulator. When the internal temperatures are higher, the external vents are open and natural convection takes place between the air in the gap and the external environment, so that the internal wall is cooled.

Fantozzi et al. (2017) concluded that this system can decrease, by at least 23%, the energy consumed for ambient heating of a house in Pisa, Italy, in winter. Actually, Pisa is located in a Europe region where, due to the weather conditions, heating systems are not so required along the year (inferior to 2000 kWh). This means that more savings can be achieved with this technology if applied to buildings in other regions of the world. Later, these authors improved their previous analysis taking into consideration different optical surface finishing for the external walls and several house orientations with respect to the sun. They concluded that the savings with the use of the device could be even superior to those previously predicted (23%).

Bellani et al. (2019) describe a loop thermsosyphon developed to be used in the solution proposed by Fantozzi et al. (2016, 2017). It is designed to be installed in walls with high solar incidence. Actually, this device provides a thermal bridge between the exterior and the interior of the building, allowing the heat to enter into the house. As already observed, the thermosyphons work as thermal diodes, i.e., the heat flux is one-directional, from the outside environment to the inside of the building, when the evaporator temperature is higher than the indoor temperature. In night or cloudy days, no heat is transported in the other direction: from inside to outside. The schematic of the tested device is shown in Fig. 11.7. The working fluid these authors used was R141b with a filling ratio of 80% (700 ml). Although this working fluid has been forbidden in many countries due to environment concerns, the thermal behavior of loop thermosyphons with such working fluids is typical and the understanding of the physical phenomena can be extended to other working fluids. The position of the liquid and vapor lines in the manifolds (horizontal tubes in the condenser) has the purpose of helping in the distribution of the working fluid in the condenser.

Models are needed for the design of such equipment. For steady state conditions, the theoretical overall thermal resistance of loop thermosyphons for solar building applications can be predicted based on a thermal circuit constructed, as depicted in Chap. 4. The effective thermal resistance of the wall-thermosyphon is, as for other two-phase devices, defined as the ratio between the evaporator and condenser temperature difference and the heat transferred. After removing the negligible resistances from the circuit, the thermal resistance can be written as the sum of the boiling resistance in the evaporator and the condensation resistance in the condenser. The coefficients of heat transfer necessary to determine these resistances can be dobtained using boiling and condensation heat transfer coefficients from the literature. There are several correlations in the literature able to predict the heat transfer coefficients involving the working fluid phase change phenomena (see Chap. 2).

It is important to note that, for the evaporator, the use of boiling correlations to predict heat transfer is obviously reasonable. However this is not so clear for the condenser, as convection and phase change may happen simultaneously. Bellani et al (2019 and 2017) observed that the temperature distribution along the condenser is

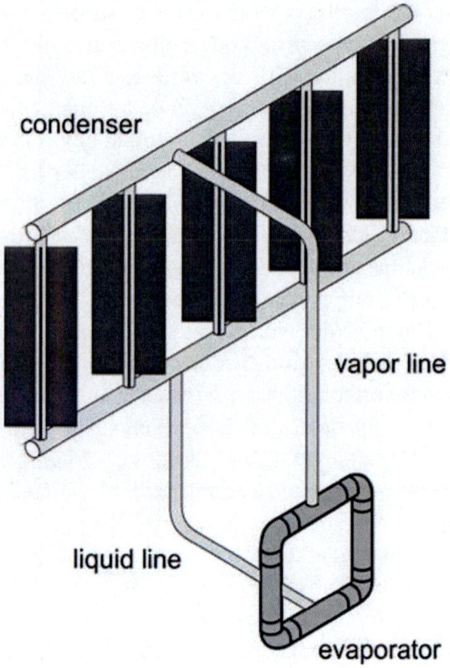

Fig. 11.7 Schematic of the wall-loop thermosyphon for the solar heating of buildings

almost uniform, varying 4% about the condenser mean temperature. This fact shows that, actually, condensation happens along all the condenser length and so phase change correlations are adequate to be used.

Bellani et al. (2019) selected the following boiling correlations: Stephan and Abdelsalam (1980), Kutateladze (see Kaminaga 1992) and Kiyomura et al. (2017) (Eq. 4.16). For the condenser, the condensation correlations were selected: Groll and Rösler (Eq. 4.53) and Kaminaga (Eq. 4.52).

As an illustration of the thermal behavior of a loop thermosyphon for this application, Fig. 11.8 shows one of the graphics provided by Bellani et al. (2019), for a loop thermosyphon which geometry is similar to the one depicted in Fig. 11.7. In this figure, the plots of the temperature and power input as a function of time, are presented. The internal surface of the evaporator did not receive any special treatment, presenting a roughness of $R_p = 1\mu\text{m}$. Power inputs, from 5 to 200 W were delivered to the evaporator by means of electrical heaters. All the device was insulated to avoid heat losses, which, for the highest power input of 200 W, was less than 14 W. At the beginning of the test, the evaporator temperature rises more quickly than the vapor line, until startup conditions are reached, in a time of approximately 17000 s, at 25 W. After this point, the device starts to work properly, with a visible decrease in the temperature difference.

Bellani et al (2019) observed that, for the same operation conditions, the startup for an evaporator, which internal surface was grounded with a course (100) grit silicon carbide paper ($R_p = 26\mu\text{m}$) happened at 20 W, lower than that for the smooth

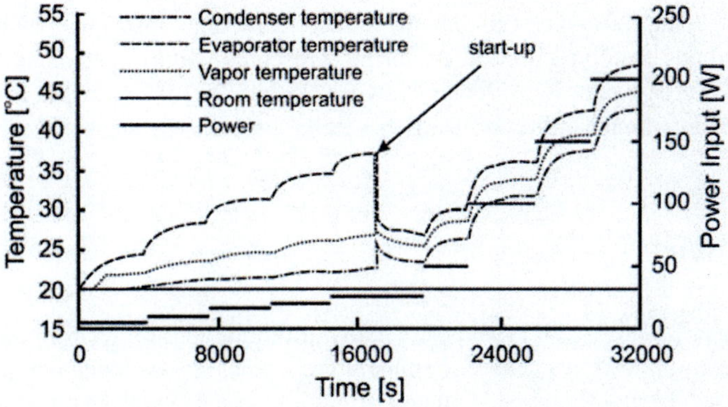

Fig. 11.8 Temperature distribution of the wall-loop thermosyphon

internal surface. This happens because the rough surface provides more nucleation sites for bubbles to start.

Figure 11.9 compares the experimental predictions and theoretical values of the effective thermal resistance of the wall-thermosyphon, which internal evaporator surface finishing is smooth ($R_p = 1\mu\text{m}$). Experimental resistances were obtained after steady state conditions were achieved.

Several combinations of the correlations selected (Stephan and Abdelsalam, Kutateladze and Kiyomura et al. for the evaporator, and Groll and Rösler and Kaminaga, for condenser) were used in the theoretical model. The combination of correlations that presented the best comparison with data, with smaller root mean square (RMS) difference, is obtained by applying Kiyomura's correlation for the condenser and Kaminaga's correlation for the evaporator. At low heat transfer rates, the combination of these two correlations under-predict the data by 50%, while, as the power input increase, the differences become smaller, of less than 15%, with a total average difference of 32%. This difference can be attributed to the fact that the correlations were obtained for liquid pools with a very different geometry from that of the loop thermosyphon evaporator.

Fig. 11.9 Comparison between model and data for the wall-loop thermosyphon

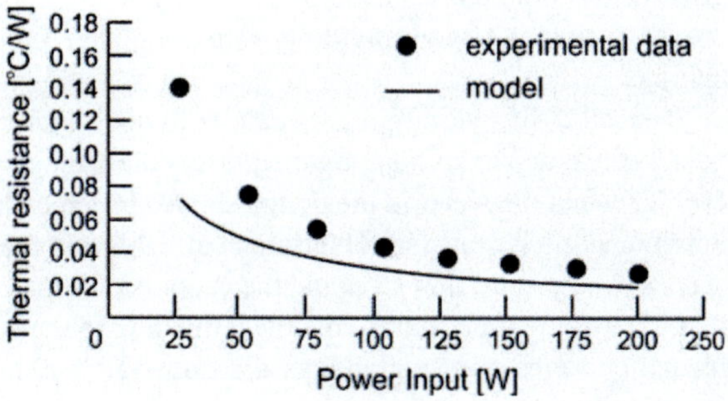

Bellani et al (2019) report that the effective thermal resistance of the wall-thermosyphon prototype that they tested presented values between 0.22 and 0.011 °C/W, for power input levels between 25 and 200 W, which is a very low value, showing the feasibility of using this technology for the present application.

11.3 Ovens

According to Mantelli (2013), there are many different types of ovens in the market, for different applications, different temperature levels, etc. An oven can be simply described as a heated chamber, insulated from the external environment, where some material receives some kind of heat treatment. Usually, the heat is delivered to the oven by means of electricity or combustion of gas (natural or liquefied petroleum gas, LPG). Other more sophisticated technologies can also be employed for heating, such as microwave, infrared radiation, etc. The cavity can be at atmosphere pressure (most applications), or filled with air or other gas or even pressurized or in vacuum.

Ovens were among the very first applications of thermosyphons. Britannic army used portable bread baking ovens in the century XIX. In the XX century, thermosyphons were also used to transfer heat from wood or oil furnace to the baking chamber in large static industrial ovens. Although until now some of this ovens are still operational, some reported accidents, usually explained by the unintentional overheating of the thermosyphons due to the lack of control systems, have caused the discontinuation of this technology along the years.

However, in the last two decades, more controlled new technologies involving thermosyphon application in ovens for bread baking, fruit drying and herb dehydration have been developed, due to the efficient use of basically any source of energy and the high quality of the heat treated product.

The Bread Baking Process

To design an efficient bread baking ovens, it is important to understand the baking process. Depending on the type of bread to be made, the major steps are the following: mixing of ingredients (dry ingredients are blended and hydrated), fermentation (yeast growth, development of flavors and aromas and modification of dough handling properties), make-up (dough dividing, rounding, sheeting and/or molding) and finally, baking.

Basically, the baking process can be divided into two steps: the bread, ready to bake (grown and at ambient temperature) enters the hot oven and heats up to around 100 °C, when the water in the dough start to be vaporized and to be spread along the dough making it grow (see Hatasani et al. 1992). In the second step, the temperature stays at water saturation level and the vapor is eliminated by the external bread layers that, after some time, start to overheat, leading to the crust formation. With the crust formation, temperature gradients are observed within the bread as the outer hard layer hinder the vapor flow out of the bread.

Domestic and industrial baking ovens usually work at temperatures ranging from 150 to 250 °C (Santos and Mantelli, 2009). Much of the energy employed in cooking is used to transform the water contained in the dough in vapor, at around 100 °C. The cooking chambers of many bread baking ovens present a parallelepiped shape. Trays (from 4 to 20) are disposed vertically, where the bread to be baked must be located.

Heating Sources for Bread Baking Ovens

Baking ovens are basically feed by two kinds of energy sources: electrical and combustion. In electrical ovens, few electrical resistances, usually in cartridge shapes, are located close to the lateral walls of the baking chamber. Electrical energy is delivered directly inside the cooking ambient, minimizing the heat losses; however, as the electrical resistances are not in large number, they dissipate high amount of concentrated energy, which makes the temperature distribution inside the chamber not homogeneous. Fans are used in an attempt to mitigate this temperature distribution problem, by increasing the convection heat transfer inside the cooking chamber. However, due to the radiation coming from the high temperature electrical resistances, the breads located in the oven close to the electrical resistances, tend to be darker. The lack of temperature uniformity along the baking chamber forces some regions to be overheated to guarantee that all the volume inside the oven can be used for baking. Besides, the fan wind can be a problem to some bread finished with grated cheese or dry herbs, for instance.

Considering the baking ovens heated by gas, liquefied petroleum gas (LPG) or natural gas (NG) are the most common fuels. In most domestic gas powered ovens, the gas burner is located bellow the cooking chamber. The gasses resulting from combustion passes through small lateral holes, being mixed with the chamber air. As the chamber bottom plate is hotter than the rest of the cooking chamber, the food is heated by natural convection. Fans located in the rear region of the cooking chamber may also be applied, improving the convection. The radiation comes from the heated walls, especially from the bottom one, which is hotter than the others. Burning in the lower surface of the food may occur, if food receives direct radiation from the bottom surface (with the use of glass trays, for instance). Due to the contamination of the air with burning products, this oven configuration is in being vanished especially for industrial applications, but still is very common in domestic ovens in many countries.

In another configuration for gas bakery ovens, hot combustion gases pass through tubes (heat exchangers) located at the rear position of the oven. The cooking chamber air is blown against the heated tube by fans. Usually hot air resulting from gas combustion air flows through the tube by natural convection. Energetically this is a not a good solution because, to deliver the necessary heat to the cooking chamber, the hot gases need to flow through the tubes (by natural convection) at considerable velocities, to allow the necessary convection heat exchange with the internal tube walls. This is possible if the hot gas leaves the tubes to the environment at high temperature levels. Another solution is to use electrical fans to improve the air velocity within the tubes, but this design also increases the use of energy.

To improve the thermal efficiency to such ovens, two strategies can be applied. First to avoid the use of fans, as they consume electrical energy and, in some applications, the resulting wind can disturb the baking process. Second to improve the temperature distribution along the oven baking chamber. Both requirements are fulfilled by the thermosyphon technology (see Silva and Mantelli, 2003). Thermosyphon condensers can provide uniform temperature distribution and rational use of energy, as the heat delivered in the evaporator can be transferred directly to the cooking chamber with thermal resistance close to zero. In the ovens presented in this chapter, water is used as the working fluid.

Tree Configuration Thermosyphons

As stated by Mantelli (2012), only a small share of energy applied to the oven is used for the baking process. Most of the energy is applied to heating up the oven structure and walls and keep them at the needed elevated temperature. Therefore, to increase the efficiency of the baking ovens, besides the use of good external thermal insulators, it is advisable to deliver the heat directly to the cooking cavity filled with air, instead of using the chamber walls to conduct heat.

Silva and Mantelli (2003), Milanez and Mantelli (2004, 2005, 2006a and b) and Mantelli et al. (2005) propose the use of tree configuration thermosyphons. In this configuration, a common evaporator (trunk) is connected in its upper region to several condensers (branches). The evaporator upper regions actually works as a vapor manifold. Vertical branches (Fig. 11.10 left) are recommended for vertically oriented chambers. For horizontally oriented cavities, these branches must present two inclinations: a small vertical section, to allow the gravity to act over the condensate liquid, and a very close to horizontal region, where heat is delivered to the cooking chamber (see Fig. 11.10 right).

The operation principles of tree configuration thermosyphons are very similar to the conventional ones: for the vertical oriented thermosyphon (see Fig. 11.10, left), heat, delivered to the evaporator at the bottom external face, evaporates the working fluid, forming a vapor layer in the upper region of the evaporator. As the evaporator

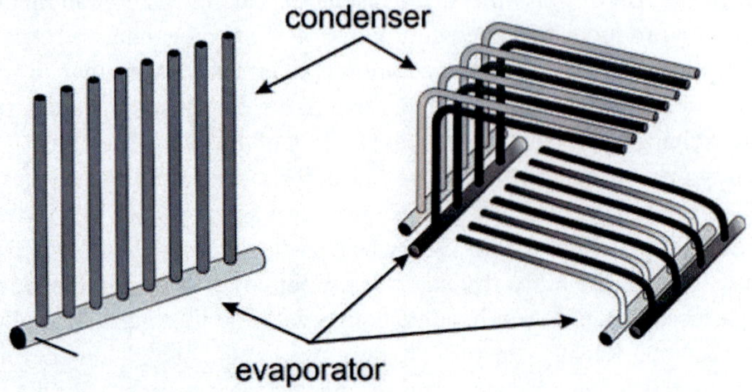

Fig. 11.10 Tree configuration thermosyphons. Left: vertical heating. Right: horizontal heating

is connected to the condenser tubes, the vapor formed is evenly distributed among them. Heat is removed from the condensers and the vapor condensates, returning to the evaporator by gravity action. The volume of the working fluid must be carefully controlled so that liquid does not block any vapor connection with the condensers. The main feature of this configuration is that any point of any condenser has basically the same temperature in all condenser tubes. The tree configurations horizontally oriented thermosyphon, which operates similarly to the vertical one, can substitute the use of heat pipes in horizontal heating applications, if a minimum height difference between evaporator tube and condensers is provided. Leaning the vertical tree thermosyphon to close to horizontal position may also work to heat horizontally oriented volumes.

The analogy between electrical and thermal circuits is applied to model the steady state operation of the tree termosyphons. Figure 11.11 shows a schematic of the equivalent thermal circuit adopted in this work. In this circuit, $R_{ex,e}$ is related to external heat transfer and is determined from the correlation of Churchill e Bernstein (see Eq. 11.2), and $R_{cond,e}$ and $R_{cond,c}$ are related to the heat conduction through the tube wall in the axial direction (evaporator and condenser regions); $R_{lv,e}$ is related to liquid evaporation and $R_{vl,c}$ is related to vapor condensation inside tube walls and are determined using the correlations presented by Groll and Rosler (Mantelli 2013).

Thermosyphon Assisted Bread Baking Ovens

For vertical oriented baking ovens, it is recommended the use two vertical tree configuration thermosyphons close to the lateral walls, to transport, with negligible thermal resistance, the heat generated in the combustion chamber (separated from the baking chamber by a horizontal wall) to the baking environment. Although other heat sources

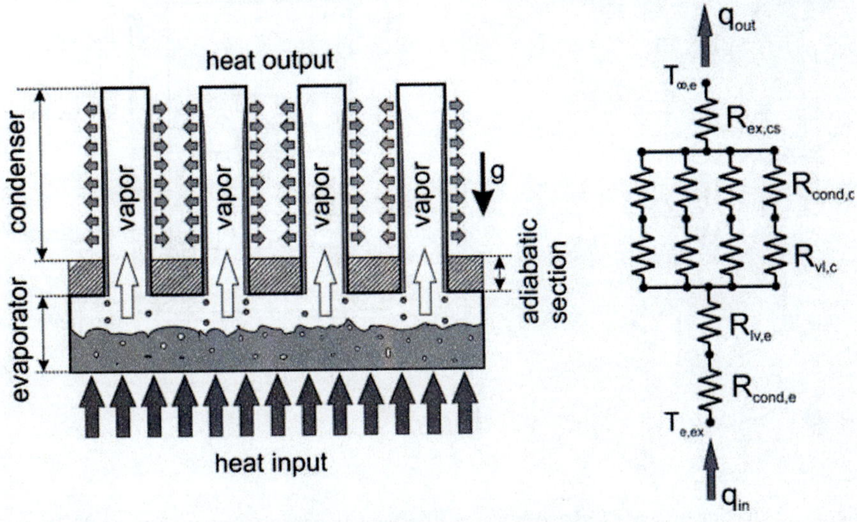

Fig. 11.11 Thermal circuit of the tree configuration thermosyphon

can be used (with the assistance of a heat exchanger) this configuration is very convenient to be associated to flute type gas burners, which are located in the bottom surface of the evaporator. In this case, the combustion heat is uniformly delivered inside the cooking chamber by the several condensers. A fan is also employed inside the cooking cavity to promote convection. The fan is necessary because, due to the thermosyphons, the bottom and the lateral vertical walls of the oven are at very uniform temperatures, and no natural convection is promoted.

Figure 11.12 shows a drawing of the bread baking oven assisted by thermosyphon technology, highlighting the location of the evaporator, condensers and gas burners. The quality of the bread cooked in the electrical ovens varies according to the position of the bread inside the cooking chamber: overcooked when close to the electrical resistances and other hot regions and undercooked in colder regions. On the other side, the bread produced in thermosyphon assisted ovens has uniform cooking quality. Experiments also showed that the heating up time is considerable smaller when compared to other conventional gas burner technologies. Besides, the energy consumption of gas is around only 50% compared to the traditional gas ovens (see Milanez and Mantelli 2006a, b).

Conveyor Belt Baking Ovens

In large scale production of bread and similar products, electrical conveyor belt oven is usually employed. It basically consists of a long electrical heated chamber, insulated in their external surfaces, within which a conveyor belt transport, at a

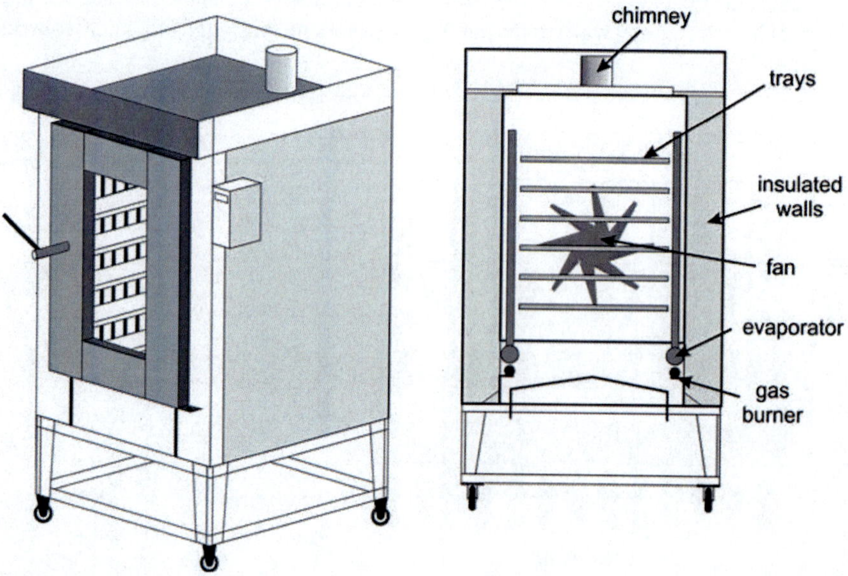

Fig. 11.12 Baking oven assisted by vertical tree configuration thermosyphons close to the lateral vertical walls

controlled velocity, the food to be cooked. The longer the oven, the faster the conveyor speed and the larger is the production. Some industrial ovens (biscuit factories, for instance) may have 50 m of length. Baking time depends on the food: cookies may take around 20 min to bake, while cakes can take up to one hour. The cooking time is controlled by the belt velocity. Although electrical resistances are spread over and above the belt, the resulting temperature distribution within the cooking chamber is not uniform, as noticed by the variation of the food quality, as the product may present different blushes depending on its position over the belt.

Direct heated gas driven conveyor belt ovens are not usual, as the combustion gases may contaminate the cooking chamber. On the other hand, indirect heated gas ovens may not be thermally efficient. Thermosyphon technologies, however, are very suitable for gas ovens, as they are able to efficiently transfer heat from the combustion chamber to the cooking cavity, without contamination by combustion gases.

In the left side of Fig. 11.13, a schematic of a thermosyphon assisted conveyor belt oven, made of four similar modules, is presented. Details of one of the modules, with the thermosyphons highlighted, are shown in the right side of this figure. A pair of two parallel water-stainless steel thermosyphons, in tree configuration, are employed (see Fig. 11.10, for thermosyphon geometry details). Flute type burners, located just below each pair of evaporators are recommended, although other solutions such as industrial gas burner may be applied. The two thermosyphons in parallel are necessary to provide the evaporator external area, able to collect the heat produced by the burner. As shown in Fig. 11.13, a pair of thermosyphons are located above and another pair below the belt (left and right sides of the belt, respectively). As the belt is usually made of metallic wire screens, hot air can cross the belts, delivering heat by natural convection to the food. In normal applications, most of the baking energy comes from these thermosyphons located below the belt. The above belt thermosyphons provide radiation heat, normally used to blush the food.

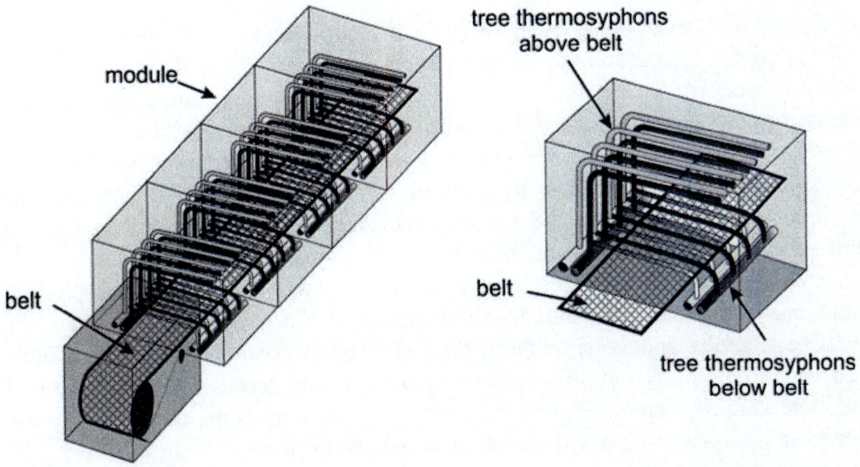

Fig. 11.13 Conveyor belt oven assisted by tree thermosyphons

According to Mantelli (2013), an advantage of modular configuration is that the oven is expandable, being adjustable for the specific application. Besides, the temperature of each module can be controlled independently, by means of the burning gas control, so that the oven can accommodate different cooking process at different temperature levels. Fans can also be used to improve thermal convection. Besides, usually the failure of one of the thermosyphons is not critical, due to the many thermosyphons in parallel.

An important feature of this technology is that the temperature distribution of the thermosyphon conveyor belt oven is very uniform. As for the bread baking oven, the gas consumption is lower than that of the usual gas driven equipment available in the market (more than 30% of energy savings, see Mantelli 2013).

Herbs and Fruit Dryers

According to Cisterna et al. (2016), in herb drying process, the consumption of energy is directly related to the time that the herb remains in the drying chamber. The major part of this energy is used to convert water contained in the leaves to vapor. Long times and low temperatures levels are associated with medicinal herbs. In many countries, the medicinal herb drying processes are still very precarious, with firewood being used as the fuel, due to its availability and low cost. Firewood heating presents many drawbacks, among them: poor energetic efficiency, contamination of the herb with the combustion gases, not uniform drying, lack of temperature control and lack of hot gases flow control.

Drying is the main method used to preserve the active principles (essential oils) of medicinal herbs (Cisterna et al. 2016). Too fast drying may cause degradation of the active ingredients, while too slow drying may cause the reproduction of undesirable microorganisms. The importance of temperature and relative humidity control in medicinal herbs drying is a subject of many literature works (Park et al. 2002, 2008; Muller and Heindl 2006; Argyropoulos et al. 2012 and Argyropoulos and Muller 2011).

Drying can be provided by means of convection heat transfer of a preheated air that flows over the solid to be dried, evaporating moisture and carrying out the formed vapor. The drying conditions can be controlled by temperature and humidity of the heated air (Mujumdar 2006). Kiranoudis (1998) proposed a mathematical model to predict a conveyor dryer operating parameters (similar to the one described in the last section, but without thermosyphons), where each drying chamber has an individual heating unit and a fan for circulating the air through the product. In this scheme, mixing of ambient and process air is provided, in order to control the relative humidity of the air and to save energy. Holowaty et al. (2012) proposed a mathematical model that simulates the behavior of a convective dryer, where the relative humidity and temperature of the heated air is controlled. They concluded that, due to the mass and heat exchange processes, the drying air loses its quality by increasing its humidity during the drying process. Therefore, the use of several layers of conveyors, one above the other, should be avoided.

As for bakery and food processing ovens, thermosyphons can be used in association with gas in herbs and fruit drying. This technology allows for the rational and efficient use of fuel, with much better control of the applied thermal power, of the heated air flow distribution and of the drying temperature level, providing clean drying processes and improving the quality of the final product. Heat is delivered to the drying chamber by means of the tree configuration thermosyphon condensers, which consists of vertical tubes located near the oven vertical lateral walls. In the drying chamber, there are several trays in vertical arrangement to accommodate the product to be dried.

To dry a fruit or herb, water needs to reach the surface in contact with the hot air where it is evaporated. The surface water removal causes a diffusive liquid flow inside the material do be dried. The water vapor from the drying herbs or fruits, is absorbed by the drying hot air, which must have its relative humidity controlled. Therefore, to avoid the air saturation, it is necessary to renew a fraction of the drying air. Thus, atmospheric air is ventilated to the chamber through a centrifugal fan. The air jet entering the chamber is directed to the center of the mixer, allowing a rapid mixture with the inside hot chamber air, resulting in uniform humidity distribution along the drying chamber. Figure 11.14 shows a schematic of the thermosyphon assisted drier, with the renewal air system.

The heat used for drying leaves and fruits is a necessary parameter for the design of driers. Mass and energy balances were performed for the control volume encompassing a drying chamber. The total mass inside the chamber was considered constant along the time. The vapor mass fraction of water in the air is also considered uniformly spread along the drying chamber. Actually, these hypotheses were verified by experiments (see Cisterna et al. 2016).

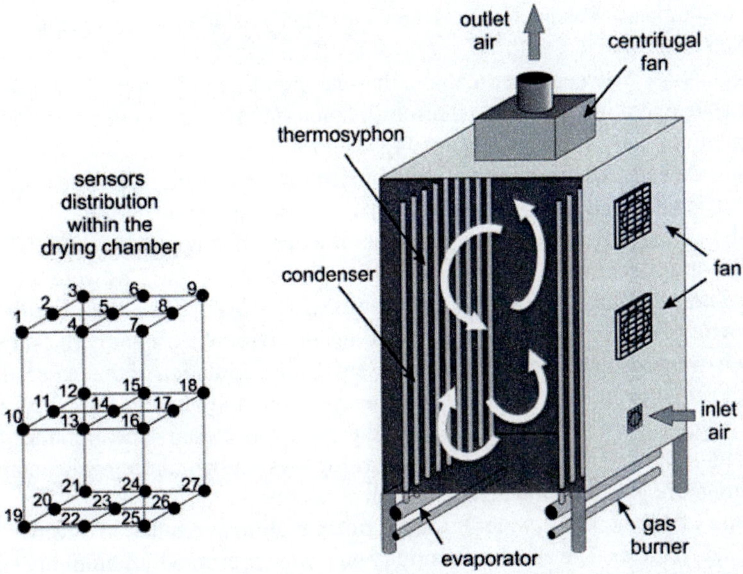

Fig. 11.14 Fruits and herbs thermosyphon assisted drier

From the mass balance, it is possible to obtain the following expression for the vapor mass fraction in the drying chamber, which varies with time:

$$X(t) = X_v + (X_0 - X_v)e^{-X_{air}(t)} \tag{11.19}$$

where X_0 is the is water mass fraction of the ambient air:

$$X_0 = \frac{m_v}{m_{air}} = \frac{\omega_{air}}{1 + \omega_{air}} \tag{11.20}$$

m_{air} is the total mass of air in the drying, m_v is the vapor mass and ω_{air} is the absolute humidity of air. Besides:

$$X_v = \frac{(\dot{m}_0 X_0 + \dot{m}_v)}{\dot{m}_0 + \dot{m}_v}, \quad X_{air}(t) = \left(\frac{\dot{m}_0 + \dot{m}_v}{m_{air}}\right)t \tag{11.21}$$

where $\dot{m}_0$ and $\dot{m}_v$ are respectively the mass flow rate of renovation air and the vapor release rate in the chamber due to drying.

The total thermal energy consumption during the drying process can be predicted by:

$$q_t(t) = \frac{1}{2}\{\dot{m}_v[(1 - X(t))h_{ch} + h_{lv}] + \dot{m}_0[(1 - X(t))h_{ch} - (1 - X_o)h_0]\} \tag{11.22}$$

where h_{ch} and h_0 are respectively the specific enthalpies of the air inside the chamber at a fixed time.

The analogy between electrical and thermal circuits can be applied to model the steady state operation of the tree thermosyphon used in the drier oven, resulting in a thermal circuit very similar to that of Fig. 11.11.

Cisterna et al. (2016) performed experiments for the validation of the above mathematical models. They constructed a drying oven similar to the one depicted in Fig. 11.14 and studied the drying behavior at a typical temperature of 80 °C, for a mass flow rate of the renewal air of 0.00218 ± 0.0004 kg/s. The fans were set so that the resulting Reynolds number of hot air was around $Re = 1.6 \cdot 10^5$. In this case, water in trays were dried, mimicking the drying behavior of a fruit or herb. At regular times (10 min) the oven was open to represent the manipulation of the material to be dried in actual conditions. This causes a temperature and humidity decrease in the drier chamber that increased again after the oven was closed. The amount of heat delivered to the drying chamber by gas burning was set by an electronic controller, releasing more gas to burn when the chamber is cold.

Figure 11.15 shows the water mass fraction variation as a function of time within the drying chamber. The absolute humidity ω_{air} was measured by a humidity sensor. The vapor mass fraction prediction model does not consider the door openings and

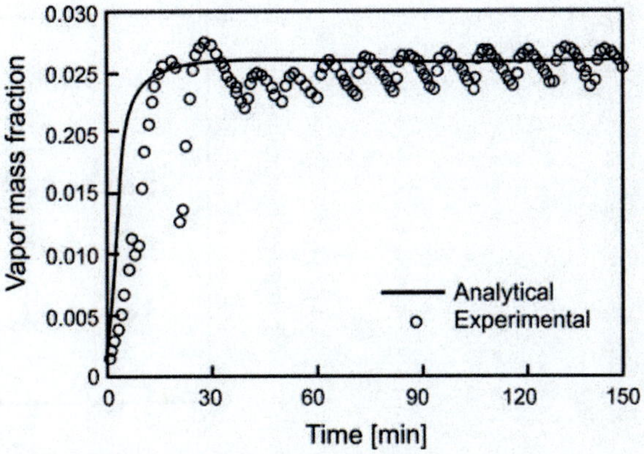

Fig. 11.15 Vapor mass fraction of the heated air inside the drier chamber

so the experimental data oscillates around the analytical predictions. However, the comparison is quite good.

Figure 11.16 shows the comparison between the experimental and theoretical thermosyphon condenser temperature. In this case the model is able to capture the temperature oscillations, with the controlled power used as an input parameter. The comparison between model and data is quite good. The evaporator temperature behaviors, of both model and data, are similar to the ones shown in Fig. 11.16 and are not presented.

The objective of the fans installed in the drier vertical walls were to provide better convection heat exchanger conditions between the condenser tubes and the hot air and to provide uniform temperature and humidity distributions along the drying chamber. Wet small 3 mm thick sponges, with $45 \times 45 \text{mm}^2$ of area, were located in several trays of the drier. In this case, the chamber was kept at 55°C and the mass flow rate renewal air was the same as before ($0.00218 \pm 0.0004 \text{kg/s}$). The fans caused air fluxes inside the chamber with Reynolds numbers varying between $0 < Re < 2.7 \cdot 10^5$. Experiments (see Cisterna et al. 2016) shown that the vapor released by the sponges

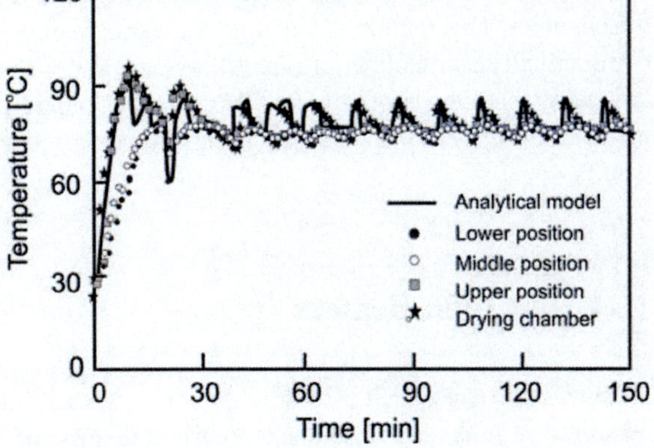

Fig. 11.16 Comparison between model predictions and data for the condenser of the tree thermosyphon temperature as a function of time

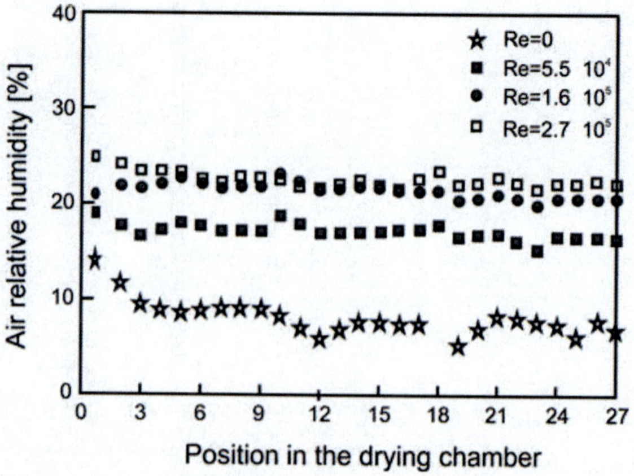

Fig. 11.17 Relative humidity along the drying chamber

(drying process) presents a close to linear mass flow rate behavior as a function of the Reynold number. Also the uniformity of the temperature and humidity increase with the increasing hot air Reynolds number up to a level, when the temperature and humidity distributions became almost uniform and constant. Figure 11.17 shows the humidity distribution along the condenser for the monitored positions (1 to 27, shown in Fig. 11.14), for several levels of the hot air Reynolds number, for a same power input and for a constant air renovation rate: the relative humidity level increases with the Reynolds number as well as the uniformity. The temperature distribution (not shown here) present similar behavior but opposite trend: the temperature level and uniformity increase as the Reynolds number decreases.

Finally, Cisterna et al. (2016) tested the performance of the dryer in actual conditions, by drying medicinal mint herb (Mentha x Villosa Huds). The temperature of the drying chamber was set to around 55 °C and the renewal air mass flow rate was maintained constant at 0.00218 ± 0.0004 kg/s. The Reynold number inside the drying chamber was kept at $Re = 1.6 \cdot 10^6$. The total mass to surface area ratio of the herb "in natura" over the tray was of 2.5kg/m^2. The uniform temperature and humidity distributions characteristics lead to a uniform drying process, with the dried herb with the same characteristics of humidity, color and smell everywhere inside the chamber. This feature is the largest advantage of this dryer in comparison to other commercially available equipment. Besides, as for the thermosyphon bakery oven, the energy savings is of at least 50%, when compared with the conventional equipment. Experiments were also successfully performed with fruits, with very similar results.

11.4 Oil Tank Heaters

In some industries, such as petrochemicals, oils and other viscous products need to be stored in tanks. At ambient temperature, some of these materials may solidify, being difficult to handle, requiring constant heating to be at the liquid state while stored. This is the case of asphalt (or bitumen), a highly viscous liquid (or semi-solid form of petroleum).

To stay in liquid state, asphalt needs to be above its melting temperature, of 68 °C (see Lucena et al. 2003). Usually, asphalt storage tanks can be large and so is the energy used for keeping the asphalt at the appropriate temperatures. However, higher temperature levels are only necessary when it needs manipulation. Actually, some countries regulate the temperature in which the asphalt must be delivered to the costumer (filling of trucks, for instance), at around 140 °C, so that the client can control the actual volume (or mass) of asphalt purchased. Considering steady state conditions, all the energy used to heat the asphalt is ultimately lost to the environment, therefore, much energy can be wasted if the asphalt is stored at higher than the adequate temperature levels. The ideal condition is the one in which asphalt is stored at around 70 °C, being heated to 140 °C only when it is about to be dumped to trucks.

The temperature distribution of the asphalt along the tank is also an important aspect, as chemical degradation is observed when temperatures overpass 230 °C (see Costa et al. 2013). Also, the temperatures must be known, especially at the location of the hose mouth, from which asphalt is removed from the tanks and dumped in trucks.

Probably the most usual way to heat stored asphalt in tanks is by steam (Read and Whiteoakthe 2003), which flows through a few (one to three) serpentines, located in the bottom of the tank. These serpentines are designed to occupy most of the bottom area of the tank, to provide distributed heat and supposedly provide even temperature distribution. The number of serpentines depends on the total volume of the asphalt to be heated and on the temperature of the available steam. Vapor is usually generated away from the tank and, to reach the tanks, the vapor has to travel along long tubes, passing through many purges, where, to reduce undesirable pressure levels, some vapor is released to the environment, decreasing the efficiency of the industrial plant.

It is important to note that vapor is a valuable industrial input and its use in secondary processes, like asphalt heating, should be avoided. Besides, vapor in industrial plants is available only in a predetermined pressures and temperatures, being hard to use is to control efficiently the asphalt temperature. Other technologies such as gas or oil burners could be used, with dedicated burners to the asphalt heating. However, combustion gases can easily reach temperatures of the order of 1000 °C, and so the direct contact of the asphalt with very high temperature surfaces must be avoided (see Costa et al. 2013).

In an attempt to join the good characteristics of the serpentine and the gas burner technologies, thermosyphon heat exchangers have been considered to be employed to heat asphalt stored in tanks (Milanez and Mantelli 2006a, b and c; and Costa et al. 2013). The idea is to design a controllable burner (gas, oil or other heat source) that would transfer heat to the thermosyphon evaporator section (the working principles of thermosyphons are discussed in Chap. 1).

The thermosyphon tubes, in parallel array configuration, may have an inclination of at least 7°. The evaporator, located externally to the tank, receives heat from the burner, and the condenser, located internally, delivers this heat to the asphalt. Figure 11.18 shows the schematic of the asphalt tank heater configuration, highlighting the concentrated thermosyphon array (parallel and inclined thermosyphons).

The loop thermosyphon configuration can also be applied, as shows the schematic of Fig. 11.19.

The heater configuration composed of the array of thermosyphons can be simple to design and construct and have the advantage of safety: the effect of the damage of one of the thermosyphons could be minimum, with negligible reduction of the heat transfer capacity of the device (depending on the total number of thermosyphons) and with the release of a small amount of water (or other working fluid) in the asphalt. On the other hand, the evaporator of loop thermosyphons can be designed in serpentine geometry, increasing the heat exchange area with the asphalt. However, if there is a failure in one of the tubes, the whole serpentine heater fails, with all the working fluid lost to the asphalt.

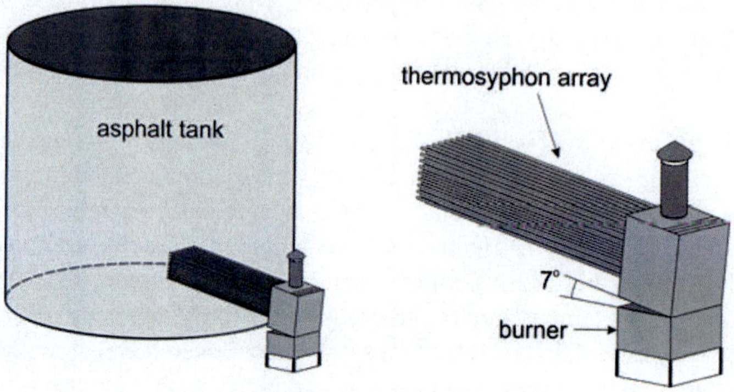

Fig. 11.18 Oil storage tank assisted by a thermosyphon array heated by gas burner

Fig. 11.19 Oil storage tank assisted by loop thermosyohon, heat by gas burners (not to scale)

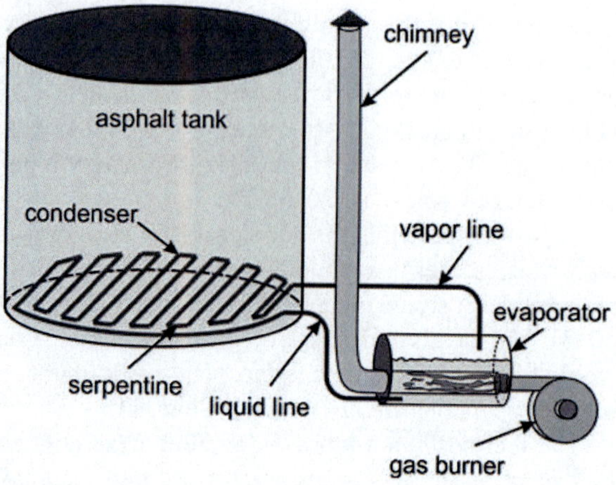

The major difference between these configurations is in the concentration of heat within the asphalt tank: higher for the tube array and lower for the serpentine configuration. To decide the best thermosyphon heater configuration, the thermal behavior of the asphalt needs to be accessed.

For this purpose, a numerical study was performed (Costa et al. 2013). To save computational efforts, a radial slice of the tank was simulated. Differently from studies that deal with well-characterized medium (e.g., air, water, etc.), the main difficulty in simulating the thermal and hydrodynamic behavior of asphalt is the proper determination of its thermophysical properties, some of which are highly influenced by temperature. Costa et al. (2013) also conducted experimental work in a small scale stainless steel asphalt tank that was heated by cartridges heaters, in concentrated (array of cylinders) and spread (serpentine) configurations. Their numerical results and experimental data showed a very good comparison. The most important conclusions is that the temperature distribution of the asphalt, for both concentrated and spread heat sources, is uniform, within 5 °C. This means that the asphalt temperature, in steady state conditions, is uniform, despite of its very viscous characteristics and of the location of the heat source.

As straight parallel thermosyphon array configuration is more reliable and less expensive to construct, Costa et al. (2013) proposed this last configuration as the most effective one.

Gas Heating in City-Gate Stations.

According to Ângelo et al. (2007) two-phase thermosyphon heaters can be used to raise the temperature level of natural gas in city-gate stations. A city gate station is a coupling point between natural gas transport system and the local gas delivery company, where the gas is sold to the client.

Usually, the delivery pressure established by the contracts is smaller than the pressure used to pump the gas along the transport pipelines. The pressure reduction may cause a sensible decrease of the natural gas temperature, resulting in potential problems, like the material embrittlement and the hydrate formation. Thus, the city-gate stations demand heating systems in order to guarantee an acceptable delivery gas temperature range.

Indirect heating is the most usual way to heat hydrocarbon fuel. The water-baths are the most common type of natural gas heater on city-gate stations. This simple heater (see schematic in Fig. 11.20) is composed by a cylindrical horizontal vessel filled with water, where a pipe (usually in serpentine shape) is sunk. Through this pipe, hot flue gas, resulting from natural gas burn, flows. The bath water is heated by this tube. Above the heating tube and within the heated water, another serpentine pipe is drowned, within which the natural gas to be delivered flows. The heated water, by natural convection, heats the gas to be delivered, to reach the appropriate temperature level. In most cases, the hot gas tube is dimensioned to keep the water at around 70 °C. Although reliable and almost completely independent of electricity, this water-bath system requires constant water reposition, due evaporation process. This can be a problem as these stations can be located at remote sites with difficult access.

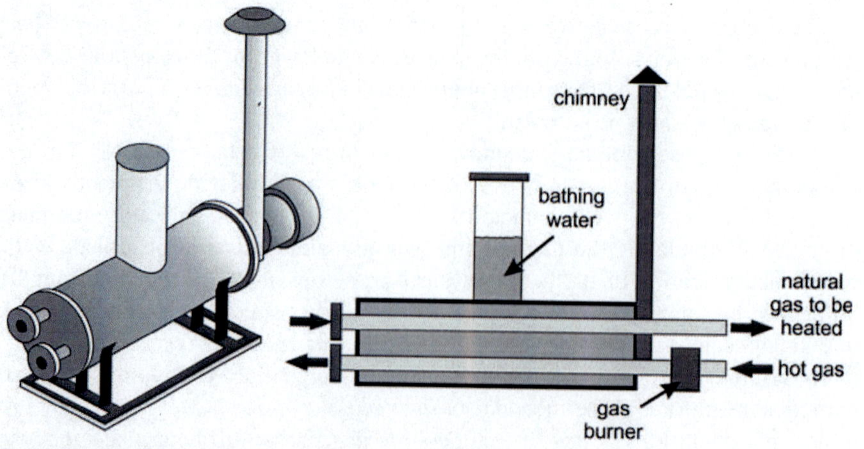

Fig. 11.20 Water bath natural gas heater in city-gates

Ângelo et al. (2007) proposed the use of thermosyphon technology to heat the natural gas in these city-gates. The main idea is to transfer latent heat between water (or other working fluid) and the external surface of the tube where the natural gas, to be heated, circulates. The original cylinder (water tank) geometry can be kept and partially filled with water, also heated by tubes where hot flue gas flow, as for the original equipment. When heated, the water evaporates. The tubes, within which the natural gas to be heated flows, are located in the vapor region of the cylinder, removing the heat delivered by the hot tube. The vapor condensates and the resulting liquid returns to the liquid pool by gravity. Actually, this configuration works as an inverted thermosyphon, where the heat source and sink are located internally to the cylinder. Figure 11.21 shows the schematic of this thermosyphon assisted natural gas heater. Different from the original configuration that works at atmosphere pressures, the proposed configuration can work below the atmosphere pressures. Due to its geometry, cylindrical tanks subjected to low pressures can collapse and so they need to be mechanically designed, following pressure vessels regulations, usually resulting in thick wall vessels.

Loop thermosyphon is another configuration that can be used in this application: the water heated cylinder is used as the evaporator and the natural gas heating cylinder as the condenser of this circuit, as shown in Fig. 11.22. The selection of the water level, and so its volume, is an important parameter, as it can determine the start-up time of the device. Besides, by controlling the liquid volume, the undesirable presence of Geyser boiling, which decreases the efficiency of the condenser (see Chap. 3 for Geyser boiling discussion) may be avoided. A steam-water separator is provided to mitigate the effects of liquid over the condenser region. Besides, the equipment must be designed to accommodate the presence of NCG, in a tank, placed at top of the condenser drum to collect eventual gases that might be in the circuit.

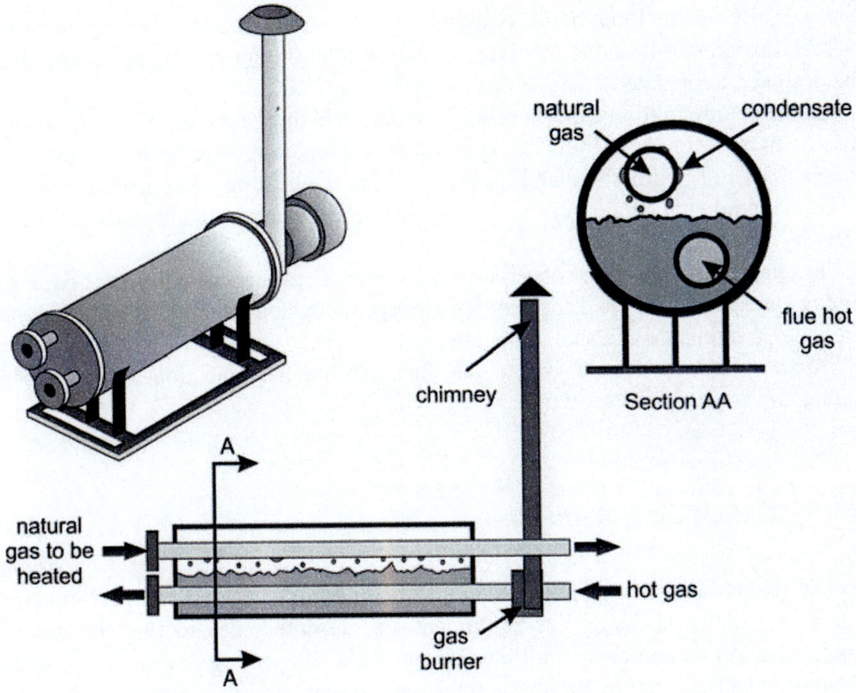

Fig. 11.21 Inverted thermosyphon natural gas heater

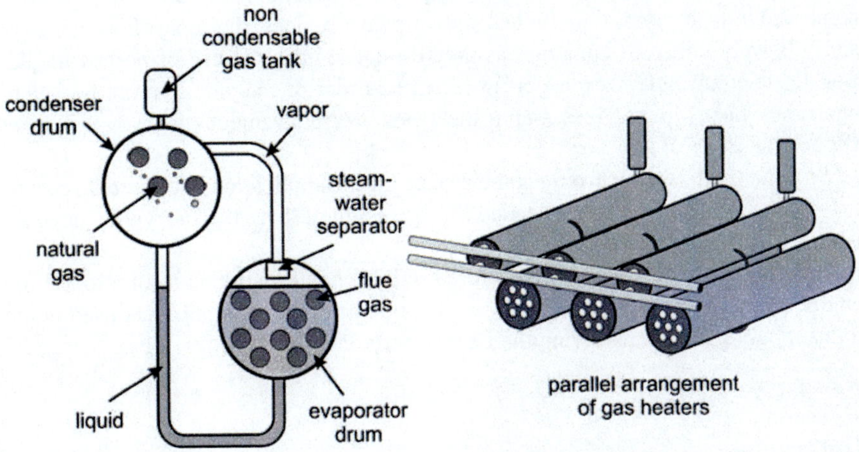

Fig. 11.22 Loop thermosyphon configuration for natural gas heater

Specially for the loop thermosyphon heater, to avoid the need of using pressure vessels design norms and regulations, several smaller thermosyphon equipment can be designed to operate in parallel, as shown in Fig. 11.22.

In both configurations, the working fluid (water in this case) can be heated by the use of an array of inclined thermosyphons. The evaporators are heated by flue hot gases from a gas burner (usually located far from the heating equipment, for safety reasons), while the condensers are located inside the water to the heated (as for the heating of storage oils tanks, see last section).

In general terms, it can be observed that the proposed thermosyphon assisted technology presents lower cost, are electricity independent, does not require replenishing of water, are reliable and safe to operate.

Models for loop thermosyphons developed for other applications, as described along this book, is also valid here.

11.5 Concluding Remarks

In this chapter, discussions about the improvement of the heat transfer performance of several industrial large size equipment are presented, considering the use of thermosyphon technology. In all the thermal solutions proposed, more reasonable designs, which also allow for considerable energy savings, are achieved.

Many times, two-phase technologies is considered not cost effective for industry. Thermosyphons can easily fulfill this gap. They present many advantages, when compared to heat pipes. Gravity driven two-phase devices, thermosyphons are easy and relatively cheap to construct as they do not require the use of porous media. Besides they are able to transport at least one order of magnitude more heat than equivalent heat pipe devices, being therefore very convenient for large industrial applications.

Specific models, concerning some aspects of the application itself, are discussed in this chapter. Other models, needed for the design of these thermosyphon devices, are presented along the book.

It is expected that the material offered in this chapter work as inspiration to the engineers to use this technology to improve other traditional equipment used in the industry, especially considering their energy efficiency and compactness.

References

ASHRAE Handbook Fundamentals Inch-Pound System, ASHRAE (2001)

Akachi, H., Structure of a heat pipe, Patent US4921041A (1990)

Al-Bassan, E., Maheshwari, G.P.: A New Scheme For Cooling Water Conservation in Arid-Zone Countries. Energy **36**, 3985–3991 (2011)

Argyropoulos, D., Müller, J.: Effect of Convective Drying on Quality of Lemon Balm (Melissa Officinalis), 11th International Congress On Engineering And Food. Procedia Food Science **1**, 1932–1939 (2011)

Argyropoulos, D., Alex, R., Kohler, R. and Müller, J., Moisture Sorption Isotherms and Isosteric Heat of Sorption of Leaves and Stems of Lemon Balm (Melissa Officinalis L.) Established by Dynamic Vapor Sorption, Lwt - Food Science And Technology, **47**, 2, 324–331 (2012)

Ashby, M.F.: A First Report on Sintering Diagrams. Acta Metall. **22**, 275–289 (1974)

Asvapoositkul, W., Kuansathan, M.: Compartive Evaluation of Hybrid (Dry/Wet) Cooling Tower. Appl. Therm. Eng. **71**, 83–93 (2014)

Bedingfield, C.H., Drew, T.B.: Analogy between Heat Transfer and Mass Transfer. Ind. Eng. Chem. **42**(6), 1164–1173 (1950)

Bellani, P., Milanez, F., Mantelli, M., Filippeschi, S., Mameli, M., Fantozzi, F.: Theoretical and Experimental Analyses of the Thermal Resistance of a Loop Thermosyphon for Passive Solar Heating of Buildings. Interfacial Phenomena and Heat Transfer **7**(1), 57–68 (2019)

Bellani, P., Milanez, F. H., Mantelli, M. B. H., Filippeschi, S., Mameli,, M. and Fantozzi, F., Experimental Studies on a Loop Thermosyphon for Solar Heating of Buildings, Proc. of 9th World Conf. on Experimental Heat Transfer, Fluid Mechanics and Thermodynamics, Iguazu Falls, Brazil (2017)

Bergman, T.L., Lavine, A.S., Incropera F.P. and Dewitt, D.P.: Fundamentals of Heat and Mass Transfer, 7th edn. Wiley and Sons (2016)

Bhattacharya, A., Calmidi, V.V., Mahajan, R.L.: Thermophysical Properties of High Porosity Metal Foams. Int. J. Heat Mass Transf. **45**, 1017–1031 (2002)

Brennan, P. J., Thienen, L., Swanson, T., Morgan, M.: Flight Data for the Cryogenic Heat Pipes (CRYOHP) Experiment", AIAA Paper No. 93–2735 (1993)

Castro, F. R., Mantelli, M. B. H., Silva, A. K.: Use of Two-Phase Thermosyphons in Cooling Towers: An Experimental Evaluation of Water Recovery, AIAA Int. J. Heat Mass Transf. (2021)

Chien, K.H., Lin, Y.T., Chen, Y.R., Yang, K.S., Wang, C.C.: A Novel Design of Pulsating Heat Pipe With Fewer Turns Applicable to All Orientations. Int. J. Heat Mass Transf. **55**, 5722–5728 (2012)

Churchill, S.W., Bernstein, M.: A Correlating Equation for Forced Convection From Gases and Liquids to a Circular Cylinder in Crossflow. J. Heat Transfer **99**(2), 300–306 (1977)

Cisterna, L. H. R., Caldas, D. M. M., Milanez, F. H. and Mantelli, M. B. H., Theoretical and Experimental Analysis of an Herbs Dryer Assisted by Two-Phase Thermosyphons, Joint 18th International Heat Pipe Conference and 12th International Heat Pipe Symposium, Korea (2016)

Costa, C.A.S., Mantelli, M.B.H., Milanese, F.H., Da Silva, A.K., Rucker, C., Furlan, L.T.: Experimental and Numerical Study of an Asphalt Storage Tank in a Reduced Scale. Appl. Therm. Eng. **56**, 101–109 (2013)

Costa, C.A.S., Miranda, V., Mantelli, M.B.H., Silva, A.K., Modenesi, C.R., Furlan, L.T.: Experimental Study of Flexible, Unstructured Metal Foams as Condensation Structures, Experimental Thermal and Fluid Science, **57**, 102–110 (2014)

Cremasco, M.A.: Fundamentos de Transferência de Massa, 3rd edn. Editora Edgard Blücher Ltda, São Paulo (2015).(in Portuguese)

Czubinski, F.F., Mantelli, M.B.H., Passos, J.C.: Condensation on Downward-Facing Surfaces Subjected to Upstream Flow of Air-Vapor Mixture. Exp. Thermal Fluid Sci. **47**, 90–97 (2013)

Delforge, P., Whitney, J., Data Center Efficiency Assessment, Technical Report, Natural Resources Defense Council, New York (2014)

Fantozzi, F., Filippeschi, S., Mameli, M., Nesi, S., Cillari, G., Mantelli, M.B.H., Milanez, F.H.: An Innovative Enhanced Wall to Reduce the Energy Demand in Buildings. J. Phys: Conf. Ser. **796**, 12043 (2017)

Fantozzi, F, Filippeschi, S., Mameli, M.; Mantelli, M. B. H. and Milanez, F. H., How a wall thermosyphon can enhance the energy savings in a prefabricated house in Italy. In: 18TH IHPC AND 12TH IHPS. Jeju, Korea: Jun. 2016

Florez, J.P.M., Mantelli, M.B.H.: Thermal Model for Sintered Cylindrical Evaporators of Loop Heat Pipes. J. Thermophys. Heat Transfer **32**, 1–13 (2016)

Fuller, E.N., Schettler, P.D., Giddings, J.C.: New Method For Prediction Of Binary Gas-Phase Diffusion Coefficients. Ind. Eng. Chem. **58**(5), 18–27 (1966)

Gerasimov, Y.F., Maydanik, Y.F., Shchogolev, G.T.: Low-temperature heat pipes with separate channels for vapor and liquid. Journal of Engineering Physics **28**, 683–685 (1975)

Ghosh, R., Ray, T.K., Ganguly, R.: Cooling Tower Fog Harvesting in Power Plants–A Pilot Study. Energy **89**, 1018–1028 (2015)

Hasatani, M., Arai, N., Harui, H., Itaya, Y., Fushida, N., Hori, C.: Effect of the Drying on Heat Transfer of Bread During Baking in Oven. Drying Technology an International Journal **10**, 623–639 (1992)

Hashimoro, H., Kaminaga, F.: Heat Transfer Characteristics in A Condenser of Closed Two-Phase Thermosyphon: Effect of Entrainment on Heat Transfer Deterioration. Heat Transfer-Asian Research **31**(3), 212–225 (2002)

Hensley, J.C. (ed) in: Cooling Tower Fundamentals, 2nd ed., SPX Cooling Technologies, USA (2009)

Hisrchfelder, J.O., Curtiss, C.F., Bird, R.B.: Molecular Theory of Gases and Liquids. Wiley, New York (1954)

Holowaty, S.A., Ramallo, L.A., Schmalko, M.E.: Intermittent Drying Simulation in Deep Bed Dryer Of Yerba Mate. Journal of Food Enginnering **111**, 110–114 (2012)

Hu, Z., He, W., Ji, J., Zhang, S.: A Review on the Application of Trombe Wall System, Buildings. Renew. Sustain. Energy Rev. **70**, 976–987 (2017)

Incropera, F.P., DeWitt, D.P.: Fundamentos de Transferência de Calor e de Massa. LTC, Rio de Janeiro (2008)

Jasper, J.J.: The Surface of Pure Liquid Compounds. J. Phys. Chem. Ref. Data **1**, 841–1010 (1972)

Kim, B.H., Peterson, G.P.: Analysis of the Critical Weber Number at the Onset of Liquid Entrainment in Capillary-Driven Heat Pipes. Int. J. Heat Mass Transfer **38**(8), 1427–1442 (1995)

Kiranoudis, C.T.: Design and Operational Performance of Conveyor Belt Drying Structures. Chem. Eng. J. **69**, 27–38 (1998)

Koh, J.C.Y., Sparrow, E.M.: Hartnett, The Two Phase Boundary Layer in Laminar Film Condensation. Int. Journal Heat Mass Transfer **2**, 69–82 (1961)

Kutateladze, S.S.: Semi-Emprirical Theory of Film Condensation of Pure Vapours. Int. Journal Heat Mass Transfer **25**, 653–660 (1982)

Lindahl, P.A.: Plume Abatement. Power Engineering, Pennwell Publishing Corporation **113**(11), 28–30 (2009)

Lindahl, P.A., Jameson, R.W.: Plume Abatement and Water Conservation With The Wet/Dry Cooling Tower, Proceedings of Cooling Tower Institute Annual Meeting, New Orleans (1993)

Lindhal Jr, P. A., Mortensen, K.: Plume Abatement – The Next Generation, Proceedings of Cooling Technology Institute Annual Conference, Houston (2010)

Lohse, E., Schmitz, G., Inherently Safe Looped Thermosyphon Cooling System for Aircraft Applications Using Dielectric Fluid H-galden, International Refrigeration and Air Conditioning Conference at Purdue, Purdue University, USA (2012)

Lucena, M. C. C., Soares, S. A., Soares, B. J. and Leite, L. F. M., Reologia de Asfaltos Brasileiros Puros e Modificados por SBS, in XVIII Transport Research and Teaching Congress, Florianopolis, Brazil (2004) (in Portuguese)

Ma, J., Fu, X., Hu, R., Luo, X.: Effect of Inclination Angle on the Performance of a Kind of Vapor Chamber. Journal of Solid State Lighting **1**, 1–9 (2014)

Maezawa, S. Gi, K., Ishihara, H., Thermal Characteristics of Gas-Loaded Two-Phase Closed Thermosyphon with Gas Reservoir, 4th International Heat Pipe Symposium, Tsukuba, Japan (1994)

Mantelli, M.B.H.: Development of Porous Media Thermosyphon Technology for Vapor Recovering in Cross-Current Cooling Towers. Appl. Therm. Eng. **108**, 398–413 (2016)

Mantelli, M.B.H., Milanez, F.H.: A Loop Thermosyphon for Asphalt Tank Heating, 8th International Heat Pipe Symposium. Kumamoto, Japan (2006)

Mantelli, M. B. H.: Thermosyphon Technology for Industrial Applications. In: Vasiliev, L.L. and Kakaç, S., Heat Pipes and Solid Sorption Transformations - Fundamentals and Practical Application, 411–464, CRC Press, Florida (2013)

Mantelli, M. B. H., Milanez, F. H., Mielitz, G.. Tree Configuration Thermosyphon Study, Aiaa 38th Thermophysics Conference, Toronto (2005)

Mantelli, M.B.H., Study of Closed Two-Phase Thermosyphons for Bakery Oven Applications, Proceedings of the 33rd National Heat Transfer Conference, ASME-NHTC99–205, Albuquerque (1999b)

Mantelli, M. B. H. and Milanez, F. H., A Loop Thermosyphon for Asphalt Tank Heating, 8th International Heat Pipe Symposium, Kumamoto, Japan (2006)

Mantelli, M. B. H., Milanese, F. H. . Thermal Performance Theoretical Prediction of an Enclosure Heated by Aligned Thermosyphons, 43th Aerospace Sciences Meeting and Exhibit (2005)

Milanese, F. H., Mantelli, M. B. H., A New Methodology for Measuring Heat Transfer Coefficients - Application to Thermosyphon Heated Enclosures, 13th International Heat Pipe Conference, Shanghai (2004)

Milanez, F.H., Mantelli, M.B.H.: Analytical Model for Thermal Performance Analysis of an Enclosure Heated by Aligned Thermosyphons. J. Thermophys. Heat Transfer **20**(2), 267–275 (2006a)

Milanez, F.H., Mantelli, M.B.H.: Thermal Characteristics of a Thermosyphon Heated Enclosure. Int. J. Therm. Sci. **45**, 504–510 (2006b)

Moreira Junior, J., Cisterna, L.H.R., Mantelli, M. B. H., Milanez, F. H., Development of Numerical Tools for Shell-and-Shell Thermosyphon Heat Exchanger Design, 10th Minsk International Seminar of Heat Pipes, Heat Pumps, Refrigerators, Power Sources, Minsk, (2015)

Mujumdar, A.S., Handbook of Industrial Drying, 3rd Edn, CRC Press, Taylor and Francis, Boca Raton (2006)

Munson, B.R., Young, D.F., Okiishi, T.H., Huebsh, W.W.: Fundamentals of Fluid Mechanics, 6th edn. John Wiley & Sons, Massachusetts (2009)

Müller, J., Heindl, A., Drying of Medicinal Plants. In: Bogers, R J, Craker, L. E. and Lange, D., Medicinal And Aromatic Plants - Agricultural, Commercial, Ecological, Legal, Pharmacological And Social Aspects Medicinal and Aromatic plants, 237–252, Springer (2006)

Panjeshahi, M.H., Ataei, A., Gharaie, M., Parand, R.: Optimum Design of Cooling Water Systems For Energy And Water Conservation. Chem. Eng. Res. Des. **87**, 200–209 (2009)

Park, K. J., Vohnikova, Z., Brod, F. P. R., Evaluation of drying parameters and desorption isotherms of garden mint leaves (Mentha crispa L.), Journal of Food Enginnering, 51, 193–199 (2002)

Park, K. J. B., Park, J. K, Cornejo, F. E. P. and Dal Fabbro, I. M., Considerações Termodinâmicas das Isotermas, Revista Brasileira de Produtos Agroindustriais, Campina Grande, **10,** 1, 83–94 (2008)

Park, K., Chhatre, S. S., Srinivasan, S., Cohen, R. E. and McKinley, G. H.: Optimal Design of Permeable Fiber Network Structures for Fog Harvesting, Langmuir ACS Publications, **29**, 43, 13269–13277 (2013)

Peterson, P.F., Tien, C.L.: Gas-Concentration Measurements and Analysis for Gas-Loaded Thermosyphons. ASME Journal of Heat Transfer **110**, 743–747 (1988)

Pozzobon, J. C., Mantelli, M. B.H and SILVA, A. K., Experimental Study of Unstructured Porous Media Inserts for Water Recovery in a Reduced Scale, Crossflow Cooling Tower, Applied Thermal Engineering, **96**, 632–639 (2016)

Read, J., Whiteoak, D.: The Shell Bitumen Handbook, 5th edn. Thomas Telford Publishing, London (2003)

Reay, D. A., Kew, P. A and Mcglean, R. J., *Heat Pipes: theory, design and applications*, 6th edn. Elsevier (2014).

Reyes, M., Alonso, D., Arias, J.R., Velazquez, A.: Experimental and Theoretical Study of a Vapour Chamber Based Heat Spreader for Avionics Applications. Appl. Therm. Eng. **37**, 51–59 (2012)

Robinson, A. J., Smith, K., Hughes, T. and Filippeschi S., Heat and Mass Transfer for a Small Diameter Thermosyphon With Low Fill Ratio, International Journal of Thermofluids, **1–2**, (2020)

Rohani, A.R., Tien, C.L.: Steady Two-Dimensional Heat and Mass Transfer in The Vapor-Gas Region of a Gas Loaded Heat Pipe. ASME Journal of Heat Transfer **95**, 377–382 (1973)

Rohsenow, W. M.: Boiling. In: Rohsenow, W.M. and Hartnett, J.P (eds) Handbook of Heat Transfer, pp.13.1–13.73. Mc Graw-Hill Inc, New York (1973)

Saadatian, O., Sopian, K., Lim, C.H., Asim, N., Sulaiamn, M.Y.: Trombe Walls: a Review of Opportunities and Challenges in Research and Development. Renew. Sustain. Energy Rev. **16**(8), 6340–6351 (2012)

Sarno, C., Application of Phase Change Systems in Avionics, 16th International Heat Pipe Conference, Lyon, France (2012)

Shen, X., Lu, W., Sun, Y., Zhang, Y., Deng, N.: Experimental study on heat pipe assisted heat exchanger used for industrial waste heat recovery. Appl. Energy **169**, 177–186 (2016)

Silva, A.K., Mantelli, M.B.H.: Thermal Applicability of Two Phase Thermosyphons in Cooking Chambers - Experimental and Theoretical Analysis. Appl. Therm. Eng. **24**, 717–733 (2004)

Srinivasan, V., Khandekar, S.: Thermo-Hydrodynamic Transport Phenomena in Partially Wetting Liquid Plugs Moving Inside Micro-Channels". Sadhana - Indian Academy of Sciences **42**, 607–624 (2017)

Stephan, K., Abdelsalam, M.: Heat Transfer Correlations for Natural Convection Boiling. Int. J. Of Heat and Mass Transfer **23**(1), 73–87 (1980)

Sundararajan, T., Ayyaswamy, P.S.: Hydrodynamics and Heat Transfer Associated With Condensation on a Moving Drop: Solutions for Intermediate Reynolds Numbers. J. Fluid Mech. **149**, 33–58 (1984)

Susheela, N., Sharp, M.K.: Heat Pipe Augmented Passive Solar System for Heating of Building. Journal of Energy Engineering **127**(1), 18–36 (2001)

Thewsey, D.J., Zhao, Y.Y.: Thermal Conductivity of Porous Copper Manufactured by the Lost Carbonate Sintering Process. Physica Status Solidi. **205**(5), 1126–1131 (2008)

Tyagi, S.K., Pandey, A.K., Pant, P.C., Tyagi, V.V.: Formation, Potential and Abatement of Plume From Wet Cooling Towers: a Review. Renew. Sustain. Energy Rev. **16**, 3409–3429 (2012)

Wayner, P.C., Jr., Kao, Y.K., LaCroix, L.V.: The Interline Heat-Transfer Coefficient of an Evaporating Wetting Film. J. Heat Mass Transfer **19**, 487–492 (1976)

Zaghdoudi, M.C., Tantolin, C., Sarno, C.: Experimental Investigation on the Use of Flat Mini Heat Pipes for Avionics Electronic Modules Cooling. International Review of Mechanical Engineering **5**(4), 770–883 (2011)

Zhang, Z., Sun, Z., Duan, C.: A New Type of Passive Solar Energy Utilization Technology - The Wall Implanted with Heat Pipes. Energy and Buildings **84**, 111–116 (2014)

Zhang, P., Wang, B., Shi, W., Li, X.: Experimental Investigation on Two-Phase Thermosyphon Loop With Partially Liquid-Filled Downcomer. Appl. Energy **160**, 10–17 (2015)

Zhang, P., Wang, B., Wu, W., Shi, W., Li, X.: Heat Recovery From Internet Data Centers for Space Heating Based on An Integrated Air Conditioner With Thermosyphon. Renew Energy **80**, 396–406 (2015)

Zhang, P., Zhou, D., Shi, W., Li, X., Wang, B.: Dynamic Performance of Self-Operated Three-Way Valve Used in a Hybrid Air Conditioner. Appl. Therm. Eng. **65**, 384–393 (2014)

Zhang H., Shi Z., Liu K, Shao S., Jin T., Tian C., Experimental and Numerical Investigation on A CO2 Loop Thermosyphon for Free Cooling of Data Centers, Applied Thermal Engineering 1983–1090 (2017c)

Zuber, N.: Nucleate Boiling, The region of Isolated Bubbles and the Similarity with Natural Convection. Int. J. Heat Mass Transfer **6**, 53–78 (1963)

Ângelo, W., Mantelli, M. B. H. and Milanez, F. H., Design of a Heater For Natural Gas Stations Assisted by Two-Phase Loop Thermosyphon, 14th International Heat Pipe Conference, Florianópolis, Brazil (2007)